Reverse Engineering

Can a physicist visualize an electron? The electron is materially inconceivable and yet, it is so perfectly known through its effects that we use it to illuminate our cities, guide our airlines through the night skies, and take the most accurate measurements. What strange rationale makes some physicists accept the inconceivable electrons as real while refusing to accept the reality of a Designer on the ground that they cannot conceive Him?

WERNHER VON BRAUN
in a letter to the California State Board of Education, September 11, 1972

Dr. Wernher von Braun (1912–1977), German rocket scientist, aerospace engineer, and space architect, directed the U.S. space program from the Marshall Space Center of NASA, Huntsville, Alabama, from 1950 to 1972.

Cover: The background image is the most recent diagram depicting the gears that constituted the Antikythera mechanism, discovered in the wreck of a first century BC Greek cargo ship off the small Greek island of Antikythera in the Aegean Sea in 1900. Deduced by mathematician Dr. Tony Freeth and his associates as part of the Antikythera Mechanism Research Project for which Dr. Freeth is a principal, this refinement on the gear system originally put forth by Professor Derek de Solla Price in 1974 in his *Gears of the Greeks* was the result of clever reverse engineering from high-resolution x-ray images taken of a large corroded lump of brass, together with a great deal of mathematical expertise. Permission for use of the © 2012 image was granted by Tony Freeth, Managing Director, Images First Ltd.

Reverse Engineering

Mechanisms, Structures, Systems, and Materials

Robert W. Messler, Jr.

New York   Chicago   San Francisco
Athens   London   Madrid   Mexico City   Milan
New Delhi   Singapore   Sydney   Toronto

Cataloging-in-Publication Data is on file with the Library of Congress

Copyright © 2014 by McGraw-Hill Education. All rights reserved. Printed in the United States of America. Except as permitted under the United States Copyright Act of 1976, no part of this publication may be reproduced or distributed in any form or by any means, or stored in a data base or retrieval system, without the prior written permission of the publisher.

1 2 3 4 5 6 7 8 9 0   QVS/QVS   1 9 8 7 6 5 4 3

ISBN 978-0-07-182516-0
MHID 0-07-182516-9

Sponsoring Editor
Michael Penn

Editing Supervisor
Stephen M. Smith

Production Supervisor
Pamela A. Pelton

Acquisitions Coordinator
Amy Stonebraker

Project Manager
Virginia E. Carroll,
North Market Street Graphics

Copy Editor
Virginia E. Carroll,
North Market Street Graphics

Proofreaders
Chris Crocamo, Virginia Landis,
and Stewart Smith,
North Market Street Graphics

Art Director, Cover
Jeff Weeks

Composition
Mark Righter,
North Market Street Graphics

Printed and bound by Quad/Graphics.

McGraw-Hill Education books are available at special quantity discounts to use as premiums and sales promotions or for use in corporate training programs. To contact a representative, please visit the Contact Us page at www.mhprofessional.com.

This book is printed on acid-free paper.

Information contained in this work has been obtained by McGraw-Hill Education from sources believed to be reliable. However, neither McGraw-Hill Education nor its authors guarantee the accuracy or completeness of any information published herein, and neither McGraw-Hill Education nor its authors shall be responsible for any errors, omissions, or damages arising out of use of this information. This work is published with the understanding that McGraw-Hill Education and its authors are supplying information but are not attempting to render engineering or other professional services. If such services are required, the assistance of an appropriate professional should be sought.

To all of those who taught me by their example and by mentoring me in the practice, ethics, and art of engineering, your names reside in my mind and your spirits in my heart.

ABOUT THE AUTHOR

Robert W. Messler, Jr., Ph.D., FASM, FAWS, is Emeritus Professor of Materials Science and Engineering at Rensselaer Polytechnic Institute in Troy, New York. He spent 16 years in industry, 11 years at Grumman Aerospace and 5 years at Eutectic-Castolin, and then returned to Rensselaer, where he earned his degrees, to serve as Technical Director and Associate Director of the Center for Manufacturing Productivity. Dr. Messler later became a tenured professor and was appointed Associate Dean for Academic and Student Affairs. He is the author of six books, including *Engineering Problem-Solving 101* (McGraw-Hill Professional, 2013). He has also written more than 140 papers in diverse areas of materials processing, joining, design, manufacturing, and engineering education. Dr. Messler is a Fellow of ASM International and of the American Welding Society for career accomplishments and contributions.

Contents

Preface xv

CHAPTER 1 Introduction 1
    1–1 Human Beings Are a Naturally Curious Species 1
    1–2 Taking Things Apart to Learn 3
    1–3 Learning from Experience 3
    1–4 The Fundamental Approaches of Engineering 6
    1–5 The Critical Role of Dissection 9
    1–6 Summary 12
    1–7 Cited References 13
    1–8 Thought Questions and Problems 13

CHAPTER 2 The Status and Role of Reverse Engineering 15
    2–1 The Status of Reverse Engineering in References 15
    2–2 Reverse Engineering Defined 16
    2–3 Motivations for Reverse Engineering 18
    2–4 Engineering Design and the Engineering Design Process 19
    2–5 Types of Design 22
    2–6 Uses for and Benefits and Risks of Reverse Engineering 23
    2–7 Summary 26
    2–8 Cited References 27
    2–9 Thought Questions and Problems 27

CHAPTER 3 History of Reverse Engineering 29
    3–1 The Likely Emergence of Reverse Engineering 29
    3–2 Reverse Engineering in the Middle Ages 33
    3–3 Reverse Engineering during the Industrial Revolution 35
    3–4 Reverse Engineering during World War II 42
    3–5 Reverse Engineering in the Cold War and Beyond 47
    3–6 Summary 50
    3–7 Cited References 51
    3–8 Recommended Readings 51
    3–9 Thought Questions and Problems 51

CHAPTER 4	The Teardown Process..55
	4–1 The Purpose of Teardown 55
	4–2 Observation 57
	4–3 Measurement 60
	4–4 Experimentation 64
	4–5 Other Specific Forms of Teardown 65
	4–6 Summary 66
	4–7 Cited References 66
	4–8 Thought Questions and Problems 66

CHAPTER 5	Methods of Product Teardown...69
	5–1 The Product Teardown Process Revisited 69
	5–2 The General Procedure for the Teardown Process...................71
	5–3 Teardown Analysis or Value Analysis Teardown.....................73
	5–4 The Subtract-and-Operate Procedure 74
	5–5 Force Flow Diagrams (or Energy Flow Field Design) 76
	5–6 Functional Models...79
	5–7 Illustrative Example of a Product Teardown 82
	5–8 Summary 87
	5–9 Cited References 87
	5–10 Thought Questions and Problems 88

CHAPTER 6	Failure Analysis and Forensic Engineering..............................89
	6–1 Introduction to Failure Analysis 89
	6–2 Sources of Failures in Mechanical Systems 92
	6–3 Mechanisms of Failure in Materials 95
	6–4 The General Procedure for Conducting a Failure Analysis............104
	6–5 Two Exemplary Failure Analysis Cases 106
	6–6 Forensic Engineering..111
	6–7 An Exemplary Forensic Engineering Case 111
	6–8 Summary 113
	6–9 Cited References 113
	6–10 Thought Questions and Problems 114

CHAPTER 7	Deducing or Inferring Role, Purpose, and Functionality during Reverse Engineering 117
	7–1 The Procedure for Reverse Engineering 117
	7–2 Knowing versus Identifying versus Deducing versus Deferring.........119
	7–3 The Value of Experience 120
	7–4 Using Available Evidence, Clues, and Cues 124
	7–5 Using Geometry..126
	7–6 Using Flows of Force, Energy, and/or Fluids 128
	7–7 Using Functional Units or Subsystems from a Functional Model 130

	7–8 Summary 139
	7–9 Cited References 140
	7–10 Thought Questions and Problems 140
CHAPTER 8	The Antikythera Mechanism 143
	8–1 The Discovery 143
	8–2 The Recovery 144
	8–3 The Suspected Device 146
	8–4 Operation of the Mechanism 148
	8–5 Reverse-Engineering Investigations and Reconstructed Models 152
	8–6 Proposed Planet Indicator Schemes 159
	8–7 Similar Devices, Possible Predecessors, and the Possible Creator 159
	8–8 Speculation on Role, Purpose, and Functionality 161
	8–9 Summary 162
	8–10 Cited References 162
	8–11 Thought Questions and Problems 162
CHAPTER 9	Identifying Materials-of-Construction 165
	9–1 The Role of Materials in Engineering 165
	9–2 The Structure-Property-Processing-Performance Interrelationship 166
	9–3 Material Properties and Performance 169
	9–4 A Primer on Materials 172
	9–5 A Primer on Material Properties 177
	9–6 Relationships for Material Properties in Material Selection Charts 181
	9–7 Identifying Materials by Observation Only 186
	9–8 Laboratory Identification Methods 192
	9–9 Summary 194
	9–10 Cited References 194
	9–11 Recommended Readings 194
	9–12 Thought Questions and Problems 195
CHAPTER 10	Inferring Methods-of-Manufacture or -Construction 199
	10–1 Interaction among Function, Material, Shape, and Process 199
	10–2 The Role of Manufacturing or Construction 201
	10–3 The Taxonomy of Manufacturing Processes 203
	10–4 Process Attributes 207
	10–5 Inferring Method-of-Manufacture or -Construction from Observations 209
	10–6 A Word on Heat Treatment 218
	10–7 Summary 221
	10–8 Cited References 222
	10–9 Recommended Readings 222
	10–10 Thought Questions and Problems 222

CHAPTER 11   Construction of Khufu's Pyramid: Humankind's Greatest Engineering Creation 227

11–1 Herodotus Reveals the Pyramids to the World 227
11–2 The Great Pyramid of Khufu 229
11–3 Theories on the Purpose of the Pyramids 234
11–4 Theories on the Location of the Great Pyramid 238
11–5 Theories on the Construction of the Great Pyramid 244
11–6 Deducing the Likely Reality of Construction by Reverse Engineering 252
11–7 Summary 257
11–8 Cited References 258
11–9 Recommended Readings 258
11–10 Thought Questions and Problems 258

CHAPTER 12   Assessing Design Suitability 261

12–1 Different Designs, Different Role, Purpose, and Functionality ... 261
12–2 Form, Fit, and Function 265
12–3 Using Observable Evidence and Clues to Assess Form, Fit, and Function 267
12–4 Summary 278
12–5 Thought Questions and Problems 278

CHAPTER 13   Bringing It All Together with Illustrative Examples 283

13–1 Proverbs Make the Point; Pictures Fix the Lesson 283
13–2 Conair Electric Hair Blow-Dryer 285
13–3 An Automatic Electric Coffeemaker 292
13–4 Toro Electric Leaf Blower 299
13–5 Skil Handheld Electric Circular Saw 304
13–6 Lessons Learned 312
13–7 Summary 313
13–8 Cited References 313
13–9 Thought Questions and Problems 313

CHAPTER 14   Value and Production Engineering 317

14–1 Manufacturability 317
14–2 Design for Manufacturability 318
14–3 Value Engineering 322
14–4 Production Engineering 325
14–5 Summary 331
14–6 Cited References 331
14–7 Recommended Readings 331
14–8 Thought Questions and Problems 332

Contents **xiii**

CHAPTER 15	Reverse Engineering Materials and Substances 335
	15–1 Flattery or Forgery..335
	15–2 Motivations for Reverse Engineering Materials and Substances 337
	15–3 Finding Substitute and Replacement Substances and Materials........344
	15–4 Creating Generic Materials (Generics) 345
	15–5 Synthesizing Natural Materials and Substances: Biomimicry 348
	15–6 Imitating Natural Materials....................................352
	15–7 Summary 357
	15–8 Cited References 357
	15–9 Thought Questions and Problems 357

CHAPTER 16	Reverse Engineering Broken, Worn, or Obsolete Parts for Remanufacture 361
	16–1 Necessity Is the Mother of Invention.............................361
	16–2 The Motivation for Reverse Engineering for Remanufacture 362
	16–3 Reverse Engineering Broken Parts for Remanufacture................364
	16–4 Reverse Engineering Deformed or Worn Parts for Remanufacture......365
	16–5 Reverse Engineering Obsolete Parts for Remanufacture..............367
	16–6 Summary 369
	16–7 Cited References 369
	16–8 Thought Questions and Problems 369

CHAPTER 17	The Law and the Ethics of Reverse Engineering........................373
	17–1 Without Morals and Ethics, Laws Mean Nothing 373
	17–2 Legal versus Ethical...375
	17–3 The Legality of Reverse Engineering 376
	17–4 The Ethics of Reverse Engineering 377
	17–5 Summary 381
	17–6 Cited References 381
	17–7 Thought Questions and Problems 381

CHAPTER 18	The End of a Book, the Beginning of a New Story: Closing Thoughts 385
	18–1 The First Design 385
	18–2 Imperfect Humans Need Reverse Engineering 387
	18–3 Order from Chaos, Light from Darkness, Knowledge from Knowledge .388
	18–4 Learning from the Old to Create Anew: Four Opportunities..........393
	18–5 Final Words...403
	18–6 Cited References 405
	18–7 Recommended Readings......................................405
	18–8 Thought Questions and Problems 405

	Appendix A List of All Material Classes and Major Subtypes, and Major Members of Each 407
	Appendix B Comprehensive List of Specific Manufacturing Methods by Process Class.......................................409
	Index 417

Preface

There are those, especially engineers involved in the design of new mechanisms, structures, systems, or materials, who resent and would argue with the saying extracted from Ecclesiastes 1:9, " . . . there is nothing new under the Sun." But with some honest reflection, most would soon probably agree with it, particularly as they gained experience and grew older and, hopefully, wiser. It is practically impossible to create something *completely new,* with *nothing like it* under the Sun. As intelligent physical beings living in a physical world, we are immersed in our environment, where our brains are inundated with inputs to our five senses. Our minds, from which come all our thoughts, are shaped by our experiences. How would one create something from a vacuum of nothing, whether physical, sensual, emotional, or intellectual?

One of the potentially most powerful problem-solving techniques available to engineers is *reverse engineering.*[*] In reverse engineering, one takes apart one thing to create another thing, whether identical, similar, marginally or peripherally related, or seemingly quite different. That "taking apart," or dissection, may be quite literal and physical or figurative and conceptual. By seeing what made up the original, and intuiting, deducing, or inferring how it did what it did, we derive ideas on how to make an exact copy or a totally different creation—or anything in between—based on what was learned.

Too often seen as a lazy person's shortcut to copy the good idea of another, or even oneself, reverse engineering is—or should be—much more than mindless mimicry or artless cloning. This book is intended to place reverse engineering where it should be placed, at the top of the list of methods for learning from our experiences. The book is unique in the breadth and depth of treatment of the method of reverse engineering. Nothing like it exists elsewhere.

Join this engineer, who chose to teach after years of carefully observing what can be done if one keeps one's eyes and mind open, in what is intended to be more than a textbook. The engineer-teacher has attempted to take readers on a journey through the technical details, the practical techniques, and the long and honorable history of reverse engineering that will be engaging beyond informative. With any luck, the author may even succeed in making engineering education fun, enlightening, and inspiring. Come see what reverse engineering *really* is, what it has accomplished, and what it is capable of accomplishing.

I hope you enjoy the journey!

Robert W. Messler, Jr.

---

[*]Check out the author's previous book with McGraw-Hill: *Engineering Problem-Solving 101: Time-Tested and Timeless Techniques,* 2013.

Reverse Engineering

CHAPTER 1

Introduction

1-1 Human Beings Are a Naturally Curious Species

An old, anonymous proverb says: "Curiosity killed the cat." Which it may or may not have, depending on what the cat got into.[1] After all, another anonymous proverb says: "What you don't know won't hurt you." Which at least some nonstereotypical cats might heed. But the great humorist Mark Twain modified the latter proverb to read, "It's not what you don't know that hurts you, it's what you know that just isn't so." Confused? Or just curious about where all this is leading?

Curious thing this curiosity exhibited by human beings, as well as by some of our friends in the animal kingdom, most notably those with whom Darwin, at least, believed we share the closest kinship, that is, the apes. As but two examples, the reader is referred to YouTube to view the initial 40 seconds of "Curious Ape" posted by *independentconceptz* and, also, about the first 29 seconds of "Curious chimpanzee approaches video camera" posted by *Goualongo*. Not that either of these endearing distant relatives, who bear some resemblance to at least some of us, do much to get to the bottom of the object that fascinates them so, but it shows how primitive curiosity may be to our own species. And, of course, there is Curious George®, the eyeglass-wearing, intelligent-looking chimp with whom most of us are familiar as a children's book character that got into adventure after adventure for that trait which we humans share with him (Figure 1–1).

There's also the quote many of us have misquoted as: "This is becoming curiouser and curiouser!" Knowingly or unknowingly, we were quoting—actually misquoting—from Lewis Carroll's *Alice in Wonderland,* Chapter 2. The full and proper quote goes: " 'Curiouser and curiouser' cried Alice (she was so much surprised that for the moment she quite forgot how to speak

---

[1] The original proverb, of unknown authorship or origin, goes on to add: "And satisfaction brought it back." It seems the full proverb implies that unnecessary inquisitiveness and prying are risky but potentially worthwhile.

Figure 1-1 Iconic image of Curious George®, who epitomizes the curiosity of human beings' closest kin. (*Source:* Illustration of Curious George by H. A. Rey. Copyright © 2013. Reprinted with permission of Houghton Mifflin Harcourt Publishing Company. All rights reserved. The character Curious George®, including the character's name and character's likenesses, are registered trademarks of Houghton Mifflin Harcourt Publishing Company.)

good English). 'Now I'm opening out like the largest telescope that ever was! Good-bye feet!' (for when she looked down at her feet they seemed to be almost out of sight, they were so far off)."

There is no doubt about it. We human beings are a naturally curious species, even though curiosity is not unique to our species alone. But we are alone in several respects. Defined as "marked by a character that is eager to learn more," our *curiosity* runs deeper than simply wondering *what* something is. As a species, we seem to be alone in wondering: *Why* are we here? *How* did we get here? *When* did we come into being? *Where* are we going? A few quotes on curiosity make the point far better than any words the author can come up with here, including:

"Curiosity is lying in wait for every secret" (Ralph Waldo Emerson)
"The important thing is not to stop questioning. Curiosity has its own reason for existing" (Albert Einstein)
"Only the curious have something to find" (Anonymous)

The dictionary[2] defines a *journalist* as "a person whose occupation is journalism," which leaves one curious about what journalism is. Fortunately, the curious will find a satisfying definition for

---

[2] Throughout this book, definitions presented are from the online dictionary at www.thefreedictionary.com by Farlex).

journalism as "the collection, writing and distribution of news and other information," which would make authors journalists of a sort. The young, apprentice journalist is mentored by his or her editor to dig into a story to answer the questions "Who?," "What?," "When?," "Where?," "Why?," and "How?" In fact, James Glen Stovall, professor of journalism at the University of Alabama, wrote what is considered to be the "bible" for prospective journalists, namely, *Journalism: Who, What, When, Where, Why, and How* (ref. Stovall). Known colloquially as "the Five Ws" or, more precisely, "the Five Ws and one H," the answers to these questions are considered basic in information-gathering, whether for journalists, police investigators, or researchers. For the pursuit of some knowledge, they are the basic questions that need to be answered by engineers also!

1–2 Taking Things Apart to Learn

It is fairly well known that it is not unusual at all for little boys to take things apart. Not because the things need to be taken apart, but simply because it's fun and it's the best way to help a curious young mind figure out how things work—when, in fact, they worked just fine prior to being taken apart. Really young boys, as toddlers, begin by finding more joy in knocking down the towers (as a quick way to take something apart!) their mother or father or uncle or grandfather built from wooden blocks or Legos than by building towers themselves. But, soon, most find joy in building their own structures, however unrecognizable it may be what these are meant to be, if anything. Next, though, comes taking apart toy cars and trucks or trains to see what makes them work, then pulling off the head of the farmer in their Fisher-Price Little People Animal Sound Farm to see what held it on, then "dissection" to find what makes the cow moo. From this point, the taking apart progresses to Big Wheels, trikes, bikes, skateboards, then cell phones or laptops that no longer work. Somewhere between taking apart a Razor kick scooter and a Motorola Droid RAZR MAXX device, it's dissecting a frog in high-school biology class. The goal (in all but the rarest cases of sheer destructiveness) is usually the same: to find out how things work.

Little girls, too, take things apart: Barbie dolls get undressed and dismembered; dollhouses get dismantled; and, eventually, electric hair blow-dryers get disassembled in the hope of finishing preparations for a big date. Some girls—much to the disappointment of some boys—even find the dissection of that frog in biology class less disgusting than fascinating, and, perhaps, some get the first hint that they may want to become a nurse or a doctor or a biomedical engineer.

There is much to learn from taking things apart, and, as will be seen later, there is often even more to be learned from putting them back together, whether one is an engineer or a surgeon.

1–3 Learning from Experience

If "a picture is worth one thousand words," how many pictures—no less, how many words—is an experience worth?[3] "Experience is the best teacher," while apropos to the purpose of the technique that is the focus of this book, unfortunately, is also misquoted, the original saying being: "Experience is the hardest lesson, as it gives the test before the lesson." This will prove to be very apropos to our exploration of reverse engineering.

---

[3] The original "Chinese proverb" from a streetcar advertisement was wrongly translated. The literal translation is "A picture's meaning can express ten thousand words." The point about the value of experience is still valid, even in mis-translation.

So, what is all this about anyway? It's this: Human beings learn best from experience. What the author knows, having worked as an engineer before returning to academe to teach engineering, is that obtaining a degree from an engineering school is necessary but not sufficient for one to become an engineer. *Real engineering,* while built on theory, is learned in practice, and proficiency grows with experience. This fact is embedded in how we refer to what engineers do, even though it seems to be overlooked by most; that is, they *practice* engineering. So, too, do the other professionals, as doctors *practice* medicine, lawyers *practice* law, and dentists *practice* dentistry. In fact, doctors, lawyers, and dentists have "practices." Practicing, in this context, does not mean starting from scratch and hoping to get it right. Rather, it means working and constantly striving to do that work better! After all: "Practice makes perfect."

All this about *practice* has to do with physically *doing* something, not simply mentally pondering it. It is practice that helps one gain experience, and it is experience that has given rise to the most effective learning. But more about this later.

The importance and value of experience took on a new significance in the early 1970s, when David A. Kolb (born 1939), an American psychologist and educational theorist, together with a colleague, Ron Fry, developed the *Experiential Learning Model* (ELM).[4] The ELM is composed of four key elements to be used iteratively, as follows:

1. Concrete experience
2. Observation and reflection on that experience
3. Formation of abstract concepts based upon the reflection
4. Testing the new concepts
   [repeat]

These four elements are the essence of a spiral of learning that could, conceivably, begin with any of the four, but typically begins with a concrete experience. Kolb named his model to emphasize links to ideas formulated by others, including John Dewey, Jean Piaget, and Kurt Lewin, who all wrote about the experiential learning paradigm.

Prior to Kolb's model, visual, auditory, and kinesthetic (tactile) styles of learning were proposed and occasionally were adopted by conscientious educators, at all levels, to attempt to appeal to the different ways in which different students learn best.[5] What these represent, of course, are the ways by which information from outside can enter one's mind, that is, as a sensory input. Thus, rather logically, the three learning styles are sight ⇒ visual, sound ⇒ auditory, feel ⇒ kinesthetic. This is not, by the way, to say that one cannot learn from the other two senses, smell and taste. We all surely do learn from these; for example, the head-clearing/eye-watering/nose-burning smell of ammonia that teaches us to sniff lightly, the smell of cookies baking in the oven that reminds us of our grandmother or mother, the way we know how hot chocolate will warm us on a cold day, and how our first too-large-a-taste of wasabi brought tears to our eyes and caused us to gasp for breath. In any case, it should be clear that experiential learning comes from sensory input too, so the experiential learning model is not so far removed from the V-A-K model. But it was Kolb who really launched the learning style movement in the early 1970s, and his model has become the most influential yet developed.

---

[4] David A. Kolb is Professor of Organizational Behavior at Case Western Reserve. He and Ron Fry were with the Weatherhead School of Management at Case Western Reserve when they came out with the ELM.
[5] Visual, auditory, and kinesthetic learning styles are often referred to by their abbreviations, i.e., VAK.

David Kolb may have said it best in 1984, when he said: "Learning is the process whereby knowledge is created through the transformation of experience. Knowledge results from the combination of grasping experience and transferring it."

Kolb proposes that experiential learning has six main characteristics, as follows:

1. Learning is best conceived as a process, not in terms of outcomes [by which he means, the process of *doing* is often the source of greater learning than are the results].
2. Learning is a continuous process grounded in experience [by which he means, we grow only by doing—and experiencing—more and more].
3. Learning requires resolution of conflict, as learning is, by its very nature, full of tension [by which he means, obtaining the answer to a question inevitably gives rise to other questions].
4. Learning is a holistic process of adaptation to the world [by which he means, we learn from everything we experience—as an integrated whole].
5. Learning involves transactions between the person and the environment [by which he means, to learn we have to relate what we observe in our environment to ourselves and we have to project ourselves onto and into our environment].
6. Learning is the process of creating knowledge that is the result of the transaction between social knowledge and personal knowledge [which goes back to the previous quote, and means, growth of knowledge involves give and take between ourselves and others].

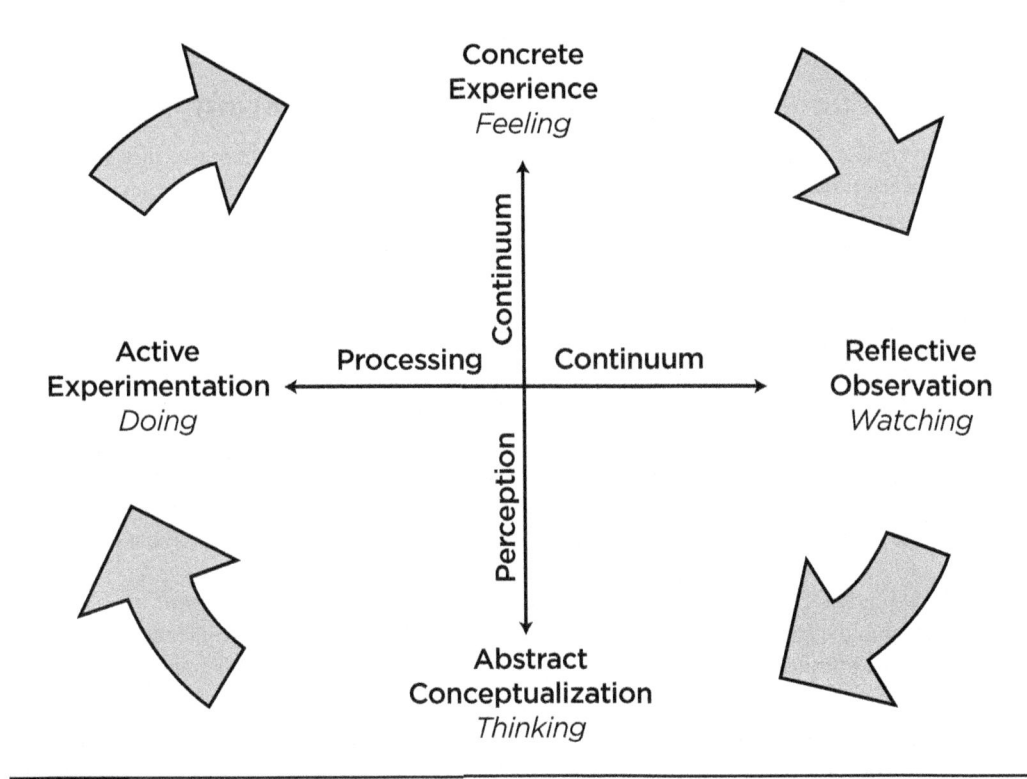

Figure 1-2 Schematic of Kolb's Experiential Learning Model (ELM). (*Source:* http://www.nwlink.com/~donclark/, with permission from Don Clark.)

Kolb's experiential learning theory sets out four distinct learning styles based on a four-step learning (iterative) cycle (Figure 1–2). His model differs from others by offering a way to understand individual learning styles (which he named the "Learning Styles Inventory" or "LSI") *and* an explanation of a cycle of experiential learning that applies to all learners. Who among us doesn't learn from doing? In fact, Lao-tse (Laozi in Chinese), sixth century BC philosopher in ancient China, and founder of Taoism, said: "Give a man a fish, feed him for a day. Teach a man to fish, feed him for a lifetime." The point: Learn by doing and really learn!

In short, the schematic in Figure 1–2 shows *immediate or concrete experiences* that provide a basis for *observations and reflections*. These observations and reflections are assimilated and distilled into *abstract concept(s)* to provide input for action which can be *actively tested,* in turn, creating new experience(s). The cycle is: Feeling ⇒ Watching ⇒ Thinking ⇒ Doing.

A final anecdote before moving on . . .

As a young engineer, having worked for only a couple of years in advanced materials and processes development, the author was told by the department manager: "You've spent more than enough time thinking. It's time to act. Do something! Make *something*! By acting—by doing—you'll learn more about what you're only thinking about, and you'll end up with more to think about. Thoughts in one's head—no matter how clever—mean nothing if they are not turned into something real." A very wise man indeed!

So, let's move on our way toward doing something!

1-4 The Fundamental Approaches of Engineering

Those who choose to become engineers need to learn and become proficient at two things, and they usually learn these two in their formal engineering education as follows: First, they need to learn how to analyze a problem in order to find a solution; that is, they need to learn *analysis.* Second, and, hopefully, in short order, or, ideally, along with analysis, they need to learn to create or synthesize something from its components; that is, they need to learn *synthesis*. In a proper preengineering curriculum, some of both are taught, with the culmination being embodied in the ABET-required senior capstone design experience.[6] Here, students, as teams, are expected to conduct up-front analysis on the way to solving a real-world engineering problem, but the end product of their effort is to be a workable—and, preferably, working—design which they synthesized. The intent is multifold: to teach the formal process of engineering design, to force integration of technical knowledge obtained in discrete courses, and to teach the importance of teamwork. But, more than anything else, the all too often implicit, versus explicit, purpose is to provide students with an experience in design as the most effective way of having them learn to design.

Math and physics courses tend to focus on analysis: analyzing how the process of differentiation in the calculus allows one to find maxima and minima and how integration allows summation over infinitely small increments, and how, in physics, free-body diagrams allow one to

---

[6]ABET is the Accreditation Board for Engineering & Technology. It accredits engineering degree programs that meet a well-conceived set of learning outcomes seen as vital for graduates to be able to succeed in the practice of engineering, as part of a process of continuous quality improvement at the college or university.

determine how an object acted upon by a number of forces will react or respond and how one form of energy can be transformed into another form. In physics and chemistry laboratory sessions, on the other hand, lectures that focus on analyzing what happens to a block at rest on a shallow-angle incline as the angle of the incline is slowly increased or what happens when one mixes an acid with a base in the presence of an indicator solution like phenolphthalein, take on new and added meaning as new knowledge is gained by a process of synthesis. The component of force acting on the block along the plane of the incline is measured to rise as the incline is made steeper, until, finally (and rather repeatedly), the block eventually slides down the incline. The addition of a base, like sodium hydroxide, to an acidic solution (like hydrochloric acid in water) turned pink in the presence of phenolphthalein becomes progressively less intensely pink, until it becomes clear (i.e., the solution has been *titrated*), at which point, if the solution is evaporated by heating, a residue of salt (for this example, sodium chloride) remains.

More is usually learned from hands-on experiences (such as physics or chemistry experiments) than from words—whether spoken and heard or seen and read.

As students progress through their undergraduate education in engineering, they are, hopefully, being taught the critical role of both analysis *and* synthesis. In fact, they should be learning— and should explicitly be being taught—that one without the other leaves too many unanswered questions. An engineer would have little success—without extraordinary good luck—in creating (i.e., synthesizing) a design for a new device, product, material, system, or process without having initially, before beginning to design anything (e.g., during the stage of problem formulation), done some analysis. In fact, the process of analysis must be repeated all through the design process, constantly analyzing precisely where the design stands.

Analysis is required during the stage of formulating the problem (i.e., during problem formulation) to be addressed by the design, in an unambiguous statement. Analysis is required during the stage of concept generation, or *conceptual design,* to quickly assess technical feasibility—or infeasibility—of each new concept. Analysis is required during the stage of down-selecting a preferred design, known as *embodiment design,* during trade-off studies to determine that needed functionality will be obtained, along with some idea of the estimated performance that can be expected and costs that will be incurred. And detailed analysis is required as the design is refined during the stage of *detail design* to determine precisely what the final design's functionality will be, to what level it will perform, with what reliability, and at what costs for materials, energy, manufacturing or construction processes, labor, information resources, and so forth. Proper synthesis demands proper analysis, usually performed in an iterative fashion.

Table 1–1 summarizes the four stages of engineering design.

Another way of expressing the fundamental approach of engineering is this: An engineer knows what he or she knows from having conducted analysis (often, but not solely, involving mathematics) or from having made measurements (usually from experiments and/or from the use of physical models) or by having proof-positive that what is under question has been successfully accomplished or demonstrated before (e.g., by reference and citation). Some would add to these three that a modern engineer could know what will happen by using computer-based models. There is a caveat here, however—that is, a physical model is one thing, a computer-based model (or simulation) is another! A physical model, for which there are a number of types, in terms of level of sophistication and intended function (see my previous book *Engineering Problem-Solving 101: Time-Tested and Timeless Techniques,* Chapters 18, 19, 20, 21, and 22, pp. 141–164), is

TABLE 1-1 The Stages of Engineering Design with the Engineering Approaches Used

	Stage 1	
Analyze needs, constraints, and goals; synthesize problem statement	**Problem Formulation**	Entirely a mental activity
	Stage 2	
Synthesize ideas into design concepts; analyze concepts against the problem	**Conceptual Design**	Physically model for proof of concept(s); measure ability of concept(s) to function; conduct mathematical analyses
	Stage 3	
Analyze alternative concepts within trade-off studies; synthesize rank-ordered list to down-select best concept(s)	**Embodiment Design**	Experiment to find best concept(s); measure basic performance; construct and test experimental model(s); mathematically analyze test results
	Stage 4	
Analyze "best" concept test results; synthesize refined "best" design; analyze final design prototype	**Detail Design**	Construct test model; measure performance and mathematically analyze to allow optimization of design; construct, test, and analyze prototype

intended to be tested and for some measurements to be taken. It is what it is—a physical model. If it has captured the essential feature(s) to be assessed, the engineer(s) will know if that (those) feature(s) work(s) or not. A computer-based model—or simulation—on the other hand, is only as good as the modeler or software programmer. What the simulation predicts may be built into the simulation. In short, at the risk of sounding old-fashioned (but not from being a Luddite), hardware never lies; software can.[7]

Table 1–1 also summarizes the role of analysis and of synthesis at each stage of design, along with the role of mathematical analysis, use of experiments to allow measurements to be made, and physical models to allow measurements to be made.

Hopefully, it is clear from the preceding discussion how proper engineering demands both thinking and doing, both analysis and measurement (using well-conceived experiments and/or one or more of several types of physical models) (ref. Messler), both analysis and synthesis. Usually, these pairs of actions are done in an iterative fashion until what is required as an output has been achieved to the level of detail and degree of precision that is needed or appropriate.

---

[7]The Luddites, named after Ned Ludd, a youth who allegedly smashed two stocking frames 30 years earlier, were nineteenth-century English textile artisans who violently protested against the machinery introduced during the Industrial Revolution that made it possible to replace them with less-skilled, low-wage laborers, leaving them without work. The term now refers to one who fears technology—or new technology—as they seem pleased with things as they are.

1-5 The Critical Role of Dissection

So by now it ought to be clear that engineers need to know how things work in order to design and build things that work better. Furthermore, it ought to be clear that the best way to know how something works is to take it apart; with many fledgling engineers-to-be getting an early start as a natural part of growing up in a world filled with wonders tugging at the curiosity that seems to be such an integral part of us being human. As for taking things apart to learn, Galen did it, Leonardo da Vinci did it, and modern pathologists still do it.

Aelius Galenus or Claudius Galenus (AD 129–ca. 200), better known as Galen of Pergamos (modern Bergama, Turkey) was, in many ways, the "father of modern medicine" (Figure 1–3). While of Greek ethnicity, he is best remembered as a Roman physician, surgeon, physiologist, pathologist, pharmacologist, and philosopher. He contributed immeasurably to the practice of medicine, being the most accomplished, by far, of all medical researchers of antiquity, and he did so through the vivisection of animals (Figure 1–4) and dissection of deceased humans (many of whom came out of the gladiatorial arena of Pergamos). Leonardo da Vinci, more correctly, Leonardo di ser Piero da Vinci (AD 1452–1519), was an Italian Renaissance painter, sculptor, architect, scientist, engineer, inventor, anatomist, mathematician, geologist, botanist, cartographer,

Figure 1-3 Galen (here referred to as "Claude Galien") was born in Pergamos in Asia Minor in AD 129. After receiving medical training in Smyrna and Alexandria, he gained fame as a surgeon to the gladiators of Pergamos, eventually being summoned to Rome to serve as physician for Emperor Marcus Aurelius. He spent the rest of his life at the Court writing an enormous body of medical works until his death ca. AD 200. (*Source:* Wikipedia Creative Commons, a lithograph by Pierre Roche Vigneron, ca. 1865, contributed by Mgoodyear on 1 September 2008.)

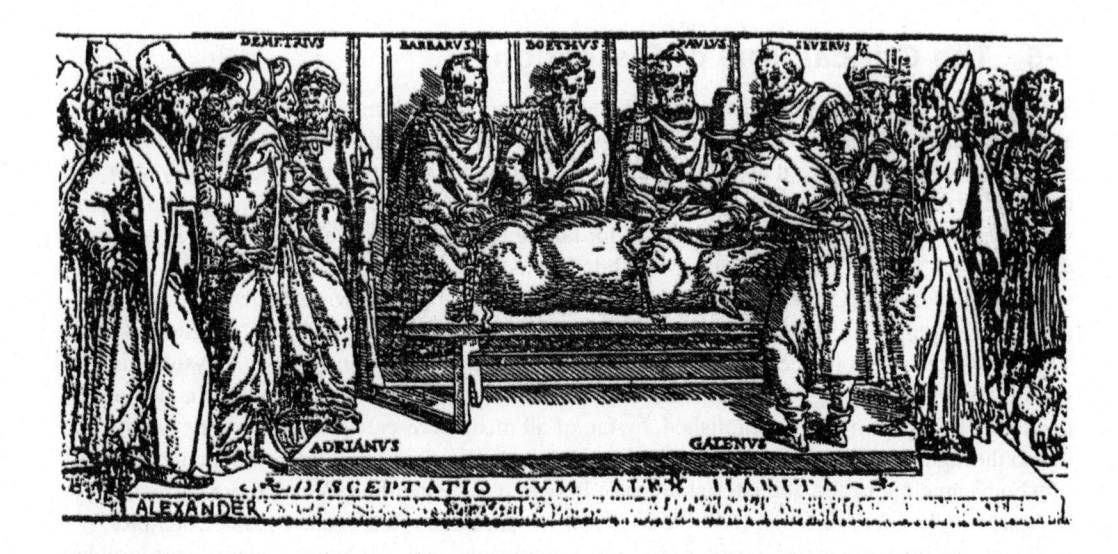

Figure 1-4 A print from the 1541 Junta edition of *Galen's Works* from the Yale University Harvey Cushing/John Hay Whitney Library showing Galen demonstrating with vivisection of a pig. (*Source:* In the public domain, but from Wikipedia Creative Commons, contributed by Rswarbrick on 8 April 2008.)

musician, and writer (Figure 1–5). To say he was a genius, may, in his case, be an understatement! He used vivisection of animals and dissection of both deceased animals and human cadavers extensively to perfect his art and satisfy his insatiable curiosity (Figure 1–6). *Gray's Anatomy,* the seminal work on the topic used in medical schools around the world to teach medical doctors, is an unparalleled collection of the results of dissection that serves us all (Figure 1–7).

Figure 1-5 A painting from the early sixteenth century on a wooden panel used as part of a cupboard in a private home in Lucan, Italy, where it was discovered by the homeowner. The image bears a striking resemblance to Leonardo da Vinci and is believed to be either a self-portrait or a work by minor sixteenth-century artist Cristofano dell'Altissimo. (*Source:* In the public domain, but from Wikipedia Creative Commons, contributed by Murray Menzies on 21 November 2010.) **Don't miss the color version of this figure, available at www.mhprofessional.com/ReverseEngineering.**

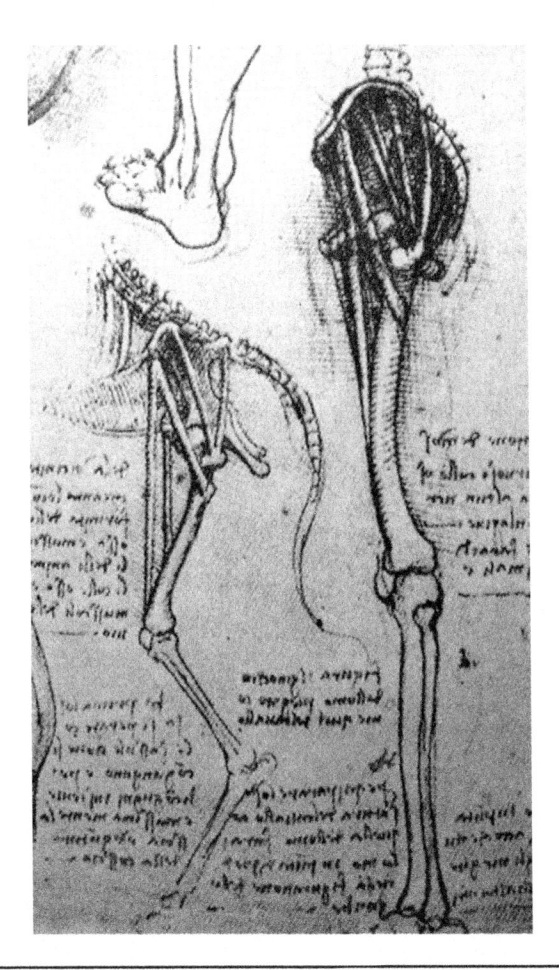

Figure 1-6 Leonardo da Vinci executed innumerable sketches showing anatomy, which, from the accuracy of details, he obviously studied via dissection. Many can be found online, but most are the property of the Royal Collection of Her Majesty Queen Elizabeth II. The one shown here, comparing the anatomy of the legs of an adult man and a dog, is representative, and has been found to be in the public domain in the United States and elsewhere where copyright ends 100 years after the author dies. No attempt has been made here to circumvent copyright law. (*Source:* Wikimedia Commons, contributed by OldekQuill on 14 February 2005.)

Dissection is the systematic process by which things are taken apart to aid in understanding those things, whether once alive or never alive. As such, it is an essential tool for learning in fields beyond medicine.

This book is devoted—for the remainder of its entirety—to reverse engineering, the process, the technique, and the procedure. There may be no better way of satisfying an engineer's natural curiosity or helping an engineer grow in knowledge than by this process involving "mechanical dissection"—taking apart and analyzing one thing to enable the synthesis of another thing.

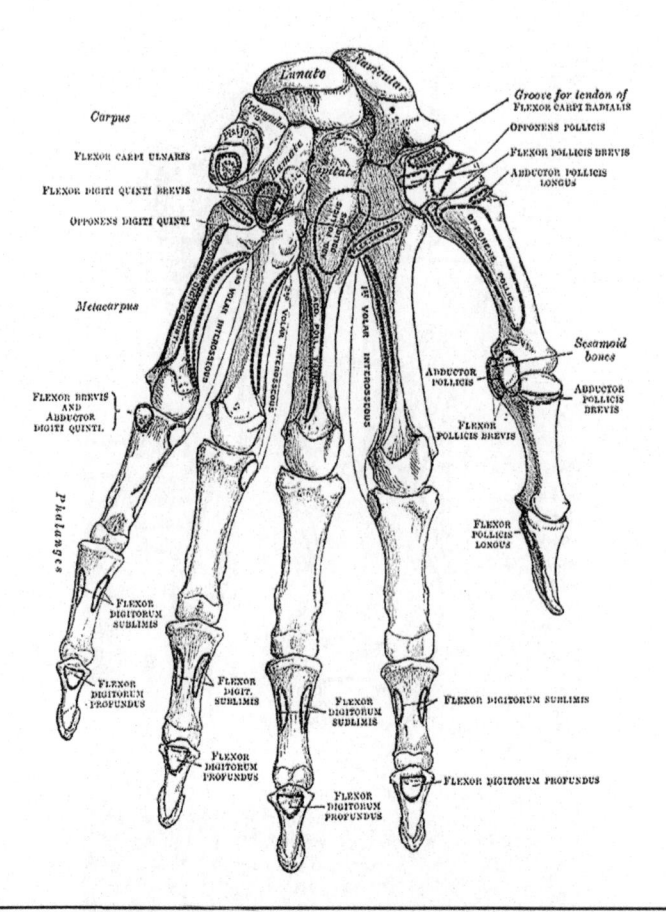

Figure 1-7 Anatomy of bones of a human right hand from the 1918 edition of *Gray's Anatomy of the Human Body* (often shortened to simply *Gray's Anatomy*). Henry Gray (1827–1861), an English anatomist, worked 18 months with his colleague Henry Vandyke Carter to create the central body of illustrations for this seminal work, dissecting bodies obtained from the morgues of workhouses and hospitals with the passage of the Anatomy Act of 1837. Initial publication was in 1858. (*Source:* Wikipedia Creative Commons, contributed by Tene on 17 March 2007.)

1-6 Summary

Human beings are innately curious. Cats and rats and elephants may be curious, but only about what is going on or about what something is.[8] We humans are uniquely curious about more than what; we are also curious about who, where, when, why, and how. From our earliest days, we take things apart to learn. Modern learning theory proposes that we learn best from our experiences and, perhaps, in no other way. We learn best through a repetitive cycle of Feeling ⇒ Watching ⇒ Thinking ⇒ Doing.

---

[8] "... cats and rats and elephants ..." is from the lyrics of "The Unicorn Song" made famous by the Irish Rovers. It goes: "Some cats and rats and elephants, but sure as you're born, you're never gonna see no unicorn!"

Our curiosity causes us to analyze, which then leads us to create through thoughtful design. We first analyze and then we synthesize. Great thinkers (e.g., ancient philosophers like Laozi) and great doers (e.g., ancient physicians like Galen) and the epitome of the Renaissance man (i.e., da Vinci) all knew that action, more than thought, changes the world for the better.

Reverse engineering is a powerful technique, process, method, and means for creating a design—maybe better, maybe less expensive, maybe hardly different at all, perhaps barely recognizable as a derivative of the original. It is, quite simply, mechanical dissection or "teardown" of mechanical, electrical, electromechanical or mechatronic, and, occasionally, biological entities.

The remainder of this book looks into the basic concepts, the history, the varied uses, the variety of methods of teardown, the identification of materials-of-construction, the inference of methods-of-manufacture or -construction, specific application areas (e.g., value engineering, methodizing, productionizing, repair), the legal and ethical ramifications, and some wonderful examples from the ancient past to modern times. The focus is on mechanical mechanisms, structures, systems, and materials, with reference to electrical and biological systems in passing only for completeness.

Throughout the book are what are hoped will be enjoyable tangents intended to make more interesting the journey. Enjoy this first and only comprehensive and practical treatise on reverse engineering. And enjoy being an engineer!

1-7 Cited References

Messler, Robert W., Jr., *Engineering Problem-Solving 101: Time-Tested and Timeless Techniques,* McGraw-Hill, New York, 2013.

Stovall, James Glen, *Journalism: Who, What, When, Where, Why, and How,* Allyn & Bacon/Pearson, Boston, 2005.

1-8 Thought Questions and Problems

1-1 Using the Internet, look up "visual, auditory, and kinesthetic learning styles" or "VAK."
   a. Briefly describe the overall theory *and* each style, specifically, using your own words.
   b. Think of and write down *two* example instances from your own experiences where each style seemed to work to facilitate your learning.
   c. Which style of learning do you feel appeals to and is most effective for you?
   d. Think of and write down *two* examples of how the use of smell and of taste, individually, or together, helped you learn or evokes particularly strong memories from your past.
   e. Briefly explain why inputs to multiple senses (e.g., sight and hearing; sight, sound, and touch, etc.) might be even more effective for imparting learning than an input to only one sense.
   f. What do your responses to (c) and (e) suggest as far as how you should engage in new experiences where learning is important?

1-2 Using the Internet, look up "Kolb's Experiential Learning Model" or "ELM."
   a. Briefly describe the model in your own words.
   b. Think of and write down *two* examples from your own experiences of how the four key elements of (1) concrete experience, (2) observation and reflection on that experience,

(3) formulation of abstract concepts based upon that reflection, and (4) testing the new concept helped you learn something new.

   c. Which of the four experiences shown in Figure 1–2 (i.e., feeling, watching, thinking, and doing) do you feel has the greatest positive impact on your learning? Explain your answer.

   d. Think of and briefly describe a situation in which you found it was essential for you to *do something,* not just *think about something* in order to learn.

   e. Are there situations where doing something, rather than just thinking about it, could be dangerous to your well-being? Give an example.

1-3 While it is essential that an engineer think through a problem needing solution before proposing a solution, it is also often important for an engineer to conduct tests or experiments and make measurements before locking onto a final solution.

   a. Think of an example for which and briefly describe how and why testing or experimenting and measuring were important for the solution of a problem in your own past or from what you find on the Internet.

   b. Think of and briefly describe a situation or example for which it was useful, if not essential, for a physical model to have been created on the path to solving a problem.

1-4 Taking things apart is an extremely valuable and effective way to learn how things work. *Biological dissection* as well as *mechanical dissection* have been used for centuries to advance understanding and learning. Think of or find *two* examples of how "dissection" has been or is used to advance understanding and learning in each of the following areas:

   a. Artistic painting
   b. Artistic sculpting
   c. Medicine
   d. Engineering
   e. How does the use of dissection relate to (e.g., support or refute) the VAK?
   f. How does the use of dissection relate to (e.g., support or refute) ELM?

1-5 One obviously needs to carefully consider the morality, humanity, and ethicality of *vivisection*. This said, there are situations where it is arguably important and arguably can be done humanely.

   a. Provide a definition, in your own words, of each of the terms *morality, humanity,* and *ethicality.*

   b. Look up how neurosurgeons have learned and continue to learn about the function of the human brain using legal, ethical, and humane means.

   c. Where do you stand on the use of vivisection? For example, what limits would you impose on its use?

CHAPTER 2

The Status and Role of Reverse Engineering

2-1 The Status of Reverse Engineering in References

Widely used by practicing engineers (for far longer than most would imagine, as the next chapter makes clear), *reverse engineering* has been sadly neglected as an explicit topic in the majority of the large number of engineering design textbooks and references that are available. The first issue is that the overwhelming majority of books on the topic of "reverse engineering" focus on the technique's use with software and *not* hardware. Worse yet, in the opinion of the author, some of the titles of these books on reverse engineering software—not to mention the abstracts—make one wonder about the propriety, no less the legality, of the information being provided.[1] A second issue is that, more often than not in books on the design of hardware, the method is hinted at implicitly in the guise of looking at the design of an earlier entity as a guide to improving the design of *that* entity. The term *reverse engineering* is not even used, and, surely, the method would not be traceable by searching the index of the book. When the technique is explicitly addressed, it is often only in a few sentences or paragraphs, almost as if in passing.

A third issue is that, in three books dealing with the design of hardware where reverse engineering is explicitly found in the title, the technique is relegated to very narrow applications. In one, a superb book on product design—versus broader, and generally more demanding, engineering design—the technique of reverse engineering is only addressed for creating a new product based on an older product (ref. Otto and Wood). In fact, herein, in this book, the author treats the technique of reverse engineering for its full potential, without restrictions on use other

---

[1] Titles of books devoted to the reverse engineering of computer internals, operating systems, assembly language, and software include: (1) *Reversing: Secrets of Reverse Engineering*, Eldad Eilam, Wiley, 2005; (2) *Design for Hackers: Reverse Engineering Beauty*, David Kadavy, Wiley, 2011; (3) *Hacking: The Art of Exploitation*, John Erickson, No Starch Press, 2008; and (4) *Hacking the Xbox: An Introduction to Reverse Engineering*, Andrew "Bunny" Huang, No Starch Press, 2013.

than it being for ethical motives and within legal restrictions of the prevailing authority. In another reference, reverse engineering is specifically defined as "essentially the development of the technical data necessary for the support of an existing production item developed in retrospect as applied to hardware systems" (ref. Ingle). Nebulous, yes? It is subsequently defined (three pages later) as "a four-stage process for the development of technical data to support the efficient use of capital resources and to increase productivity." Better, perhaps, but narrow. Again, the book you are presently reading covers the full potential afforded by this essential—and ages-old—problem-solving technique. Well beyond for the purpose of developing data, what is covered herein develops knowledge and promotes learning, as there is far more to learning than data. In yet a third book, reverse engineering is promoted in the title, via use of a colon, thus: ". . . [the] Technology of Reinvention" (ref. Wang). This is, in this author's opinion, a sad testament to engineering ingenuity and creativity. It implies that reverse engineering plays a role in reinvention but not in invention, which is totally incorrect! Hopefully, the present work will make it very clear that reverse engineering has greater value for creating anew than for simply re-creating! It is—or could and should be—a stimulus for ingenuity and creativity, not a shortcut for laziness or complacency or mindless mimicry.

A final recent addition is an edited book that is a collection of *essays* on reverse engineering (ref. Raja and Fernandes). It may be good; it may not be good. It all depends on whether the essays fit together to present a full picture of reverse engineering or whether there is a particular essay that addresses the reader's concern and interest. In either case, collections of essays—each with different authorship—seldom, if ever, present a complete and flowing treatment of a topic, which a topic as important as reverse engineering needs and deserves.

Believe it or not, it is *not* the intention of what has just been written to demean these other works. They absolutely have a niche. The point intended is this: Reverse engineering is far more than a niche technique or technology. There is a *really* big void in the design literature that needs filling. Reverse engineering is a much more versatile and valuable design tool than has been suggested so far in published books, implicitly or explicitly.

2-2 Reverse Engineering Defined

Within the context of physical hardware (as opposed to software, which is not addressed in this book), *reverse engineering* is the process for discovering the fundamental principles that underlie and enable a device, object, product, substance, material, structure, assembly, or system through the systematic analysis of its structure and, if possible, its function and operation.[2] Usually, it involves taking the aforementioned apart and analyzing its makeup and workings, subsystem or subassembly by subsystem or subassembly, part by part, and feature by feature, until the entire entity has been analyzed and is understood. The most common objective of the process is to understand the device, object, product, substance, material, structure, assembly, or system well enough to allow a new one capable of doing essentially the same thing or fulfilling essentially the same role (albeit, perhaps better or, perhaps, not to the same degree but at a lower cost)

---

[2] Reverse engineering has applicability in software, as well as hardware, but use for software is not treated in this book because it is well treated in several other works, and, quite honestly, frequently seems to drift over the line of ethical use. Example references include *Exploiting Software: How to Break Code,* Greg Hoglund and Gary McGraw, Addison-Wesley, 2004; *Reversing: Secrets of Reverse Engineering,* Eldad Eilam, Wiley, 2005; and *Design for Hackers: Reverse Engineering Beauty,* David Kadavy, Wiley, 2011. The interested reader needs to be discerning and adhere to ethical conduct.

without using or simply duplicating (without fully understanding) all or some key portion of the original. However, in the remainder of this chapter it should be clear that reverse engineering has the potential—and, thereby, affords the opportunity—for much more varied purposes and goals. The technique can be—and is—useful for mechanical or electrical, or electromechanical or mechatronic, entities, but can also be applied to a chemical substance (which has been tried with Coca-Cola innumerable times!) or to biological entities, often with the goal of mimicking all or some key aspect of the chemical or biological entity (e.g., flavor of a Coca-Cola look-alike or functionality in an artificial joint or artificial organ).

Quite simply put: Reverse engineering is, most often, and quite literally, a process of "mechanical dissection," fully analogous to its biological counterpart discussed in Section 1–5.

Another somewhat underrecognized, underplayed, and underappreciated aspect of reverse engineering is that it is a superb example and embodiment of backward problem-solving. In *backward problem-solving* (ref. Messler, Chapter 34, pp. 237–243), one begins with the end result of a design or some other problem-solving activity and attempts to discover the starting conditions, as well as, often, the path(s) from beginning to end. In other words, one begins with the solution and attempts to discover the method by which the problem can be solved. This is distinctly different from the more conventional technique of forward problem-solving predominantly taught in engineering schools wherein starting conditions (materials, energies, forces, information) are given and the outcome is sought, often via the solution of mathematical equations.

The quandary is: The practice of engineering is generally involved with obtaining a known or desired outcome or goal and trying to find out a way—if not the best way—to get there. Said another way: Forward problem-solving takes you where it takes you. So, unless you are very clear about where you want to end up—and check progress along the way—there is risk of not ending up where you need to be. Backward problem-solving, on the other hand, takes you where you know you want to end up because where you want to end up is where you begin solving the problem. Finally, forward problem-solving (in forward engineering) involves mostly synthesis, while backward problem-solving (which is the key to reverse engineering—or vice versa) involves mostly analysis.

Figure 2–1 schematically illustrates the differences between forward and backward problem-solving.

The method by which reverse engineering is done, from a cognitive standpoint, is deduction. *Deduction* is defined as "the deriving of a conclusion by reasoning; inference in which the conclusion about particulars follows necessarily from general or universal premises." One looks at what is there to see and derives a conclusion (or conclusions) simply by reasoning why it looks as

In forward problem-solving:

Model parameters or observations (inputs) $\Rightarrow$ **data or knowledge (outcomes)**

...relying mostly on synthesis

In backward problem-solving:

Data or knowledge (outcomes) $\Rightarrow$ **model parameters or observations (inputs)**

...relying mostly on analysis

Figure 2-1 Schematic illustrating the difference between forward and backward problem-solving techniques.

it does or is as it is. It is, in fact, a challenging and, often, exciting combination of solving a puzzle or mystery and decoding a secret. And who doesn't like decoding a secret?

Hopefully, it is already clear that reverse engineering is not simply blindly copying—or duplicating or reproducing—what has been done before. Rather, it involves seeing—experiencing—what has been done before as a means for learning what is possible in the future, which, hopefully, is much more than developing a new product from an old product, reinventing a tired invention, or collecting data as opposed to gathering knowledge and learning from experience.[3]

2-3 Motivations for Reverse Engineering

There are many motivations for reverse engineering that go well beyond—and are far more varied than—what existing references on the topic suggest and address. The motivation(s) for reverse engineering (of hardware) include the following:

- *Military or commercial espionage,* with the goal of learning about an enemy's or competitor's latest research or development by stealing or capturing a prototype and dismantling it, often to develop a similar product. (As discussed in Chapter 3, the former is a necessary evil in a world plagued by evil, while the latter is *not* an ethical use—and may be an illegal use—of reverse engineering!)
- *Competitive technical intelligence,* with the goal of understanding what competitors are actually doing, versus what they say or imply they are doing. (It is important, and can be perfectly legitimate, to assess a competitor's capability versus one's own capability.)
- *Product security analysis,* with the goal of examining how a product works to determine its specifications, estimate it cost(s), and, perhaps, determine whether there has been potential infringement on a patent or, alternatively, whether there is a way to remove patent or copy protection and/or circumvent access restrictions. (This often involves acquiring sensitive data by disassembling and analyzing the design of a device, object, component, material, substance, structure, or entire assembly or system, and could, easily, cross ethical, if not legal, lines.)
- *Improve documentation shortcomings,* with the goal being to fully document a system for its design, production, operation, maintenance, or repair when shortcomings exist (all the way to no documentation existing!) and the originator is no longer available to offer improvements. (See Chapter 16.)
- *Academic/learning purposes,* with the goal being to understand the key issues of an unsuccessful design and, subsequently, improve the design. Within this category could be trying to understand what an obsolete—and undocumented—device or system was intended to do (see Chapter 8).

An illegitimate motivation (considered in Chapter 15), not endorsed here, is:

- *Creation of unlicensed/unapproved duplicates,* with the unethical (if not also illegal)

---

[3] Before there were engineering schools, and before there was any degree in engineering, there was what anyone would recognize as an engineer. They were there when the pyramids were built, when the siege machines of the Middle Ages were conceived and built, and, as is apparent in a wonderful book about four generations of *Treasure Island* author Robert Louis Stevenson's ancestors, entitled *The Lighthouse Stevensons* (Bella Bathurst, Harper-Collins Publishers Ltd., 2000), in late-eighteenth-century/early-nineteenth-century England to build some of the most formidable guardians of sailors ever created. These engineers learned engineering by an *apprentice system*; working alongside a "master engineer," learning from observation and experience. Reverse engineering was a vital part of how they learned (see Chapter 3).

TABLE 2-1 Motivations for Employing Reverse Engineering (with Examples)

Military espionage	German jerry can by British and U.S.; German V2 rocket by U.S. and U.S.S.R.
Industrial espionage	Attempted copying of Coca-Cola formula
Competitive technical intelligence	One automobile company watching all others through their production models; smartphone manufacturers watching competitors
Product security analysis	Some clones of original IBM desktop computers; Intel processors
Improve documentation shortcomings	Replacement of broken or worn parts on U.S. Navy ships when the original manufacturer is no longer in business (via RAMP, see Chapter 18)
Academic/learning purpose	Understanding the *Antikythera* mechanism (Chapter 16)
Creation of unlicensed/unauthorized duplicates	Rolex look-alikes, Coach knockoff handbags, etc.

goal of producing a look-alike or knockoff misrepresented to be the original. (Cloning of IBM's desktop computers by companies anxious to enter the marketplace in the 1980s, by "mechanically dissecting" and reverse engineering IBM's product, was neither illegal nor unethical, as the originals were purchased on the open market. However, duplicating the original and misrepresenting it as an original—as was, and is still, done by some foreign countries—is neither legal nor ethical!)

In Section 2–6, the potential value of reverse engineering for stimulating entirely new ideas by analogy or similarity, even when remote, is discussed.

Table 2–1 summarizes these motivations for using reverse engineering.

Various of these reasons or motivations are discussed elsewhere throughout the book, in specific contexts.

2-4 Engineering Design and the Engineering Design Process

Before moving on to study the use of reverse engineering as a powerful design tool, it is important to have a proper sense of what engineering design is and what the engineering design process involves.

The verb *design* is defined in several ways, with those deemed most appropriate to the goal of this book being: "to conceive or fashion in the mind"; "to plan out in a systematic, usually graphic, form"; "to create or contrive for a particular purpose or effect." The keys are these:

1. The process of *design begins in the mind,* which is that portion of the human brain in which thought originates that cannot be found by dissection any more than a human soul can be found, yet both surely exist.

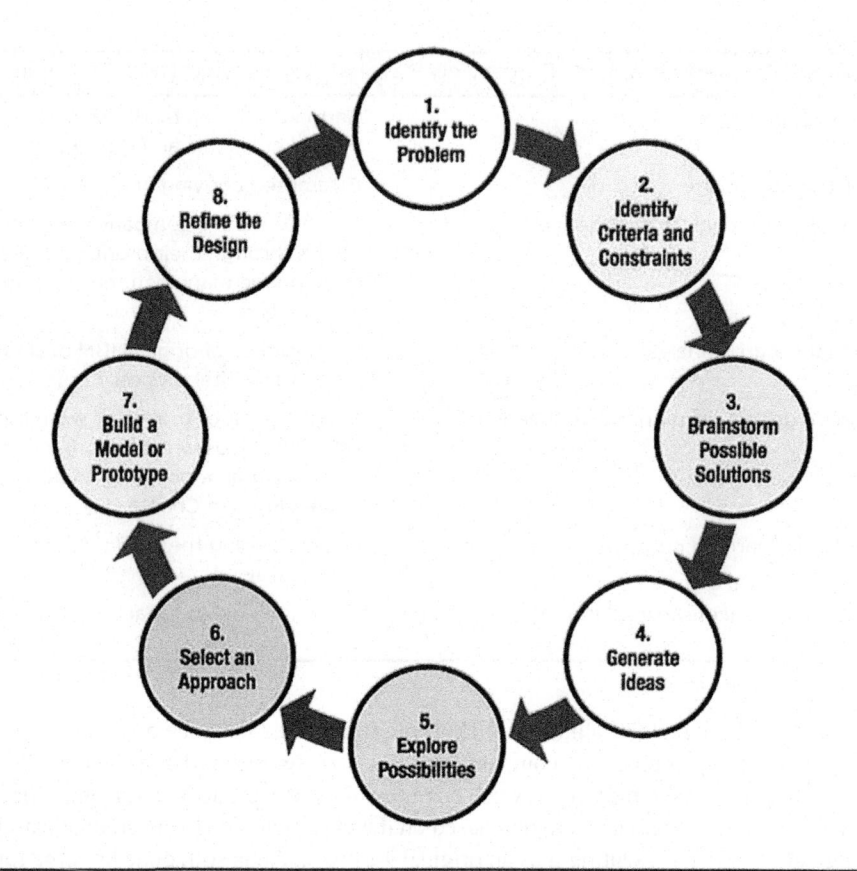

Figure 2-2 Schematic depiction of the engineering design process. (*Source:* NASA.)

2. What begins as *an abstract thought or idea is transformed into a physical* (often initially, graphic) *form* for others—or one's self—to see and execute.
3. The *design is a creation with a purpose,* with the purpose being known, and the design process being the pathway chosen via a backward problem-solving technique involving analysis.
4. Once the initial idea forms in the mind of the designer, *the process* by which it evolves and becomes real *involves synthesis.*

The result of the action of design is *the* design. The process of *engineering design,* which is—or should be—particularly rigorous, involves both analysis *and* synthesis, neither to the exclusion of the other and, usually, performed iteratively (i.e., repeatedly by completing a loop—or iteration).

Figure 2–2 is a schematic depiction of the engineering design process from the U.S. National Aeronautics and Space Administration (NASA). The first six steps are largely mental—as opposed to physical—although Steps 5 and 6 could—and often do—involve the use of physical models for proof of concept in Step 5 and experimentation and testing in Step 6 (ref. Messler, Chapters 18, 19, and 20, pp. 141–158). Step 7 definitely involves the creation of a physical model, often, by the

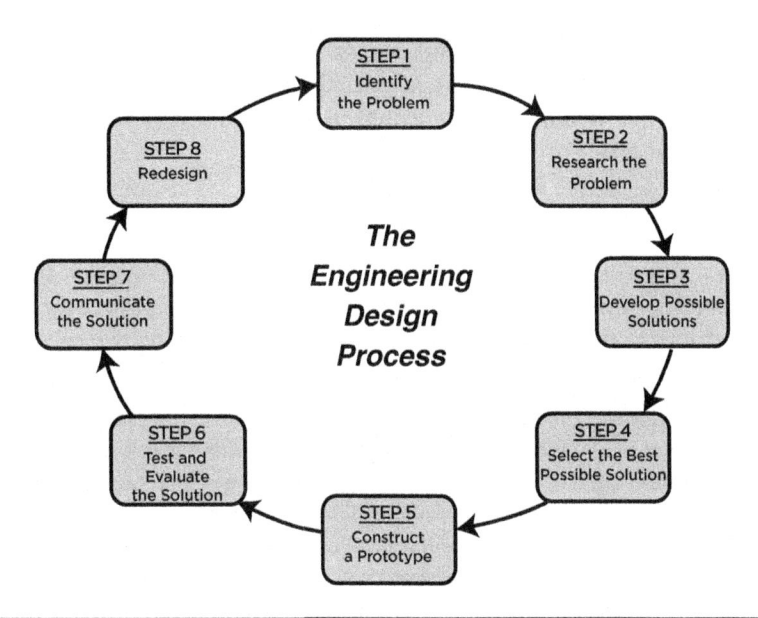

Figure 2-3 Schematic depicting the more traditional stages involved in a systematic and rigorous engineering design process.

time it is acknowledged by most to be a "prototype," having had technical feasibility demonstrated with proof-of-concept models, unproven or questionable technologies or concepts having been proven with experimental models, and functionality and performance having been assessed with test models. Step 8, which involves refining the design, is, again, largely a mental process that derives its motivations and goals from results with models, and is, in fact, an iterative step. Inputs are the outputs of earlier steps, and these inputs are used to create a new (hopefully, improved) output, going back to at least Step 5, and perhaps Step 4, to run through the process again and again until the output meets needed or desired requirements and goals.

Figure 2–3 is a more traditional schematic depiction of the steps involved in a systematic and rigorous engineering design process. A look at the steps makes clear the importance of moving back and forth between analysis and synthesis (e.g., Step 2 to 3, then Step 3 to 4), the reliance on experimental measurements beyond mathematical analysis (e.g., Steps 5 and 6), the importance of communicating (Step 7), and the importance of iterating until the target has been sufficiently converged upon (Step 8).

What sets engineering design apart from other forms of design are several things, including: (1) the inherent need for rigor, which most often derives from mathematical analysis, and (2) a model-centric and systems approach. Use of a *model-centric approach* is often either overlooked or ignored, but it is extremely important. An engineering design *must* work. To work, it *must* have been proven. For it to be proven, it *must* have been modeled, and not simply as a computer-based simulation (which is, at best, a mathematical analysis, and, at worst, is—heaven only knows!), but as one or more of several types of physical/mechanical models (ref. Messler, Chapters 18–22, pp. 141–164). If one seeks to create a physical entity, one must employ physical models on the way. By *systems approach* is meant: considering the entire entity as the integrated sum of all of its parts. It involves always keeping the end goal in sight and the overall goal in context, that is, never losing sight of the "big picture." At the same time, it involves meticulous attention to details, as "the Devil is in the details."[4]

2-5 Types of Design

Engineering design is commonly classified into types as follows: (1) *original design* or *new design* (or, in some contexts, *inventing*), (2) *adaptive design,* and (3) *variant design* (or *modification*). The author, and others, tend to prefer to subdivide adaptive design into two subclassifications): (a) *adaptive design* and (b) *developmental design.*[5] Here's what these denote:

- *Original design* involves creating a solution to a problem for which no design previously existed, at least not in any recognizable form. One starts with a blank sheet of paper or a blank computer screen and creates or invents a solution to a given or perceived problem. The resulting design may be totally new—with no previous design ever having existed—or it could be just so radically changed in the way the particular problem being addressed was ever solved that it is unique. (This risk is that the world may not know it needs the invention!)
- *Adaptive design* (which some, confusingly, call *synthesis,* as synthesis is the cognitive process involved in the process of every engineering design) involves evolving—or adapting—a known design or solution to a changing task or need, often, but not only, to resolve some shortcoming. The key for the subclassification of this type as "adaptive design" is that the motivation for the changed design is a market pull.
- *Developmental design,* as a subclassification within adaptive design, as an existing design is being changed, is motivated and enabled by a technology push. A new material, a new method-of-manufacture, -construction, or -processing, or a new technical capability is the driver for an adaptation, not a market pull. The user of the original design might be quite satisfied but is unaware of the opportunity for some dramatic improvement in capability, performance, reliability, or reduced cost that is possible because of a new technology.
- *Variant design* (or *modification*) involves varying the parameters (size, geometry, material properties, control parameters, etc.) of certain aspects of an existing design to develop a different design. In most cases, variant design involves primarily scaling (ref. Messler, Chapter 10, pp. 87–90).

Table 2–2 gives examples of the types of engineering design just described.

Reverse engineering has applicability to all of the types of engineering design described.

---

[4] An idiom that derives from an earlier phrase "God is in the details," which expresses the idea that whatever one does should be done thoroughly. It sounds a lot like what virtually all fathers tell their children as the children are maturing: "If it is worth doing, it is worth doing right!"

[5] Some references in design refer to *product development* as something akin to, but different from, engineering design. This author respectfully disagrees, feeling that product development most often involves adaptive or development design, depending on whether the driver for the new product is the need *or* opportunity for some improvement. This said, product design/development, right or wrong, is becoming, more and more, focused on ergonomics and aesthetics and is more often than not done by experts in computer-aided design (CAD) who are not degreed engineers. Hence, rigorous analysis is frequently lacking.

TABLE 2-2 Examples of the Different Types of Design in Engineering

	Original Design	Adaptive Design	Developmental Design	Variant Design
Televisions	Small-screen analog B&W CRT	Larger screens; analog color CRT	Digital; HD; plasma; LCD	Larger screens; mini-screens
Telephones	Crank analog; operators	Dial analog; automated	Digital push-button; wireless	Size variations; BlueTooth
Automobiles	Carbureted gasoline	Fuel-injected gasoline	Electric and hybrid	Different engine sizes (power)
Airplanes	Fixed-wing; single-engine; propeller	Multi-engine; propeller; commercial	Jet engine(s); swept-wings; morphing wings	Different sizes of same basic aircraft; stretch bodies
Home energy	Fire for heat; fossil fuel for light	Gas lighting; central heating	Electrification; solar; wind; nuclear	Increased power capacities in homes

2-6 Uses for and Benefits and Risks of Reverse Engineering

By now, it should be clear that there are many uses for reverse engineering in engineering design, as well as in product design or development, at least in terms of the overall motivations (Section 2–3). Before leaving this general introduction to reverse engineering, however, it is important that the reader, as a new or potentially new user of the technique, have a fuller understanding and appreciation of the full potential of the technique, beyond the aforementioned motivations, that is, of the potential benefits. As for any new experience, however, it is equally important, along with potential benefits, to alert one to the potential risks, as there is almost always, in engineering, a trade-off between risks and benefits.

Intentions and motivations are one thing; action and outcomes are another. Good intentions alone cannot save the world. Good actions might. In a similar vein, outcomes are usually more important than motivations, whether the motivations were good but the outcome a failure, or the motivations may not have been perfectly pure or worthy but the outcome proved useful or valuable or worthy.

If this sounds too philosophical, it is not entirely intended to. Rather, it is to try to separate for the reader (and the author) the difference between the *potential motivations* for employing reverse engineering and the *potential uses* of reverse engineering. Section 2–3 considered the potential motivations. This section considers the potential uses.

In reflecting on the uses and compiling a long list, it seemed appropriate to attempt to divide potential uses into logical areas. As this book is all about using the technique of reverse engineering to accomplish something meaningful and worthy, it seemed most logical to divide potential uses into categories that have significance in engineering components, parts, devices, objects, products, structures, assemblies, and/or systems. For the purposes of this discussion, "engineering" is being used in its broadest context to rely upon and interact with market need/demand (i.e., marketing), design, manufacturing/production/construction, and quality assurance. When all of these are addressed, two things become apparent: (1) some potential uses span more or fall into more than one category, and (2) there may be a few overarching uses that merit special mention and attention. Admittedly, not all potential uses, like not all potential motivations, involve marketing, but it is important to engineering, in general, to include marketing here.

So here goes!

Potential Uses for Marketing

It is most important in business (as well as in national defense) to know where one stands relative to one's competition (or enemies). This requires *benchmarking* one's design/product against all others. This is the first, and potentially most important, use of reverse engineering. Knowing where one stands compared to one's competition, and finding one is not up to that competition, one needs to take action. Two possible actions where reverse engineering can be used to advantage are: (1) as a way to make a comparable design/product (in terms of functionality and performance) at less cost (which might equate to lower price) and (2) as a way to obtain improved or enhanced functionality and performance without increasing cost (and, thus, price). Both of these goals clearly impact engineering, as engineering responds to marketing in commercial companies.

Two other potential uses relating to marketing are: (1) to make a defective or deficient design/product better and (2) to uncover an opportunity for an entirely new, but often logically related, design/product. The former use clearly relates to quality assurance, addressed later.

Potential Uses in Design

The most important, if not the most common, potential use of reverse engineering is to uncover any uncoordinated features of a design/product to correct these and improve the design/product in the process. These may be incompatible features or, simply, improperly or incompletely integrated features.

Relating to both design and manufacturing/production/construction is the potential use of reverse engineering to replicate or re-create parts for which tooling no longer exists or is unknown. In such cases, the part is used, more or less, as a pattern from which the tooling must be designed, using appropriate corrections for shrinkage during casting or molding, and so forth. Tooling, in this context, includes dies, molds, and fixtures.

Potential Uses in Manufacturing/Production/Construction

The two greatest uses are to lower costs of the entity itself and to improve the efficiency of the process of manufacture, production, or construction. The former is generally part of what is known as *value engineering*, while the latter involves one or more processes of *methodizing* or *producibility*. These topics are treated in Chapter 14.

It is almost always a goal to reduce the time to market, which involves moving more efficiently and smoothly from marketing to design, from design to manufacturing, and from manufacturing to market. Reverse engineering can play a role in any or all of these transitions.

Potential Uses in Quality Assurance

Where not so long ago (i.e., pre-1980s) the accepted mantra was "Quality costs money," it became clear after the 1980s that the mantra should be "Lack of quality costs money." After all, a customer lost to a competitor due to a problem with quality is very difficult to ever get back. Hence, reverse engineering finds several potential uses to help improve (or, at least, maintain) quality. The first key use is to troubleshoot a defective or deficient design/product to make it right. A second potential use is to increase an organization's ability to maintain a manufacturing capability at its peak performance via improved documentation (ref. Ingle).

Overarching Uses

Finally, as two overarching uses for reverse engineering, there are these: First, to discover new technologies or technological principles in a design/product that can have positive ramifications far beyond the design/product analyzed. The alert engineer is always vigilant about finding new technologies or technological principles, as we are, after all, *technologists*. Second, reverse engineering can be used to great advantage to discover new concepts in the design and manufacture of a part, component, device, object, product, structure, assembly, or system. As can be seen from a quick scan of the table of contents of this book, properly done, reverse engineering attempts to understand a design/product from the geometry of its parts and the arrangement of these parts or details (Chapter 7), the materials used in construction of parts or details (Chapter 9), and the processes used in fabrication of parts and details and processing of materials (Chapter 10) used to create the entity.

Table 2–3 summarizes the potential uses of reverse engineering.

TABLE 2-3 Summary of the Potential Uses for Reverse Engineering

For marketing-drive needs or opportunities:
- Benchmarking
- Reducing cost without reducing functionality and/or performance
- Improving functionality or performance without increasing cost
- Making a defective or deficient design better
- Uncovering an opportunity for a new design or product

For design-driven needs or opportunities:
- Uncovering any uncoordinated features of a design or product
- Re-creating missing tooling to allow production

For manufacturing-driven needs or opportunities:
- Reducing direct costs for materials, purchased items, and processes, as part of value engineering (done in cooperation and coordination with design)
- Improving efficiency through processes of methodizing (to properly identify and sequence necessary operations) and producibility (to make parts and their assembly easier)
- Reducing time to market (in cooperation and coordination with marketing and design)

For quality-driven needs or opportunities:
- Troubleshooting a defective or deficient design or product or a deficient process
- Increasing the ability to maintain manufacturing capability at peak performance via improved documentation of a design/product

Overarching Uses:
- Discovering new technologies or technological principles to allow innovative new designs/products
- Discovering new concepts for geometry (of parts or structures), arrangement (of parts in an assembly or elements in a structure), material-of-construction, and method-of-fabrication or -processing.

The benefits of reverse engineering are simple:

- More intimate *knowledge* of what has already been accomplished in the past as a guide for what might be accomplished in the future.
- More rapid development of *experience* for one's self by seeing what others have accomplished.
- Greater *appreciation* of and *insight* into the *heritage* left to us by our forebearers.

The risks from using reverse engineering are few but profound. The greatest risk is that one can become lazy—and/or complacent—leaving ingenuity and creativity and hard work to others. Using reverse engineering can stymie innovation, if one just copies what one sees, making few, if any, improvements. There is also the risk of copying from a fool, in which case bad design is perpetuated. (This has happened in industry more than once, with the "fool" sometimes being the company itself when a flawed design is blindly copied!)

The other risks discussed in other books (ref. Ingle) seem to this author to be associated more with misuse than with use of the technique. A caution that "only 1 out of 10" exercises in reverse engineering might result in a viable new design/product is unfounded. The statement or statistic may be true if one uses reverse engineering as a crutch instead of as a tool. Without meaning to be insensitive: A tool helps the able, while a crutch helps the disabled. Engineers need to be able. If they are, there's little risk. Properly done, by good engineers, reverse engineering becomes second nature. Not every new design/product results from the actual "mechanical dissection" of an old design/product, but every design/product should, somehow, be influenced by the intellectual awareness of what has been done successfully before!

2-7 Summary

Published books addressing reverse engineering applied to hardware are a rarity. Even with explicit mention of the technique (by name), treatment is usually done in several sentences or in a few paragraphs more in passing than as any serious attempt to relate the tremendous potential of the technique in engineering design. When treated explicitly, reverse engineering is relegated to a few narrow areas of application, greatly un-representing (or perhaps understanding) the full potential. The goal of this book is to change this misconception forever.

Reverse engineering is, quite simply, "mechanical dissection," involving the taking apart of hardware to see what it does or did, to see how it works or might have worked, and to see from what and how it was made. The technique is a backward problem-solving technique in which the end goal is known and the path to that goal, as well as the proper starting point, are sought by deduction.

There are many motivations for reverse engineering, some of which are quite legitimate, as well as legal and ethical, but others are ripe for abuse.

Engineering design is a rigorous, step-by-step process and, more often than not, requires iteration to achieve optimization. Four types of design—original, adaptive, developmental, and variant—are all amenable to reverse engineering.

The uses of reverse engineering are many, whether driven by marketing demands or opportunities, design proper, manufacturing, or quality assurance. Benefits are also many, while risks are few but profound—the greatest being stagnation of imagination and innovation.

2-8 Cited References

Ingle, Kathryn A., *Reverse Engineering*, McGraw-Hill, New York, 1994.

Messler, Robert W., Jr., *Engineering Problem-Solving 101: Time-Tested and Timeless Techniques*, McGraw-Hill, New York, 2013.

Otto, Kevin, and Kristin Wood, *Product Design: Techniques in Reverse Engineering and New Product Development*, Pearson/Prentice-Hall, Upper Saddle River, NJ, 2000.

Raja, Vinesh, and Kiran J. Fernandes, *Reverse Engineering: An Industrial Perspective*, Springer, New York, 2010.

Wang, Wego, *Reverse Engineering: Technology of Reinvention*, CRC Press, Boca Raton, FL, 2010.

2-9 Thought Questions and Problems

2-1 Most discussions of *reverse engineering,* which, regrettably, appear in only a small fraction (maybe one-fourth) of textbooks devoted to design, relegate this potentially powerful problem-solving technique to incrementally improving earlier design(s) of one's own product or a competitor's product by largely copying what was done before and making a few subtle modifications or one or two more significant changes. Worse is the fact that all but two books devoted to the topic (ref. Ingle; Wang) focus exclusively on certain aspects of computer hardware (Eilam) or software (Hogland and McGraw). Worst of all, several books on reverse engineering of software raise serious questions as to the ethicality, if not the legality, of their treatment (Kadavy, Erickson, Huang). [Titles by Eilam, Kadavy, Erickson, and Huang are documented in Footnote 1 in the chapter.]

Briefly express your opinion on the following uses of reverse engineering:
a. "blindly," "mindlessly," or "artlessly" *copying* another's design
b. "blindly," "mindlessly," or "artlessly" *copying* one's own design
c. "hacking"
d. "exploiting"

Be sure to consider how you would feel if it was your design that was being "copied," "hacked," or "exploited"!

2-2 a. Briefly discuss how *reverse engineering* is a "backward problem-solving technique."
b. Give and briefly discuss *two* examples of your own creation, experience, or finding (e.g., on the Internet) where one actually finds it necessary or especially useful to solve a problem backward, i.e., using a known outcome or effect to determine or discover an initial condition or cause.

2-3 There are many motivations for engaging in *reverse engineering* (see Table 2–1 for the list), indicating that the technique can be used for more than just product development (e.g., ref. Otto and Wood) or artless copying (ref. Wang).

From your own experience or work, or by using the Internet, give and briefly describe *two* examples for each of the following motivations:
a. military espionage
b. commercial espionage
c. competitive technical intelligence gathering

d. product security analysis
e. improving documentation shortcomings
f. academic/learning purposes

2-4 Engineering design is commonly classified into types as follows:
1. original or new design
2a. adaptive design
2b. developmental design
3. variant design

From your own experience or work, or by using the Internet, give and briefly describe *two* well-known examples of a design that represents each of the types in the preceding list.

2-5 There are several potential *uses* (as differentiated from *motivations*) for reverse engineering given in Section 2–6.

From your own experience or work, or by using the Internet, give and briefly describe *two* examples for each of the following potential uses:
a. for marketing
b. in design
c. in manufacturing, production, or construction
d. for quality assurance

CHAPTER 3

History of Reverse Engineering

3-1 The Likely Emergence of Reverse Engineering

Heniunu, son of Prince Nefermaat and his wife, Itet, grandson of former Pharaoh Sneferu, and nephew of Sneferu's oldest son and present Pharaoh Khufu, as architect of Khufu's "Horizon" (to be known as the Great Pyramid of Khufu, or Cheops, in Greek), looked up in shock from the scaffolding just below the ceiling of the King's Chamber, in which the worldly body of Khufu would lie safely awaiting the journey of his immortal soul to the afterlife over the horizon in the West. He was able to insert the tip of a reed he held in his hand into a crack that had formed in the immense 50-ton granite lintel (horizontal beam) that spanned from wall to wall. The roof of the chamber was in danger of collapsing, even before the Great Pyramid was finished. Something had to be done. The design had to be modified. He had to adapt.

It was almost 2560 BC, and reverse engineering was to be used, if not for the first time, surely for one of its greatest moments. Saving the Great Pyramid of the beloved pharaoh would leave the world its oldest and most magnificent engineering achievement of all time, the oldest and sole-surviving member of the seven wonders of the ancient world.

While no archeologist or Egyptologist speaks of "reverse engineering," nor does any historian, this was surely what took place. Even those who have written about reverse engineering as a valuable process for improving designs and for reinventing, often on the path to a new product, miss this seminal event. Virtually everyone attributes the emergence of the backward problem-solving technique of reverse engineering to much more modern, albeit uncertain, origin. Some will say the technique emerged with the emergence of the Industrial Revolution (AD 1740–1850). Others will say much later, perhaps World War II (AD 1939–1945). But they are *all* wrong! You know it, as the author knows it. It is only logical. And, as engineers, we are nothing if we are not logical.

An important engineering design—maybe the greatest design ever—was failing, even before it was finished. The first great (and still largest) pyramid was at risk! Fortunately, the likely architect (a student of the great Imhotep, "Father of the Pyramid") was intimately involved with

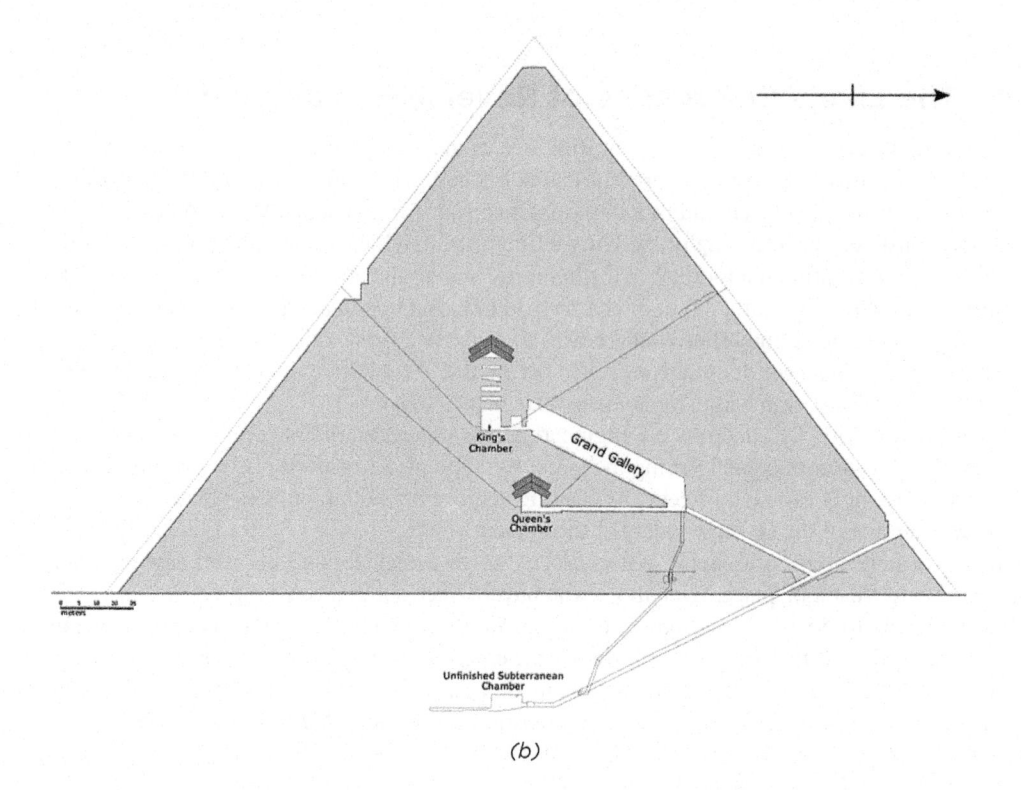

(a)

(b)

History of Reverse Engineering **31**

Khufu

5.24m 11.48m 5.33m

1. King's Chamber
2. Antechamber with Portcullises
3. Sarcophagus
4. Air Shaft
5. Horizontal Granite Beams
6. Supporting Limestone Beams
7. Gabled Roof
8. Stress-relieving Chambers
9. Grand Gallery

5.24m

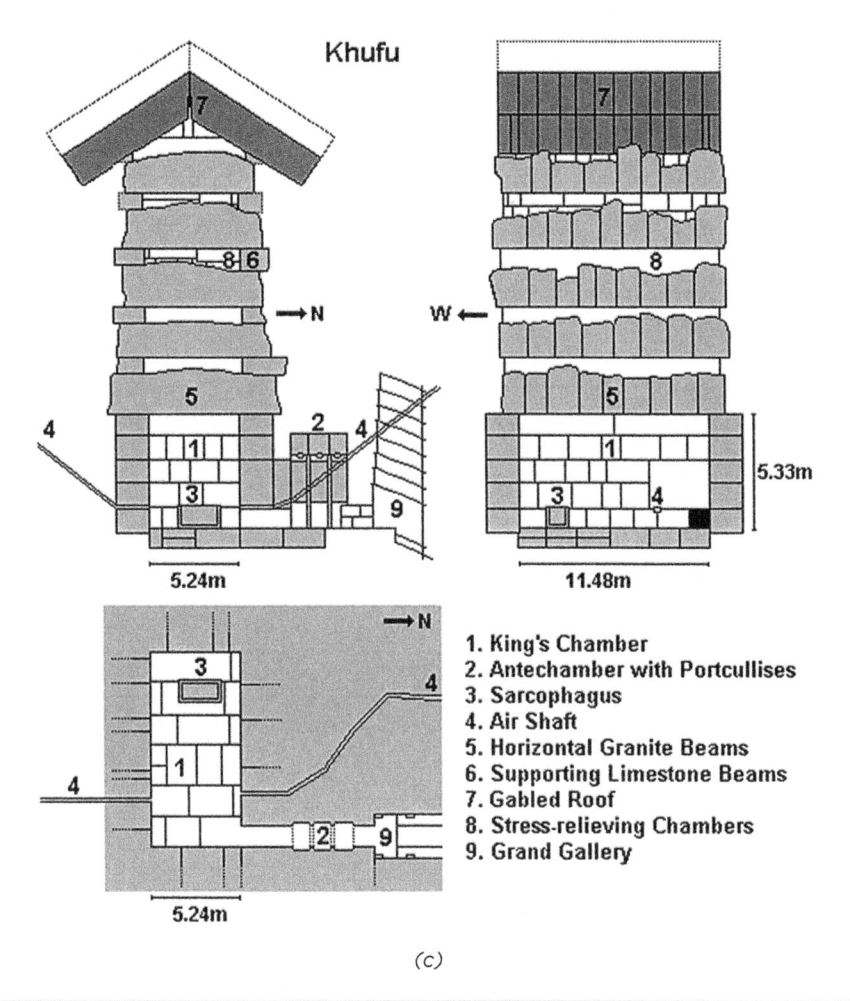

(c)

Figure 3-1 Khufu's Pyramid (known in ancient Egyptian as "Khufu's Horizon," in reference to the soul passing into the afterlife over the horizon in the West), oldest and largest of the three Great Pyramids of Giza (a). **Don't miss the color version of this figure, available at www.mhprofessional.com/ReverseEngineering.** The interior of Khufu's Pyramid showing the location of the King's Chamber (b). Schematic illustration of the King's Chamber in Khufu's Pyramid, showing special "relief chambers" added above the original stone-slab roof to deflect weight to the side walls, away from the beam's midspan (c) (*Sources:* Wikipedia Creative Commons, contributed by Minto on 20 August 2005 and modified by A. Parrot on 7 May 2010 [a]; Wikipedia Creative Commons, contributed by Jeff Dahl on 14 November 2007 and modified by Hardwigg on 22 July 2012 [b]; www.cheops-pyramid.ch, created by Franz Löhner at www.khufupyramid.ch and Teresa (Zubi) Zuberbühler at www.starfish.ch, with permission of Teresa Zuberbühler [c].)

the pyramid's construction. He devotedly overlooked every step of the construction—for almost 20 years already—by observing what happened as the great structure grew, and reacting to every flaw in the design. Some would say Heniunu was engaged in real-time engineering, which he certainly was. But, most significantly, he was learning with each new experience. He was adapting his design as required. And he was using observation of that which worked and that which was not working, correcting the latter "on the fly" and repeating the former with growing confidence. He was employing reverse engineering in its purest form: learning from the past to improve the future.

Some might argue that no "mechanical dissection" was being employed, no one was taking anything apart, as a means of learning what worked and what didn't work. No one had to! The structure was taking itself apart!

Heniunu was learning structural engineering as he constructed, from the structure. It's rather elegant, really. In seeing the great stone beam in the process of cracking, he knew he had to react. The unimaginable weight of the structure above the King's Chamber was causing the beam to fail. He needed to find a way to relieve the weight at the midspan of the beam. So what he did, according to James Spencer, deputy keeper of the British Museum's Department of Ancient Egypt and Sudan, was "deflect the weight of masonry over the core of the pyramid away from the roofing beam and out to the supports at each side of the chamber." The result was an ingenious modification and a major contribution to engineering—that is, reverse engineering (Figure 3–1).

TABLE 3-1 Summary of Design Changes in the Pyramids of Ancient Egypt

Mastabas ("Eternal House")		Volume/angles
■ Single pit in sandy desert; lined with reed mats	Narmer (3100–3050 BC)	
■ Mud brick construction	ca. 2900 BC	
■ Stone construction; two false doors on eastern wall for pharaoh's soul	Mastabet el-Fara'un (~2840 BC)	
■ Giant Mastaba	Shepseskaf (2510–2502 BC)	148,271 m^3/ 70°
Pyramids		
■ 1st pyramid; 4-level tomb of stepped mastabas		
■ 6-step pyramid	Djoser (2667–2648 BC)	331,400 m^3
■ Pyramid at Meidum (step pyramid)	Sneferu (2613–2589 BC)	638,783 m^3/51°50'35"
■ Bent Pyramid at Dashur (transition from step)	Sneferu	1,237,040 m^3/43°22' upper/ 54°50'35" lower
■ Red Pyramid at Dashur	Sneferu	1,694,000 m^3/43°22'
■ Great Pyramid at Giza	Khufu (2589–2566 BC)	2,583,283 m^3/51°50'40"
■ Pyramid at Giza/Great Sphinx	Khafra (2566–2542 BC)	2,211,096 m^3/53°10'

Modifying the design of the roof for the King's Chamber of Khufu's pyramid was but one example that reverse engineering was being used during the design and building of pyramids by the ancient Egyptians, who built 138 pyramids over a period of nearly 2000 years, ending around mid-600 BC. Other examples appear in the evolution of the pyramids' shapes, as they evolved from piles to tiered or stepped pyramids to true pyramids, and as the angle of the faces changed with refinement born of experience with what worked and what did not work. Table 3–1 summarizes some of these changes.

3-2 Reverse Engineering in the Middle Ages

Between the eleventh and sixteenth centuries in Europe, scores upon scores of great stone cathedrals were built by medieval architects and engineers (Figure 3–2). Carved from more than 100 million pounds (50 million kilograms) of stone, the largest of these still stand, although on the brink of future disaster without restoration (ref. *Nova*). Remarkable as these creations of humans to honor their God are, more remarkable still is the fact that the creators were not formally educated engineers. In fact, no one that practiced what most would recognize as engineering was *formally* educated (at a university) until the late seventeenth century (in Europe).[1] The builders of the great cathedrals were stonemasons, master builders trained in a rigid system of apprentice under a master. Fewer than 40 percent of these masons could write their own names, fewer still could read, and none knew any formal mathematics. But they did know how to create marvels with a square, a level, a plumb line, and a compass!

One learned from another, who had learned from another, ad infinitum. But each continued to learn from experience, using trial and error employing guided empiricism (using what one knew worked one place, for one purpose, in some other place for some other—albeit somehow related—purpose) (ref. Messler, Chapter 23, pp. 165–168).

These master builders, some believe, may have been descendants of the builders of ancient times, going back to at least King Solomon (970–931 BC), and, perhaps, to ancient Egypt and the builders of the pyramids. Whether they were or weren't, they shared something in common with the earlier builders. Like their forebearers, they walked the walls and halls of their constructions. In doing so, they practiced real-time engineering, using real-time feedback. They saw what worked and what did not work, and, like Heniunu before them, they reacted, changing what didn't work on the fly and employing what did work with growing confidence. And, not incidentally, they were most certainly employing reverse engineering.

A superb example of reverse engineering arose early in medieval cathedral building. The huge—immensely heavy—stone edifices incorporated vaulted stone ceilings. The vaulted ceilings produced huge lateral forces that acted to push supporting exterior walls outward. Seeing the problem, the builders had to react. To resist these forces and not have to abandon the vaulted ceiling design that appealed to the architect for its symbolism of "reaching toward heaven," the "engineers" (stonemasons) added reinforcing walls at right angles to the main load-bearing walls. Because of their resemblance to wings, perhaps, these were called "flying buttresses" (Figure 3–3).

---

[1] There is little doubt that the ancient and medieval world was blessed with real engineers, even if they weren't formally educated. Imhotep ("Father of the Pyramid" and inventor of the column), Archimedes (inventor of the screw and machines of war), da Vinci (inventor of the helicopter, tank, submarine, etc.), and Michelangelo (builder of great domes in cathedrals), to name a few, were engineers in every sense except having a degree in engineering.

34 Chapter Three

Figure 3-2 Cathedral of Notre Dame de Paris (AD 1163–1345; French Gothic style) (*a*), the Dom of Cologne, or Köln, (AD 1248–1473, when stopped; resumed in mid-1880s, completed 1880; Gothic style), Germany (*b*), and St. Peter's Basilica at Vatican City (AD 1506–1626; Renaissance and Baroque styles) (*c*). (*Source:* Wikipedia Creative Commons, contributed by Sanchezn on 25 November 2007 [*a*], Tetraktys on 13 November 2007 [*b*], and Tkgd2007 on 19 May 2008 [*c*].) **Don't miss the color version of this figure, available at www.mhprofessional.com/ReverseEngineering.**

(a)   (b)

Figure 3-3 Early flying buttresses (ca. 1170) at the eastern end of the Basilica of St. Remi in Reims, a short distance from Notre Dame Cathedral in Paris, France (a) and the Basilica of Santa Maria del Flore, the Cathedral of Florence, Florence, Italy, popularly known as "the Duomo," for which Michelangelo designed and oversaw the construction as a predecessor—and learning experience—for his great dome at St. Peter's (b). (*Source:* Wikipedia Creative Commons, contributed by Raggatt2000 on 6 February 2010 [a] and Bouncey2k on 26 June 2006 [b].) **Don't miss the color version of this figure, available at www.mhprofessional.com/ReverseEngineering.**

These initially real-time fixes for a problem that emerged as construction progressed soon became part of the design of later great stone cathedrals.

The flying buttress was but one of many design changes that had to be incorporated into these great structures as they grew in size and complexity through architectural styles of Early Christian, Byzantine, Gothic, Renaissance, and Baroque. Another famous one, also a result of reverse engineering, was executed by Michelangelo when he observed problems building a dome for the Florence Cathedral, more correctly the Basilica of Santa Maria del Flore and more popularly "the Duomo," as a predecessor of the greater dome he was to design for the Papal Basilica of Saint Peter at the Vatican (St. Peter's Basilica). The dome at St. Peter's rises 448.1 feet (136.5 meters) above the floor and has a diameter of 136.1 feet (41.47 meters), making it the largest dome of its kind in the world. To lighten the structure, having observed problems from the heavy weight of his earlier dome with uniformly thick sections in each of an inner and outer dome separated by stiffening ribs, Michelangelo designed reliefs into the dome's undersurface, providing stiffness where it was needed and a series of smaller and smaller, deeper and deeper stepped pockets to end up with a thinner and lighter shell where it was not. He died as only its base had been completed, around 1564.

Once again, reverse engineering was an active part of design and redesign early in medieval times, and, almost certainly from the time of the Great Pyramid of Khufu.

3-3 Reverse Engineering during the Industrial Revolution

It was 1701 in rural Berkshire County in England, north and west of London, and Oxford-educated Jethro Tull (1674–1741) had already improved agriculture by crumbling up the soil prior to

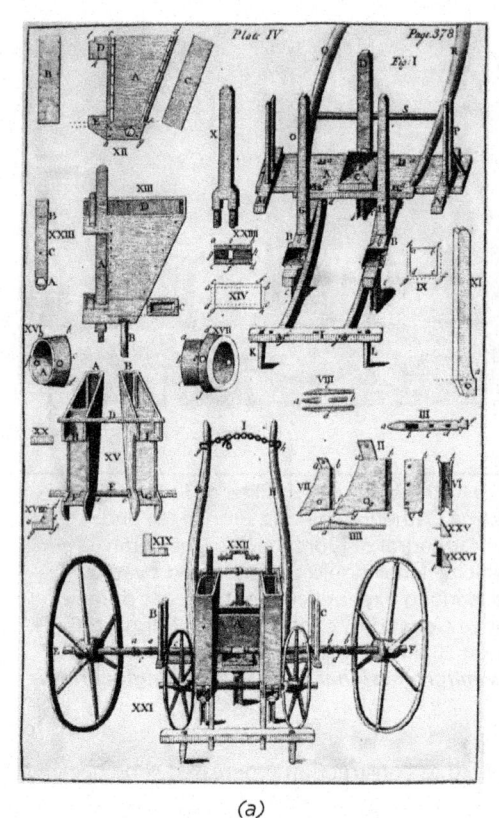

36 Chapter Three

Figure 3-4 Early British agriculturalist Jethro Tull's "seed drill" (a) that is credited by many historians as starting the Industrial Revolution in Europe in 1701, and a modern planting machine that employs the same basic concepts (b), albeit after many intermediate modifications via reverse engineering. (*Source:* Wikipedia Creative Commons, contributed by Bwwm on 15 November 2008 [a] and Mahlum on 6 May 2007 [b].)

sowing seed and by having learned that rotating crops extended the fertility of soil.[2] He was now to unknowingly start a revolution. By inventing a horse-drawn cart containing a "seed drill" and dropper device (Figure 3-4), he would begin the mechanization of agriculture and dramatically increase productivity henceforth.[3] More significantly, he paved the way for a host of quick-to-follow "machines" to hoe, till, plow, mow, reap, bale, and harvest, and, without knowing it, he began what was forever after known as the *Industrial Revolution*.

---

[2] *Jethro Tull* is better known, by many, as a British rock band formed in Luton, Bedfordshire, in 1967. The band initially played blues rock, but subsequently incorporated elements of classical music, folk music, jazz, hard rock, and art rock. The band has sold more than 60 million albums worldwide over a career spanning more than 40 years. The group took its name from the man.

[3] The Sumerians used a single-tube seed drill ca. 1500 BC, but the invention never reached Europe. Multitube iron seed drills were actually invented by the Chinese in the second century BC. These transformed agriculture in ancient China!

The period from around early 1700 (many would say 1740 or so) to about the 1830s or 1840s, is widely known as the Industrial Revolution,[4] although Jethro Tull's seed drill and subsequent horse-drawn hoe really got things going. With a start in Great Britain, it quickly spread to North America, then Japan, and then the rest of the world. Robert Emerson Lucas, Jr. (born 1937), an American economist at the University of Chicago, a Nobel Prize recipient in 1995 in Economic Science, and consistently ranked among the top 10 economists in the world, said: [With the Industrial Revolution,] "for the first time in history, the living standards of the masses of ordinary people have begun to undergo sustained growth. Nothing remotely like this economic behavior has happened before."[5] In fact, in the 200 years since, per capita income in the world increased ten times, and the population increased six times. People ate better and lived better and, apparently, for better or worse, procreated more.

With its start in agriculture, the role of mechanization and machines spread to textiles with the cotton-spinning water frame by Richard Arkwright (1732–1792) in 1768, and from this as a beginning point, through reverse engineering, to the "spinning Jenny" of James Hargreaves (1720–1779) in 1769 and then to the wedding of the spinning Jenny and the waterwheel in the "spinning mule" of Samuel Crompton (1775–1827) in 1779 (Figure 3–5). From textiles, where power came from moving water and waterwheels, a major leap forward happened with the invention of the steam engine, the history of the evolution of which is a superb example of one engineer learning from another and of using reverse engineering, with or without "mechanical dissection."

The earliest example of the use of steam for creating motion was a reaction-engine by Hero of Alexandria (AD 10–70) known as an *aeolipile* (Figure 3–6). No actual work was performed with the engine, but an idea was born! The first rudimentary steam turbine was built by Taqi al-Din in 1551. Jeronimo de Ayanz y Beaumont in 1606 and Giovanni Branca in 1629 built steam-powered lifting devices, while it took Thomas Savery to commercialize the first steam machine in 1698. Thomas Newcomen is widely credited, however, with inventing the first true steam engine in 1712, for which a wonderful animation can be found in Wikipedia under "Thomas Newcomen." A two-cylinder steam engine was invented by Jacob Leupold in 1720 (Figure 3–7).

Shortly following the development of a much-improved piston sealing system by James Smeaton ca. 1770, no doubt inspired by reverse engineering of less-than-robust steam engines, James Watt (1736–1819) invented a steam engine that improved greatly on the Newcomen engine. Newcomen's engine required that cold water be injected into the steam cylinder to cool and condense the steam and retract the piston for the next power-stroke. This caused all of the energy in the steam to be lost each cycle. Watt used a system of valves with a governor to allow two-way motion in the piston without cooling the steam and losing valuable energy (Figure 3–8). In 1775, Watt licensed Matthew Bolton to build his steam engines commercially, and a major business was born. The single greatest advance, however, was yet to come during the Second Industrial Revolution, which most historians would say began around 1850.

In 1884, Charles Algernon Parsons (1854–1931) linked the best available steam engine (invented as a refinement on a Boulton and Watt's engine by Gustav de Laval) to an electric-generator or dynamo, creating the first compound steam turbine (Figure 3–9). The age of steam

---

[4] Actually, it is now often referred to as the First Industrial Revolution, and it was to be followed by the a Second Industrial Revolution that began around 1850 in Europe and North America, and was even more profound, as it was based on steam engines, and then steam engine–electric dynamo hybrids, more than on waterwheels.

[5] Quoted in "Industrial Revolution: Past and Future," *2003 Annual Report to Federal Reserve Bank of Minneapolis,* May 1, 2004.

38 Chapter Three

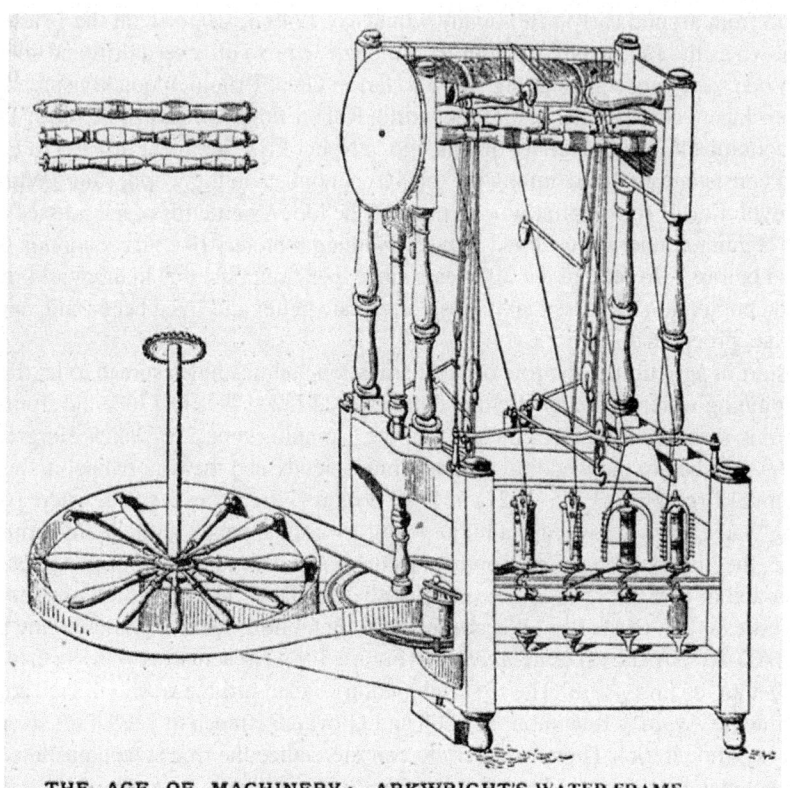

THE AGE OF MACHINERY: ARKWRIGHT'S WATER FRAME

(a)

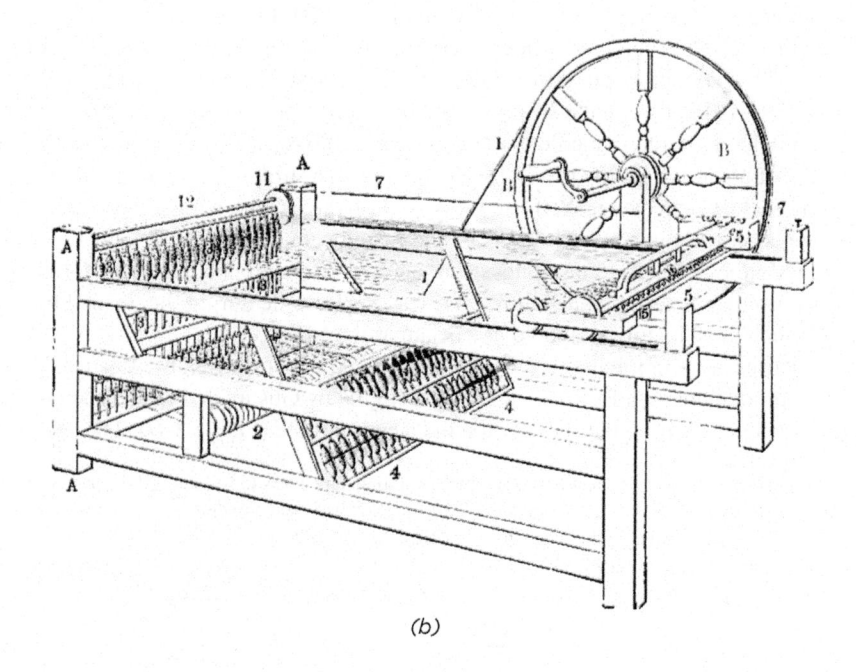

(b)

History of Reverse Engineering **39**

(c)

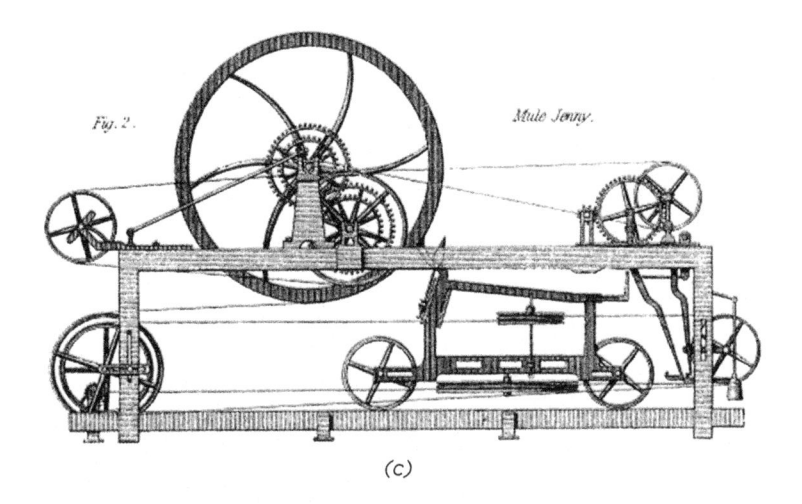

Figure 3-5 Depictions of Arkwright's "water frame" (a), Hargreaves's "spinning jenny" (b), and Crompton's "spinning mule," in which a spinning jenny was connected to a waterwheel (c). (*Source:* www.pixnet.co.uk, Wikipedia Creative Commons, from ClemRutter on 27 April 2009.)

Figure 3-6 The aeolipile steam reaction-engine of Hero of Alexandria. (*Source:* Wikipedia Creative Commons, contributed by DazB on 20 November 2009.)

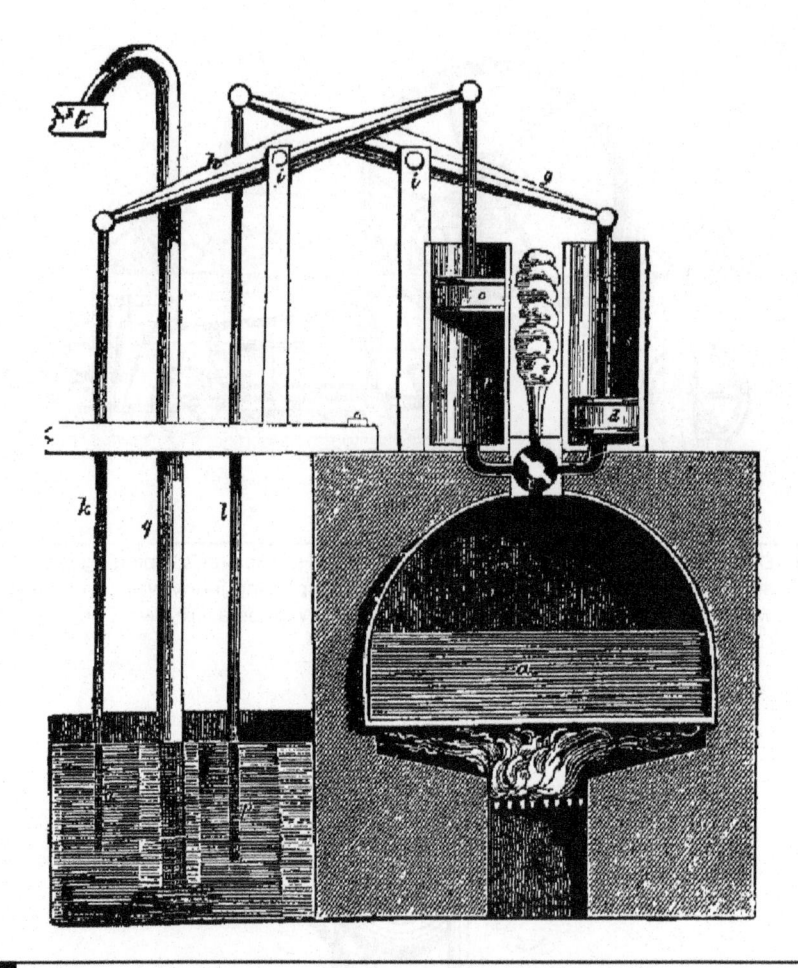

Figure 3-7 The two-cylinder steam engine designed and built by Jacob Leupold in 1720. (*Source: Wikipedia Creative Commons, contributed by Williamgelhart on 26 January 2012.*)

was truly begun, and steam changed everything! Now power no longer had to come from a kinetic energy source to be used immediately. Rather, energy could now be stored—as electrical potential energy—to be used upon demand. To this day, except for wind, hydroelectric, and photovoltaic sources, more than 85 percent of the electric power generated in the world comes from steam-driven generators, whether the steam is made from the heat of combusting a fossil fuel or biomass or from heat created by the absorption of fast-moving thermal neutrons produced by nuclear fission or, eventually, fusion reactors.

The steam engine went on to power the First Industrial Revolution with invention after invention—largely via reverse engineering—to aid in coal, mineral, and ore mining, transportation, and so on, and was a major driver (no pun intended, but rather apropos) to the Second Industrial Revolution.

Reverse engineering was now very apparent as a major tool for making engineering advances.

History of Reverse Engineering   41

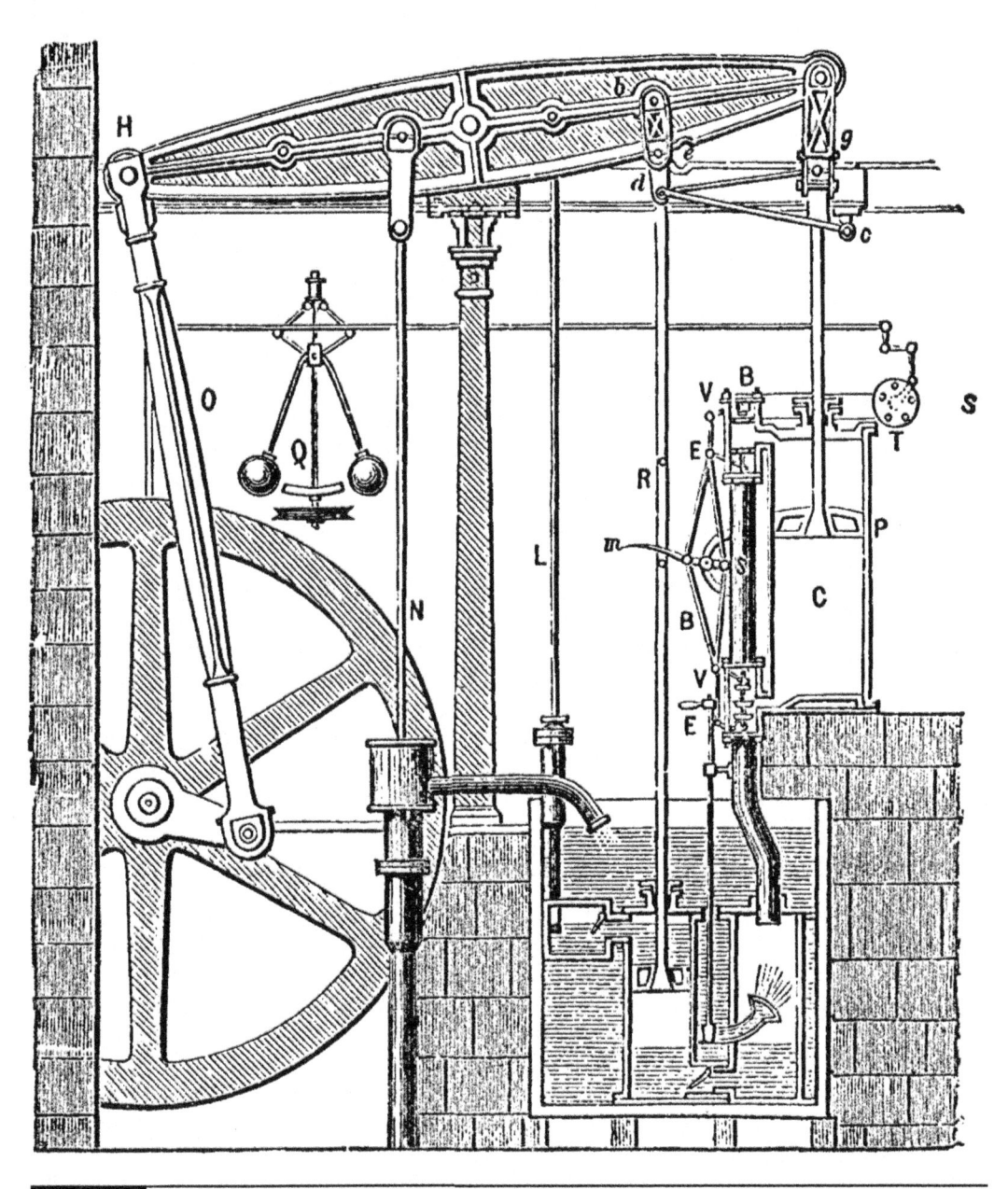

Figure 3-8 James Watt's steam engine of 1775, in which he employed reverse engineering to improve upon the Newcomen steam engine by adding a valve and governor to avoid wasting valuable energy. (*Source:* Wikipedia Creative Commons, contributed by Lidingo on 31 January 2008 and modified by Ariadacepo on 8 February 2010.)

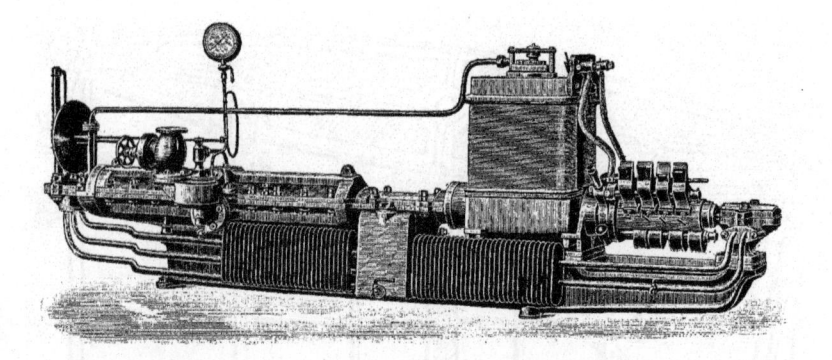

Figure 3-9 The first compound steam turbine, by Charles Parsons in 1884, linked an engine refined from a Boulton & Watt design by Gustav de Laval to an electric generator or dynamo previously powered by a waterwheel. Parson's steam turbine transformed the world forever after, making electric power generation practical. (*Source:* Wikipedia Creative Commons, contributed by Tagishsimon on 20 March 2006.)

In fact, it led to a radical difference between the Industrial Revolution that had begun and flourished in Europe—and then North America—and its manifestation in Japan. In Europe, the great majority of the technology that drove the transformation of industry, transportation (e.g., steam locomotives), power generation (e.g., gas lighting and then electrification), and society had been created in Europe, by Europeans, as creative and innovative original ideas. Likewise in America. Japan's industrial revolution, on the other hand, was, without question, largely the result of the reverse engineering of the European and, later, American inventions. Japan, in particular, and other societies in Asia, were propelled into the twentieth century by reverse engineering.

3-4 Reverse Engineering during World War II

Nowhere is the use of reverse engineering more overt, more widespread, or more apparent than in the creation—or re-creation—of weapons of war. The practice has, almost certainly been going on since the dawn of weapons, which is a long time ago. The creation—and re-creation—of what was (and is) popularly known as "Greek fire" (also known, in ancient times as "sea fire," "Roman fire," "war fire," "liquid fire," and "manufactured fire") is a good example. First documented in the ninth century BC, "Greek fire" was an incendiary weapon developed and used by the Byzantine Empire, typically to great effect during naval battles, as it could continue to burn while floating on water (hence, "sea fire") (Figure 3–10). With the original formulation a closely guarded secret of the Byzantines, enemies, as well as other armies not in conflict with the Byzantine Empire, tried to copy—i.e., reverse engineer—the weapon for centuries. The exact formula is still not known with certainty, although there are several theories based on the weapon's characteristics (which is part of how reverse engineering is done).

Whatever the formula, modifications directly or indirectly based on the original, emerged through centuries as flamethrowers, incendiary projectiles, and incendiary grenades.

The use of reverse engineering for military espionage is much more well known—and documented—than for industrial espionage or less overt commercial technical intelligence-

Figure 3-10 An ancient illustration from an illuminated manuscript showing the use of "Greek fire" during a naval battle. The inscription reads "the fleet of the Romans setting blaze the fleet of the enemies." (*Source:* Wikipedia Creative Commons, originally contributed by Mats Halldin on 31 August 2005 and modified by Amandajm on 14 September 2011.)

gathering. For this reason—that is, easy identification and documentation—what is presented here is all related to weapons.

Reverse engineering is often used by the military of one sovereign nation to copy another nation's weapons, technology, or sensitive military information. In most cases, this is enabled by the capture of weapons by regular forces in the field or by covert intelligence-gathering operations, but it is occasionally aided by a sheer stroke of good luck. Not surprisingly, activity increases dramatically during wartime, epitomized by activity during the Second World War (World War II, September 1, 1939, to September 2, 1945), as well as during the prolonged period of tension between the United States (U.S.) and the Union of Soviet Socialist Republics (U.S.S.R.) known as the "Cold War" (1945–1991). Some examples are so well known that many engineers, not to mention some authors of books or chapters on reverse engineering, believe reverse engineering, at least as a formal technique to aid design, began with World War II and reached a peak during the Cold War. Hopefully, readers here now know this is not really the case.

Well-known examples, but surely not all examples, from World War II follow.

On July 29, 1944, "Ramp Tramp," serial number 42–6256, a B-29–5-BW "Super-fortress strategic, long-range bomber, was unable to return to its base after a raid in Manchuria, being forced to land in Vladivostok in Siberia, in the Soviet Far East. On November 11, 1944, "The General H. H. Arnold Special," serial number 42–6365, damaged during a raid against Omura on Kyushu, was forced to divert to Vladivostok and land. On November 21, 1944, "Ding How," serial number 42–6358, also force landed in Vladivostok. Since the Soviet Union was not at war with Japan until August 1945, it confiscated all American aircraft that made emergency landings

Figure 3-11 The U.S. B-29 Superfortress long-range strategic bomber (in flight) (a) and the U.S.S.R.'s Tu-4 "Bull," an exact copy created by reverse engineering the "Ramp Tramp" impounded in Russia when forced to land in Vladivostok on July 29, 1944 (b). (*Source:* Wikipedia Creative Commons, contributed by Rottweiler on 23 December 2006 [a] and Ntmo on 11 May 2007 [b].)

in Russia prior to that time. In January 1945, the crews of the three B-29s were quietly returned to the United States via Teheran, Iran, but their aircraft were impounded by the Russians and stayed behind in the U.S.S.R.

After flight-testing of "Ramp Tramp," Premier Joseph Stalin launched a program to exactly duplicate the B-29 on June 22, 1945. Engineers working under Soviet aviation pioneer Andrei Tupolev dissected the "Ramp Tramp" "rivet by rivet", making exact copies of 105,000 parts, which they used to create the B-4, later to be renamed the Tu-4 "Bull" (Figure 3–11).

A Russian air force general was quoted as calling the event *dar Bozhii*—"a gift from God"—as it completely changed the Soviet's standing in the postwar world. In fact, it unknowingly helped set up the Cold War that followed, pitting the U.S. and the U.S.S.R. against one another for almost five decades!

Beginning in September 1944, the German Luftwaffe launched more than 3000 V2 short-range, liquid-fueled rockets, known by the German call-name "Aggregat-4" or "A-4," and built by forced laborers, against allied targets, mostly London but also Antwerp. This devastating new weapon of war was responsible for more than 7250 deaths of military and civilian personnel and also cost more than 12,000 forced laborers their lives. Using captured technical documents, hardware, and German scientists and engineers involved with the rocket program led by 34-year-old Dr. Wernher von Braun at Peenemunde, the United States, with von Braun at the helm, in "Operation Paperclip," the Soviets in "Operation Osoaviakhim," and, to a lesser extent, the British in "Operation Backfire" reverse engineered the V2 to create their own rockets. The result of the U.S. effort was the U.S. Army Redstone, while that of the Soviets was their R-1 (Figure 3–12). Perhaps—perhaps—neither the United States nor the U.S.S.R. realized that this effort would begin the "space race" as well as the development of long- and medium-range intercontinental ballistic missiles that could be launched from land bases or nuclear-powered submarines that would forever change the power structure of—and risk to—the entire world!

History of Reverse Engineering   **45**

(a)

(b)

(c)

Figure 3-12 The German-developed V2 (*a*) used against London, England, and Antwerp, Belgium, during World War II was reverse engineered to create the U.S. Army Redstone (*b*) and Soviet R-1 (*c*) using a combination of captured parts, unused rockets, plans, and captured German rocket engineers. Dr. Wernher von Braun, who led the group at Peenemunde, surrendered to the United States and went on to lead the U.S. rocket program and eventual "space race." (*Sources:* Wikipedia Creative Commons, originally contributed by BarchBot on 3 December 2008 and modified by Miracet on 17 December 2008 [*a*]; originally contributed by CarolSpears on 7 October 2007 and modified by Redstonesoldier on 22 November 2007 [*b*]; and www.russianspaceweb.com, unable to contact despite repeated attempts; no intent to circumvent any copyright that might exist [*c*].)

Figure 3-13 The crew of the *Apollo 11* mission that landed a man on the Moon for the first time on July 20, 1969. This event may have culminated the "space race" that began between the United States and the U.S.S.R. when each used captured technical documents, hardware, and scientists and engineers from Germany's V2 program to reverse engineer rockets and missiles for their own use. (*Source:* Wikipedia Creative Commons, originally contributed by Timon on 9 April 2001 and modified by Craigboy on 27 March 2011.)

The U.S.S.R. won the initial leg of the space race by launching and orbiting the first artificial satellite of Earth, *Sputnik 1*, on October 4, 1957. The United States won the second leg of the space race during Apollo 11, safely landing astronauts Neil Armstrong and Buzz Aldrin on the Moon on July 20, 1969, at 20:17:40 UTC, and bringing them, as well as command spacecraft pilot Michael Collins, safely back to earth (in the North Pacific Ocean) on July 24, 1969, at 16:50:35 UTC (Figure 3–13). As for who won the "missile race"—no one wins at that race, which is ongoing, albeit with a new cast of characters!

A final, far-less-well-known example of reverse engineering resulted when the United States captured what was intended to be Japan's "super weapon" to win World War II. Known as their Sen Toku I-400 class, the I-400, I-401, and I-402 (the only three built) were submarine "aircraft carriers." Larger than any submarine ever built and deployed (until the launch of nuclear ballistic missile submarines of the 1960s), besides conventional torpedoes and guns, each carried three

Figure 3-14 Photograph of the Japanese aircraft-carrying "super sub" known as their Sen Toku I-400 class, which the United States captured before it could be used in a planned attack and, from it, reverse engineered its own first nuclear ballistic missile submarines of the 1960s. (*Source:* Wikipedia Creative Commons, originally contributed by World Images on 20 December 2005 and finally modified by Hohun on 16 March 2013.)

Aichi M6A1 Seiran seaplane bombers. The plan was to use the three "super subs" in a coordinated sneak attack on New York City, Washington, D.C., and Los Angeles, dropping massive bombs in each densely populated city and bringing the war to America's homeland.

Fortunately, two of the subs were captured by the United States before they could ever be used. These were secretly tested and fully technically documented—using reverse engineering—before they were intentionally sunk in a secret location in the Pacific Ocean by the U.S. Navy, to keep the Soviet Union from ever seeing or getting their hands on them.

Figure 3–14 shows a photograph of the "super submarine." A wide, figure-eight-shaped hull was to provide stability against roll as aircraft were launched. This design provided essential information and technology invaluable to the U.S. effort to build its first missile-launching submarines, the hulls of which were more than remarkably similar.

Interested readers should view the fascinating YouTube video located under "Submarine Aircraft Carrier—Japanese Super Sub."

3-5 Reverse Engineering in the Cold War and Beyond

The Cold War between the United States and the Soviet Union resulted in innumerable situations in which reverse engineering came into play to clone weapons. One instance occurred when a Taiwanese AIM-9B Sidewinder obtained from the United States hit a Chinese-operated Soviet MiG-17 without exploding. The top-secret missile became lodged within the airframe, and the pilot returned to base with what Russian scientists would describe as "a university course in missile development." Soon to appear was the Soviet reverse-engineered copy: the K-13/R-3S NATO call-name "AA-2 Atoll."

But the greatest use of all involved nuclear espionage, wherein, in most cases, state secrets regarding a nuclear weapon were purposefully given to other states without authorization, allowing unintended nuclear proliferation. Examples abound, but only one example will be cited here, as follows.

A 1999 report of the U.S. House of Representatives Select Committee on U.S. National Security and Military/Commercial Concerns with the People's Republic of China, chaired by Rep. Christopher Cox, revealed that U.S. security agencies believed there was ongoing nuclear espionage by the People's Republic of China (PRC) at U.S. nuclear weapons design laboratories, especially Los Alamos National Laboratory, Lawrence Livermore National Laboratory, Oak Ridge National Laboratory, and Sandia National Laboratories. The *Cox Report* claimed "stolen classified information on all of the United States's most advanced thermonuclear warheads" since the 1970s had been stolen by the PRC. These weapons included designs for miniaturized thermonuclear warheads for multiple warheads, MIRV, missiles, the neutron bomb, and "weapons codes" which allow for computer simulations of nuclear testing (allowing the PRC to advance their weapon development without actually testing any devices themselves). The United States was apparently unaware of this activity until 1995.

These investigations led to the arrest of Wen Ho Lee, a scientist at Los Alamos initially accused of giving weapons information to the PRC, but the case eventually fell apart, and he was only charged with mishandling of data. Other people and groups were arrested and fined, but none were related to the theft of the actual nuclear designs. The damage, however, was done, as little doubt exists that the PRC has advanced nuclear weapons and missile delivery systems.

Without going into any detail here, a simple look at a few examples of look-alike advanced aircraft should convince all but the greatest skeptic of the degree to which reverse engineering has come into play in the modern, post–Cold War world (Figures 3–15 to 3–19).

These examples show "copying" of U.S. aircraft by Russia and of Russian, American, and Israeli aircraft by the PRC, as both production and experimental air weapons.

There are fringe groups that even believe that alien spacecraft (i.e., UFOs) that crashed on Earth

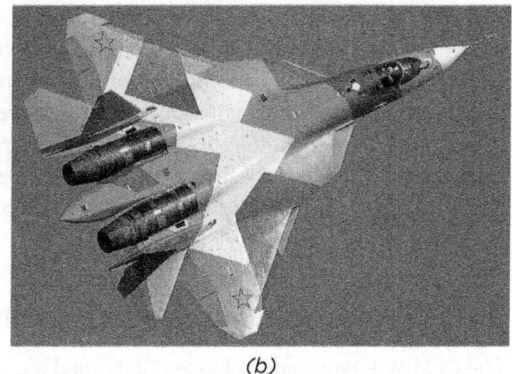

(a)   (b)

Figure 3-15 The U.S. Air Force F-22 Raptor designed by Lockheed-Martin and cobuilt by Boeing, appeared in December 2005 (*a*), while the Russian's Sukhoi PAK T-50 appeared in January 2010 (*b*), the latter exhibiting most of the features, no less geometry, of the former. (*Source:* Wikipedia Creative Commons, originally contributed by Magnus Manske on 4 April 2005 and modified by Jovianeye on 21 June 2010 [*a*], and by Nockson on 14 June 2011 [*b*].) **Don't miss the color version of this figure, available at www.mhprofessional.com/ReverseEngineering.**

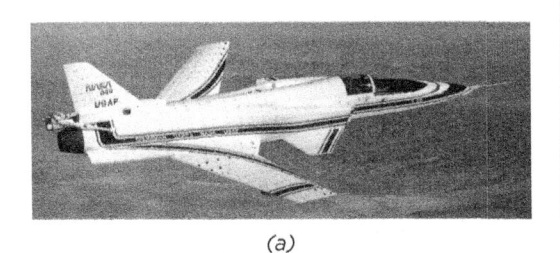

Figure 3-16 Grumman X-29 forward-swept-wing experimental aircraft built for DARPA/NASA and flown in late 1984 (a) and Russian Sukhoi Su-47 prototype forward-swept-wing aircraft first seen in late 1989 (b). (*Source:* Wikipedia Creative Commons, contributed by Stahlkocher on 30 April 2005 [a], and by High Contrast on 23 August 2011 [b].) ***Don't miss the color version of this figure, available at www.mhprofessional.com/ReverseEngineering.***

Figure 3-17 Russian Sukhoi Su-27 NATO code-name "Flanker" as it appeared in 1984–85 (a) and Chinese (PRC) Shenyang J-11 that first appeared in 1995 (b). (*Source:* Wikipedia Creative Commons, contributed by Nockson on 15 November 2011 [a], and by Orlovic on 1 April 2007 [b].) ***Don't miss the color version of this figure, available at www.mhprofessional.com/ReverseEngineering.***

were reverse engineered, giving humankind its remarkable jump in technology post–World War II (ref. King). In fact, if all we humans were able to get by reverse engineering alien spacecraft that came to Planet Earth "from galaxies far, far away" are atomic energy, jet aircraft, and some rockets that went to the Moon and Mars, we didn't learn much!

Reverse engineering of U.S. and European commercial products has, without question, been used—and abused—by Japan, China, Korea, and others. The growing appearance of look-alike products is one thing, but, worse yet, is the appearance of knockoffs that misrepresent themselves as the real thing (see Chapter 15)!

Figure 3-18 The U.S. Lockheed F-35 Lightning II, a fifth-generation multirole fighter that made its first flight on December 22, 2006 (a) and the PRC's Shenyang J-31 that appeared in late 2010 (b). Official reports indicate the top-secret plans for this aircraft were stolen by the PRC from the undersecured computers at BAE, Britain's largest military aircraft manufacturer and partner with Lockheed. The J-31 is said to look like a hybrid of the USAF F-15 from the back and F-35 from the front. (*Sources:* Wikipedia Creative Commons, contributed by Marcus Qwertyus on 22 August 2011 [a]; and www.militaryphotos.net, originally posted by Einhander, with permission [b].) **Don't miss the color version of this figure, available at www.mhprofessional.com/ReverseEngineering.**

Figure 3-19 PRC Chengdu J-10 that appeared in late 2004 (a) and Israel Lavi IAI B-2 prototype fighter that appeared around 1990 (b). (*Source:* Wikipedia Creative Commons, contributed by Retxham on 11 January 2009 [a], and by Bakvoed on 3 March 2006 [b].) **Don't miss the color version of this figure, available at www.mhprofessional.com/ReverseEngineering.**

3-6 Summary

Reverse engineering, in its broadest sense, as treated in this book, is far older than most believe but, with a little reflection, will come to understand. There is evidence in the evolution of pyramids in ancient Egypt that the architects and builders were learning from mistakes as well as successes, as cone-shaped piles or primitive pyramids evolved into tiered or stepped designs and, finally, into true pyramids, epitomized by the Great Pyramid of Khufu. Complications with earlier face angles and the effects of the unimaginable weight above, and force on, the roof of the King's Chamber were adapted to on the fly.

The builders of the great cathedrals during medieval times also employed reverse engineering, as evidenced by the evolution of those designs with the incorporation of flying buttresses to bolster exterior walls from the immense outward loads from vaulted ceilings and with sculpted domes to lessen weight.

Reverse engineering was a major driver of the rapid advances of power sources and machines during the Industrial Revolution in Europe—and North America—and it was *the* way in which that revolution made progress in Asia, notably in Japan.

But the heyday of reverse engineering was the Second World War, persisting into the Cold War that followed. The demand for better weapons of war drove one nation to keep a watchful eye on its enemy nations, ever ready to steal and copy what it could to catch up, if not leapfrog ahead.

With tensions in the world showing no signs of easing, and the balance of power shifting yet again, reverse engineering is still being actively employed in weapons development and for economic development.

3-7 Cited References

"Building the Great Cathedrals," *Nova,* PBS, December 26, 2012; available on video/DVD at www.pbs.org.

King, Thomas, *UFOs That Crashed to Earth: Reverse Engineering of Alien Spacecraft Mankind Creates the Atomic Bomb UFO Enigma Solved,* AuthorHouse, Bloomington, IN, 2011.

Messler, Robert W., Jr., *Engineering Problem-Solving 101: Time-Tested and Timeless Techniques,* McGraw-Hill, New York, 2013.

3-8 Recommended Readings

Icher, Francois, *Building the Great Cathedrals,* Abradale Books, New York, 2011.

Jackson, Kevin, and Jonathan Stamp, *Building the Great Pyramid,* Firefly Books Ltd., Richmond Hill, ON, 2013.

Kennedy, Gregory, *Germany's V-2 Rocket,* Schiffer Publishing Ltd., Atglen, PA, 2006.

Romer, John, *The Great Pyramid: Ancient Egypt Revisited,* Cambridge University Press, Cambridge, UK, 2007.

Sekaido, Henry, Gary Nila, and Koh Takaki, *I-400 Japan's Secret Aircraft-Carrying Strike Submarine—Operation Panama Canal,* Hikoki Publications, Crowborough, UK, 2006.

3-9 Thought Questions and Problems

3-1 The solution of problems as they appeared in a new design during its execution, as well as improvements upon earlier designs that somehow fell short of expectations, is actually enabled by *reverse engineering* through observation, experimentation (or testing), and measurement to gain needed understanding and knowledge. A great example is presented in Section 3–1 for the real-time modification of the of the King's Chamber in Khufu's Pyramid.

There are undoubtedly countless other examples that have occurred in the constructions of ancient civilizations and earlier societies over the ages.

Use the Internet to find another example (different from any presented in the book) where the design of a great structure had to be—or should have been—adaptively modified, as problems with the original design were encountered during construction. Be sure to consider ancient times, medieval times, and more recent times (e.g., the nineteenth and twentieth centuries), choosing only one area.

3-2 Another valuable use of *reverse engineering* that occurred well before most mentions of the technique recognize or acknowledge is to aid in the understanding of an ancient or very old design for which there are no written records of the purpose of the structure or, alternatively, the method by which it was built or manufactured.

A few examples where reverse engineering has helped or could help with understanding are:

- Stonehenge (2600–2400 BC)
- Tunnel of Samos or Eupalinos (sixth century BC)
- Hadrian's Wall (AD 122–128)
- Fountains of Villa d'Este (1550–1970)
- A 12-cylinder tractor engine assembled from the bottom or lower crankshaft upward, as the heads were integral to the block, precluding the usual assembly from above (ca. 1897–1898). [This could be *tough* to find but is a real puzzler, which Ford Motor Company pondered for years after seeing a patent!]
- Aswan High Dam (1960–1970)

For one of these, briefly describe the structure, the particular challenge(s) it posed to the builders, and any questions it may have posed to those who pondered its purpose, method-of-construction (or, for the tractor engine, manufacture), etc.

3-3 The rate of advancement of power sources, manufacturing equipment and processes, etc., during the Industrial Revolution (ca. 1740–1840 in Europe; later, in the New World and elsewhere) was staggering. Learning had to—and did—occur very quickly to allow such rapid advancement. *Reverse engineering* often played a key role.

From your personal knowledge or work, or using the Internet, choose an example area in which reverse engineering almost certainly played a major role. Possible areas could be (but are not to be limited to) agriculture, textile manufacture, power generation or production, or transportation. Prepare a brief but thoughtful one- to two-page write-up or essay about your chosen example.

3-4 For better or worse, *reverse engineering* has been frequently and widely applied to advance the design of weapons of war, from ancient to modern times.

Choose an example from among the following (based on your personal interests or knowledge) and prepare a brief but thoughtful one- to two-page write-up or essay on how reverse engineering was used to advance one or another civilization's or nation's weapons *or,* as appropriate, help modern historians (or weapon designers) understand what an ancient civilization or modern rival might have done to create a weapon for which they are known. Examples are:

- Archimedes' mirror
- Archimedes' claw
- "Greek fire"

- Medieval siege machines (e.g., trebuchets)
- The "Enigma Machine" of World War II
- "Die Glocke" ("The Bell") of Nazi Germany in World War II

3-5 More espionage has been involved with stealing secrets related to the atomic bomb (or more modern nuclear weapons) than any other weapon system or technology. Early secrets relating to the atom bomb developed by the top-secret Manhattan Project at the Y12 Plant at Oak Ridge, Tennessee, were allegedly passed to the Soviet Union by Julius and Ethel Rosenberg in the early 1950s. While there have been those who question their actual guilt, they were convicted of spying and treason and executed by electrocution on June 19, 1953, at the Sing Sing Correctional Facility in Ossining, New York, on behalf of the federal government.

Use the Internet to look into espionage relating to atomic or nuclear weapons and/or technology, with several examples being:

- Klaus Fuchs
- Theodore Hall
- David Greenglass
- Ethel and Julius Rosenberg
- Wen Ho Lee
- Peter Lee
- Dr. Abdul Qadeer Khan

Prepare a brief but thoughtful one- to two-page report on the incident you choose. Be sure to identify what was allegedly stolen, how it was allegedly stolen and transferred, and, particularly, how it might have changed the balance of power in the world—for better or worse.

CHAPTER 4

The Teardown Process

4-1 The Purpose of Teardown

Next to building things (shelters, tools, weapons, etc.), the most natural thing for human beings to do would seem to be taking things apart. Some might even argue that we are a destructive species by our nature and, thus, are more enamored by taking things apart than by putting then together. Certainly, we are the only species in control of our own destiny—having, or having been granted, free will—and, thus, of making our planet better or destroying it.

Philosophy aside, hopefully the case was made (in Chapter 1) that we human beings:

- Are innately curious
- Are uniquely motivated to know more than the "what" but also the "who," "where," "when," "why," and "how" concerning things
- Learn best from experience
- Analyze and synthesize

Taken together, these traits lead us to take things apart to understand *how* they were put together and *how* they work; in other words, they naturally give rise to the extremely powerful problem-solving technique of *reverse engineering* (ref. Messler, Chapter 16, pp. 127–134).

Reverse engineering was previously defined (in Chapter 2) as "the process for discovering the principles that underlie and enable a device, object, product, substance, material, structure (construction), assembly, or system through the systematic analysis of its structure and, if possible, its function and operation." The technique comes into play—and has utility—for far more than simply reinventing an existing invention or developing a new product from a prior product, whether motivated by need or opportunity for improvement (i.e., market pull, as in adaptive design, or technology push, as in developmental design). It is the premise of this book that reverse engineering is—or should be—a mind-set, philosophy, or guiding principle, as much or more than

56 Chapter Four

a problem-solving technique or tool to aid in design or a redesign tool. It has been for far longer than it might seem from the narrow view of product redesign (in Chapter 3).

If something falls apart or begins to fail during the process of its creation, production, or construction, engineers (as those responsible for such things) seek to determine *why,* in order to correct the problem. If, on the other hand, something survives creation, production, or construction (as is usually the case), engineers, not uncommonly, take it apart (at some point) in a systematic process of *teardown*. They do so for either of two primary purposes:

1. To learn from something that works or worked well what makes or made it successful

or

2. To discover why something no longer works as it was intended to work

To make it easier (on readers and on the author), the remainder of this chapter will focus discussion of the teardown process on teardown of a "product" in the broadest sense, by which could be meant part (e.g., automobile sparkplug or shock absorber or tire), component (e.g., machine cam or spring or bolt), structural element (e.g., beam or truss or arch), object (e.g., soccer ball or golf club or lightbulb), device (e.g., p-n junction diode or rectifier or computer microprocessor unit), product (e.g., vacuum cleaner or iPod or smartphone), substance (e.g., Coca-Cola or proprietary adhesive or coating on a consumable arc welding electrode), material (e.g., metallic alloy or engineered polymer or synthetic composite), structure (e.g., pyramid or bridge or dam), assembly (e.g., self-winding Swiss watch or aircraft landing gear or populated printed circuit board), or system (e.g., automobile or laser-guided missile or LCD TV). Based on this, *product teardown* is the process of taking apart a *product* to understand it. But, as is further described later, product teardown also allows one to understand something about how the product's creator (e.g., inventor, company, civilization, or Mother Nature) made the product succeed.

Product teardown serves three central (or key) purposes, as follows:

1. Dissection and technical (as well as cost) analysis
2. Experience and knowledge for one's own personal (or for an organization's) database
3. Benchmarking

Some references (ref. Otto and Wood) include modifiers in the first and third purposes in the preceding list that make the meaning more narrow than this author wishes to do. For example, the first purpose is written (in their superb reference on the subject of "product design," incidentally!) as "dissection and analysis during reverse engineering." This implies that reverse engineering (by their definition) is part of the process of product teardown, which it could be for some purposes. But this author feels the greatest utility of product teardown is as part of the broader and/or higher-level process of reverse engineering. "Teardown" involves mechanical dissection (except for biological entities, for which it involves physical dissection) and analysis, for sure, but as a part of "reverse engineering," not vice versa. Also, "analysis" in the context of the preceding must include analysis of all *technical details* (e.g., part geometry, orientation, arrangement, interaction; material-of-construction; and method-of-fabrication, processing, and assembly), as well as *costs* (e.g., raw material costs, material processing costs, machine and human labor costs, design costs, maintenance costs, energy costs, ultimate recycling costs)—in other words, initial and life-cycle costs.

Otto and Wood use the term *competitive benchmarking* in the third purpose of their list of purposes for product teardown. This author chooses to use *benchmarking* without this modifier, in its broadest sense; that is, comparing where one stands relative to others, whether those "others" are competitors, enemies, or simply a culture or society or civilization, current or ancient. One automobile manufacturer would be foolish to not benchmark its vehicles against all other competitive automobile manufacturer's vehicles. The United States of America, on the other hand, needs to benchmark its defense capability against declared or perceived immediate or potential enemies, as well as against its own capability at other times in its history, perhaps. Modern engineers need to—if for no more than respect for our heritage—benchmark their knowledge and capability relative to the knowledge and capability of our forebearers (ancient Egyptians, ancient Romans, medieval cathedral builders, the Wright brothers, Henry Ford, IBM founder Thomas J. Watson, Apple founder Steven Jobs, etc.).

To give credit to Otto and Wood, it is only appropriate to include their views on the ramifications of the first and third of the purposes listed previously.

Dissection and analysis, relative to actual (or true) product development are key to the evolution of a product, they say, and this author agrees. A product cannot evolve to its "next generation" if the current version of the product is not fully understood from a formal, systematic process of dissection and analysis. Analysis and understanding must include "intended" *and* "latent" (unrecognized or unintended, but still present) functions, operative technologies, and design, manufacturing (or construction), and life-cycle strengths and weaknesses.

Relative to "competitive benchmarking" (per Otto and Wood), it is essential, in order to remain competitive, to compare one's own design (or emerging design concept) to that/those of any/all competition. What a competitor does better as well as what it doesn't do as well are important data points, not just at some point in time but in terms of any trend. Are they getting better or worse at what they do? Or: Are you getting worse or better at what you do?

A very wise man (or woman) said: "Never underestimate your competition!" Later, the author, on more than one occasion, makes the point that modern technologists—unlike archeologists—should never underestimate (or underappreciate) the capability of our ancestors. There are—and have been—a lot of very smart people on this planet!

A final word or so on the second purpose of product teardown, that is, to gain "experience and knowledge for an individual's [or an organization's] personal [or organizational] database." By understanding "how things work," we learn! By dissecting products, one gains "kernels of information" (to quote Otto and Wood) on how to achieve desired function. Otto and Wood said it best: "The more we dissect technology, the larger our knowledge base of concepts grows to solve and synthesize solutions to new problems." The author couldn't say it better or agree more! As engineers—and as vital, albeit temporary, citizens of Planet Earth—we have an obligation to know what has been done before so we can do at least as well in the future.

While we are here—on Earth—we need to observe, measure, and experiment.

4-2 Observation

Thomas Huxley (1825–1895), the English biologist and anatomist, said: "Science is simply common sense at its best, that is, rigidly accurate in observation, and merciless to fallacy in logic."

But it took an American poet, Wallace Stevens (1879–1955) to say it best: "The accuracy of observation is the equivalent of accuracy of thinking." And it took Yogi Berra to say it so we are all sure to get it: "You can observe a lot just by looking."[1]

Reverse engineering, perhaps more than anywhere else in engineering, demands meticulous *observation*. To paraphrase the recruiting slogan of the U.S. Army: "You must see all there is to see!"[2] In tearing down a product (again, in the broadest sense of that word given earlier), it is critical that *everything* there is to be seen is seen—and is thoroughly documented.

At the highest level, it is very important—and may prove to be very informative—to look at the overall workmanship of the product. Workmanship reflects more than simply the care taken by the worker(s), although it surely does that. It reflects the value the creator of the product placed on creating it. Care in workmanship generally reflects care in every other aspect of the product's creation, from design to marketing to customer support and service. Design has a particularly significant impact, however, on the ability to create a physical entity from what began as an abstract concept. It has been said that decisions made during the design stages of a product have the most profound and lasting impact on the product's life-cycle cost—from "womb to tomb," as it were. It may be unfair and/or untrue to say a product that looks good (in terms of its workmanship) is good, but it is more true than not that a product that looks bad will probably be bad. What caring designers—or engineering design organization—would tolerate poor workmanship in making their design a reality? Surely, a bad design cannot be saved by the best workmanship, but a great design can be ruined by poor workmanship!

You get the point!

At the next level, prior to taking anything apart, one needs to get a sense of the product. What does it look like it might do? How old—or new—is it? How expensive does it look? Does it appear that its cost (from details of its overall appearance) reflects its intended purpose? For example, wouldn't one expect a power tool (e.g., a circular saw) intended for use by a professional (e.g., a carpenter) to look better (i.e., more robust, greater attention to details, more expensive materials) than one intended for a home do-it-yourselfer?

As teardown begins, one should observe how the product was assembled (e.g., using standard fasteners, specialty or custom nonstandard fasteners, visible or hidden integral attachment design features such as "snap-fits" commonly used in assembling polymer/plastic parts). Much more is said about this in Chapter 7.

One must look at every part that makes up the product. The parts should be laid out to reflect how they went together in the assembled product, that is, to present an "exploded view" of the product (Figure 4–1). One should observe each part's shape, size, finish, orientation to and fit with each mating part, and arrangement of each within the whole. Clues that help identify the material-of-construction for each part (metal, ceramic, glass, wood, polymer/plastic, rubber, or composite; aluminum alloy or steel; etc.) should be sought (color, relative density, etc.) (Chapter 9). Telltale details should be sought that help identify the method-of-construction or -fabrication and/or -processing of each part (machining, casting, plastic injection molding, etc.) (Chapter 10).

Later, as we delve more deeply into each step in the teardown process and the intended goal of

---

[1] Lawrence Peter "Yogi" Berra (1925–) was a catcher, outfielder, and manager with the New York Yankees from 1946 to 1965, where he was a frequent MVP, a dependable hitter in a clutch, and a renowned speaker of the obvious—e.g., "It ain't over 'til it's over!" and "Deja vu all over again!"

[2] "Be All You Can Be" was the recruiting slogan of the United States Army for over 20 years, from 1980 to 2001. E. N. J. Carter created it while at the advertising firm N. W. Ayer & Son. He received the Outstanding Civilian Service Award for his efforts.

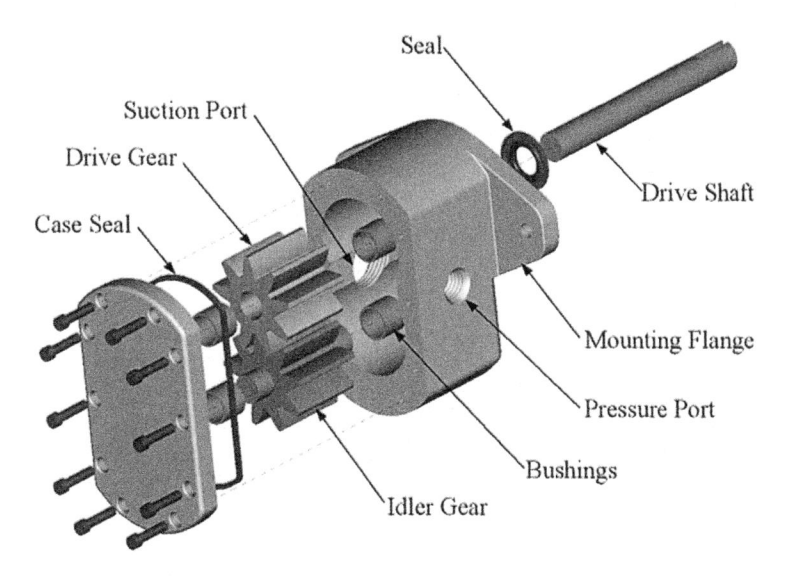

Figure 4-1 An exploded-view drawing (here, a computer-aided solid model rendering) of a gearbox, showing the detail parts and their arrangement in the device. (*Source:* Wikipedia Creative Commons, contributed by Duk on 11 January 2005 and modified by Pngbot on 24 January 2007.)

reverse engineering (Chapters 7, 9, and 10), we will use each of these observations to guide our deductive reasoning as to the purpose and logic of each part as well as of the overall product.

A few final points concerning *observation* before we move on:

First: "Nothing is inconsequential, as you don't know what will eventually be consequential" (Messler, 2013). Write down everything you observe! (This is particularly important in the conduct of a failure analysis or investigation, and is *essential* in known or potentially litigious cases, in other words, during *forensic engineering*.)

Second: Observation should employ all senses and include all sensory clues. Obviously, how things look is usually paramount. But, also, perhaps: How things feel (e.g., smooth or rough, dry or slippery, heavy or light, cool to the touch relative to other things known to be at the same [equilibrated] temperature), how they smell (e.g., burned, oily, chemical, pungent, dirty), how they sound when tapped (e.g., solid or hollow, metallic or nonmetallic). One should probably refrain from tasting anything one is not sure about—for a host of reasons.

Third: "Always turn the body over." If you don't, you may not have seen all there was to see—and nothing is insignificant! A short anecdote will make the point.

One of the author's uncles (Uncle Bill) was a police officer who, for the last 15 years of a 35+-year career, worked as a homicide detective. At family gatherings, Uncle Bill always had great stories for the teenage boy the author was at the time. One Christmas, during the author's senior year in engineering school, there was the story of the body of "a middle-age, 50-ish, white male" found at dawn lying on his back on the sidewalk along the main street through a large town of around 40,000 people. Naturally, along with the police, homicide detectives were dispatched to see if there had been any foul play.

When the county medical examiner arrived and finished examining the body—supposedly first time it had been touched—he said to the author's uncle (as lead detective for the case), "There's no problem here. Death was by natural causes." Uncle Bill replied, "What 'natural cause' would that be?" to which he received the response, "Heart attack. You can leave. The body will be transported to the county morgue for identification and notification of next of kin. There won't be any autopsy." Uncle Bill asked, "How did you come to that conclusion, 'heart attack'?" to which the somewhat perturbed doctor responded, "Well, he's a 50-ish male, he's 10 to 15 pounds overweight, he smokes cigarettes, as indicated by the yellow nicotine stains on the fingers of his right hand, and there's an open pack of Marlboros in the left inside pocket of his suit coat, and he has a high-stress job, as indicated by his expensive three-piece glen plaid suit and the expensive attaché case lying near his body. Also, there are characteristic signs of sudden cardiac arrest—ruptured blood vessels in his eyes and cyanic coloration around his mouth and lips. Heart attack! No crime!"

As the doctor began to walk away, Uncle Bill asked, "Doc, did you turn the body over? I think you should." When, together, they rolled the body up on its side, a 3- to 4-inch-diameter pool of dark blood was on the sidewalk and a matching stain was on the left upper back of the dead man's suit. Probing the 2½-inch-long slit in the suit coat revealed to the doctor "a hard, reflective object inside a deep wound." Uncle Bill said, "Look under his left breast," upon which, after rolling the body onto its back, the doctor noticed that the skin was "tented outward," held there by "a hard object." The doctor turned to the detective and said, "Bill, you've got a murder here!" A large Bowie knife had been plunged into the poor man's back, the blade broken off at the hilt by the force and violence of the blow, and the newly formed wound plugged against bleeding by the implanted blade. The tip of the long knife tented out the skin under the man's left breast, having passed completely through his chest and heart. No blood and sudden cardiac arrest had misled the doctor.

"Neat story" was the fascinated boy's response. "It's more than a 'neat story,' there's a valuable lesson for a young engineer-to-be," said Uncle Bill. "Always turn the body over—or you haven't seen all there is to see."

The author has abided by his Uncle Bill's advice for over four decades during the conduct of more than 1400 failure investigations. Be observant—meticulously observant. No observation is insignificant. And always turn the body over, or you may not see all there is to see!

4-3 Measurement

H. James Harrington (1929–) is an American author (of more than 35 books), engineer, entrepreneur, and consultant in performance improvement. In 1979, it was he who originated IBM's Internal Benchmarking Procedure (see Section 4–1). Among many things, Harrington wrote: "Measurement is the first step that leads to control and eventually to improvement. If you can't measure something, you can't understand it. If you can't understand it, you can't control it. If you can't control it, you can't improve it."

An early lesson learned in the conduct of failure analysis, in general, and an absolutely critical lesson learned in forensic engineering (Chapter 6), in particular, is the concept of taking careful and specific measurements of everything measureable.[3] Measurement is extremely important—following initial observation—during product teardown.

---

[3] The importance of measurement, if not obvious from H. James Harrington's statement within the past few decades, was made patently clear for forensic engineering from forensic anthropology. To begin to understand a person or a people (e.g., ethnic culture), anthropologists took specific measurements to determine the age of the human, and other specific measurements to determine the gender, race, stature, etc.

As discussion of the specific *methods* by which product teardown can be accomplished is presented in Chapter 5, two broad categories of measurements become apparent: (1) measurement of geometry and (2) measurement of function (when possible). It will be seen that this bears a striking resemblance to the techniques used to advance the early practice of medicine (see Section 1–5).

Measurement of geometry (i.e., *geometric measurement*) is straightforward. Every aspect of the geometry of a product (in the broadest sense) needs to be measured or quantified on some appropriate dimensional scale. This includes dimensional measurements that would allow reproduction of the overall product, as well as each and every detail part or component in the product. Measurement of length, width, thickness, hole diameter, hole depth, hole location, radii of curvature are all obvious, but, for 3D objects or features with compound curvature (i.e., curvature in two directions) or with nonuniform curvature, the task is somewhat more involved. Modern computer-based coordinate measuring systems make this task much simpler than it once was, however. By digitizing points on a surface, that surface can be reproduced with whatever accuracy, in terms of resolution, is desired or required. The points, after all, define the surface.

In making measurements, there are a number of criteria for appropriate devices, as shown in Table 4–1. What is presented there is self-explanatory and, so, is not addressed any further here.

Before leaving measurement of geometry, it is worthwhile pointing out the need for collecting enough measurements to fully and unambiguously characterize the particular product, object, part, or structure.[4] Ashby refers to the need for characterizing the critical information needed to fully define an object as *information content* (ref. Ashby), although what he presents was developed long ago, at least as part of what was known as "group technology."[5]

Two examples will suffice to make the point, one only slightly more geometrically complex than the other.

If one wished to provide the dimensional information necessary *and* sufficient to allow the creation or reproduction of a solid sphere made from a machinable steel (say, AISI-SAE 1212),[6] here's what would be needed (Figure 4–2a):

- ✓ Diameter (or radius) of the sphere, say 40.00 mm
- ✓ Dimensional tolerance (accuracy) for the diameter, say ±0.03 mm
- ✓ Surface finish of the sphere as Ra,[7] say Ra = 0.1 mm

---

[4] It is equally important, however, not to take superfluous measurements that overconstrain the design. For example, the center of a hole to be drilled or bored in a flat surface on one face of a rectangular block can be precisely located with two dimensional measurements, one from the long (longitudinal) edge and one from the short (transverse) edge. To provide dimensions (from measurements) from both opposite ends of the long length and both opposite sides of the short length, as well as providing dimensions for the long and for the short lengths, overconstrains the design, as tolerance stack-up will lead to problems. Interested readers should look into "dimensioning and tolerances."

[5] Group technology (GT) is a manufacturing technique in which parts having similarities in geometry, manufacturing process, and/or function are grouped together using a string of digits and/or letters. GT was first proposed in 1925 by Flanders, was adopted in Russia by Mitrofanov in 1933, and was actively promoted by Jack Burbidge in the United Kingdom in 1977.

[6] AISI-SAE 1212 is a plain carbon steel containing a nominal carbon content of 0.12 wt.%, but with intentionally added phosphorus (P) and sulfur (S) to promote chip formation during machining. The alloy is assigned a value of "100%" and is used as the basis for indexing every other steel alloy's machinability as a lower percentage.

[7] *Surface roughness* (or simply "roughness") is a measure of the texture of a surface. It is quantified by the vertical deviations of a real surface from its ideal form. The typical approach involves taking statistical root-mean-square (rms) values, which the interested reader should look into on the Internet.

62 Chapter Four

TABLE 4-1 Various Criteria for Devices Used to Make Measurements

Criteria	Explanation
Accuracy	The deviation of measured values from actual values
Range	The span of possible readings, from minimum to maximum, for which the device has applicability and accuracy
Repeatability	The degree by which multiple measures of the same thing vary
Dynamic accuracy	Frequency range over which measurements of cyclic phenomena are accurate
Calibration	Comparison of instrument's readings with a known standard
Mass, size, and power	Constraints on a device's physical limits
Safety	The relative risk to health, injury, or death posed by a device when properly used
Utility	The ability of a device to interface with and measure a variety of aspects of physical entities
Cost	The combination of purchase price and lifetime operating costs for a device (i.e., life-cycle cost)
Output	The form of data display, e.g., visual read-out or gauge, data recorder, etc.
Ergonomics	The ease of use of a device
Robustness	The durability of the device for normal use as intended
Nondestructiveness	The degree and nature of any discernable detrimental change to the entity being subjected to measurement

The preceding list represents three essential bits of information, which turns out to be the least information content needed to produce any object.

If instead of a solid sphere, one wished to provide the dimensional information necessary and sufficient to allow a solid rectangular block having a "blind" (i.e., not "through") hole in one face to be made, here's what would be needed (Figure 4–2b):

- ✓ length L of the long side, to a ± tolerance ⇒ 2 bits of information
- ✓ Width W of the short side, to a ± tolerance ⇒ 2 bits of information
- ✓ Thickness t, to a ± tolerance ⇒ 2 bits of information
- ✓ Angle α (=90°), to a ± tolerance ⇒ 2 bits of information
- ✓ Angle β (=90°), to a ± tolerance ⇒ 2 bits of information
- ✓ Angle γ (=90°), to a ± tolerance ⇒ 2 bits of information
- ✓ Surface finish of each of the six faces ⇒ 6 bits of information
- ✓ Location of hole from one end, L', to a ± tolerance ⇒ 2 bits of information
- ✓ Location of hole from one side, W', to a ± tolerance ⇒ 2 bits of information
- ✓ Hole diameter D, to a ± tolerance ⇒ 2 bits of information

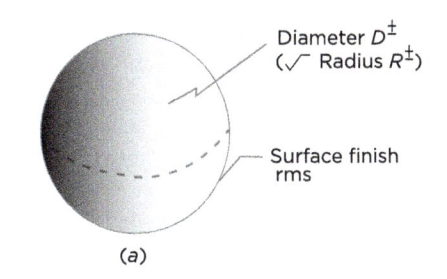

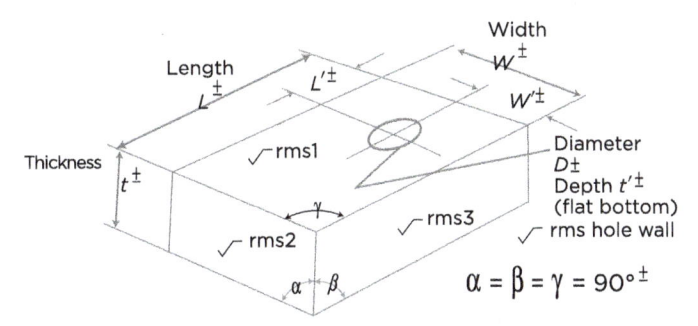

Figure 4-2 Schematic Illustrations of two simple objects showing the dimensions or other features that are required to fully characterize the object and allow its reproduction; a solid steel sphere (a) and a solid steel rectangular block containing a single round, blind hole in one face (b).

- ✓ Depth of hole T', to a ± tolerance ⇒ 2 bits of information
- ✓ Normality of hole to face, given by two angles, to a ± tolerance ⇒ 4 bits of information
- ✓ Surface finish of the hole wall ⇒ 1 bit of information
- ✓ Surface finish of the hole bottom ⇒ 1 bit of information.

This represents 32 essential bits of information, that is, an information content of 32 (versus 3 for the solid sphere). That's a lot more information for such a simple object. Image how much information content is required to fully characterize a four-tine dinner fork, a stainless steel teapot, a Skil electric-powered circular saw, a Toyota Prius, or a Boeing 787 Dreamliner!

When, during product teardown to reverse engineer a product, it is also desired to capture information to characterize the known—or deduce the uncertain—function of a product (i.e., its use) and performance (i.e., its level of function), one needs to make appropriate measurements where possible. The general procedure for *functional measurement* involves decomposing the product into a set of "elements" that represent all of the key functions, for which a set of measurements can then be defined from each function to allow the overall product's function to be quantified (i.e., measured). To do this, one must first list the known or presumed (or deduced) functions of the product from what is known as a *function structure* or *functional structure* (ref. Messler, Chapter 16, pp. 127–134). In some cases, this list can change from *predicted functions* before product decomposition to *actual* (or *deduced*) *functions* after product decomposition.

With a list of functions in hand, appropriate measurements can be taken wherever possible, or, where not possible, some enumeration of function by estimation must be attempted (see Chapter 7). One of the *methods* for accomplishing product teardown actually involves taking apart a functioning product to observe how—and from where—function arises.

Since a proper functional analysis of a product provides a complete representation of the product, a more accurate set of measurements can be developed by examining each function, one by one. In this context, *function* includes customer needs, operating ranges, and flows of energy, material, and information (or signal).

4-4 Experimentation

Claude Bernard (1828–1878), a French physiologist called "one of the greatest of all men of science" by I. Bernard Cohen (1914–2003), Victor S. Thomas Professor of the History of Science at Harvard University, said about *experimentation*: "Observation is a passive science, experimentation an active science." After *observation,* as the first key step in product teardown and reverse engineering, and *measurement,* as the second key step, it is sometimes possible and appropriate to conduct experimentation. If a product to be subjected to teardown is currently operational, particularly (but not only) if the purpose and function of the product is known, *experimentation* with the product can provide valuable information, knowledge, and understanding. Examples where experimentation with a product can prove useful include:

- Benchmarking of one's product against a competitor's (or competitors') product (or products)
- Benchmarking one's new product concept (e.g., as a prototype) against one's earlier product
- Troubleshooting one's product that falls short of expectations or is experiencing problems in the marketplace
- Reverse engineering a weapon system (as part of military espionage) or a product in the marketplace for which your organization has no counterpart (as part of industrial espionage)
- Learning about an unknown object (e.g., the Rosetta stone), device (e.g., the Antikythera mechanism), structure (e.g., Stonehenge), or product[8]

A good example of experimentation as part of reverse engineering is the Soviet flight-testing of the "Ramp Tramp" Boeing B-29 Superfortress bomber impounded after forced landing in Vladivostok on July 29, 1944, returning from a mission in Manchuria, and used by the Soviets to create their Tupelov Tu-4 copy (Section 3–4). Widespread use of experimentation, as part of reverse engineering to understand ancient devices, is by experimental archeologists. One example

---

[8] The *Rosetta stone,* rediscovered by a French soldier (Pierre-Francois Bouchard) in 1799, is an ancient Egyptian grandodiorite stele (large fragment of black granite) inscribed with a decree issued at Memphis in 196 BC by King Ptolemy V. Written as three parallel texts in ancient Egyptian hieroglyphs (at the top), Demotic script (in the center), and Ancient Greek (at the bottom), it provided the key to understanding Egyptian hieroglyphs. The *Antikythera mechanism* is the subject of Chapter 7 of this book. *Stonehenge* (built between 3000 and 2000 BC) is a prehistoric monument located in the English county of Wiltshire, 2 miles (3.2 kilometers) west of Amesbury and 8 miles (13 kilometers) north of Salisbury. Consisting of a ring of standing stones within earthenworks, it is one of the most famous ancient sites in the world. It is still being studied to understand its purpose.

is an effort to estimate whether it was feasible for the ancient Romans to construct a 6-mile- (10-kilometer-) long section of Hadrian's Wall in just 15 weeks.[9] An experiment using a typical detachment of men (i.e., Roman soldiers) to build a short-length replica proved the feat was possible (ref. "Hadrian's Wall," History Channel).

Experimentation in the context being discussed here has two goals: (1) to determine function by experiment and (2) to quantify function by measurement (Section 4–3). In both instances, what experiments should be performed needs to be carefully considered and planned to maximize the value of output. A few important criteria (from Table 4–1) for assessment by experimentation during reverse engineering include:

- Suitability to purpose, by which is meant, can it do what it was intended to do?
- Performance capability, by which is meant the peak and sustainable level of function (e.g., the duty cycle of a photovoltaic solar panel)
- Universality of the product, by which is meant use beyond the obvious and ability to interface or interact with other products
- Ergonomics, by which is meant ease of use in terms of human factors and human interface or interaction
- Robustness, by which is meant durability for intended service

Not surprisingly, experimentation involves both observation and measurement, as will all be discussed in Chapter 5.

4-5 Other Specific Forms of Teardown

There are actually several specific forms or types of teardown besides product teardown, which has been discussed at length herein, including:

- Dynamic teardown
- Cost teardown
- Material teardown
- Matrix teardown

Dynamic teardown applies the principle of comparative analysis to the assembly process. The focus is the examination of all the design features that specifically (directly or indirectly) contribute to the time and cost of assembling the product during production. *Cost teardown* has the specific objective of assessing the total cost to bring a product to market, excluding general overhead. In this method of teardown, product comparisons and differences are specifically identified and measured with cost estimates. *Material teardown* focuses on saving on direct material costs and labor costs brought about by the particular material. The goal is to identify which materials could be changed to reduce such costs, as well as to have less adverse impact on the environment. Finally, *matrix teardown* deals with the comparison of a company's own products with an eye toward standardizing and communizing wherever possible. A specific goal is to reduce part count and prevent the intrusion of new part numbers that are not absolutely necessary. Process

---

[9] *Hadrian's wall* was a defensive fortification in Roman Britain, begun around AD 122 and completed in less than six years. The entire fortification spans 80 Roman miles (73 statute miles/120 kilometers), is up to 16 to 20 feet (6 meters) high and approximately 10 feet (3 meters) thick, with a system of parallel ditches, berns, and implanted spikes, and a system of forts.

teardown is similar to matrix teardown, except that it is focused on standardizing a simplifying internal fabrication and assembly processes.

4-6 Summary

Product teardown is the intentional dissection and analysis of a product (in the broadest sense, to include parts, components, objects, devices, structures, substances, materials, assemblies, and systems) in order to gain experience and knowledge to add to a database and/or for benchmarking. When used for products that have failed either prematurely or catastrophically, product teardown is a key part of *failure analysis* (or, for the purpose of forensic engineering, to solve crimes or support cases of litigation). When used for products that work or didn't necessarily fail, it is part of the higher-level technique of reverse engineering.

The first key aspect of product teardown is *observation,* which is seeing—or, more properly, sensing—all there is to see (or sense). Effective observation means that (1) nothing is inconsequential and (2) one must "always turn the body over" to see all there is to see.

Measurement and *experimentation* are two other important activities that can and might be involved in proper product teardown and reverse engineering. Measurement involves quantification of geometry and function, while experimentation involves assessment (often involving measurement) of functionality and performance.

There are actually several different forms or types of teardown, with different types having different foci, but all, like general process teardown, aimed at reducing product cost, albeit by addressing and attacking different aspects of the product and its production.

4-7 Cited References

Ashby, M. F., *Materials Selection in Mechanical Design,* Pergamon Press, Oxford, UK, 1992.

"Hadrian's Wall," History Channel, YouTube.

Messler, Robert W., Jr., *Engineering Problem-Solving 101: Time-Tested and Timeless Techniques,* McGraw-Hill, New York, 2013.

Otto, Kevin, and Kristin Wood, *Product Design: Techniques in Reverse Engineering and New Product Development,* Prentice-Hall, Upper Saddle River, NJ, 2000.

4-8 Thought Questions and Problems

4-1 *Product teardown*, like *reverse engineering,* involves taking things apart to see how they work. The motivations for the two techniques related by methodology, however, are usually quite different, with those for product teardown being (1) much more narrow and (2) usually totally consumer oriented.

Write a brief but thorough one- to two-page report on how product teardown *differs* from reverse engineering. As part of your response, be sure to give some common applications for product teardown. Contrast these with the dissimilar applications for reverse engineering.

4-2 There are many wonderfully educational examples of *product teardown* on the Internet, including, but not limited to, the following:

- Apple's 8GB iPod Nano

- Toyota's Prius
- Sony's OLED TV
- Gibson's self-tuning guitar
- Optical mouse

A very extensive list of products that have been subjected to product teardown appears at www.electronicproducts.com under a search for "What's inside electronic products?"

Choose any product in which you have interest and prepare a one- to two-page report on the teardown. Be sure to state the initial motivation for the effort and the resulting output and its hoped-for impact.

4-3 No skill may be more important for an engineer to have or to develop than the skill of *observation*. While often presumed to refer to—or be limited to—visual observation, this is *not* true. Real and complete *observation* involves taking in everything there is to be sensed: appearance, feel, sound, smell, taste.

Think of an event from your life and reflect on all the *observations* you probably made at the time, as well as which of those *observations* most elicit your memory of the event. What sensory input created the most vivid memory? Are there things you remember because of multiple sensory inputs? What lessons are there in these to guide you as an engineer who needs to *observe* the world? How does the use of observations, experiments or tests, or measurements fit with VAX (see Section 1–3).

4-4 *Measurement* often needs to accompany observation during product teardown (as well as during reverse engineering). Measurement can be of static things, such as the geometry of a product, but dynamic things (e.g., operation, function, performance, behavior) may require active intervention to allow measurement.

One important method for making measurements is to conduct experiments or to *experiment or test*. (In fact, it is worth pointing out that engineers use one of four different approaches to learn about, define, or describe something in the physical world; i.e., they analyze it mathematically, they measure it physically, they experiment with it actively, or they simulate it virtually.)

In a brief but thoughtful one- to two-page narrative report or essay, discuss the role of *measurement* in engineering in general, and in product teardown (and reverse engineering) specifically. Be sure to explain when experiments must be performed to allow measurements.

4-5 There are actually several specific forms or types of *teardown*. These include (per Section 4–5) dynamic teardown, cost teardown, material teardown, and matrix teardown.

Briefly but thoroughly describe each of the following types of teardown:
a. dynamic teardown
b. cost teardown
c. material teardown
d. matrix teardown

CHAPTER 5

Methods of Product Teardown

5-1 The Product Teardown Process Revisited

A product teardown process (introduced in Chapter 4) can be defined as a formal approach to learning about and modeling the physical components and the functional behavior of a product. In most references, including those found at various sites on the Internet (ref. M3Design), the use of the term *product* is in the narrow sense of a literal commercial or consumer product. Without getting into a battle of semantics here, the process can be summarized as shown in Table 5–1.

TABLE 5-1 Summary of the Purposes of the Product Teardown Process
To dissect and analyze a product as it is changed for any reason (to overcome a shortcoming, evolve, etc.)   ■ To evaluate the status of a product   ■ To understand the functions, components, materials, fabrication, and technologies employed   ■ To identify strengths, weaknesses, and opportunities for evolution of new products, spin-off products, or radically new products
To conduct benchmarking against a competitor's or one's own product   ■ To establish a baseline in terms of understanding and representing the competence and capability of a producer (whether a competitor, a sister company, or one's own organization)   ■ To establish a baseline against which new conceptual designs can be compared
To gain experience and knowledge   ■ To grow engineering knowledge so as to be better equipped to improve, advance, elicit, or generate new concepts   ■ To learn the strengths and weaknesses of previously attempted or employed concepts   ■ To provide the basis for transferring solutions to analogous problems

While the common use of the term (*product teardown*) and process tends to emphasize use for consumer products to aid customers, users, and consumer evaluation and rating organizations and publications, the process is also used by hobbyists to create realistic scale models. When used with consumer products, the process aids in understanding how the product works to identify and promote innovative design features or possible design shortcomings, but can also be used to evaluate a manufacturer (e.g., for evaluation of its stock by a financial agency) by assessing its methods and attention to quality. For consumer products, the most common way of disseminating the results of a product teardown is in an article rich with photographs and component lists (as a bill of materials) so that others can make use of the information without having to disassemble the product themselves.

Teardowns have even been performed in front of live audiences in a studio at the Embedded Systems Conference (ESC). ESC is a conference and exposition that takes place at six locations year-round around the world each year, including in the United States, the United Kingdom, India, and France. Systems architects and design engineers attend ESC to learn design techniques and best practices from leading experts in industry. It provides attendees with product demonstrations, speeches by industry experts, technical training classes, and accreditation opportunities. The first one was held at Silicon Valley (California) in 1989. The first live teardown took place in San Jose in April 2006, when a Toyota Prius was torn down.[1] Since then, innumerable other popular consumer products have been "dissected" at ESC.

Readers interested in the use of the process with consumer products, outside the area of improving design, are encouraged to search out and visit websites on the topic. Two particularly nice examples of the use of the process and the general procedure are found at www.M3design.com under Google searches for "product teardown," one for a Waterpik Model WP-100 and another for an optical mouse.[2]

Recalling the purpose of this book (i.e., to come to know, accept, and fully understand reverse engineering in its broadest sense), this chapter addresses the general teardown method or procedure, a more sophisticated approach known as *teardown analysis,* as well a more unusual approach employed while a product is actually operating known as the *subtract-and-operate procedure* (SOP). In addition, some details of the method of product teardown with utility to reverse engineering are presented, including the creation of models. These include: (1) product assembly and geometric models, (2) force flow diagrams, and (3) functional models. A final illustrative example should tie everything together.

So let's begin!

---

[1] A fast (time-lapse) video showing the teardown of a Toyota Prius can be found at www.EETimes.com, using a search of "Fast teardown."

[2] The websites by M3 Design are seen as valuable despite what the author strongly feels is a misrepresentation or misunderstanding of reverse engineering as it is being covered in this book. To say, as the M3 Design website does, that "reverse engineering is for copying" and "product teardown" is for understanding and learning is simply incorrect. At the very least, it represents a very narrow view of a very versatile and valuable technique. In fact, *product teardown* is a physical process not necessarily involving any attempt to influence engineering design for the better, while *reverse engineering* is a mental exercise involving mechanical dissection with the specific immediate or long-term objective of influencing engineering design for the better. Enough said about this!

5-2 The General Procedure for the Teardown Process

The general procedure for the common teardown process involves the following five steps:

1. List the design issues of interest (i.e., the purpose for the teardown).
2. Prepare for physical teardown or mechanical dissection of the product.
3. Examine the product distribution and installation through accompanying or available product documentation.
4. Disassemble, measure, analyze data for, and model the product's overall assembly and major subassemblies.
5. Create system models for the product.

Let's look at each step.

Step 1: List the Design Issues or Purpose for the Teardown

Before beginning a product teardown, it is important (as it is with any design-related effort) to identify and articulate the purpose of the effort, that is, the needs, goals, and expectations, or, as done as the first step in a proper design process, to formulate and articulate the problem being addressed. A proper effort here preconditions one's mind as to what to look for, so when it is seen (or not seen), it will be recognized. In preparation for proper documentation, a data sheet should be created for capturing all of the information observed, measured, and discovered (e.g., by experimentation). Information in the data sheet should include at least the following:

- Names or other identifications of components
- Names or other identifications of major subassemblies or subsystems
- Shape (i.e., geometry) and dimensions of components (including dimensions to allow proper orientation and arrangement of components in major subassemblies or subsystems, and within the overall product)
- Tolerances
- Weights of components
- Material(s) used to create each component
- Manufacturing (i.e., fabrication, processing, and assembly) processes
- Finishes on components
- Function(s) of each and every component and any and all subassemblies or subsystems
- Estimated costs (for materials, processing, and assembly)
- Notes of interesting observations

A thorough job here will pay off later.

Step 2: Prepare for Teardown

It is important to gather all of the tools needed to take the product apart and collect the desired information (e.g., dimensions, weight, finish) identified in Step 1. These tools should include appropriate metrology instruments or tools, weighing scales, and so on. If functional information is to be gathered by actual operation of the product (as in the subtract-and-operate procedure described in Section 5–3), appropriate meters may also be required. Obviously, a digital camera and simple ruler (to show scale) are key for quickly capturing component geometry, orientation and arrangement within the assembly or major subassemblies, and the like, as well as for capturing rough dimensions when those will suffice. In most cases, a set of screwdrivers, English and metric wrenches (including Allen wrenches or keys), various pliers, and so on, will suffice. Hammers are seldom needed or recommended!

Step 3: Examine Distribution and Installation

In this step of traditional product teardown, one looks at how the product is packaged and what is involved in its installation or setup, as indicated by included documentation, as both are important factors in the product's development. The cost effectiveness and liability risks associated with the installation and operation of the product should also be assessed in this step of the procedure, using information provided in the product documentation literature.

This step is seldom part of mechanical dissection for reverse engineering in most of the contexts covered in this book, but it is clearly important as part of product teardown for consumers.

Step 4: Disassemble, Measure, Analyze Data, and Model by Assemblies

This step is the "meat and potatoes" of the procedure. First, the overall (complete) product should be photodocumented, analyzed, and measured to provide data identified as important in Step 1. Disassembly should be done carefully, slowly, and methodically, and in consideration of planned and needed measurement, experimentation, and eventual modeling. The overall product should be taken apart as nondestructively—and deliberately—as possible, so that, ideally, the product could be reassembled to operate as it had (if it operated at the start).[3] Every component and subassembly or subsystem, as well as the overall product prior to disassembly and in an exploded view, should be carefully and fully photodocumented. All measurements required to complete the data sheet developed in Step 1 should be taken, being sure that data and photographs are referenced to the data sheet.

Step 5: Create System Models

In situations where photographs alone do not provide sufficient detail, it may be necessary—and valuable—to create *geometric models* as part of the teardown learning procedure. Such models are especially valuable for understanding functionality, particularly from the standpoint of what each component does to allow the overall product to operate. It is also important as part of this step to generate *force flow diagrams*. These diagrams track the movement of forces through a product and are useful for exposing opportunities for combining components (to reduce part count, which is always desirable) or highlight other opportunities to improve the product. Force flow diagrams are addressed in detail in Section 5–5.

An extremely important output of a properly executed product teardown is the creation of *functional models*. Functional models show how the product (or major subassemblies or subsystems constituting the product) transform or transfer energy, material(s), and/or information

---

[3] In fact, conducting mechanical dissection in such a way that the product could be reassembled—whether this is intended or not—is an excellent way to capture all of the information that is available. Many people who conduct a product teardown or mechanical dissection for the broader purpose of reverse engineering say: "I learned more about the product trying to put it back together again than I did by taking it apart in the first place." That's because taking something apart can be done rather "brainlessly," which is *not* what should be occurring here!

from an input state to a desired output (or function) (ref. Messler, pp. 245–248). More is presented on generating functional models in Sections 5–4 and 5–6.

The procedure the author prefers for mechanical dissection associated with reverse engineering when the specific objectives relate to assessing the design for understanding, improvement, competitive assessment or benchmarking, or possible utility of knowledge for other analogous designs, is reflected by the following:

Step 1: Identify and articulate the purpose for dissection as part of a reverse engineering activity.

Step 2: Mechanically dissect (i.e., physically disassemble), observe, measure, and analyze the data obtained, component by component (or physical element by physical element), subassembly (or subsystem) by subassembly (or subsystem), and for the overall product (or system).

Step 3: Deduce or infer the role, purpose, and functionality of each and every part, component, or structural element, as well as each subassembly or subsystem, on the way to identifying the purpose and functionality of the overall object, device, product, structure, system, or material, if not known at the outset (see Chapter 7).

Step 4: Attempt to identify the material(s) used to fabricate each part, component, structural element, or, for an electronic product or system, device. Identification can be general, as: metal $\Rightarrow$ aluminum alloy $\Rightarrow$ heat-treated by aging.

Step 5: Attempt to deduce or infer the method or methods by which each part, component, structural element, or device was fabricated and, if possible, processed, as well as the method(s) by which the product, structure, or system was assembled.

Step 6: Attempt to assess the suitability of the overall design and design details based on acceptability of cost, robustness for service, service environment, expected life, etc.

Other than Step 1, identifying and articulating the purpose for dissection, the intent of which is obvious, and Step 2, which was discussed earlier (as Step 4) of the product teardown process, details of Steps 3 through 6, here for reverse engineering, are discussed, step by step, in subsequent chapters (Chapters 7, 9, 10, and 12, respectively).

5-3 Teardown Analysis or Value Analysis Teardown

A particularly comprehensive form of product teardown was invented by Yoshihiko Sato, known as "the Father of Japanese Teardown."[4] The technique is known as *teardown analysis* but is sometimes referred to as *value analysis teardown* (see Chapter 14 on value engineering). Compared to most teardown methods used in the Western world, the Japanese teardown analysis method is much more complex and detailed. Teardown activity is intended to completely analyze all aspects of the cost of a product compared to competitors'. Attention to detail is a major consideration in this method.

---

[4] Yoshihiko Sato launched into a brilliant career with his creative work as a production manager at Isuzu Motors in Japan in the 1970s. Having seen the teardown process at General Motors, he developed the value-engineering (VE) tinted teardown process that came to be known as teardown analysis or, more descriptively, as value analysis teardown. The process was quickly adopted by all Japanese carmakers, electronic manufacturers, and others. It has been and is increasingly making its way outside Japan, including in the United States.

Readers interested in the teardown analysis method are referred to an article by James A. Rains, Jr., and Yoshihiko Sato entitled "The Integration of the Japanese Tear-down Method with Design for Assembly and Value Engineering" (which can be found at www.valuefoundation.org), as well as a superb book by Sato (ref. Sato).

5-4 The Subtract-and-Operate Procedure

The usual procedure by which product teardown is accomplished uses a top-down approach. The procedure starts with the overall, intact product (and that product's overall function or functions) and then systematically dissects the product piece by piece (and decomposes the functions subfunction by subfunction). The analogy between mechanical dissection involved in this approach and dissection used in botany, biology, and medicine is direct. *Dissection* is defined as "the process of disassembling and observing something to determine its internal structure and as an aid in discerning the functions and relationships of its components." While dissection is usually applied to the examination of plants and animals, there is nothing in the definition that excludes dissection of mechanical or electrical entities.

There is another approach, however, that is a striking analog of another technique used in ancient and early biology and medicine, in particular, but is still used today, albeit less commonly and, hopefully, more humanely. That technique is vivisection.

Vivisection (from the Latin *vivus,* meaning "alive," and *sectio,* meaning "cutting") is defined as "surgery conducted for experimental purposes on a living organism, typically animals with a central nervous system, to view living structure." As disturbing as it is to many people with sensibilities, as opposed to top-down dissection, vivisection is a bottom-up approach. As used by ancient Greek medical practitioners, such as Herophilus of Chalcedon and Erastistratus of Chios in the third century BC and, later, by the famous physician of ancient Rome, Galen (AD 129–ca. 200/ca. 216), surgery progressively removed tissue (i.e., structures or organs) from living animals (monkeys and pigs, in particular) to see what function was lost as a means of learning from where and what various functions arose. During the vivisection of a pig (shown in Figure 1–4), Galen demonstrated that the recurrent laryngeal nerve rendered an animal voiceless when cut out—mercifully for the observers, although not for the pig. The analogous approach for mechanical and electrical systems is the subtract-and-operate procedure.

The *subtract-and-operate procedure* has particular utility for developing a function or *functional diagram* or *functional tree*. The procedure begins by considering the least important or lowest-level functions of a product that can be isolated, as well as the smallest components in the product believed responsible for that function. Lowest-level functions are those that cannot easily be decomposed, if they can be decomposed at all, into further subfunctions. For each such lowest-level function, the engineer removes the component(s) or feature(s) believed to supply the given function and then attempts to operate the product. Actual (or literal) operation may not be done if doing so is deemed a safety hazard. Instead, operation is considered conceptually. In either case, literal or conceptual, operability is assessed so that the engineer can establish the removed component's or feature's contribution to the overall product's function.

Lowest-level functions can be combined into a function tree, progressively working down the tree toward the root, which represents the primary function of the product. This usually involves following the assembly structure of the product itself from lowest to highest level. The subtract-

TABLE 5-2 Summary of the Subtract-and-Operate Procedure (To Assess and/or Develop a Product and Create a Function Tree)

Step 1: Disassemble and remove (i.e., subtract) one component or feature of the assembly: Start from the lowest-level functions (i.e., those that cannot be further decomposed easily or at all). Other components that must be removed simply to gain access to the component of interest should be reassembled or reinstalled, if at all possible. If reassembly or reinstallation of surrounding components is not possible, effort should be made to replicate their function in some way.

Step 2: Attempt to operate the altered product. The product should be tested for operability through its full range required by (or promised to) users once a selected component has been subtracted. Any effect(s) of subtraction on structural, kinematic, dynamic, ergonomic, and other customer requirements should be noted and, to the extent possible, quantified.

Step 3: Analyze the effect of subtraction. The effect(s) of having subtracted a component should be analyzed in terms of impact on operability, performance, safety, etc.

Step 4: Deduce the function of the subtracted component. The function (or subfunction within higher-level function) of the subtracted component must be deduced from Step 3. Particular attention must be paid to any change in the degree of freedom (DOF) of the product (looseness, slop, backlash, loss of balance, vibrations, etc.) within a major subassembly or subsystem during operation. Such changes might represent a critical issue (e.g., risk) in determining component functionality.

Step 5: Replace the component and repeat the procedure for each and every other component, one by one. Before proceeding to subtract another (i.e., different) component, the formerly removed component should be reinstalled. Each time a different component is subtracted, the effects must be documented, often in tabular form (i.e., as an effects table). It is often useful, if not necessary, to analyze a product according to subassemblies or subsystems.

Step 6: Translate the collection of subfunctions into a function tree. With data collection (by subfunction via Step 5) grouped into sets with common or closely related functionality, each set becomes a higher-level functional description node in a function tree for the product. The process is repeated until the higher-level functions in the tree (albeit, often still lower-level functions in the product) converge into the overall product function as a single node at the root of the tree.

Source: Based upon input from Kevin Otto and Kristin Wood's *Product Design: Techniques in Reverse Engineering and New Product Development,* Prentice-Hall, Upper Saddle River, NJ, 2000, p. 161, with extensive modification.

and-operate procedure should be worked for each subsystem until the entire system has been addressed and modeled as a *functional structure* or *functional model.*

As this particular approach is more common to product teardown of consumer products than to reverse engineering in the broader sense being addressed in this book, the summary of the subtract-and-operate procedure shown in Table 5–2, based on a summary in Otto and Wood's excellent book on product design, but greatly modified, is considered to suffice (ref. Otto and Wood).

At least for product teardown, some references (e.g., Otto and Wood) recommend using both the traditional top-down teardown approach and the bottom-up approach independently, and comparing the results of the two approaches to merge into a suitable functional model.

5-5 Force Flow Diagrams (or Energy Flow Field Design)

The subtract-and-operate procedure described in Section 5–4 is a method used to expose the functionality of a product, first of individual components, then of subassemblies of components (e.g., subsystems), and, finally, of the overall assembly, system, or product. The method is especially useful for identifying opportunities for eliminating superfluous components. Force flow diagrams focus on how components interact or combine to provide function(s) that contribute to the overall assembly's, system's, or product's function(s). The diagram models the design and the resulting model is then analyzed to understand (and possibly improve) the design, in what can be known as *energy flow field design*.

Force flow diagrams (alternatively known as *energy field flow diagrams* or *energy flow field design* in some references or contexts) represent the transfer of force(s) (or energy) through a product as it performs its intended function(s). In the diagram being generated, components are generally symbolized as nodes using circles or squares, and the force(s) operating between components are represented by arrows that connect the components involved in force transfer, with the arrow pointing in the direction of the transfer during normal operation.

A simple, but nice, example used by Otto and Wood (ref. Otto and Wood) is the three-piece binder clip shown in Figure 5–1a. To operate this clip (which, in most cases, holds bundles of papers but is routinely adapted to other uses), the user's hand (usually thumb and forefinger) applies (and transfers) force to each of the stiff formed steel wire arms and, in turn, these lever arms transfer force to the one-piece steel spring clip, as shown in Figure 5–1b. A corresponding *force flow diagram* would show a force from the user's "hand," via the thumb and forefinger, being transferred to each lever arm (generally balanced between the two), which, in turn, jointly transfers force to the spring clip, forcing the clip to open to grip whatever is to be gripped using the elastic recovery force.

The force flow diagram is a "map" or a model that shows the movement of force (i.e., the force flow or "energy field flow") through the assembly of components that constitute a product that is easy to analyze to help identify opportunities for combining components and, in the process, reduce the part count in the assembly or product.[5] Once the model has been generated, the first step to allow analysis is to label each arrow between two components that move relative to one another in the assembly with an "R." By then dividing the diagram into groups separated by those with an "R" and those without an "R," consideration can be given to combining two components into a single component where there is no "R," or relative motion. A caveat here, however, is that combining components is possible only if (1) the materials used in each are the same (based on the material being selected for a particular set of properties) and (2) combining components does not give rise to issues relating to complications with assembly (for production) or disassembly (for service, repair, or replacement).

Going back to the example of the binder clip, the formed stiff steel wire arms do not need to rotate relative to the steel spring clip, for either Lever Arm 1 or Lever Arm 2. Thus, based on there not needing to be relative motion, the lever arms and clip could be combined into one piece. This would, of course, create a one-piece binder clip that would bear striking resemblance to the

---

[5] Part count is one consideration in the cost of a product, as higher counts complicate assembly, whether accomplished manually or via automation. Higher part count also complicates inventory and the logistics of manufacturing, which also adds to a product's cost.

Methods of Product Teardown **77**

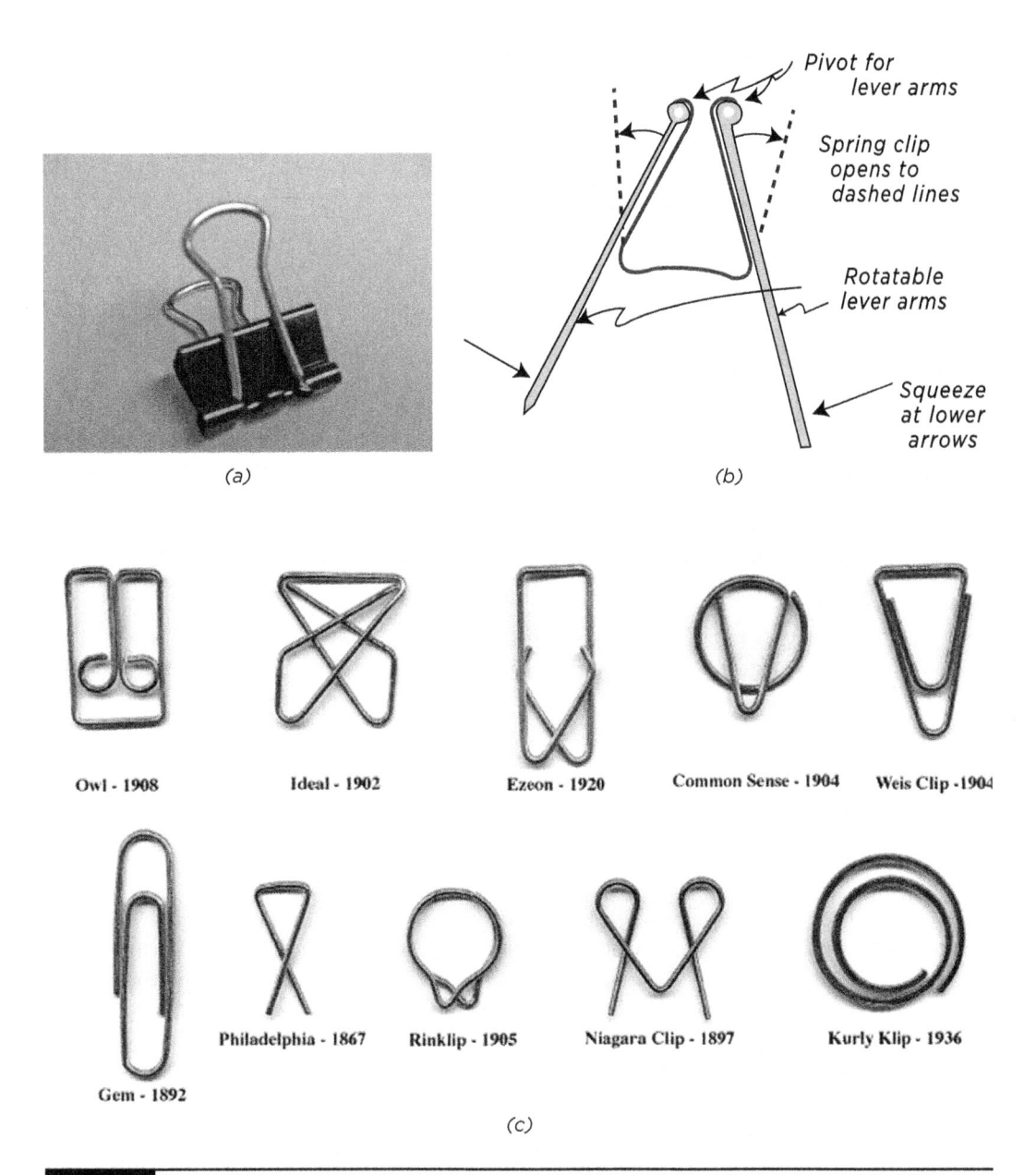

(a)

(b) Pivot for lever arms / Spring clip opens to dashed lines / Rotatable lever arms / Squeeze at lower arrows

Owl - 1908   Ideal - 1902   Ezeon - 1920   Common Sense - 1904   Weis Clip - 1904

Gem - 1892   Philadelphia - 1867   Rinklip - 1905   Niagara Clip - 1897   Kurly Klip - 1936

(c)

Figure 5-1 A common three-piece steel binder clip (a), along with a schematic showing the operation and the forces that operate such a three-piece binder clip (b). Variants of the common one-piece metal wire paper clip, including the most commonly seen type—here, a "Gem -1892" (c). (*Sources:* Wikipedia Creative Commons, contributed by HenryLi on 22 February 2007 [a]; original schematic and force flow diagram by the author [b]; and by Lynne Belluscio, "Kurly Klips," *LeRoy Pennysaver & News*, September 12, 2010, at www.leroypennysavernews.com, with permission from the publisher and Lynne Belluscio [c].)

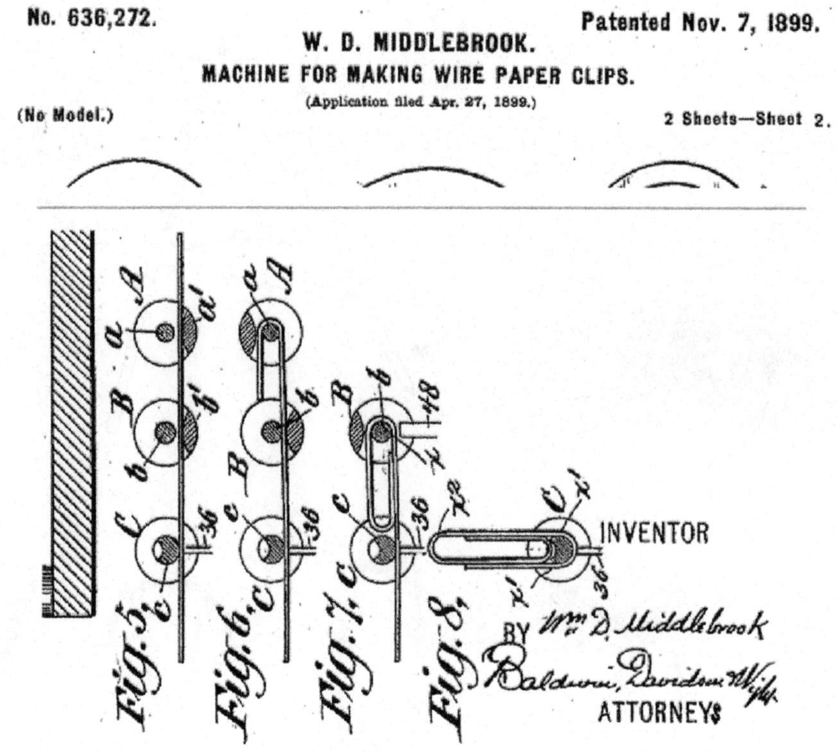

Figure 5-2 Schematic illustration used in the November 7, 1899, U.S. Patent 636,272 granted to W. D. Middlebrook for a machine for making wire paper clips. (*Source:* Wikipedia Creative Commons, contributed by Homunculus 2 on 28 October 2011.)

common metal paperclip and certain variants (Figure 5–1c), for which credit is given to William Middlebrook of Waterbury, Connecticut, as he received a patent for a machine to make them on April 27, 1899 (Figure 5–2).[6] Of course, none of the common one-piece paper clip designs offer the key feature possessed by the binder clip, which is the ability to open very wide to accommodate a thicker bundle of papers. However, to open wide, the length of lever arms would have to be long (to allow a great enough force to be applied), and without the ability to move relative to the spring clip, they would snag with other items in a file. So for compactness as well as the ability to open wide, the binder clip must consist of three pieces.

Table 5–3 summarizes the procedure for creating a force flow diagram.

For more details on creating force flow diagrams, the interested reader is referred to Otto and Wood or appropriate websites searchable on the Internet. An illustrative example is given in the final section of this chapter, however.

---

[6] There is evidence of the existence of many other patents for a similar clip, including a design by Samuel B. Fay in 1867 and 50 other designs before 1899, although none are reminiscent of the paper clip known today.

TABLE 5-3 Summary of the Procedure for Creating Force Flow (or Energy Field Flow) Diagrams

Step 1: Identify and trace the force (or energy) flow from external source(s) to and through each component of the product until the flow exits to a real or imaginary "ground" or output. Exercise particular care when force (or energy) flows split (to become parallel) through components.

Step 2: Document or map the results of Step 1 in a force flow diagram (or energy field flow diagram), in which nodes represent components and arrows represent transfer paths of connections for forces (or energy terms).

Step 3: Analyze the diagram, labeling paths between components for which there is relative motion using an "R."

Step 4: Separate or decompose the diagram into groups separated by "R"s, placing these in a box.

Step 5: Deduce the subfunctions and user needs that are affected for each group.

Step 6: Develop creative conceptual designs to combine components that do not involve relative motion.

Step 7: Repeat Steps 3 through 6 for each force (or energy) flow.

Source: Based upon input from Kevin Otto and Kristin Wood's *Product Design: Techniques in Reverse Engineering and New Product Development,* Prentice-Hall, Upper Saddle River, NJ, 2000, p. 219, with extensive modification.

5-6 Functional Models

Many mechanical, electromechanical (or mechatronic), and structural systems consist of multiple subsystems, each of which, in turn, may consist of multiple components, devices, or structural elements. The same is true for many processes, which may consist of multiple subprocesses and/or operations or steps. To deal with such complex systems or processes, usually as part of an effort to design a new product, system, or process, engineers typically employ the technique of *functional analysis* (ref. Messler, pp. 245–248). In its most basic form, *functional analysis* considers the activities or actions that must be performed by each subsystem and component or subprocess or operation in order to achieve the desired outcome at each level, step, or stage, and, in turn, via proper integration, in the overall system, structure, or process. *Functional analysis,* quite simply, identifies the transformations necessary to turn available inputs of materials, energy, and/or information into the desired output as material(s), energy, and/or information.

Complex *technical systems* (e.g., a commercial airliner, an automobile, or a laptop computer) are commonly decomposed into major assemblies, subassemblies, or subsystems and then further into components within each assembly or subassembly or subsystem.[7] The breakdown for a generic technical system is shown in Figure 5–3*a,* while an example for the breakdown of a modern airliner is shown in Figure 5–3*b.*

The manner of decomposing a complex technical system is useful for analyzing an existing

---

[7]The term *technical system,* as used here, denotes a mechanical, electrical, electromechanical, or structural system. It is used instead of simply "system" to differentiate it from organizational systems (governments, corporations, businesses, etc.).

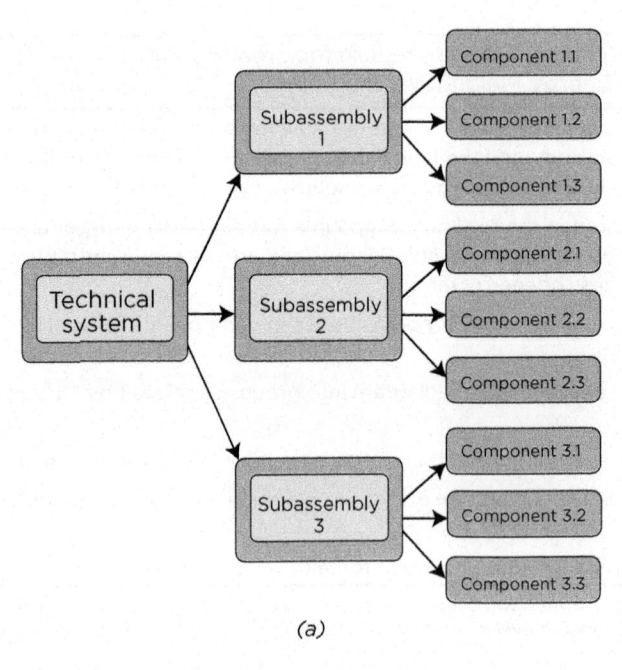

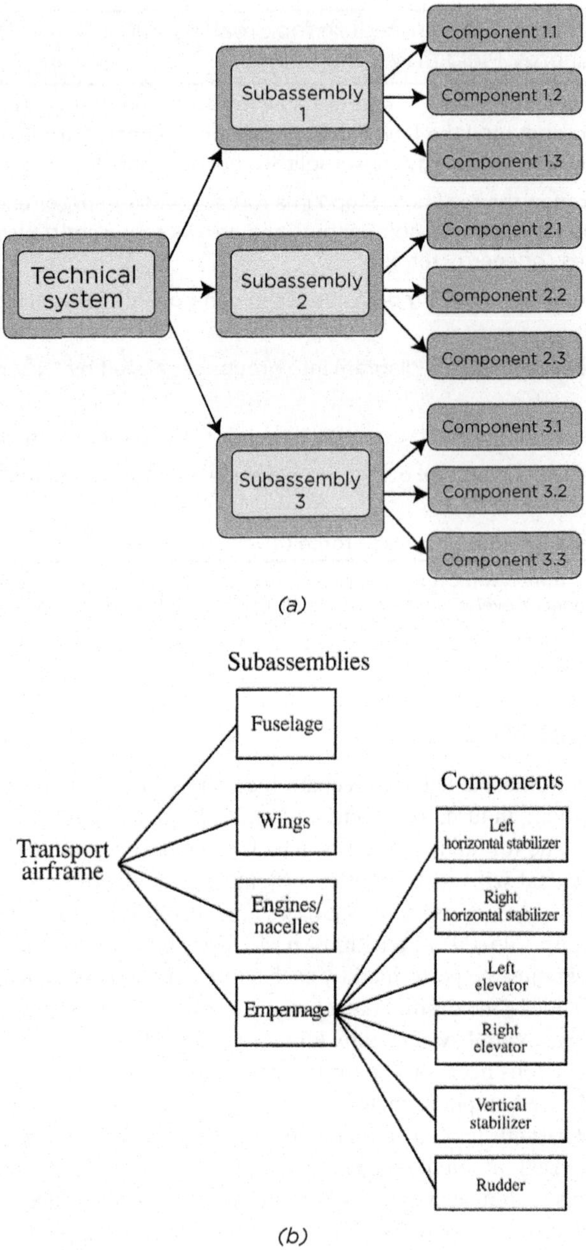

Figure 5-3 A generic breakdown for a complex technical system into subassemblies and, within subassemblies, into components (a), and the breakdown structure of a modern airliner into subassemblies and components within the empennage assembly (i.e., vertical and horizontal stabilizers section) (b). (*Source:* Reproduced from Michael F. Ashby, *Materials Selection in Mechanical Design,* 3rd edition, Elsevier, New York, 2005, page 14, Figure 2-6; used with permission [a], while [b] is original.)

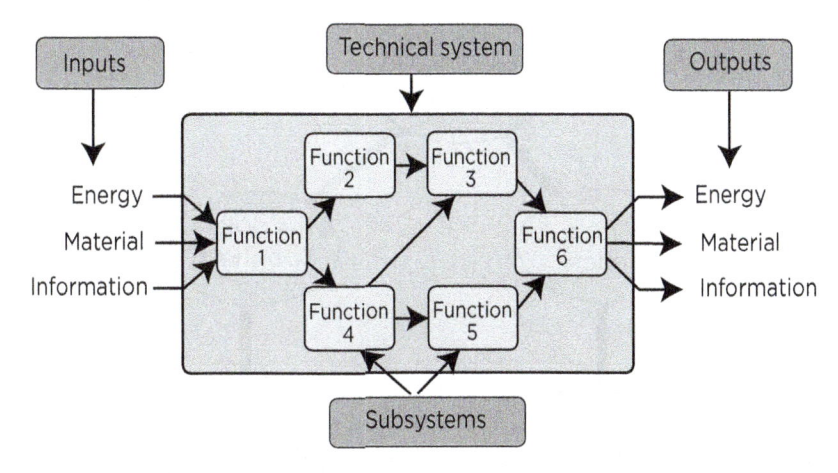

Figure 5-4 A generic breakdown for a complex system into key functions within systems engineering using functional analysis to create a functional structure or model. (*Source:* Reproduced from Michael F. Ashby, *Materials Selection in Mechanical Design*, 3rd edition, Elsevier, New York, 2005, page 14, Figure 2-7; used with permission.)

product, physical entity, or physical process to allow synthesis of a new design and is based on the idea of *systems analysis* using the technique of functional analysis. *Functional analysis* creates an arrangement known as a *function structure* or *functional model* of the system. To develop a *functional model,* the technical system is decomposed into subsystems. For each of these subsystems, the engineer thinks about the inputs, flows, and outputs. Within each subsystem (often treated as a "black box," without internal details), transformations take place which convert inputs of materials, energy, and/or information into desired or needed outputs of material(s), energy, and/or information. The outputs for a particular subsystem may become the inputs to another subsystem in the next stage of the overall, integrated system or might be the desired outputs for the overall system.

The functional model initially obtained from functional analysis is abstract from the standpoint that there are usually not—or need not be—details about how each subsystem function actually accomplishes the transformations. At the initial stage, all that is being sought is to better understand what is needed in the system at a fairly high level. Obviously, during product teardown, details as to what is actually needed in each subsystem (or functional box) must be determined as part of the teardown process and analysis or, for a new design not involving teardown, by the designer.

Figure 5–4 shows a generic breakdown of a complex system into key functions within systems engineering using functional analysis to create a functional structure or functional model.[8]

The next section develops both the force flow (or energy field flow) diagram and the functional model for a product as an illustrative example of the overall process of product teardown.

---

[8] The term *functional tree* tends to be used when a functional model is laid out in a vertical format, i.e., with the flow of material, energy, and/or information taking place vertically, usually from top to bottom (i.e., the root of the "tree"). Figure 5–5 shows an example of a simple—albeit inverted—function tree, or functional tree.

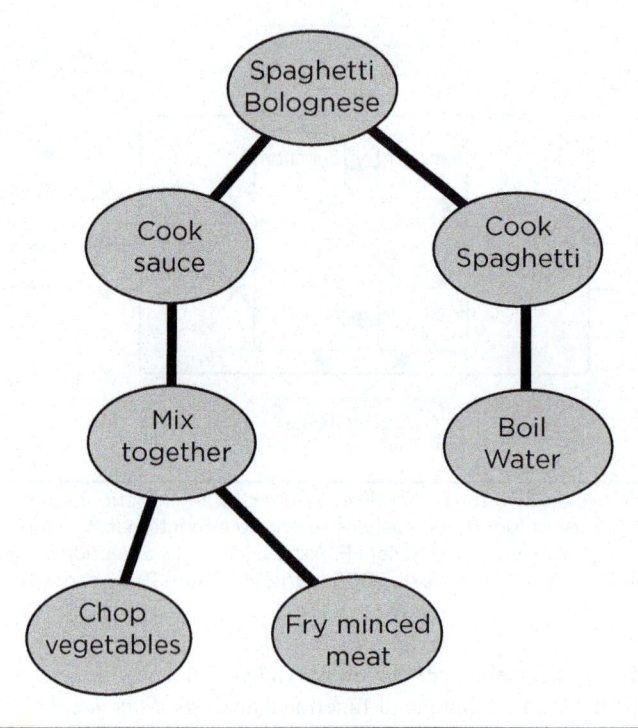

Figure 5-5 A simple function (or functional) tree for the preparation of spaghetti Bolognese. In this particular tree the flow is upward to the "root." (*Source:* Wikipedia Creative Commons, contributed by McSquirrel on 11 July 2009.)

5-7 Illustrative Example of a Product Teardown

A wonderful example of product teardown found with a Google search of "product teardown examples" is that performed by three undergraduate engineering students as part of a sophomore course in design at the University of Idaho.[9] The product subjected to teardown was a SureBonder light-duty, low-temperature (10-watt) hot-glue gun, Model LT-160.[10]

Figure 5–6a shows the product before disassembly, while Figure 5–6b shows the product disassembled with a ruler placed within the field of view to show the scale of details, as is good practice. Figure 5–7 shows the disassembled glue gun as an exploded view, reflecting the arrangement of parts within the assembly, with each and every part labeled. Close-up views of the trigger assembly and the heating element for the glue gun are shown in Figure 5–8a and b, respectively. Again, a ruler is included in the view to show scale.

---

[9] The student team, "The Rocket Avengers," consisted of Paul Sowinski, Tyler Merritt, and William Kramp. The project was performed during spring 2008, with their report dated March 3, 2008, as it was posted on the senior design website http://seniordesign.engr.uidaho.edu/processdocs/teardown.pdf. Permission to use selected photographic images, figures, and information from the report was kindly granted by Dr. Jay McCormack, Assistant Professor, Mechanical Engineering, University of Idaho, Moscow, Idaho.

[10] SureBonder mini glue guns, along with many other products, are marketed by FPC Corporation, 355 Hollow Hill Drive, Wauconda, IL 60084.

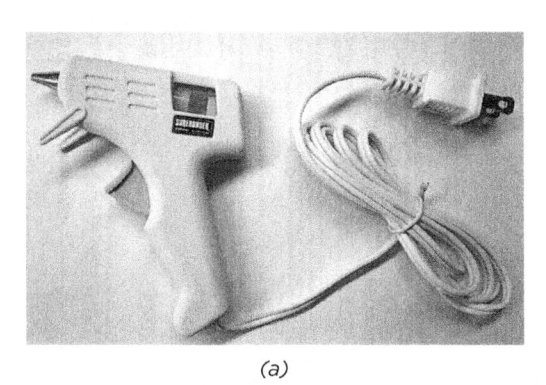

Methods of Product Teardown 83

(a)

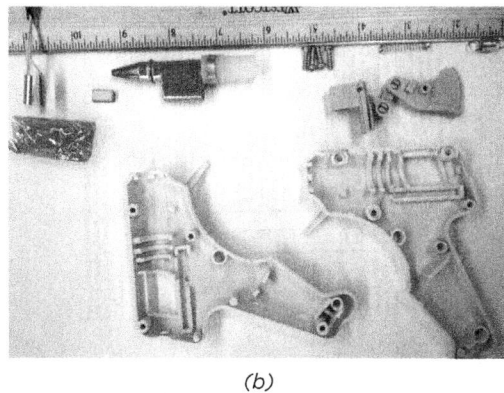

(b)

Figure 5-6 Photograph showing the SureBonder Model LT-160 light-duty, low-temperature hot-glue gun prior to disassembly (*a*) and after disassembly (*b*). (*Source:* Dr. Jay McCormack, Assistant Professor, Mechanical Engineering, University of Idaho, Moscow, Idaho; the work of students Paul Sowinski, Tyler Merritt, and William Kramp in a sophomore design course, with the kind permission of Dr. McCormack.)

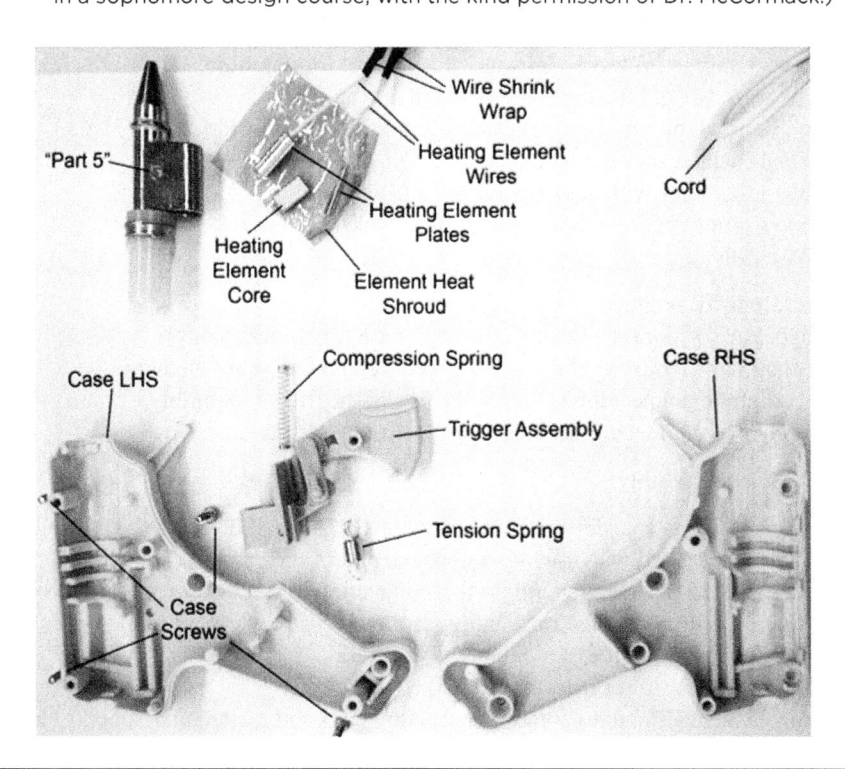

Figure 5-7 Exploded view of the disassembled hot-glue gun shown in Figure 5-6a. (*Source:* Dr. Jay McCormack, Assistant Professor, Mechanical Engineering, University of Idaho, Moscow, Idaho; the work of students Paul Sowinski, Tyler Merritt, and William Kramp in a sophomore design course, with the kind permission of Dr. McCormack.)

Figure 5-8 Close-up views of the trigger assembly (*a*) and heating element assembly (*b*) for the hot-glue gun shown in Figure 5-6a. (*Source:* Dr. Jay McCormack, Assistant Professor, Mechanical Engineering, University of Idaho, Moscow, Idaho; the work of students Paul Sowinski, Tyler Merritt, and William Kramp in a sophomore design course, with the kind permission of Dr. McCormack.)

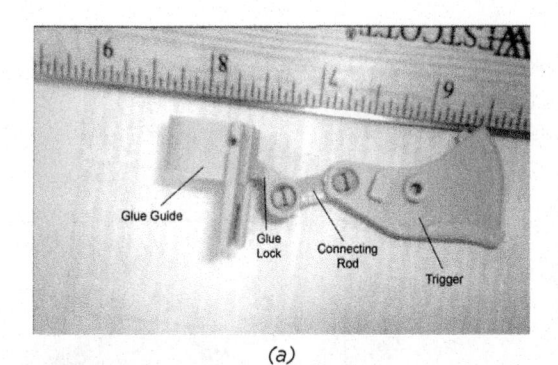

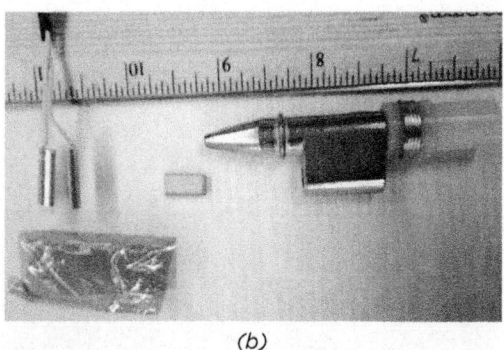

Teardown revealed the hot-glue gun consisted of 28 pieces or parts, as follows:

- 1 plastic packaging: to protect and display the product for purchase.
- 4 exterior screws: to hold the case (or body) halves together.
- 1 right case half: which acts as a handle and contains the rest of the parts.
- 1 left case half: which acts as a handle and contains the rest of the parts.
- 1 short flat spring: to provide resistance for the trigger.
- 1 long skinny spring: to provide a returning force for the trigger.
- 1 trigger assembly:

    1 trigger: to activate glue gun operation.
    1 linkage arm: to connect trigger to the rest of the trigger mechanism.
    1 glue clamp: to clamp the glue stick for feeding into the heater element.
    1 clamp shaft: to connect the clamp to the rest of the trigger assembly.
    1 glue guide: to guide the glue stick into the heating element.

- 1 heating element assembly:

    1 AC power cord: to transfer electrical energy from the wall outlet to the heating element. (The cord is wrapped with 1 white wire tie as part of the package for sale.)
    2 heat-shrink sleeves: to cover and protect connection of the AC power cord to smaller wires.
    2 small wires: to transfer electrical energy from the AC power cord to the heating pads.
    2 heating pads: to convert electrical power to heat.
    1 rectangular block: to serve as the heating element core.
    1 heating element shroud: to contain the heating pads and rectangular support block and act as a barrier between the pads and the casing.

- 1 rubber guide: to guide the glue stick into part #5 casing.
- 1 clamp: to clamp the rubber guide to part #5 casing.
- 1 metal part #5 casing:

Methods of Product Teardown   **85**

      1 check ball valve in tip: to stop glue from flowing when no pressure is applied to the trigger.
      1 valve backing plate: to hold the check ball valve in place.
      1 check ball valve spring: to return the check ball valve to its closed position.

Depending on the purpose of the teardown, one might make detailed measurements on each of these parts, as appropriate. Measurements might include dimensions to characterize component geometry and arrangement with other components in the assembly, component weights, electrical measurements for electrical or electronic components (e.g., here, the heat element's wattage), and so forth.

Figure 5–9 shows the *force flow diagram* or *energy field flow diagram* for the mini hot-glue gun. As described in Section 5–5, this diagram shows how force or energy, as appropriate, is transferred from one component to another in a product to allow the achievement of function at each subsystem and, when all subsystems are properly integrated, in the overall product. For the hot-glue gun, at the most basic level, and initially, force is transferred from the user's hand to the *case*, allowing the glue gun to be held and manipulated. *Part 5* and the *compression spring* also transfer force to the *case*, as indicated by the arrows. The next line down from the top of the diagram shows force from the user's hand being transferred to the *trigger*, which acts against a *tension spring* (to return the trigger to its rest position), as well as to *linkage arm 1*, which transfers force to the *glue stick clamp*. The *glue stick clamp* transfers force to both the *glue stick* and the *glue stick clamp shaft*, which, in turn, transfers force to the *glue stick guide*, which acts against the *compression spring*, the *check ball valve*, and the *rubber guide*. The *check ball valve* transfers force to the *check ball spring*, while the *rubber guide*, which is acted upon by *part 5*, transfers force to the *spring clamp*.

A little study of the resulting diagram, which is developed by either actually operating the product or, more likely, going through the product's operation mentally, provides complete understanding of the product—here, a mini hot-glue gun.

Analysis of the mini hot-glue gun is completed with the creation of the product functional model, shown for the exemplary product in Figure 5–10. Recall that functional models decompose a product (or other system) into its functional elements or subsystems, and consider the flow and transformation of material(s), energy, and/or information into and out of each, until the final output of the integrated system matches the material, energy, and/or information that appear or result when the product (or system) is operating properly.

In the case of the hot-glue gun, four parallel horizontal lines, with arrows to indicate the direction of flow, appear. From top to bottom, at the left, representing input to the glue gun, are: (1) mechanical energy input from the user's hand; (2) electrical energy input from an electrical outlet to the device (i.e., the glue gun); (3) material in the form of the glue stick; and (4) information in the form of "aim" to physically direct the gun's glue-dispensing tip at the target work to which glue is to be applied. Movement of mechanical energy or force from the user's hand causes movement of the entire glue gun along one path and plugs in the power cord along another path. The former also acts to deal with insertion of the glue stick, while the latter serves to allow electrical energy to flow into the device. Conversion (or transformation) of the electrical energy into heat (to melt the thermoplastic adhesive glue stick) is clear, as, too, is movement of the glue stick through the device and, ultimately, to the object(s) to be glued. It is interesting to note how heat serves to melt the hot-melt adhesive glue stick but also results in the flow of heat into the gun components and case, as unintended heat loss, so that is sensed by the user.

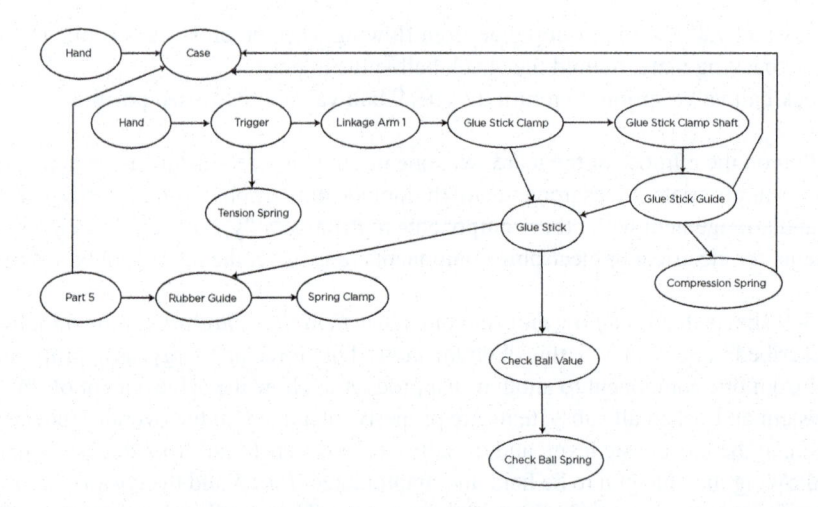

Figure 5-9 The force flow diagram developed for the hot-glue gun shown in Figure 5-6a. (*Source:* Dr. Jay McCormack, Assistant Professor, Mechanical Engineering, University of Idaho, Moscow, Idaho; the work of students Paul Sowinski, Tyler Merritt, and William Kramp in a sophomore design course, with the kind permission of Dr. McCormack.)

As two examples of what the design engineer would have to do to provide needed functionality of two subsystems to change them from "black boxes" to operating subsystems, consider (1) conversion of electrical energy to thermal energy (in the third box from the left along the second path from the top) and (2) how heat loss to the casing must be dealt with (at the upper right of the model schematic).

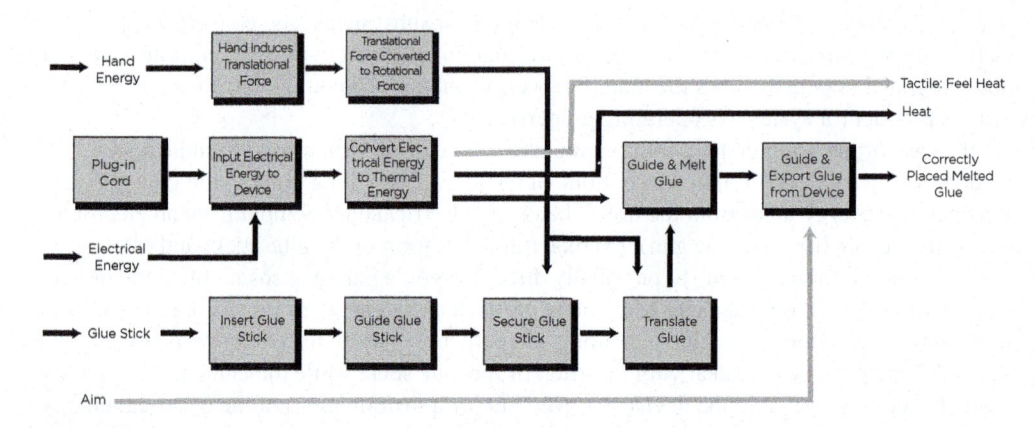

Figure 5-10 The functional model developed for the hot-glue gun shown in Figure 5-6a. (*Source:* Dr. Jay McCormack, Assistant Professor, Mechanical Engineering, University of Idaho, Moscow, Idaho; the work of students Paul Sowinski, Tyler Merritt, and William Kramp in a sophomore design course, with the kind permission of Dr. McCormack.)

Electrical energy is likely converted to heat using resistance or joule heating, almost surely using wire elements contained inside the heating element core. Obviously, this core must be designed to produce the intensity of heat desired (for this device, 10 watts) from I^2R, in which I is the current flowing in the heater element and R is the resistance of the element. This could be done a couple of ways, one using the line voltage applied as electrical energy input (say, 120 volts) and designing the wire element to have the proper resistance to generate 10 watts based on the current flow or, perhaps for safety, reducing the voltage (with a step-down transformer, not found in the actual device!) and adjusting the current and resistance accordingly.

As for the *tactile heat* that eventually gets to the user's hand, the design engineer needs to find out what can be tolerated by a human being as continuous heat and determine how much thermal insulation would be required to keep heat loss to a level less than this. Such a problem is solved using an inverse method (ref. Messler, pp. 240–242). For the subject hot-glue gun, a heat shroud was employed.

In closing, this is a nice example to study, reflecting on each detail as a mental exercise.

5-8 Summary

Product teardown plays a particularly important role in consumer products, where it is employed to provide useful information to the potential consumer, the user, and to evaluation and rating organizations and publications, as well as to engineers engaged in the design of consumer products (e.g., at ESC conferences). The general procedure for the process involves five steps to be followed, with the fourth step involving dissection, measurement, and analysis and the fifth step involving the creation of models of the product. The three models are: (1) geometric model (for assemblies for which photographs alone may not by sufficient); (2) force flow diagrams; and (3) functional models.

A particularly elaborate form of product teardown is that developed by Yoshihiko Sato known as *teardown analysis*. This technique is strongly oriented toward value engineering, hence, its other name, *value analysis teardown*.

The *subtract-and-operate (SOP) procedure* performs teardown while a product is operated—in actuality or as a purely mental effort or thought experiment—to identify the source of each function and subfunction in a product. Force flow diagrams consider the transfer of forces (or, alternatively, energy) through a product to help with its understanding. Functional models are created by decomposing a product into its requisite functional systems and subsystems, and its subsystems into its requisite components. Each function or subsystem is addressed for the transformation of material(s), energy, and/or information that takes place in a "black box" in the model.

A hot-glue gun provides an excellent example of the teardown process, force flow diagram, and functional model.

5-9 Cited References

Messler, Robert W., Jr., *Engineering Problem-Solving 101: Time-Tested and Timeless Techniques*, McGraw-Hill, New York, 2013.

M3 Design: A Product Development Firm, 575 Round Rock West Drive, Suite 100, Round Rock, TX 78681, 512–218–8858, ext. 1 (Design Studio), www.M3design.com.

Otto, Kevin, and Kristin Wood, *Product Design: Techniques in Reverse Engineering and New Product Development,* Prentice-Hall, Upper Saddle River, NJ, 2000, p. 161.

Sato, Yoshihiko, *Value Analysis Tear-Down,* Industrial Press Inc., New York, 2004.

5-10 Thought Questions and Problems

5-1 A search of the Internet results in far more hits for "product teardown" than for "reverse engineering," and definitions or descriptions differ from source to source. Worse, in the opinion of the author, many sites cast "reverse engineering" in an inferior light. While this book provides its own definitions of "product teardown" and "reverse engineering" in several places in Chapter 2, Chapter 4, and Chapter 5, it is important for students new to one or the other or both of these techniques to fix their own definitions in their minds.

After reading and rereading the author's descriptions and definitions, prepare your own concise and, most important, unambiguous definitions. (While these may change with further reading in this book, they will serve as a good starting point.)

5-2 Most references emphasize the value of *product teardown* to prospective consumers or actual users of the product. Obviously, what the prospective consumer wishes or needs to know about a product he or she is considering purchasing and what an owner/user wishes or needs to know are different.

In a brief but thoughtful one- to two-page write-up, describe the benefits that product teardown provides for each group.

5-3 a. Try to find out anything and everything you can from the Internet on the "subtract-and-operate procedure" sometimes used in *product teardown.* Assuming you try hard enough and succeed, prepare a brief but thoughtful description of the procedure.

b. Whether you are able to find out anything more on the subtract-and-operate procedure beyond what is presented in the book, describe several situations in which this approach would prove necessary or beneficial.

5-4 Choose a popular product with which you are familiar (e.g., from personal use) or simply find an interesting product for which a case study is available online (e.g., at M3 Design or the Embedded Systems Conference [ESC]) and generate the *energy field* or *force flow diagram* for it. Consult Section 5–4, Table 5–3, and Figure 5–9 for guidance.

5-5 For the same product you chose in Question 5–4, create an appropriate *functional model.* Consult Section 5–5 and Figure 5–10 for guidance.

CHAPTER 6

Failure Analysis and Forensic Engineering

6-1 Introduction to Failure Analysis

While probably not the first or only person to say it, Jim Owens, former CEO of Caterpillar Inc., said, "We actually learn more from our mistakes than we do from our success." Some might—and do—disagree, but all of us understand the gist of the saying.[1] Few of us dwell on why something succeeds, as we are too taken with the success. But when something fails, that gets our attention. Why did it fail?

Everyone who has practiced engineering long enough knows the simple—but still disturbing—fact that everything fails eventually. One of the corollaries of *Murphy's law* is surely: "Everything fails eventually, and failure will occur at the most inopportune time."[2] Failures are inevitable when one works with complex things. They may occur during early testing of a new concept or during initial production of a new product or shortly after introduction of a new product to the marketplace. But they will occur.

The issue is not so much eventual failure—which, by the way, helps keep an economy going and growing by leading to the development and purchase of new and better replacements—but

---

[1] Tim Harford, economist and columnist for *Financial Times,* wrote a book on the subject entitled *Adapt: Why Success Always Starts With Failure* (Farrar, Straus, and Giroux, New York, 2011).

[2] *Murphy's law* is an adage (or epigram) that is typically stated as: "Anything that can go wrong will go wrong." The American Dialect Society's Stephen Goransen found a version of this law on the perceived perversity of the universe in a quote by Albert Holt at an 1877 meeting of an engineering society, where he said: "It is found that anything that can go wrong at sea generally goes wrong sooner or later . . ." The contemporary form goes back to 1952. Fred R. Shapiro, the editor of the *Yale Book of Quotations,* found specific reference to the adage, by name, in a book by Anne Roe, quoting an unnamed physicist as he described: "Murphy's law or the fourth law of thermodynamics [when there are only really three!] that states 'If anything can go wrong, it will.'" Anne Roe (1904–1991) was a clinical psychologist and researcher who wrote *The Making of a Scientist* (1953), in which she interviewed more than 60 scientists.

premature failures and catastrophic failures. *Premature failures* refer to the failure of things that occur before any problems would or should be expected, based on the intended design life of the product or structure. Recognize that *everything*, however, has a design life. Nuclear power plants, like fossil-fuel power plants, have a design life of around 40 to 50 years of operation (limited mostly by pressure vessels and steam turbines). Commercial airliners (Boeing 737s, Airbus 380s, etc.) are designed for about 60,000 flight-hours, while military fighter aircraft are typically designed for 6000 to 8000 flight-hours (in both cases, being limited by accumulated fatigue damage to airframe structural components, especially in wings). Without intending to sound cynical, modern automobiles are designed to outlast their warranties—but barely (say, 100,000 to 150,000 miles/160,000 to 240,000 kilometers). Why would a company that is in the business of selling automobiles make them last forever? They'd drive themselves out of business—figuratively and literally!

Catastrophic failures are—or should be—every engineer's nightmare, as they refer to the sudden, complete failure of something, without warning (often because one is not performing proper, routine in-service inspection and maintenance), for which the consequence(s) of failure can be life threatening.[3] Catastrophic failures are, obviously, always premature!

Other than these two types, failures are generally taken in stride as part of the life of a product or structure or system. Automatic electric coffeemakers and hair blow-dryers have a finite life of two to three years for the former and a year or so for the latter. The hard drive of a computer—which is often what ends the functional life of a computer—has a finite life (if not made obsolete in the meantime!) of a few years, and surely less than a decade. Commercial buildings and bridges have a finite life, usually of 40 to 50 years, although many remain functional for much longer. Airplanes, automobiles, and petrochemical refineries all have finite lives, as do we.

The second of the two primary purposes of product teardown given earlier (Section 4–1) was: "To discover why something no longer works as it was intended to work." This may not sound as important as it really is. Unless one knows why something doesn't work, one may not know why or how it works! Confused? Don't be! Those "in the know" (and honest enough to admit it) will confess that many processes used every day in manufacturing are not fully understood. As long as things work, allowing parts and product to get made, the motto by which engineers seem to operate is "If it ain't broke, don't fix it!"[4] A young engineer quickly comes to understand the realities of manufacturing when a process for which he or she is responsible stops working as it should work and has been working. When asked by an angry manager, "Why is this *?!#@! process not working?," the thought crosses the young engineer's mind—even if not his or her

---

[3] Bridge 9340, an eight-lane steel truss arch bridge that carried Interstate 35W across the Mississippi River in Minneapolis, Minnesota, suddenly collapsed during evening rush hour, killing 13 and injuring 145 people. The root-cause of the failure was traced to undersized steel gusset plates used on girders supporting the entrance ramps. Fatigue cracking was first detected when the Minnesota Department of Transportation started inspecting the then 30-year-old bridge in 1997 but, after some minor repairs, concluded the bridge was not in any danger of failing. They were wrong! But, also, what made the DoT wait 30 years to begin inspections?

[4] The meaning of the phrase is clearly "If something is working adequately well, leave it alone." While the thought seems to have come from the Stone Age, the phrase in its present form is much more recent. It is most often attributed to Thomas Bertram ("T Bert") Lance, the director of the Office of Management and Budget in President Jimmy Carter's 1977 administration. T Bert was quoted in the May 1977 newsletter of the U.S. Chamber of Commerce, *Nation's Business,* thus: "Bert Lance believes he can save Uncle Sam billions if he can get the government to adopt a simple motto 'If it ain't broke, don't fix it.' He went on to say, 'That's the trouble with government. Fixing things that aren't broken and not fixing things that are broken.'"

lips—"I don't know why it works when it's working, so I sure don't know why it's not working now!" Engineers engaged in manufacturing are pragmatists. If it ain't broke, they don't fix it!

The reality is this: There are many processes and products that are so complex—if not also complicated—that the likelihood of some failure arising that prevents the process or product from working as it should—and once did—is near 100 percent! Things fail before they should—prematurely or catastrophically. In fact, all things fail eventually. Do they not? The real question is: What constitutes failure?, as *failure* is somewhat in the eye of the beholder.

Everyone would agree that something has failed when it no longer works at all because some key component has fractured, one component that drives another has worn, parts that should move relative to one another have seized, or the mechanism or device just stops. But for a race car driver competing in the Indianapolis (Indy) 500 on Memorial Day Weekend, his or her race car has "failed" when it can no longer run 220+ mph laps, when it did at the start of the race. The race car is on its way to failure, even though it would certainly allow any of us to go anywhere we wished as fast as we could imagine, and if the race team leader decides the driver should keep the car running, it will likely fail so that everyone can see—and agree—that it has failed.

Failure of mechanical devices or systems can be considered to have occurred when either of the following has taken place:

- The device or system has experienced a catastrophic (sudden and devastating) fracture of some critical structural member or seizure between moving parts, and the device or system has ceased to function at all.
- The device or system can no longer perform as required.

The latter criterion is the better of the two, as it often prevents a catastrophic event from ever occurring.

Failure can manifest itself in any of the following ways:

- A complete fracture has occurred in some key component.
- A key component has cracked and is on its way to fracturing completely, with either proper operation (e.g., free of excessive vibration) or safe operation jeopardized.
- A key component has experienced so much distortion (in shape and/or dimension) that operation cannot continue or is severely impeded, this distortion being either elastic and temporary, as it is fully recoverable upon unloading, or plastic and permanent, respectively.
- A key component has degraded due to mechanical wear and/or chemical corrosion to the extent that it can no longer perform its intended function in the system.[5]

Failure analysis is the systematic examination/investigation of a part, component, device, product, structural element, structure, assembly, or system that is no longer able to operate to its intended and required level of performance. The conclusion of a failure analysis is, not uncommonly, a reverse-engineered product, structure, or system, in which a new design is guided by what was learned from the prior unsuccessful design.

The process of failure analysis can be performed by any of several individuals or groups within a company or an organization, including:

---

[5] *Degradation* could be to the part or to the material constituting the part (e.g., the material losing its strength or becoming more brittle).

- Design engineering (which should always be informed, if not involved, as their participation assures feedback of unsuccessful design to those most responsible)
- Manufacturing or process engineering (which is valuable since the majority of failures actually have their origin in manufacturing-induced defects, per Figure 6–1)
- Quality assurance or quality engineering[6]

In addition to these groups within a company or organization responsible for the product can be external people or organizations, such as:

- Accident or crash investigators, including the National Transportation Safety Board (NTSB) for commercial or civilian airplane accidents or insurance companies
- Legal agencies (usually with the assistance of a consultant)
- Consultants

Who does the failure analysis is not as important as how it gets done, which must be by a rigorous and systematic procedure. Also, once a *root-cause* for a failure has been determined, it is critical to inform all stakeholders.

In the modern, global economy, remaining viable and competitive is essential to manufacturers. Quality, agility, and responsiveness are keys to success. For these reasons, manufacturers need to have failure analysis capability as a vital aid for reverse engineering and continuous quality improvement.

6–2 Sources of Failures in Mechanical Systems

American singer-songwriter Madonna (1958–), said it in her song "Material Girl," in which, in the chorus, she sang: " . . . we are living in a material world . . ."[7] Even though the context of the song is slightly different than herein, materials engineers like the author have long known and love that it's a material world. Without question, we are physical beings and we live in a physical world. The world and everything we can see or touch in it are physical, comprising physical materials. And, since engineers design, test, and build the things we all use that make our lives better, they know—or ought to know—best that it's all about *materials*.

Failures too have their origin in materials. This is *not* to say that the material is at fault—that it is the root-cause for the failure—as it seldom is, relative to other causes identified and enumerated later in this section. But no matter what the root-cause for a failure is, the failure manifests itself *in* the material by one or more of several basic mechanisms described and discussed in Section 6–3.

There are many potential sources for a failure in a mechanical part, component, device, structural element, structure, product, or system.[8] These sources are overwhelmingly *physical*

---

[6] Occasionally, sales and/or marketing personnel may get involved in a failure investigation, if only by insisting on one, when a product they have sold or promoted doesn't perform as it should and is suffering failures.

[7] The first chorus goes: "Living in a material world/And I am a material girl/You know that we are living in a material world/And I am a material girl." "Material Girl" was written by Peter M. Brown and Robert S. Rans, and was released on November 30, 1984, by Sire Records. Rights belong to EMI Music Publishing/SONY/ATV Music Publishing LLC.

[8] Failures in electrical or electronic parts, components, devices, assemblies, or systems also have their origin in the material(s) used to create the part, component, device, assembly, or system. Hence, while not the focus of this book, much of what is said about reverse engineering of "mechanisms, structures, systems, and materials" (in the title), and all of what is said about failure analysis in this chapter, applies to the electrical and electronics area as well. Readers specifically interested in failure analysis of electronic systems are referred to Sachs (ref. Sachs).

sources. The potential physical sources for a failure (as a possible root-cause) can be conveniently divided as follows:

- Design sources:
    - ✓ Improper part shape or size to meet form, fit, and function requirements
    - ✓ Improper choice (and call-out) of material to meet demands
    - ✓ Improper or incomplete identification of design requirements (e.g., loading and environment)
    - ✓ Design defects (e.g., insufficient or absent radii or other stress risers)
    - ✓ Improper selection (and call-out) of material condition (e.g., heat treatment)
    - ✓ Deficient or negligent design (e.g., improper or insufficient analysis)

- Manufacturing sources:
    - ✓ Improper part shape or dimensions (versus the design drawing)
    - ✓ Process-induced flaws (e.g., machining gouges or score marks) or defects (e.g., porosity in castings, cracks in welds)
    - ✓ Improper processing (e.g., erroneous heat treatment)
    - ✓ Improper assembly (e.g., forced fit, missing fasteners, wrong part)
    - ✓ Mishandling (e.g., handling damage)

- Service sources:
    - ✓ Use for unintended purposes (e.g., a screwdriver as a pry bar)
    - ✓ Use beyond design limits (e.g., exceeding the duty cycle, overspeed operation)
    - ✓ Abuse (e.g., bring a smartphone into a swimming pool)
    - ✓ Failure to conduct proper maintenance (including unauthorized maintenance)

Naturally, many of these physical sources involve human beings, and failures may actually be caused by the action or inaction of a human being. But it is common to consider the aforementioned as physical sources, even though humans are involved at some point. On the other hand, there are failures that occur almost entirely due to the inappropriate action or inaction of a human being or group of human beings, frequently including management.[9] These failures may result directly or indirectly from the human being, but, without question, with the likelihood that the failure, if the human had done what he or she should have done, would not have occurred. So to complete the prior list, there are:

- Human sources:
    - ✓ Negligence (in design, manufacturing, or operation)
    - ✓ Insufficient attention to quality or quality assurance measures)
    - ✓ Mismanagement (e.g., forced action or inaction involving engineers)

These sources, along with a few additional details, are summarized in Table 6–1.

---

[9] The reason is that if management didn't force rush to market—or launch of a Space Shuttle—an imminent failure may have been detected prior to launch. Other times, management is so concerned about costs, that shortcuts are forced on engineering and/or manufacturing. In either situation, failures could be averted by proper action.

94 Chapter Six

TABLE 6-1 Potential Sources for Failure in Parts, Products, Structures, or System (with Likely Failure Behavior)

Design sources	
✓ Improper part shape or size	Excessive play; seizure; malfunction.
✓ Improper material selection	Overload; excessive wear or corrosion.
✓ Improper requirements	Unexpected loading, wear, or corrosion.
✓ Design defects	Stress concentration failures.
✓ Deficient design	Static overload or premature fatigue.
Manufacturing sources	
✓ Improper part shape or size	Excessive play; seizure; malfunction.
✓ Process-induced flaws	Stress concentration; brittle fracture or fatigue.
✓ Improper processing	Wear or corrosion; overload or fatigue.
✓ Improper assembly	Excessive play or vibration; malfunction.
✓ Mishandling	Overload distortion or fracture.
Service sources	
✓ Unintended use	Any mechanism is possible.
✓ Use beyond design limits	Overload or fatigue; creep.
✓ Abuse	Overload.
✓ Improper maintenance	Any mechanism is possible.
Human sources	
✓ Negligence	Any mechanism is possible.
✓ Insufficient quality assurance	Any mechanism is possible.
✓ Mismanagement	Any mechanism is possible.

Another way of dividing the sources of failures that is useful to the company or organization responsible for it is by categories of Design errors, Fabrication/processing errors, Assembly errors, Material defects, Service/maintenance errors, Misuse, and Other.[10] Naturally, the relative contribution of each of these depends on the details and complexity of the product, the industrial sector or application area, the experience and capability of the company or organization, and other factors. Hence, "wraparound"—or generic—percentages are not nearly as meaningful as real data. This said, Figure 6–1 is a pie chart reflecting the author's overall experience as both a practicing engineer in both design and manufacturing in industry and as a materials consultant with experience in more than 1400 failure analysis investigations.

---

[10] For products manufactured in a plant, the categories of Fabrication/processing errors and Assembly errors can logically be added to account for Manufacturing-based errors. This is not true for the erection of civil structures, however, as detail parts, structural elements, or prefabricated modules are made in a plant and assembly is done outside by a completely different group of people, if not an organization.

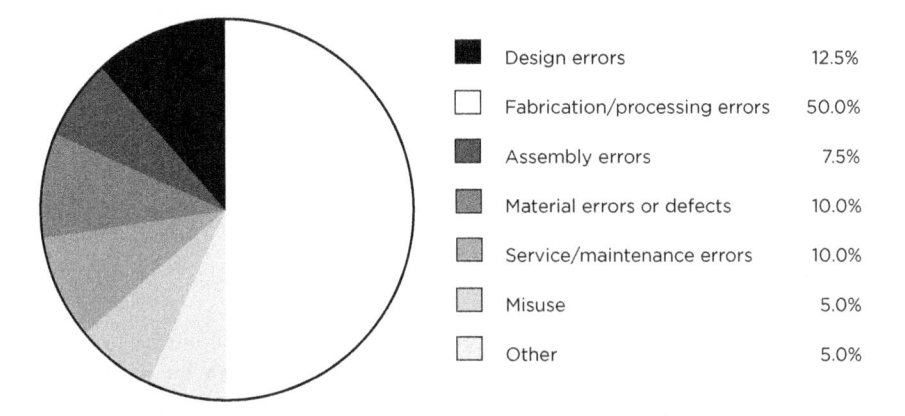

Failure Analysis and Forensic Engineering

■	Design errors	12.5%
□	Fabrication/processing errors	50.0%
■	Assembly errors	7.5%
■	Material errors or defects	10.0%
■	Service/maintenance errors	10.0%
■	Misuse	5.0%
□	Other	5.0%

Figure 6-1 Pie chart showing the generic sources of failures in mechanical mechanisms, structures, systems, and materials (as percentages) from the author's 40+-year experience and more than 1400 failure investigations. Obviously, precise statistics depend on the industry, application area, product and producer sophistication, service environment, users, and so on. **Don't miss the color version of this figure, available at www.mhprofessional.com/ReverseEngineering.**

6-3 Mechanisms of Failure in Materials

Since all physical things—like all products (in the broad sense)—as stated before, are made from materials, failures initiate in the material and propagate to destroy the product. Hence, it should come as no surprise that the clues to how a physical object or product fails are to be found in the material—actually, usually on the surface of fractures or cracks or at the surfaces and/or interfaces between parts that interact and wear or corrode. The key to successful failure analysis is finding, identifying, and interpreting the clues left behind in or on the materials constituting the parts that make up the product.

The principal mechanisms that can lead to failures in parts or components include:

- Structural overload by a static load or force (and stress) by:
  ✓ Brittle fracture
  ✓ Ductile plastic deformation (i.e., distortion) or fracture

- Fatigue cracking or fracture by cyclic loading (and stress):
  ✓ High-cycle/low-stress
  ✓ Low-cycle/high-stress

- Wear-induced loss of material due to part-to-part adhesion or foreign material abrasion, with several subtypes
- Chemical or electrochemical corrosion to cause loss of material, cracking or fracture, or property degradation, with eight subtypes
- Elevated temperature degradation of properties or permanent distortion or rupture (by creep), with several subtypes

96 Chapter Six

It is possible for two (or more) of these basic mechanisms to act together in what is known as *combined mechanisms*. Examples of the most common combined mechanisms that lead to failures are:

- Combined mechanisms:
  ✓ Wear-corrosion (e.g., erosion-corrosion)
  ✓ Wear-fatigue (e.g., fretting, which some experts and references consider fatigue-wear-corrosion)
  ✓ Corrosion-fatigue

In each case, the onset of failure is generally accelerated, as the energy contributed by each individual mechanism adds to greater energy acting to destroy or degrade the material or the part the material constitutes. After all, it takes a certain amount of energy—irrespective of the type or source—to break the bonds that hold the atoms of a solid material together.

Each of these mechanisms leaves behind telltale clues at both the macroscopic (naked-eye-observable) and microscopic (magnified) level. The reader seeking more details should consult a reputable reference on failure analysis (see "References Cited"). Table 6–2 summarizes key clues associated with each of these mechanisms in metal parts.

Just as it is possible to consider the relative contribution of various sources of failure to the overall instance of failures, it is also possible to consider the relative contribution of the various underlying mechanisms by which a failure can manifest itself, as it propagated from the material to the product. However, once again, data are only meaningful if they are specific to an industrial sector (e.g., aerospace, mining, petrochemical refining) or, better yet, an individual company or organization, as the operative mechanism is highly dependent on the types of loading (e.g., static versus fatigue) and environment (e.g., severe wear versus corrosion). Once again, for what it is

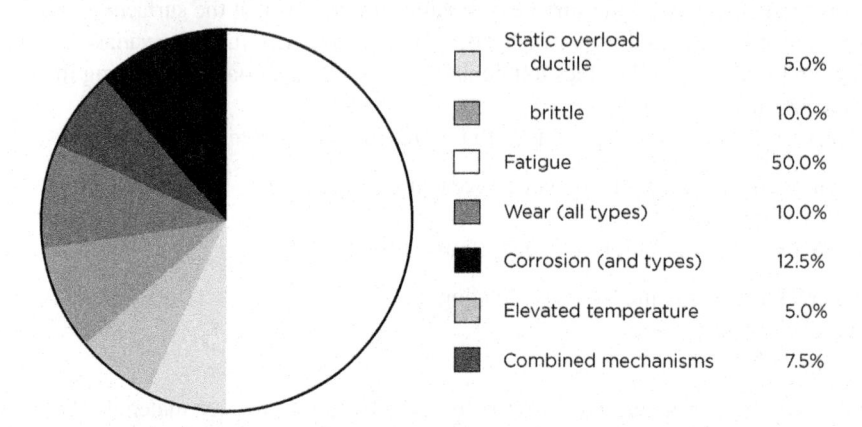

	Static overload ductile	5.0%
	brittle	10.0%
	Fatigue	50.0%
	Wear (all types)	10.0%
	Corrosion (and types)	12.5%
	Elevated temperature	5.0%
	Combined mechanisms	7.5%

Figure 6-2 Pie chart showing the relative occurrence of failures by various mechanisms in mechanical mechanisms, structures, systems, and materials (as percentages) from the author's 40+-year experience and more than 1400 failure investigations. Obviously, precise statistics depend on the industry, application area, product and producer sophistication, service environment, users, and so on. ***Don't miss the color version of this figure, available at www.mhprofessional.com/ReverseEngineering.***

TABLE 6-2 The Fundamental Mechanisms (and Subtypes) That Can Lead to Failure of Mechanical Components or Structures, along with Macro- and Microscale Clues

Mechanism	Subtype	Macrofeatures	Microfeatures
Single-event static overload	Brittle fracture	River pattern; chevrons	Cleavage fracture topography
	Ductile fracture	Gross plastic deformation	Dimpled rupture topography
Dynamic, cyclic fatigue	High-cycle/low-stress	Thumbnail crescent origin; beachmarks; overload area	Fine striations
	Low-cycle/high-stress	Thumbnail crescent origin; beachmarks; overload area	Coarse striations
	Multiple origins	Rachet marks at perimeter	—
Wear	Adhesive (metal-to-metal)	Wear scar; galling; heat tint	Metal deposit/transfer
	Abrasive	Wear scar; gouging; debris	Fine wear debris
	Erosion	Wear flow; polishing	Silt or ash
	Cavitation	Wear flow pattern; pits	Tiny pits
	Fretting	Discolored deposit	—
Corrosion	Uniform/general	Overall rust or tarnish layer	—
	Crevice	Attack in crevices	Possible corrosion residue
	Pitting	Localized pits	Preferential by microstructure
	Galvanic	Dissimilar metals in contact	Corrosion on more anodic metal
	Selective leaching	Preferential attack; severe pits	—
	Intergranular	Corroded network	Local attack of grain boundaries
	Hydrogen	Brittle fracture	Possible "fish-eyes"
	Stress corrosion	Highly branched cracks	—
Elevated temperature	Creep	Distortion or rupture	Intergranular; dimples
	Thermal fatigue	Checked crack pattern	—
	Thermal shock	Brittle fracture/cracks	—
Combined mechanisms	Wear-corrosion (erosion-corrosion)	Loss of material; polishing; corrosion pits, cracks, or residue	—
	Wear-fatigue (fretting)	Discolored deposit	Tiny pits
	Corrosion-fatigue	Thumbnail crescent origin; Possible beachmarks	Possible striations

worth, the author has observed the frequencies shown in Figure 6–2 as a general average, with deviations noted.

Fatigue (especially high-cycle/low-stress fatigue) is responsible for the majority of failures in aerospace, reaching 60+ percent (particularly in high-performance/high-demand military aircraft). Fatigue is involved in around 50 percent of the failures that occur in automobiles, machines, and railroad locomotives and cars, and is probably around 40 percent in bridges and other large structures. While also common in earthmoving, mining, and oil drilling equipment, its relative contribution (at 20 to 25 percent) is masked by the preponderance of failures due to wear in these application areas.

Static overload failures are—and ought to be—relatively rare, as designing against static loads is far easier than against fatigue loads (i.e., the required analysis is more straightforward and less statistical in nature). Brittle fracture is more common than ductile overload, as brittle fractures are exacerbated by material type (e.g., body-centered cubic metals and alloys, such as carbon and low- and medium-alloy steels), high rate of load application and low temperature (in BCC metals and alloys), geometric and metallurgical notches, thick sections, and environmental factors (e.g., hydrogen in hardened steels).

Failures due to corrosion are relatively minor in aerospace (at less than 3 to 4 percent), are moderate in modern automobiles (at 5 to 7 percent), but are, obviously, very severe in petrochemical refining (at 30 to 40 percent), pulp and paper (at around 30 percent), marine (at around 25 percent), and general chemical production and processing industries (at 20 to 30 percent).

Failures due to elevated temperatures are rare, except in heat engines (e.g., especially steam and gas turbines).

Before leaving this section on mechanisms of failure in materials, and the telltale evidence each leaves behind to allow identification, there are a few more details to discuss.

First, the orientation of a primary fracture or crack is directly related to whether the material is behaving in a ductile or brittle manner. This is important, since most people (including most engineers) have a sense that a material is, by its nature, either inherently ductile or brittle, which can be the case, with caveats. For example, is marshmallow a ductile (easily deformable, cracking-resistant) material or a brittle (crack sensitive, nondeformable) material? The answer is: It depends. If the temperature is low (i.e., the marshmallow is very cold), it behaves in a brittle fashion, cracking and fracturing into pieces if it is struck by a hammer. If it is at or above room temperature—and it is not stale (i.e., dried out and hard)—it behaves in a ductile fashion, resisting cracking and deforming to whatever shape one wishes. Finally, if the marshmallow is pulled slowly, it behaves in a ductile fashion (and stretches), but if it is pulled very quickly, it behaves in a brittle fashion (and breaks or snaps), like Silly Putty.

Some materials, such as ceramics (including cement and concrete), glasses (at ordinary temperatures), and some metal alloys (like cast iron, die-cast zinc alloys, some cast bronzes), are inherently brittle. That is, they virtually always fail by fracturing in a brittle fashion and they are prone to the formation and easy growth of microscopic cracks. Metals and alloys, on the other hand, tend to be ductile and are, without question, used in design by engineers to behave in a ductile manner. Those metals and their alloys that have a face-centered cubic (FCC) crystal structure (e.g., Al and Al alloys; Cu and Cu alloys; Ni and Ni alloys; Au, Ag, and Pt and their alloys; and austenitic stainless steels) are ductile under all conditions, not being bothered by low temperature, high strain rate, or notches or other stress concentrators. Those metals and alloys that have a body-centered cubic (BCC) crystal structure (e.g., Fe and carbon and low-alloy steels below about 1670°F/912°C; refractory W, Mo, Ta, and their alloys; and Cr) have a tendency toward increasingly brittle behavior when the temperature is low (below their ductile-brittle transition temperature, or DBTT), the strain rate is high, stress concentrations are present, sections become thick, or under certain chemical environments (e.g., hydrogen). Hexagonal close-packed (HCP) metals and alloys (e.g., Ti and beta-Ti alloys; Mg, Zn, and Sn and their alloys) tend to be ductile but can exhibit brittle behavior under extremely low temperatures or extremely high strain rates. Furthermore, it is generally true that metal or alloy parts directly produced by casting from the melt are more prone to brittle behavior than parts produced by cold-, warm-, or hot-working (wrought) processes, such as rolling, forging, drawing, or extrusion.

Polymers tend to be ductile above their glass transition temperature (T_g) and brittle below. Most exhibit less ductility as the rate at which they are strained increases. Many are embrittled by certain chemicals (e.g., chlorine, oxygen, some organic solvents), as well as by radiation, such as ultraviolet (UV) from sunlight.

This said, here's what is important:

- When cracks form and fracture occurs at 90 degrees to the principal direction of a tensile, compressive, or bending stress, the material is behaving in a brittle fashion. Also, there is generally little or no evidence of any gross (readily observable macroscopic) plastic deformation in the form of bending, buckling, wall bulging or thinning, or necking in the region of fracture.
- When there is evidence of gross plastic deformation, cracks form, and fracture eventually occurs at ±45 degrees to the principal direction of a tensile, compressive, or bending stress, the material is behaving in a ductile fashion, especially as the crack reaches the edges of a thick cross section.
- Under torsion loading (i.e., twisting, as in shafts), cracks form and fracture occurs at ±45 degrees to indicate brittle behavior and at 90 degrees to indicate ductile behavior.

This is shown in Figure 6–3.

A simple mental exercise when pondering a failure containing cracks and/or a fracture is to ask: How would glass—or a fresh pretzel—act (as inherently brittle materials)? Contrarily, how would a fresh Tootsie Roll act (as an inherently ductile material)? What we are doing when we ask such questions is going back to our experiences as physical beings in a physical, or *material,* world.

Second, ductile versus brittle fracture during overload can be verified by the topography of the fracture surface, that is, by the *fractographic evidence* during *fractographic analysis* or *fractography.* Brittle fracture evidences itself at the macroscopic (naked-eye) level by a flat fracture face with ridges that fan outward from some point to create a *river pattern* or *fan structure* (Figure 6–4a). The pattern of ridges points back to the origin or initiation site of the fracture event. Alternatively, or in addition, in some materials (like steels), *chevrons* or a *herringbone pattern* is sometimes seen (Figure 6–4b). Here, the chevrons—like arrows—point back to the fracture origin. At the microscopic level, brittle fracture is characterized by *cleavage* and a smaller-scale fan structure (Figure 6–4c). Ductile fracture, on the other hand, tends to be evidenced at the macroscopic scale by gross plastic deformation in the form of bending, buckling, or bulging (Figure 6–4d). The plane of the fracture lies at a ±45-degree angle to form *shear lips,* as opposed to lying flat (at 90 degrees to the primary stress) (Figure 6–4e). At the microscopic level, ductile fracture is characterized by *dimpled rupture,* which has a "soft" or "spongelike" appearance (Figure 6–4f).

100 Chapter Six

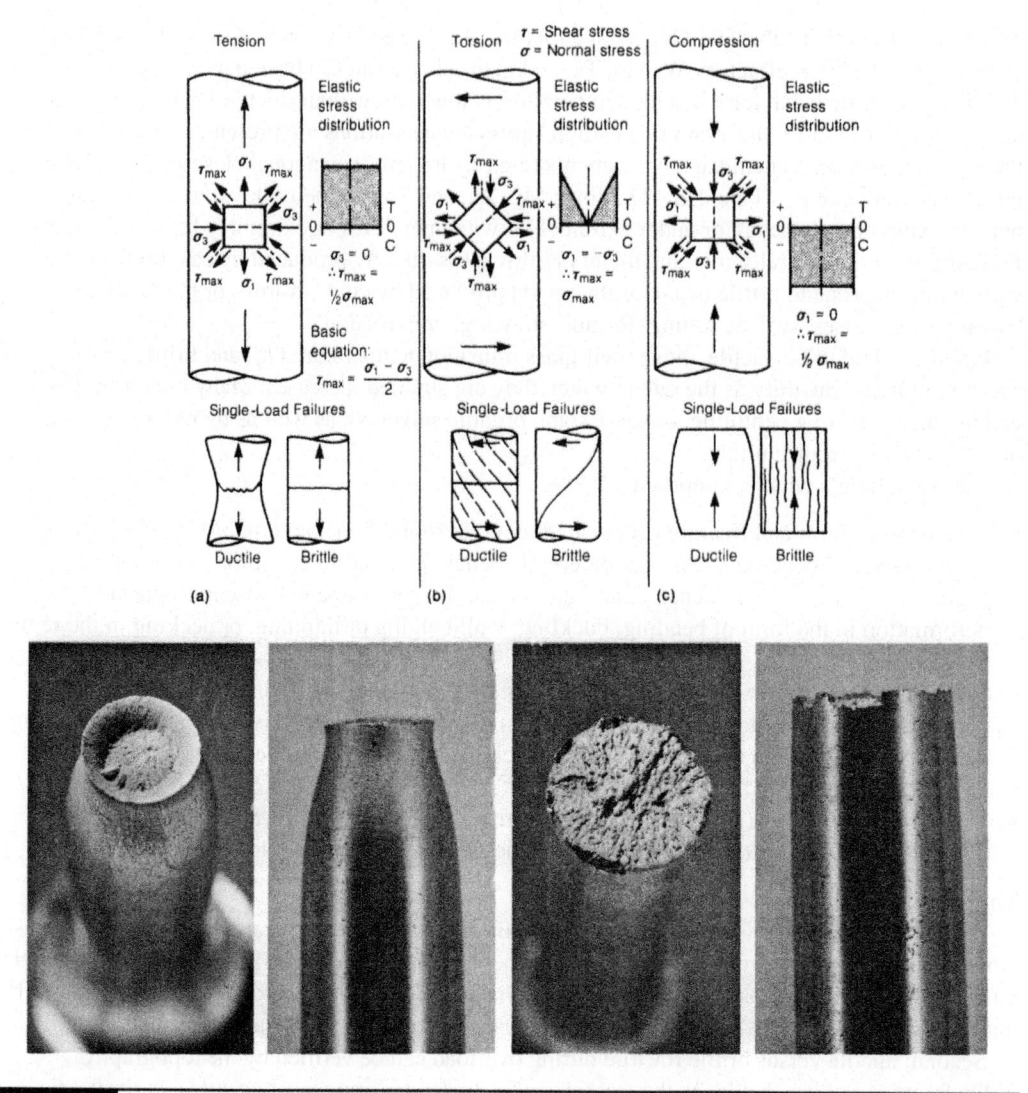

Figure 6-3 Schematic illustrations showing the orientation of fractures with brittle versus ductile behavior, depending on the type of loading (*top*) and characteristic necking and cup-cone formation with ductile fracture versus flat fracture with brittle fracture (*bottom*). In the first, the type of loading is shown in (a), (b), and (c) beneath the schematics. In the second, ductile fracture is shown by the two photographs at the left and brittle fracture by the two at the right. (*Sources:* Donald J. Wulpi, *Understanding How Components Fail,* 2nd edition, ASM International, Materials Park, OH, 1999, page 30, Figure 1, [*top*]; and R. J. Shipley and W. T. Becker, [editors], *ASM Handbook, Volume 12, Fractography,* ASM International, Materials Park, OH, 1987, page 102, Figure 22 [*bottom*]; all used with permission of ASM International.)

Failure Analysis and Forensic Engineering **101**

(a)

- Drilled hole
- Punched hole
- Drilled hole
- Punched hole (Chevron markings start here)

(b)

Figure 6-4 Macroscopic (gross, naked-eye-observable) and microscopic (magnified) overall or surface topography features, respectively, associated with and characteristic of brittle versus ductile fracture behavior. The macroscopically observable ridges referred to as *river pattern* or *fan structure* that typify brittle fracture found in many metals and alloys, as well as in ceramics, glasses, and hard polymers. The pattern points back toward the origin of fracture (*a*). Arrowlike *chevrons* or *herringbone pattern* associated with brittle fracture in some metals and allows, notably steels. The "arrows" point toward the origin of the fracture (*b*). Characteristic and typical appearance of cleavage during brittle fracture at the microscopic level, as seen in the fracture surface topography. Note that the fan structure evident at the microscopic level is like the one that manifests itself in the macroscopic fracture topography (*c*). Gross, macroscopic (naked-eye-observable) plastic deformation associated with—and evidencing—ductile fracture behavior (*d*). Characteristic *shear lips* (at top of fracture pull-out) created when fracture moves onto 345-degree shear planes during ductile fracture (*e*). The microscopic appearance of the surface of a fracture that occurred by a ductile mode. The surface topography shows *dimpled rupture,* with its smooth, soft-appearing, spongelike texture (*f*). (*Sources:* R. J. Shipley and W. T. Becker [editors], *ASM Handbook, Volume 11, Failure Analysis and Prevention,* Materials Park, OH, 2002, page 86, Figure 6 [*a*], and page 90, Figure 15 [*b*]; R. J. Shipley and W. T. Becker [editors], *ASM Handbook, Volume 12, Fractography,* Materials Park, OH, 1987, page 17, Figure 12 [*c*], page 114, Figure 43 [*d*], and page 294, Figure 336 [*e*]; and Donald J. Wulpi, *Understanding How Components Fail,* 2nd edition, ASM International, Materials Park, OH, 1999, page 110, Figure 6 [*f*]. All images used with permission of ASM International.)

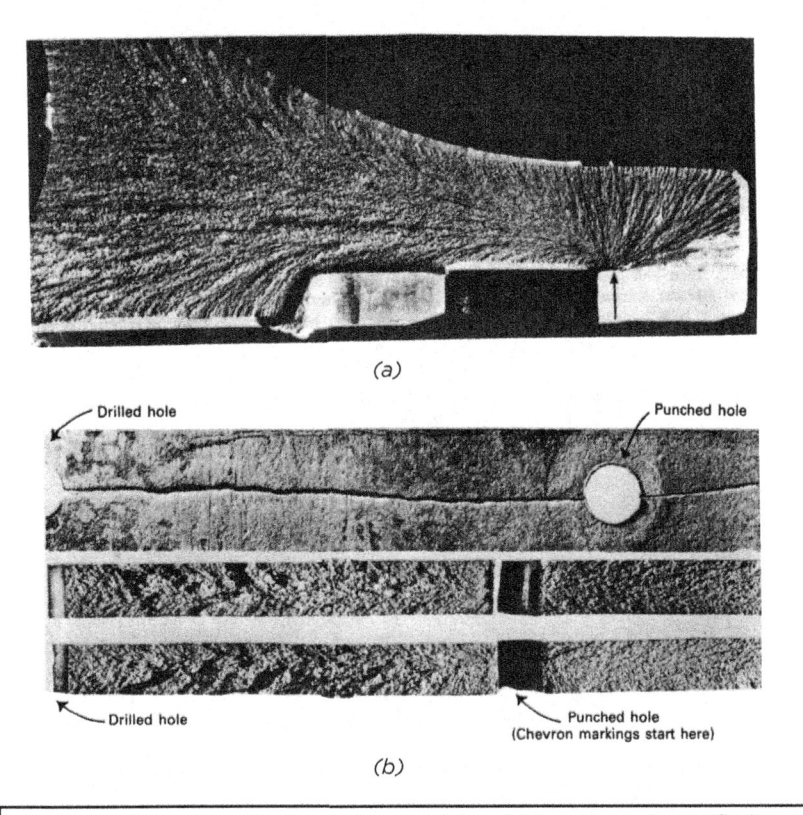

(c)

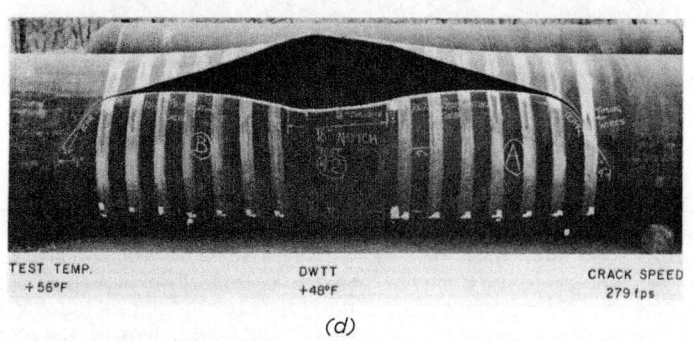

(d)

TEST TEMP.　　　　　　　　DWTT　　　　　　　　CRACK SPEED
+56°F　　　　　　　　　　+48°F　　　　　　　　　279 fps

(e)

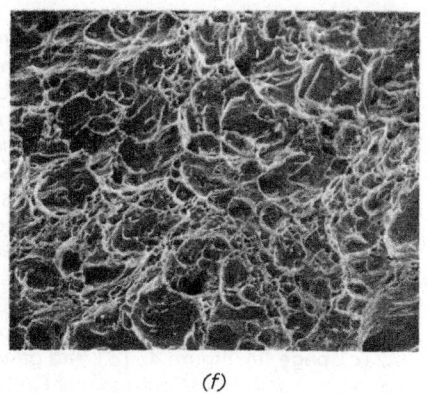

(f)

Figure 6-4 *(Continued)*

Failure Analysis and Forensic Engineering **103**

(a)

(b)

(c)

(d)

Figure 6-5 Fatigue is an inherently brittle mode of fracture. Macroscopic (naked-eye-observable) features help identify the occurrence of fatigue, but microscopic (highly magnified) striations are telltale. *Beachmarks* appear as concentric macroscopic texture bands that are associated with major load-event changes. Initially, beachmarks emanate from the fatigue initiation site or origin like ripples in water. The crack front can change direction to produce confusing patterns that are still concentric, however (*a*). *Striations* are very fine scale remnants that denote the cycle-by-cycle advance of the crack front— here, 10 coarse striations separated by 10 very fine striations due to alternating blocks of 10 cycles of high and low stress. Striations may rub away in very soft materials and may not form in very hard materials (*b*). The origin of a fatigue crack/fracture can be found by tracing normal vectors to the beachmark contours backward to the point of convergence as the origin. A small, flat, crescent-shaped *thumbnail* sometimes can be seen at an initiation site. Here there are four thumbnails associated with four initiation sites (*c*). When fatigue cracks initiate at more than one site (as they can, especially in parts with symmetrical gross geometric features, like splines on a shaft or threads on a fastener, as here), light-reflecting *ratchet marks* may be visible around the perimeter of the fractured part (*d*). (*Source:* Donald J. Wulpi, *Understanding How Components Fail,* 2nd edition, ASM International, Materials Park, OH, 1999, page 152, Figure 22 [*a*], page 122, Figure 3 [*b*], page 148, Figure 18a [*c*], and page 133, Figure 10 [*d*]. All images used with permission of ASM International.)

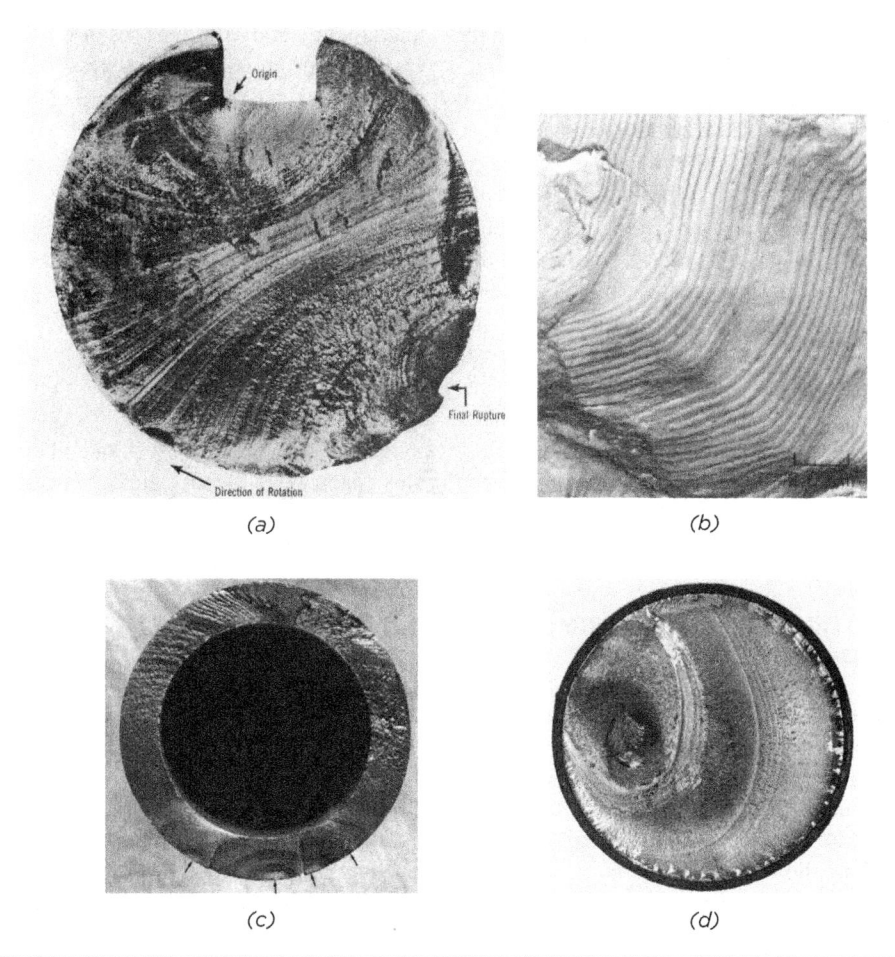

It is not unusual for fracture to occur by either mixed ductile and brittle modes or for the mode to transition from a ductile to a brittle mode as fracture progresses and the stress intensity increases and/or the crack rate accelerates. In such situations, the fractographic topography switches from dimpled rupture to cleavage over a transition range.

Wear-induced and corrosion-induced failures, which may or may not result in eventual fracture, also leave evidence behind, as indicated in Table 6–2.

Fatigue fracture always occurs by a brittle mode (i.e., it occurs at 90 degrees to a tensile, compressive, or bending stress and at ±45 degrees to a torsion load). Furthermore, it is generally evidenced by macroscopically visible beachmarks on the fracture surface (Figure 6–5a) and, unless rubbed away, by striations (Figure 6–5b) at the microscopic level.

The origin of a fatigue crack is sometimes indicated by a small, flat, crescent-shaped *thumbnail* region (Figure 6–5c), and multiple initiation sites (which can and do occur) give rise to reflective *rachet marks* around the perimeter of the fracture remnant (Figure 6–5d).

The reader interested in failure analysis more generally is encouraged to seek out other references, of which several are given at the end of this chapter (ref. Brooks et al.; Martin; McEvily; Sachs; Shipley and Becker; Wulpi).

6-4 The General Procedure for Conducting a Failure Analysis

Proper failure analysis requires a rigorous, systematic procedure to determine root-cause—*systematic* to allow repetition of the procedure by one's self or by another to obtain the same conclusions; *rigorous* to ensure thoroughness and to yield convincing conclusions.

All reputable references on failure analysis (ref. Shipley and Becker or Wulpi) list somewhere around 14 to 16 steps. The number of steps is not important (some lists combine items). What the steps require, however, is important. The general procedure recommended by ASM International follows here and is summarized with additional details in Table 6–3:

1. Collect background data and select samples.
2. Conduct preliminary visual examination of failed parts.
3. Conduct nondestructive evaluation (so as not to alter anything).
4. Conduct any mechanical testing (hardness, strength, etc.).
5. Select, identify, and preserve specimens (and compare to nonfailed parts, if possible).
6. Conduct macroscopic examination, analysis, and documentation.
7. Conduct microscopic examination (progressing from optical to electronic, including fractography) and documentation.
8. Select and prepare metallographic sections, and document observations.
9. Examine and analyze metallographic specimens.
10. Determine the mechanism of failure (see Table 6–2).
11. Conduct any chemical analysis (general, local; wear debris; corrosion residue).
12. Conduct any fracture mechanics analysis.
13. Test under simulated conditions (try to reproduce the failure).
14. Analyze all evidence, draw conclusions, compile report.
15. Follow up on any recommendations to prevent recurrence of the failure.

Failure Analysis and Forensic Engineering

TABLE 6-3 The General Procedure for Conducting a Failure Analysis Investigation (According to ASM International)

1. Collection of background data and selection of samples
2. Preliminary examination of the failed part (visual examination and record keeping)
3. Nondestructive testing
4. Mechanical testing (including hardness and toughness testing)
5. Selection, identification, preservation, and/or cleaning of specimens (and comparison with parts that have not failed)
6. Macroscopic examination and analysis and photographic documentation (fracture surfaces, secondary cracks, and other surface phenomena)
7. Microscopic examination and analysis (optical and electron)
8. Selection and preparation of metallographic sections
9. Examination and analysis of metallographic specimens
10. Determination of failure mechanism
11. Chemical analysis (bulk, local, surface deposits, residues, coatings)
12. Analysis of fracture mechanics
13. Testing under simulated service conditions (to reproduce failure)
14. Analysis of all evidence, formulation of conclusions, and writing of report (including recommendations)

Each step in the recommended general procedure should be considered for its merit in a particular failure investigation, and any exclusion of a step should be made very thoughtfully as to why evidence gathered so far suggests that a particular step can be omitted. For example, if there is an obvious design flaw (see Table 6–1), such as a sharp corner (i.e., no radius) at the bottom of a machined keyway in a shaft *and* there is no indication (e.g., from a measured proper Rockwell hardness that matches the engineering drawing call-out) *and* fracture is evidenced by beachmarks as being by fatigue, there is no reason to conduct a chemical analysis of the shaft material, as nothing suggests a problem with the material. (An even better reason for excluding chemical analysis would be having seen the appropriate microstructure from examination of a metallographic section taken from the shaft.) However, be sure to never skip a step without solid reason!

A proper failure analysis finds a root-cause from evidence left behind in the failure debris and then checks the presumed scenario and sequence of events that led to failure, deduced by a backward problem-solving technique, by starting from the operation of the part, product, structure, or system and working forward to see whether the observed failure is a logical outcome. The forward and backward paths should be the same; that is, they should include the same events but in the opposite order.

Enough on failure analysis, except to say this: The key to successful failure analysis is working the problem backward, from failure (effect) to service (cause). Systematic dissection—mechanically and mentally—is essential. From this standpoint, failure analysis is a powerful tool for reverse engineering things that failed.

But not everything fails before it becomes important or interesting to know how it works. Here, too, reverse engineering is a powerful tool.

In summary, in conducting a failure analysis:

- *Think* before you act.
- Consider every step in the general procedure for its merit.
- Be vigilant and meticulous in your observations, and remember that nothing is insignificant.
- "Always turn the body over."
- Be skeptical, trying to find why your initial impression of the cause of failure is wrong, as the truth will never change no matter how many times you tell the story or how many different ways you come at the story.
- Find the root-cause.
- Play the final story forward and backward, to see that they are consistent.
- Write a report that draws conclusions only from evidence.
- Follow up on any recommendations to prevent recurrence of the failure.

Failure analysis involves observation, measurement, experimentation, and analysis, just as reverse engineering does.

6-5 Two Exemplary Failure Analysis Cases

By now it should be clear that *failure analysis* is a powerful and essential tool for helping engineers discover (by deduction) how a part, component, device, object, structural element or structure, assembly, or system failed prematurely or, worse, catastrophically.[11] Two cases are presented here to highlight two slightly different purposes of a particular investigation.

Case 1: Failure of Tongue-and-Groove Slip-Jaw Pliers

Tongue-and-groove slip-jaw pliers are a hand tool that allows the gripping jaws of a pair of hinged handles to open unusually wide and in a manner that keeps the gripping faces parallel to one another for better grip (Figure 6–6a). The pliers shown in close-up Figure 6–6b and c fractured during use, with the gripping portion of one jaw completely separating from the handle by a catastrophic fracture. The user (a female student working in a university laboratory) alleged, "They just broke!"

Cursory examination revealed that the failed pliers were made from steel (probably a plain carbon steel with a nominal carbon content of about 0.45 to 50 wt.%), as they were strongly attracted by a magnet (i.e., the alloy was ferromagnetic) and the measured Rockwell C hardness was about 38 to 40, as it should be. The handles of the pliers were each made in one piece and were obviously strong and stiff enough (i.e., were robust enough) that it seemed impossible that anyone could squeeze them so tightly that they would elastically distort (deflect), no less fracture!

---

[11] The technique, following the same general procedure presented in Section 6–4, can obviously be applied and have relevance to determine the cause of an *eventual failure* or *ultimate failure* of something. Here, the purpose is usually to determine what ultimately ended a product's life, with one of a few goals: (1) to make possible improvements to extend life even further; (2) to allow consideration of design changes that might reduce cost and/or slightly shorten unnecessarily long life; or (3) to help learn what worked especially well, as well as what ended up being life-limiting, to aid in the more intelligent design of future products.

They also appeared relatively new, as there was little or no evidence of marring from use, except as modest wear scars along each side of the slip-slot.

The *first observation* was that the failure was the result of complete fracture, as the gripping jaw of the one handle was totally separated from the handle proper. An immediate *second observation* was that the fracture had occurred in the handle that contained the elongated slot that allowed the jaw opening to be adjusted (i.e., the jaws to be "slipped") to allow wider opening (Figure 6–6b). This was intuitively satisfying, as the cross-sectional area of this handle was less than that of the other, slot-free handle, thereby causing the stress to be higher in this handle than in the other handle. An immediate *third observation* was that the fracture also occurred where sharp-bottomed grooves (to allow plier half pivoting) amplified the bending stress (Figure 6–6c), which is also intuitively satisfying. A *fourth observation* was that the orientation of the fracture line (i.e., crack path) was essentially 90 degrees to the bending force/stress developed when the jaws gripped an object and the handles were squeezed together tightly (Figure 6–6d). (In fact, the jaw portion of the one handle was loaded as a cantilever beam, with a force concentrated near the free end of the grip jaw and a bending moment resulting where the gripping jaw blended into the handle just past the end of the slip-slot.) A 90-degree orientation of a fracture from a primary bending force/stress indicates brittle behavior in the material (here, a body-centered cubic iron-carbon alloy or steel). The immediate *fifth observation* (after seeing the fracture path) was that there was no evidence of any gross plastic deformation in the pliers' handles (e.g., bending), further supporting the conclusion that fracture occurred by a brittle mode.

As a *sixth observation,* closer examination of the fracture path indicated a distinct change in the outboard leg along the slot (based on loading by an object between the jaws prying the jaw faces outward). The crack orientation changed from 90 degrees to ±45 degrees from the primary bending force/stress as the crack propagated across this section (Figure 6–6d). This indicates that, while cracking began by a brittle mode, it seemed to change to a ductile mode near the outer edge of the handle. Reflecting for a moment on details of the loading of the grip portion of the jaws, it becomes clear that squeezing on the pliers' handles, as the jaws gripped some solid object, forced the jaw tips outward, so that fracture initiated at the inner surface of the inner leg along the slip-slot (by a brittle mode, at approximately 90 degrees), propagated through the leg, initiated at the inner edge of the second, outboard leg also by a brittle mode, and then changed to a ductile mode as the remaining cross section decreased, until fracture terminated with a path distinctly at ±45 degrees from the bending axis.[12] This observation implies that the steel used to fabricate the pliers was inherently ductile *but* was forced to behave in a brittle manner at the point of initiation.

Very close examination of the termination of the fracture path revealed (as a *seventh observation*) a small protrusion on the smaller fractured tip portion and a corresponding (i.e., mirror-image) notch in the larger, mating portion (Figure 6–6e). This feature of a fracture is known as a *compressive curl,* as it forms during fracture termination as the failing part rotates backward (or outward, in this case) to give rise to a local compressive stress. Having found the termination of the fracture, one looks to the opposite side to attempt to find the fracture initiation site or fracture origin. Here, fracture appeared to have initiated at the inner edge of the jaw in a region where not only was the cross section of the handle reduced by the slip-slot but also there were curved machined grooves that guide the movement between the two handles (and jaws)

---

[12] Cracks propagating by a brittle mode do so at the speed of sound (as an elastic wave) in the material—here, a steel, at ~20,000 feet per second or ~6000 meters per second.

108 Chapter Six

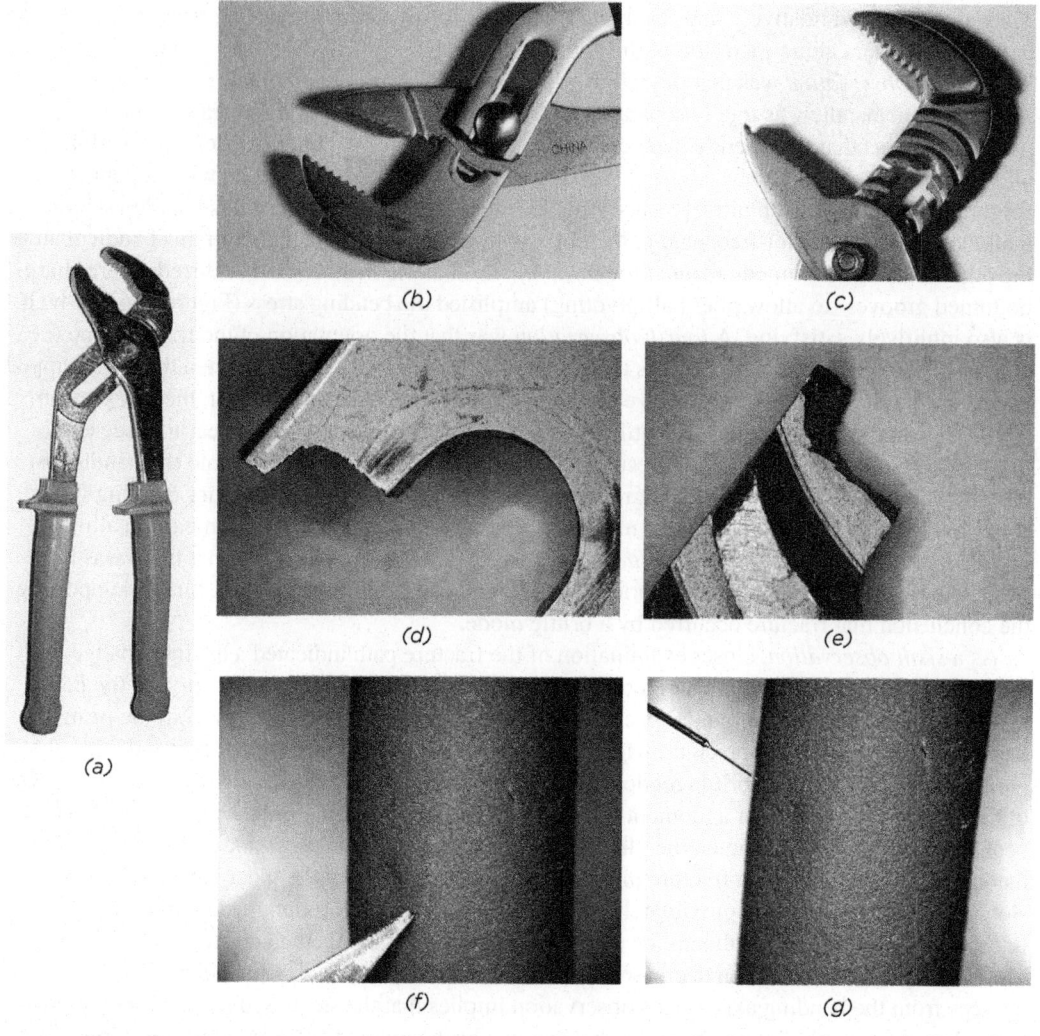

Figure 6-6 Typical tongue-and-groove slip-jaw pliers (*a*). Close-up views of the plier handle that fractured in use as seen from the handle's outer (*b*) and inner (*c*) surface. The fracture occurred where the load-bearing cross-sectional area was reduced by the slot and the resulting stress was amplified by the sharp-bottomed grooves. The fracture path can be seen to switch from 90 degees at initiation (in tension from bending) to 45 degrees near termination (in compression from bending), indicating an initial brittle mode due to stress concentration in BCC steel and final ductile mode due to the inherent ductile nature of the alloy (*d*). The very end of the fracture can be seen to have created a *compression curl*, seen as a protrusion in (*e*). A depressed dimple (*f*) and depressed line (*g*) appearing on opposite sides of the soft rubber grips on one plier half handle indicated that a pipe was used to gain mechanical advantage. (*Sources:* Wikipedia Creative Commons, contributed by Ivob on 25 October 2007 [a]; photographs are the property of the author, Robert W. Messler, Jr. [*b* through *g*].) The kind and able assistant of Jason Benyeda and Craig Galligan is gratefully acknowledged.

when the pliers is assembled, by the hinge bolt (Figure 6–6b). This region gives rise to three—multiplicative—stress concentration factors: one from the slip-slot, one from the machined groove (which, like the slip-slot, reduces the cross section and gives rise to higher stress), and a third from a nearly absent radius between the groove sidewall and bottom. Failure initiation at this location is intuitively satisfying, as failures tend to initiate (1) where the operative (applied and amplified) stress is highest and (2) where body-centered cubic materials (like steel) are encouraged to fail by a brittle mode. Thus, while the steel used may be inherently ductile (based on static stress-strain data), a sharp design feature promotes brittle fracture.

Visual examination of the fracture surfaces of the two fracture remnants, revealed (as an *eighth observation*) a reflective, granular texture (or fracture topography) that suggested a reasonably fine grain size (Figure 6–6e), supporting the reasonable presumption that fabrication was by forging, as is the preferred method.

The question still remained: How could anyone—no less a 110-pound female student—break pliers? One can reasonably assume that the engineer designs the length of the handles relative to the jaws such that, under normal use, no one could cause an overload failure, even with the designed-in stress concentrations. This suggests two possibilities for the root-cause of the failure: (1) fatigue fracture from repeated use or (2) overload due to abuse. There being no evidence of any fatigue (e.g., no thumbnail crescent initiation and no beachmarks) and the low likelihood that enough loading cycles could be applied to any pair of pliers, no less to a seemingly new pair of pliers to cause fatigue, implied the failure involved abuse and a single overload event. So it's time to look a little closer!

The obvious suspicion is that the user of the pliers employed some form of "mechanical advantage," using what is known as a *helper*. Really close examination of the red-rubber handle grips on the pliers revealed a distinct dimple, or indentation, on one rubber grip (Figure 6–6f), as a *ninth observation*, and another depressed line on the opposite side of the same handle (Figure 6–6g). The location of these "scars" suggested that the user employed a pipe to extend and enhance the mechanical advantage she could apply to the pliers' handles. The increased leverage, apparently, allowed static overload of the tool. The root-cause of the failure was "misuse," as pliers are *not* designed for use as levers!

Case 2: Warning Lightbulbs on F14 Aircraft #2

Grumman Aerospace Corporation's F14 Tomcat prototype aircraft (A/C #1) crashed during a test flight on December 30, 1970 (see YouTube, "F14 Prototype Crash in 1970"). Both the pilot and copilot survived with only very minor injuries, ejecting just 3 to 5 seconds before impact. The failure was traced to a leak at a common crossover point between supposedly independent "redundant" primary and backup high-pressure hydraulic control systems, with the root-cause being a design error in linking supposedly independent systems.

In a stroke of irony, Chief Test Pilot Bob Miller was killed on June 30, 1972, when A/C #10 crashed into the Chesapeake Bay off Norfolk, Virginia. Miller was practicing aerobatic maneuvers in preparation for an international air show during the July 4th holiday. Immediately upon takeoff, he did a series of 360-degree barrel rolls. Unfortunately, with a gray sky and gray water in the bay, he unknowingly ended his rolls with the aircraft inverted at about 2200 feet. Announcing to a fellow test pilot in an A6 Intruder chase plane taking 16mm movies that he was going to put the aircraft into a 75-degree climb under maximum afterburner (i.e., maximum thrust), he instead put the aircraft into a 75-degree dive under maximum thrust. Seconds later, the aircraft crashed into the water at over 475 knots (545 miles per hour/877 kilometers per hour), causing the aircraft to totally disintegrate. Radio communication and movie film seemed to suggest that the cause of the crash was pilot error (i.e., disorientation).

Upon recovery of the debris by the U.S. Navy, a meticulous crash investigation was conducted by Materials and Processes engineers at Grumman's Bethpage, Long Island, New York, facility to determine the root-cause of the accident. The author, as the newest and youngest engineer in the group, was disappointed when he didn't receive a big piece of the airplane to analyze as a half dozen other engineers had. The M&P manager of the F14 program saw the dejected young engineer and said, "What's your problem? Are you feeling sorry for yourself because you didn't get a piece of the airplane? Are you convinced you didn't because you are the 'new guy,' and we don't trust you?" "Yes, yes, and yes," thought the young engineer. "Well, you're wrong!" said the manager. "These are dedicated professionals who all know—as you and I know from having heard the voice recordings and seen the film—our airplane didn't kill our test pilot. He made a fatal error and crashed the plane! But take note: They are all still looking at every inch of thousands and thousands of inches of fracture surfaces to see if there is any evidence of a root-cause from a structural failure in flight. There won't be!"

He then produced three small, brown coin envelopes and said, "I have something I want you to do. It's the key to the real cause of the crash! These are tiny lightbulbs that indicate "Left Engine Fire Warning," "Right Engine Fire Warning," and "Master System Caution." I need you to tell me whether the pairs of bulbs under the transparent green covers for normal operation or red covers for malfunction, for each pair, were on or off at the time of the crash.[13] These will tell us whether there was a major malfunction in the aircraft or if, as it appears, the crash resulted from pilot error."

The young author/engineer called Sylvania's Lamp Division in New Jersey, where the bulbs were manufactured, to get advice about lightbulbs. The receptionist connected the author with the head of lamp design and the young Grumman engineer told him, "One of our F14s crashed and killed our test pilot. I need to determine whether warning indicator lights were on or off at the time of the crash." The expert responded, "I'll come to Grumman immediately to help you," to which I replied, "You don't understand. We're all essentially locked down until the crash investigation is complete." The expert said, "You get me in, and I'll help as long as I need to be there!"

To cut a long story short: The fact that lamp filaments are made of extremely fine, coiled pure polycrystalline tungsten (W) wire and that tungsten is a body-centered cubic metal that exhibits a ductile-brittle fracture transition under impact (as would occur in a crash) at around 1350°C/2450°F, if the lightbulbs were on at the time of impact the filament wires would exhibit ductile localized necking and cup-and-cone fractures, while if they were off there would be no necking and a flat surface would indicate brittle fracture. The expert and the young engineer spent two days perfecting how to remove the glass envelopes from identical bulbs without damaging the filaments and creating and photodocumenting filament failures under severe impact (replicating estimated crash g-forces) until, finally, examination of the bulbs from A/C #10 were analyzed. All of the bulbs under green covers had been on, and all of those under red covers had been off. There had been no major malfunction in the engines or aerodynamic control systems. As hard as it was to accept, the plane crashed due to an error by our chief test pilot. This finding supported the lack of any evidence of any structural abnormality in any of the recovered fragments of the aircraft.

---

[13] Bulbs were placed in pairs beneath each red and each green cover for each warning system, with no bulb ever being allowed to have operated for more than 1000 flight-hours for bulbs having a 2000-hour nominal operating life.

6-6 Forensic Engineering

Forensic science (often shortened to *forensics*) is the application of a broad spectrum of sciences and technologies to investigate and establish facts of interest in relation to criminal or civil law. The obvious analogy between failure analysis and autopsy leads to the common use of the term *forensic engineering* for failure analysis, just as *forensic medicine* is used to denote autopsy. But, to be perfectly correct, the adjective *forensic* refers to use of either procedure *only* for legal purposes, as in crime solving or litigation in cases of liability.

The first documented use of forensic science was in the "Eureka" legend told of Archimedes (287–212 BC). According to the great Roman architect, engineer, and writer Marcus Vitruvius Pollio, or simply Vitruvius (ca. 80–70 to ca. 15 BC), a votive crown for a temple had been made for King Hiero II, who had supplied the pure gold to be used. Archimedes was asked to determine whether the dishonest goldsmith had substituted some much less valuable silver. Archimedes had to solve the problem without damaging the crown, so he couldn't melt it down to cast a regularly shaped body in order to calculate its density from its measured weight divided by its calculated volume. Inspired by seeing the water level in his tub rise when he immersed himself, he jumped from the tub and ran through the streets shouting, "Eureka! I have found it!" He solved the problem by immersing the crown in water and measuring the displaced volume to calculate the metal's density from the crown's weight, that is, using the principle of buoyancy. Forensic science was born!

In any case, in modern times, forensic engineering—which includes failure investigations—is most often associated with explosions, fires, floods, and criminal cases, although use in cases of litigation for liability is also common (ref. Carper).

6-7 An Exemplary Forensic Engineering Case

A case in which the author served as a consultant closes this chapter.

Four male friends and college students were working the summer painting houses in the city where the author served as a faculty member in the Department of Materials Science and Engineering. Early one morning, the young man painting the exterior trim on a kitchen window in the back of a house owned by a pediatrician heard an odd squeal, "like a hurt cat," he said. When he stepped around the corner of the house, he saw his friend high up on an aluminum extension ladder, where he had been painting the upper trim. The young man's clothes and hair were smoking and it was clear he was being electrocuted, as the ladder had come into contact with the electric power service cable where it connected to the mast on the sidewall of the house. Despite knowing the risk of being electrocuted, too, he threw his body against the ladder, knocking it and his friend to the ground.

Hearing the clatter of the ladder, the two other young men working on the opposite side of the house ran to the scene, where one friend was staggering to his feet and the other was lying unconscious, still and smoking, on the ground. One young man ran into the house to get the doctor, who came out and immediately began CPR, while the doctor's receptionist called for EMTs from the fire department. Despite twice being resuscitated before being transported to the burn unit of a local hospital, the young victim died during the night of internal injuries from electrocution.

The author learned of the tragedy when a colleague in the Department of Electric Power Engineering asked for his assistance in examining the portion of the power cable that attached to the house. The heavy (1-inch-diameter) cable from the utility pole had been stripped of its thick black rubber insulation casing for a distance of 13½ inches from the point of connection to the mast on the house. The local power company insisted that their installers would never strip a cable that far back from a connection and surely wouldn't leave it uninsulated without multiple wraps of electrical tape. In fact, they were quick to show they had a rigid "specification" on precisely how such connections were to be made, as if having a specification meant there was no chance of negligence.

While asked only to determine whether, as the power company claimed, the "rubber insulation deteriorated from weathering" to expose the underlying bare copper twisted wires that came into contact with the aluminum ladder, the author found no such evidence. Quite the contrary, in fact, as the remaining nitrile rubber (chosen for its weather resistance as an electrical insulator) was clean, shiny black, and pliant, with no indication of any drying, cracking, or crumbling. Furthermore, distinct cut marks could be seen on the surface of the insulation's cross section and fine score marks could be seen running circumferentially across the tops of the outermost copper wires. In stripping another length of cable, the author reproduced the marks in both the rubber and on the wires. No doubt about it: The cable had been stripped for a length of 13½ inches. However, there was more! At distances of 9½ and 11 inches back from the point of connection, an odd gray, gritty material was found lying between abutting copper wires, along with some white spots on several wires in the same orientation around the cable, but at several different locations from the point of connection.

Examination of samples of both the gritty gray material and the rubbery white material (following careful photodocumentation while still on the cable wires) using energy-dispersive x-ray spectroscopy (EDS) analysis revealed distinct intense peaks corresponding to the following elements: oxygen (O), calcium (Ca), silicon (Si), aluminum (Al), iron (Fe), sulfur (S), and titanium (Ti). Eureka! The gritty gray material was Portland cement and the rubbery white material was titanium dioxide, a stark white pigment used in white paint.[14] Serendipity had come into play. The author knew about Portland cement after being forced to become an expert to support the cement-making industry with welding maintenance repair and hard-facing consumables between 1980 and 1984. He knew of titanium dioxide pigment from undergraduate chemistry taken in 1961–62.

But how could these materials have gotten onto the surface of exposed copper wires 24 feet above the ground? The answer lies in the following timeline: The doctor confirmed he had the power service to his home increased to 400 amperes when he recently moved his practice into his residence. The date of the new cable installation was only five to six weeks prior to the accident. The power company installer had told the doctor that the bricks on the chimney on the side of the house where the new power line was installed needed to be repointed with cement, as several bricks had come loose and were in danger of falling from the chimney. The doctor hired a handyman to repoint the bricks of the chimney above the roof—just above and adjacent to the new power cable installed a week before. The handyman told the doctor that the trim along the roof near the chimney was rotted, and he was authorized to replace it and prime it before the doctor was to get the entire house painted. The new trim piece was installed and primed with white paint

---

[14] Portland cement is composed of the following ingredients (after heating ground "clinker" in a rotating kiln): 61 to 67% CaO, 19 to 23% SiO_2, 2.5 to 6% Al_2O_3, 0 to 6% Fe_2O_3, and 1.5 to 4.5% sulfate (SO_4^{-2}).

a week later. The college housepainters began working below the roofline and cable three weeks after that.

So wet Portland cement dropped onto the cable and stuck between the copper wires left exposed when the power company installer stripped off the rubber insulation for a distance of 13½ inches (despite the specification that required far less stripping) and never wrapped insulating tape around the exposed wires to replace the lost covering. Spots of wet white paint from the trim had also seemingly been blown to fall onto the tops of the exposed copper wires. If the cable hadn't been improperly stripped for a length of 13½ inches by the installer and left uncovered by electrical tape, neither cement nor paint could have been deposited on it!

Presented with the evidence, the power company—at the recommendation of its attorneys—agreed to settle with the distraught family of the young man who had died from electrocution. The author put the pieces together by not ignoring evidence that was not expected. The lesson to be learned from this is to be vigilant, nothing is insignificant, and always turn the body over, or you may not see all there is to see!

6-8 Summary

Failure analysis is a powerful and essential tool for modern manufacturers to remain competitive in a global economy, where responsiveness to customers is critical. The process must follow a rigorous systematic procedure in which each step is considered for its merit and is omitted only with careful consideration and justification. The end goal of the investigation is to determine the sequence of events, as well as the prevailing conditions, under which a failure occurs, being complete only when the unambiguous root-cause of the failure has been determined so that repeat failure can be avoided. Root-cause involves identifying the operative mechanism by which failure occurred in the material(s) constituting the part(s) that failed, as well as the specific source for the failure from among design errors, manufacturing-induced errors, service errors, and possible exacerbating human errors.

Forensic engineering is a subset of failure analysis in which the purpose is to aid in the solution of a crime, the determination of the cause for an accident that involves litigation for liability, and so on.

In the context of this book, failure analysis is one reason for conducting product teardown as part of reverse engineering.

6-9 Cited References

Brooks, Charlie R., and Ashok Choudhury, *Failure Analysis of Engineering Materials,* McGraw-Hill, New York, 2002.

Carper, Kenneth L. (editor), *Forensic Engineering,* 2nd edition, CRC Press, Boca Raton, FL, 2000.

Martin, Perry L., *Electronic Failure Analysis Handbook,* McGraw-Hill, New York, 1999.

McEvily, A. J., *Metal Failures: Mechanisms, Analysis, Prevention,* John Wiley & Sons, Hoboken, NJ, 2002.

Sachs, Neville W., *Practical Plant Failure Analysis: A Guide to Understanding Machinery Deterioration and Improving Equipment Reliability* (Dekker Mechanical Engineering), CRC Press, Boca Raton, FL, 2007.

Shipley, R. J., and W. T. Becker (editors), *ASM Handbook, Volume 11, Failure Analysis and Prevention,* ASM International, Materials Park, OH, 2002.

Wulpi, Donald J., *Understanding How Components Fail,* 2nd edition, ASM International, Materials Park, OH, 1999.

6-10 Thought Questions and Problems

6-1 Everything fails eventually. *Eventual* or *ultimate failure* once a product or structure has served its intended use and becomes obsolete or just too old to be of further use is not usually considered problematic. Failures are problematic, however, if they are *premature failures* or *catastrophic failures.*
   a. In your own words, define (1) *premature failures* and (2) *catastrophic failures.*
   b. Briefly explain how these two are different.
   c. What are the most serious consequences of each type of failure? Explain your response.
   d. From your personal experience or work, or by using the Internet as a resource, give *three* examples of each type of failure—*eventual* or *ultimate failure, premature failure,* and *catastrophic failure.* Defend your choices.

6-2 Failures of products, devices, structures, or systems can manifest themselves in any of several ways, as follows:
   - *Temporary distortion* or *elastic deformation* (which is recoverable once loading is removed)
   - *Permanent distortion* or *plastic deformation* (which is not recoverable when loading is removed)
   - *Fracture* (in which some key component of the product, device, structure, or system physically breaks or cracks to preclude further function)
   - *Corrosion* or *corrosive degradation* (which leads to degraded properties and some or all loss of function or performance)
   - *Wear* (which leads to degraded properties or, more often, loss of some or all function and/or performance)

   a. Indicate which of these manifestations of failure is seemingly involved in each of the following, explaining your answer in each case:
      (1) The blade of a rotary lawn mower no longer cuts the grass cleanly, but shreds it.
      (2) The guard over the chain of a child's bicycle rubs against the chain.
      (3) The swinging gate in a picket fence rubs on the ground to cause the gate to stick when a teenager hangs on it.
      (4) The drive sprocket gear on a bulldozer fails to move the crawler treads properly after a couple of years of use.
      (5) An incandescent lightbulb "blows out."
      (6) The water pump of a water-cooled automobile engine begins to "squeal" and the engine temperature rises above "Normal."
   b. Give *two* examples of your own for each of the following failure types:
      (1) fracture
      (2) wear: *one* involving plain metal-to-metal (adhesive) wear and *one* involving wear by a hard particulate substance (abrasive wear)

(3) corrosion leading to cracking or fracture *and* corrosion leading to loss of needed properties (e.g., toughness)
(4) distortion: *one* temporary and *one* permanent
c. Different industries and/or application areas tend to favor (or promote) failures by different mechanisms (e.g., fatigue, wear, corrosion, some form of elevated-temperature failure such as creep). Indicate which of these mechanisms you believe would predominate for each industry or application area in the following pairs, and support your answers:
(1) pulp and paper mills versus lumber mills
(2) coal mining versus oil recovery (after drilling)
(3) railroad yards versus railroad cars versus locomotive engines
(4) military civil aircraft versus spacecraft
(5) automobile engines versus jet aircraft engines

6-3 Failures can arise from many possible sources, including design sources, manufacturing sources, and service sources (see Section 6–2).
a. From your own personal knowledge or work, or with assistance from online sources, give a *specific example* of how a failure could occur in some part, device, structure, or system (by application) for each of the following sources:
(1) improper part size or shape
(2) improper choice of material or material condition
(3) design defect or error
(4) process-induced defect
(5) improper assembly
(6) improper use
(7) improper maintenance
b. For each of the following situations, indicate, as specifically as possible, the likely failure mechanism from among those listed in Section 6–3:
(1) a very sharp (almost nonexistent) radius at the bottom of a machined keyway in a shaft
(2) gas pores (i.e., porosity) in the surface mount technology (SMT) solder joint between an electronic chip package and a metallized circuit board
(3) severe seismic vibrations to the welded steel framework or support structure of a 10-story office building
(4) gears in a transmission gearbox that have lost lubrication
(5) two different alloys, one of which is much more anodic than the other in a wet environment
(6) the blades on the rotating disc of a gas turbine allowed to operate at too high a temperature for too long
(7) the handle on a bench vise to which the user applied a 2-foot length of pipe for extra leverage during tightening

6-4 An essential aspect of a failure analysis investigation is to systematically and meticulously examine every detail (e.g., every inch of fracture surfaces) of each and every part of the failed entity for clues. In this way, conducting a failure analysis investigation is similar to reverse engineering. Another similarity between failure analysis investigations and reverse engineering activities is that both employ *backward problem-solving techniques.*

Prepare a one- to two-page thoughtful essay on how failure analysis can contribute to reverse engineering and, contrarily, how reverse engineering contributes to failure analysis.

6-5 Use the Internet to find a case study in failure analysis that interests you and, in a one- to two-page thoughtful essay, describe how the selected case followed the steps of the "General Procedure . . ." for conducting a failure analysis investigation in Section 6–4 and Table 6–3.

CHAPTER 7

Deducing or Inferring Role, Purpose, and Functionality during Reverse Engineering

7-1 The Procedure for Reverse Engineering

The procedure for the process or technique of reverse engineering when the specific objectives relate to assessing a design for understanding, improvement, competitive assessment or benchmarking, or possible utility of the knowledge gained for other future analogous designs follows:

Step 1: Identify and articulate the purpose or motivation for disassembly or mechanical dissection and analysis.

Step 2: Systematically mechanically dissect (i.e., physically disassemble) the object, device, mechanism, product, structure, system, or material, and, in the process, measure and, perhaps, experimentally test subassembly by subassembly (or subsystem by subsystem) and, within subassemblies (or subsystems), component by component.

Step 3: Deduce or infer the role, the purpose, and the functionality of each and every part, component, structural element, or, for an electronic product or system, of each device, as well as each subassembly or subsystem, on the way to identifying the purpose and function of the overall object, mechanism, product, structure, system, or material, if not known at the outset.

Step 4: Attempt to identify the material or materials used to create each and every part, component, or structural element, or, for an electronic product or system, each device. Identification can be general, as opposed to specific (e.g., metal $\Rightarrow$ steel $\Rightarrow$ quenched and tempered).

Step 5: Attempt to deduce or infer the method or methods by which each part, component, or structural element, or, for an electronic product or system, each device was fabricated and, if possible, processed, as well as the method(s) by which the product, structure, or system was assembled.

Step 6: Attempt to access the suitability of the overall design and design details to the intended purpose, including acceptability or appropriateness of cost, robustness for service, service environment, duty cycle and expected life, market sector, and so on.

Table 7–1 summarizes these steps for convenient reference.

The various possible motivations or purposes for conducting reverse engineering (Step 1) are addressed in Section 2–3, and are summarized in Table 2.1. Details for conducting actual physical disassembly or mechanical dissection (Step 2) are addressed in Chapter 4, as well as in Section 5–2. The remainder of this chapter, for the first time in any book, describes and discusses how one goes about deducing or inferring the purpose and function of the details that make up a product, structure, or system. Likewise, for the first time, Chapter 9 describes and discusses how to identify materials-of-construction, Chapter 10 describes and discusses how to deduce or infer methods-of-manufacture or -construction, and Chapter 12 describes and discusses how to assess design suitability.

Let's begin with deducing or inferring purpose and function.

TABLE 7-1	Summary of the Procedure for Reverse Engineering
Step 1:	*Identify and articulate the purpose or motivation* for disassembly or mechanical dissection and analysis.
Step 2:	*Systematically mechanically dissect (i.e., physically disassemble)* the object, device, product, structure, system, or material, and, in the process, measure and, perhaps, experimentally test subassembly by subassembly (or subsystem by subsystem) and, within subassemblies (or subsystems), component by component.
Step 3:	*Deduce or infer the role, purpose, and functionality of each and every part,* component, structural element, or, for an electronic product or system, device, as well as each subassembly or subsystem, on the way to identifying the purpose and function of the overall object, mechanism, product, structure, system, or material, if not known at the outset.
Step 4:	*Attempt to identify the material or materials used to create each and every part,* component, or structural element, or, for an electronic product or system, device. Identification can be general, as opposed to specific (e.g., metal $\Rightarrow$ steel $\Rightarrow$ quenched and tempered).
Step 5:	*Attempt to deduce or infer the method or methods by which each part,* component, or structural element, or, for an electronic product or system, device *was fabricated* and, if possible, *processed,* as well as the method(s) by which the product, structure, or system was *assembled.*
Step 6:	*Attempt to access the suitability of the overall design and design details to purpose,* including acceptability or appropriateness of cost, robustness for service, service environment, duty cycle, and expected life, market, and so on.

7-2 Knowing versus Identifying versus Deducing versus Inferring

A detective knows a murder has occurred with certainty if he or she finds a body with a knife in its back. He or she is able to infer there is a murderer because someone was murdered. He or she is able to deduce who is a prime suspect from gathered clues that support the three aspects of a crime needed to convince a jury of guilt: means, motive, and opportunity. Four different levels of confidence are expressed by knowing, inferring, deducing, and identifying in the order from absolute to likely certainty being knowing, identifying, deducing, and inferring. All four play a role in reverse engineering, along with the invaluable role played by experience.

The dictionary defines *knowing* as "perceiving directly with clarity or certainty. To regard as true beyond doubt." Ergo: How the detective knows someone has been murdered, as a person does not plunge a knife into his or her own back. If something is not immediately known with certainty, one of three possibilities, as paths, exists for learning what the unknown is or, at least, is likely to be. These three paths involve (1) identifying, (2) deducing, or (3) inferring the unknown.

Identifying is defined as "establishing the identity of something," with *identity* being defined as "the collective aspect of the set of characteristics by which a thing is definitely recognizable or known." Ergo: How the detective eventually comes to know a previously unknown person of suspicion using uniquely identifying fingerprints or DNA.

The second path for attempting to know what was previously unknown or uncertain is to use deduction, deducing a prime suspect from among several possible suspects. *Deducing* is defined as "reaching a conclusion by reasoning" or, in the best case, "using a systematic method for drawing conclusions that cannot be false when the premises are true." For detectives, deducing conclusions well from clues is their path to a successful career. The result of a deductive process is not a certainty, only the logically most likely. For a murder suspect, they try to deduce motive and opportunity.

The fourth, and final, path is slightly less certain than deduction, although it too relies on reasoning, albeit from somewhat less certain or obvious clues. This path involves inference, where the gerund form of the verb *infer*—*inferring*—is defined as "reasoning from circumstances." Our detective inferred that a murderer needed to be identified by logically inferring that there must be one if there is a murder victim.[1]

When an engineer employs the technique of reverse engineering, some things about the entity being examined will be known with certainty and other things will initially not be known; that is, they will be unknown. To learn or know about these things, the engineer will have to use a combination of approaches for *identifying* the unknown from unambiguous data or facts, *deducing* the unknown from logical reasoning based on evidence, or *inferring* something about the unknown from reasoning based on less clear or certain clues or evidence.

In this chapter, a combination of deduction and inference will be used to deduce or infer the likely purpose and functionality of the entity being subjected to reverse engineering.

---

[1] The ancient Athenian philosopher Socrates (469–399 BC) argued that a statue inferred the existence of a sculptor, which some people use as the argument for the existence of God. We are here on the Earth, so there must be a creator. Read the quote from Dr. Wernher von Braun at the front of this book.

7-3 The Value of Experience

Andrei Nikolayevich Tupolev (1888–1972), a brilliant 34-year-old Soviet aerospace pioneer, founded Tupolev OKB in 1922. Tupolev OKB was also known as the Tupolev Design Bureau at one time, as it focused on research, development, and design, and left manufacturing of its aircraft to other firms. Having celebrated its ninetieth anniversary on October 22, 2012, the modern official name of the company had long before become Public Stock Company Tupolev (PSC Tupolev), which the Russian government merged with the other aerospace firms Mikoyan, Sukhoi, Ilyushin, Irkut, and Yakovlev under President Vladimir Putin in February 2006. The new company is known as Joint Stock Company United Aircraft Corporation (JSC United Aircraft Corporation). The capabilities of the former PSC Tupolev included development, manufacturing, and overhaul for both military and civil aerospace products, such as aircraft and weapons systems. Large bombers and transport aircraft were its specialty. The company also built the world's first supersonic transport, the Tu-144 (Figure 7–1).

A. N. Tupolev pioneered Soviet research into all-metal airplanes during the 1920s based on earlier pioneering work already done by the German engineer Hugo Junkers (1859–1933) during World War I. Andrei Tupolev's company built the first all-metal airplane in the world, with its forte being large airplanes (Figure 7–2).[2]

It was Andrei Tupolev's unmatched experience with large airplanes that led the Soviet leader Joseph Stalin to choose him to head the effort in early 1945 to reverse engineer the impounded B-29 Superfortress "Ramp Tramp" (see Section 3–4) and quickly create the Soviet's exact copy, the Tu-4 "Bull" (see Figure 3–11). This success, enabled by A. N. Tupolev's expertise in knowing what he was looking at in the completely dissected B-29, completely changed and strengthened the Soviets' postwar status, placing them on a par with the United States, and unknowingly setting up (along with captured German rocket scientists from Peenemunde) a 50-year-long bitter—and potentially catastrophic—rivalry during the Cold War.[3]

Expertise and experience proved to be invaluable assets for reverse engineering. It is always useful for an engineer to know what he or she is looking at during a reverse-engineering procedure. With his tremendous experience and expertise in the design of large airplanes, A. N. Tupolev knew what he was looking at when he saw the dissected B-29. He recognized the role, purpose, and functionality of each and every part, how they interacted within major subassemblies to give rise to their functionality and performance, and how subassemblies and/or subsystems integrated into the overall aircraft.

The following interesting and relevant anecdote is offered. As a young materials and processes engineer at Grumman Aerospace Corporation from 1970 to 1980, the author was part of a four-

---

[2] The ANT-20, called the *Maxim Gorky* by the Soviets, was the largest airplane in the world in the 1930s, having a wingspan similar to a modern Boeing 747 airliner, and the first to have all-metal construction. With only a couple built between July 3, 1933, and April 3, 1934, corrugated steel sheet metal (originally proposed by Hugo Junkers) was used extensively throughout the airframe, as well as in the skins. The purpose of the airplane was principally for Stalinist propaganda, although it prepared Tupolev for the design of the heavy bombers for which his company became famous (e.g., Tu-14 "Bosun" torpedo bomber; Tu-16 "Badger" strategic bomber; Tu-20/Tu-45 "Bear" long-range strategic bomber; Tu-22 "Binder" supersonic medium bomber, Tu-22M/Tu-26 "Backfire" supersonic swing-wing long-range/maritime strike bomber, Tu-166 "Blackjack" supersonic swing-wing bomber).

[3] This kind of minor event at one time having major repercussions later is known as the *butterfly effect*, whereby everything that happens, no matter how seemingly slight, changes the world, if not the universe!

Deducing or Inferring Role, Purpose, and Functionality during Reverse Engineering 121

Figure 7-1 The Tu-144 "Charger" designed and built by PSC Tupolev (headed by Alexei Tupolev, A. N. Tupolev's son) as the first supersonic transport aircraft (SST) in the world, first publicly revealed in January 1962. Here, a Tu-144LL used by NASA to carry out research for High Speed Civil Transport (HSCT) is shown taking off at Zhukovsky Air Development Center (*a*) and in flight (*b*). (*Source:* Wikipedia Creative Commons, contributed by FightinGFalcoN on 7 July 2010 [*a*], and by Selefant on 18 May 2007 [*b*].) **Don't miss the color version of this figure, available at www.mhprofessional.com/ReverseEngineering.**

person team consisting of "an expert in aerodynamics, an expert in structural design, an expert in materials and processes, and any other person of the company's choice" that the U.S. Air Force invited from each major military aerospace company to inspect—separately, team by team—what turned out to be four captured Soviet aircraft held in a secret location at Wright Patterson Air Force Base in Dayton, Ohio, sometime around 1974 or 1975. Along with several MiG fighters that had

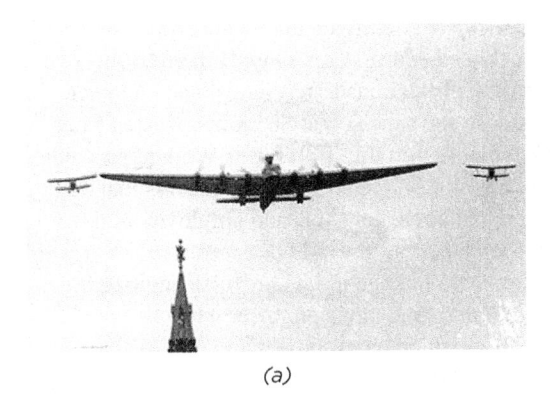

Figure 7-2 The ANT-20 *Maxim Gorky* built by OKB Tupolev between July 4, 1933, and April 3, 1934, was the first all-metal aircraft in the world and was the largest built for decades. An ANT-20 is seen with two Po-2 airplanes during a fly-by over Moscow in 1935 (*a*) and in another version later in 1935–36 (*b*). (*Source*: Wikipedia Creative Commons, contributed by REOSarevok on 26 November 2005 [*a*], and by Stahlkocher on 20 November 2005 [*b*].)

been shot down during the six-day Arab-Israeli War (June 5–10, 1967) or that had been flown into South Korea by a defecting pilot from North Korea was a Tu-144S CCCP-77102, Serial Number 01–2, the second airplane built. It had crashed at the Paris Air Show during flight demonstration on June 3, 1973, killing the Soviet crew of six and another eight people on the ground (see YouTube video "Soviet Tu-144 Crashes at Paris Air Show in 1973"). Two theories for the crash were that (1) unbeknownst to the Soviet pilots of the Tu-144, a French Mirage fighter aircraft taking video forced the pilots to take an evasive maneuver that caused the aircraft to crash, and (2) in the heat of a fierce rivalry with the French/British *Concorde,* the Soviet pilots attempted a maneuver of which the aircraft was incapable. In any case, upon examination of most of the wings, the engines, and most of the fuselage and empennage assembly, two conclusions were shared with the Air Force intelligence officer in charge of the secret visits.

First, Grumman's designer of the F14, Michael Pelehach, native of the Soviet Union until a teenager, and an expert on Soviet air weapon systems, was convinced the aircraft was actually a bomber converted to look like a civil airliner for publicity purposes.[4] (He found evidence to support his theory in cockpit instrumentation and numerous details of the airframe, which he shared with the Air Force officer.) Second, shoddy workmanship in several areas of the airframe that contrasted dramatically with superb workmanship in the engines supported information from the intelligence officer that A/C 2 had been rushed to completion after A/C 1 had crashed outside of Moscow during preparation for the Paris Air Show. Long-lead-time engines reflected actual Soviet capability, while the shoddy workmanship that supported many of our U.S.-propaganda-shaped expectations of Soviet capability was in error, being only the result of last-minute rush efforts to have the airplane at the air show (recall earlier comments on the importance of workmanship!).

This entire effort by the U.S. Air Force was focused on reverse engineering to assess the design and manufacturing capability and status of air weapons technology possessed by the Soviet Union compared to that of the United States from airplanes that had "dissected" themselves!

A case for not knowing what one is looking at during reverse-engineering activity is likely if one adheres to the theory that UFOs (unidentified flying objects) that have crashed on the Earth have, for decades, been the subject of reverse engineering (ref. King). Rumors abound of such events, with several implicating experts from the Lockheed-Martin Company. Robert "Bob" Lazar (1959–), a controversial individual who claims qualifications as a scientist and engineer, alleged that he worked on extraterrestrial technologies in the supersecret Section 4 of Area 51 at Groom Lake, Nevada, from 1988 to 1989 (do an Internet search for "Bob Lazar Worked on Alien Spacecraft Reverse Engineering: Lockheed-Martin's Senior Scientist"). The problem is that many of Lazar's credentials, including alleged degrees from CalTech and MIT, and employment have been brought into question, and legal actions have been taken on several occasions.

A more credible person, albeit with no less questionable linkage to UFO reverse engineering, is Clarence "Kelly" Johnson (1910–1990), the brilliant and innovative engineer who created the "Skunk Works" at the Lockheed-Martin Company, from which have come many remarkable aircraft. The one that led to suspicion that Kelly Johnson had access to alien technology—from where, no one says—is the SR-71 "Blackbird" (Figure 7–3).

---

[4] Michael Pelehach later became senior vice president and then president of Grumman Aerospace Corporation. At a Paris Air Show in the mid-1960s, he spotted a MiG 21, then the envy of every air force in the world for its speed, maneuverability, and firepower. Upon returning home to Grumman, he set about designing a countermeasure not just for it but for what he projected would be the capability of the next two generations to evolve from the MiG 21. That countermeasure was the Grumman F14 Tomcat, and Mike Pelehach was the "Father of the F14."

Deducing or Inferring Role, Purpose, and Functionality during Reverse Engineering **123**

Figure 7-3 The SR-71 "Blackbird" designed by Kelly Johnson under a "black project" at the Lockheed "Skunk Works" and introduced in 1966. Retired in 1998 after 32 were built, the aircraft set innumerable records, including highest cruise altitude flown and highest sustained speed. (*Source:* Wikipedia Creative Commons, originally contributed by David Legrand on 20 December 2004 and modified by Dbenenn on 16 August 2006.) **Don't miss the color version of this figure, available at www.mhprofessional.com/ReverseEngineering.**

This advanced, long-range, Mach 3+ strategic reconnaissance aircraft was developed from the Lockheed A-12 in the 1960s! It appeared from a top-secret "black project" from Kelly Johnson's "Skunk Works." Its unique Pratt & Whitney Aircraft J58 engines were hybrid turbojets inside a ramjet; it featured a unique blended body to provide additional lift, was made from titanium (which had not been used before), and was stealthy (with an extremely low visibility on radar). Over a 30-year mission life, it set innumerable records for speed and altitude, reaching the highest altitude yet achieved by any airplane at 85,069 feet (25,929 meters) and highest sustained speed of 1,905.81 knots (2193.2 mph/3529.6 kph). It was so advanced that some are convinced it could only have come from "alien technology."

For what it's worth, the author believes the technology may have been alien to other aircraft manufacturers around the world, but need not—and did not—require the reverse engineering of any UFO. In fact, as stated in Chapter 3, if an alien spacecraft that came from deep space, across light-years of distance, crashed on Earth and was reverse engineered to create the SR-71, the job was not very well done! As impressive as the SR-71 was, it was no extraterrestrial spaceship. It simply didn't exhibit anywhere near the technologies that would be required in such a craft!

So, while experience in related designs is a tremendous asset during reverse engineering, not everyone is so fortunate. In these cases, one needs to identify with certainty what one can and deduce or infer the rest as best as possible.

7-4 Using Available Evidence, Clues, and Cues

Frank Sinatra made it famous, but it was written for the Broadway musical *Finian's Rainbow*: "When I'm not near the girl I love, I love the girl I'm near."[5] Apropos to the technique of reverse engineering, the message is this: Make do with what you've got, or, if one doesn't know with certainty what one is looking at (what it is, is for, is made of, etc.) during mechanical dissection, one needs to do what one can to identify, deduce, or infer what it is, is for, is made of, and so on. Recall that identification (as the act used to identify) results in knowledge with certainty of what was previously unknown. Deduction (as the act to deduce) and inference (as the act to infer), on the other hand, result not in knowledge of what is certain but, rather, in knowledge of what is likely.[6] So the engineer who engages in reverse engineering needs to read and follow clues and cues, short of finding incontrovertible evidence.

For clarification, here are the formal definitions of these three words:

- *Evidence* is "ground(s) for belief or disbelief; data on which to base proof or to establish truth or falsehood."
- *Clue* is "something that serves to guide or direct [belief or action] in the solution of a problem or mystery."
- *Cue* is "a signal [often a word or action] used to prompt another event or action [often in a performance]."

Each and every one of these three needs to be sought during a reverse-engineering procedure, in this order: (1) seek evidence, (2) be alert to and follow clues, and (3) be alert to cues, should any arise during actions taken upon the entity being dissected (e.g., using measurement or experiment), to prompt another action. A much better understanding and appreciation of each of these are found in the illustrative cases of reverse engineering covered in Chapters 13 and 14.

Each individual's brain operates slightly differently than every other individual's brain or even from one situation or time to another in the same individual. After all, except during deep contemplation in total isolation, the brain is stimulated by input from one or more of an individual's senses and causes the individual to act or react accordingly. One sees, hears, feels, smells, or tastes something, and a cascade of memories and thoughts occur, which lead to some action or reaction. This is often obvious during the procedure of reverse engineering as the dissected entity is meticulously examined.

What one engineer sees first is not necessarily what another engineer sees first. Using a fatigue fracture during failure analysis (Chapter 6) as an example (here, Figure 7–4), one person may spot the set of concentric bands first, indicating crack propagation under repeated (cyclic) loading in the form of *beachmarks*. A second person (or the first person on another occasion) may spot the

---

[5] *Finian's Rainbow,* story and book by E. Y. Harburg and Fred Saidy; the song "When I'm Not Near the Girl I Love," lyrics by E. Y. Harburg, words by Burton Lane, 1947.
[6] Going back to the earlier analogy of the detective, while the good detective uses inference and deduction as essential tools in solving a crime, without evidence that is incontrovertible, backed by solid identification or admission, what is deduced or inferred is often considered circumstantial and is rarely certain.

Figure 7-4 Fracture surface of a steel shaft containing a machined keyway in which fatigue initiated at the lower-left-hand corner of the keyway, where there was virtually no radius, and propagated across the shaft until terminal overload occurred in the small region near the 4:30 clock position. (*Source:* Donald J. Wulpi, *Understanding How Components Fail,* 2nd edition, ASM International, Materials Park, OH, 1999, page 152, Figure 22.)

keyway cutout first, indicating a severe stress concentration site at the bottom of the keyway due to the near absence of any discernable radius. A third person (or one of the other two on another occasion) may spot the dark featureless region, devoid of concentric bands, near the lower right, at the 4:30 clock position; indicating fracture termination by overload as the remaining sound cross section became so small that the next loading cycle gave rise to a stress that exceeds the material's static fracture strength.

Each "first" observation of topological evidence on the fracture surface (i.e., fractographic evidence) provides a clue as to what the observer should look for next to deduce the sequence of the fracture on the way to seeking a root-cause for the failure. Upon seeing the concentric bands, the first person might try to deduce their origin—looking to the sharp radius and stress concentration at the root of the keyway—by tracing the path of the bands upward. Alternatively, the first person might look to see where the concentric bands end up—by tracing the path of the bands downward and to the right—to identify the region of terminal overload. The second person, having spotted a stress concentration in an obvious geometric feature (i.e., the keyway), might next spot the concentric bands, recognize that fatigue occurred from that sharp radius, and then look to the opposite side of the fracture surface in an attempt to identify the region of fracture termination. The third person, having spotted the overload region first, might not recognize it as evidence of termination but, instead, think it is the initiation site (or origin) for the obvious concentric bands that appear to be emanating from it. However, by tracing the path of the concentric bands (using normals to the contours as vectors), the third person ends up at the lower-left-hand corner of the sharp keyway and realizes this is a more likely site for fatigue crack initiation.

The point is this: In conducting reverse-engineering dissection, one needs to see all there is to see, using one bit of evidence as a clue for what to look for next.

In the remainder of this chapter, the focus is on making *observations*! This means seeing all there is to see, being sure to "turn the body over," and using what there is to be seen as hard evidence or, if not hard evidence, as a clue as to what to look for next. Recall (from Section 4–2) that when experiments are conducted to allow measurements to be made, use any cues from one action to prompt the next action.

To accomplish the end goal of learning about and understanding what the entity being examined was intended to do, one needs to think about and—short of identifying with certainty—deduce or infer what the role, purpose, and functionality of each and every detail was (or is). As used here, these three words have the following subtly different meanings, which will become more and more clear as this chapter continues:

- *Role,* as a noun, is "a usual or customary function," in other words, the part something *actually* plays in the whole.
- *Purpose,* as a noun, is "the reason for which anything is done, created, or exists," in other words, the part something is *intended* to play in the whole.
- *Functionality,* as a noun, is "the quality of being functional," by which is meant "designed for or adapted to a particular use," in other words, *how* the part works.

The clues and cues to be used may come from one or more of the following sources: (1) geometry; (2) flow of forces, energy, and/or fluids; and (3) major functional entities (e.g., structure, energy or power source, controls, sensors, actuators).

Let's look at these one by one, with this being the first and only reference to address this essential step in reverse engineering.

7–5 Using Geometry

The term *form, fit, and function,* sometimes called FFF or F3 in design and manufacturing, is a description of an item's identifying characteristics. Together, form, fit, and function uniquely identify an item's physical, functional, and performance characteristics or specifications. When the specifications for form, fit, and function for a particular item are met, the item is generally considered interchangeable with other items with the same specifications or requirements. An assessment of the impact on form, fit, and function is commonly used to determine whether a proposed change to the item's design or manufacture will be "minor," with little or no impact on the form, fit, and function, or "major," with a significant effect on the form, fit, and function. Definitions of these three terms from www.businessdictionary.com follow:

- *Form* is "the shape, size, dimensions, mass, and/or other parameters that uniquely characterize an item, that is, its 'look.'"[7]

---

[7] Color is generally not considered in "form," except when it has a specific functional meaning and/or impact.

- *Fit* is "the ability of an item to physically interface or interconnect with or become an integral part of another item or assembly."
- *Function* is "the action(s) that an item is designed to perform," which generally sets the reason for the item's existence.

This section focuses on the observation of evidence (see Section 4–2), as well as clues or cues, relating to the *form* of an item, and its component parts, being subjected to reverse engineering.

There is, without question, evidence, or there are clues, which can be obtained—by deduction—from the geometry of a part, component, device, or structural element. Overall shape may afford a clue to function. *Prismatic geometry* (i.e., three-dimensional, rectilinear shape) is usually (but not only) to simplify manufacture and either enclose details that have to be attached to the prismatic unit or help with the function of a structural element or structure (e.g., to carry loads along a straight line or in a plane, perhaps to minimize unwanted bending). *Curvilinear geometry* (e.g., spherical, cylindrical, toroidal, or compound curvature) is usually (but not only) to provide aerodynamic or hydrodynamic streamlining (to aid fluid flow) or aesthetics (line, style, etc.) or ergonomics (e.g., human factors).

Dimensionality often affords another set of clues, with 3D shapes used to allow containment of details or payload (i.e., to provide volume), fill volume, or provide a degree of structural solidness (stability, load-carrying capacity, structural stiffness, etc.). On the other hand, 2D shapes are often used to facilitate easy fabrication, handle low and in-plane loads, minimize volume and weight, and may, perhaps, indicate the absence of any need to contain details (except when 2D panels are used for creating an enclosure). Finally, 1D shapes are usually used to operate in tension, as a cable stay or tie or linkage to an actuator.

Geometric shape is extremely important when it relates to the *cross section* of a structural member or element, for example. While the cross-sectional shape has no effect on the load-carrying capability for tensile loads or forces (where stress is simply load or force/area), it is extremely important for structural members or elements loaded in compression, bending, or torsion (i.e., twist). For these three types of loading, singly or in combination with each other and/or with tensile loads, the moment of inertia (I) is extremely important, as it affects structural stiffness against unwanted or intolerable elastic (i.e., temporary and recoverable) or plastic (i.e., permanent and unrecoverable) deflection, elastic or plastic buckling (e.g., of a column, strut, or truss), or elastic or plastic twist. Since the moment of inertia for a shape increases as the distribution of mass around the center of mass or center of gravity (i.e., centroid) of the section around some axis increases, more complicated cross-sectional shapes generally indicate an attempt to obtain a higher value of I and greater structural stiffness, often at minimum weight.[8]

Geometric symmetry (circles, equilateral triangles, squares, hexagons, etc.) often indicates some need for symmetry of mechanical properties and/or some functional requirement or behavior. However, overall symmetry may have other purposes. Contrarily, asymmetry may indicate the need for different properties and/or different functions or functionality in different directions.

There are other clues that geometry can suggest, but what is presented here, while not complete, covers the most significant clues.

Table 7–2 summarizes the clues geometry may imply.

---

[8] The smallest value of I occurs for the cross section of a solid round. All other shapes (a square, oval, hollow round, I- or H-shape, etc.) have a higher value of I. It is common to compare the value of I for a nonsolid round to that of the solid round as a ratio, i.e., $I_{shape}/I_{round} > 1$.

TABLE 7-2 Summary of Clues Geometry May Imply in a Dissected Technical System

Aspect of Geometry	Possible Implications
General Geometry/Overall Shape:	
Prismatic (3D rectilinear)	■ Enclose details
	■ Facilitate detail attachment
	■ Favor in-line or in-plane loading
	■ Simplify manufacturing
	■ Simplify assembly/joining
Curvilinear (simple)	■ Equalize external or internal pressure
	■ Provide rotational symmetry
	■ Facilitate assembly by joining
Curvilinear (compound)	■ Provide streamlining
	■ Shape fluid flow
	■ Enclose details
	■ Impart aesthetics
Dimensionality (3D vs. 2D vs. 1D):	
Gross 3D (solid)	■ Provide high load-carrying capability
	■ Provide structural stiffness
	■ Provide stability
	■ Provide internal volume or space
Gross 2D (planar)	■ Handle low, in-plane loads
	■ Minimize volume and weight
	■ Facilitate mounting of details
	■ Simplify manufacture
Gross 1D (linear)	■ Favor tensile loading
Cross Section of Members:	
Solid or hollow round	■ Suggests tensile loading
	■ Favored for rotating shafts
Nonround	■ Suggests need for moment of inertia
	■ May suggest compressive, bending, or torsion loading
Symmetry:	
Symmetrical	■ May reflect need for symmetry of mechanical properties
	■ Often reflects need for symmetry of functional requirements or behavior
	■ Imparts aesthetics
Asymmetrical	■ May reflect need for aerodynamic or hydrodynamic streamlining
	■ May imply different loading in different directions

7-6 Using Flows of Force, Energy, and/or Fluids

Flows can be used as clues to the operation of a product or system or other item or technical entity being subjected to dissection as part of a reverse-engineering effort. Flow (or transfer) of force, flow (or transport) of energy, and/or flow (or movement) of fluids are all potential indicators of or clues to the role, purpose, and functionality of a technical entity.

The *flow* or, more familiarly, *transfer of force,* through an assembly of parts, components, or structural elements is discussed in Section 5–5, on force flow diagrams. *Force flow diagrams*

TABLE 7-3 Summary of the Potential Uses of the Flow of Forces, Energy, and/or Fluids as Evidence or Clues in Dissected Technical Systems

Flow of Force(s):
- For the transfer or transmittal of force(s), load, stress, or pressure
- May be to apply a holding force (to develop friction) or provide grip
- May be to cause motion or movement
- May be to cause needed elastic deflection

Flow of Energy:
- May imply the presence of power source, external or internal
- May imply the conversion from one form to another form
- May indicate the need for insulation (for electrical or thermal energy) or damping (for mechanical energy)

Flow of Fluids:
- May be to move or remove heat (i.e., for heating or cooling)
- May be to provide thrust, gust, force, jet, or spray
- May be to activate a mechanism (e.g., hydraulic or pneumatic) or flex a diaphragm

are routinely used during consumer product teardown but also can prove useful during reverse engineering for other design-oriented end goals. Whether or not an actual force flow diagram is created and constructed for a product, device, mechanism, structure, or system being subjected to reverse engineering is not particularly important. What is important is for the engineer(s) to consider the transfer (or flow) of mechanical forces through such entities, as this can aid in deducing role, purpose, and functionality of details on the way to overall functionality in the system.

There occasionally is evidence of the flow of forces, as with the presence of wear scars, metal transfer (from the softer to the harder material), and/or heat tint between mating parts that move relative to one another. Otherwise, the flow of force(s) must be extracted logically from the way in which details interact.

Flow of energy (or *energy transport*) is also important, regardless of the form of that energy (e.g., mechanical, electrical, or thermal).[9] Likewise, sources of energy (or power) and points (as devices or subsystems) where energy is converted from one form to another are important to identify or deduce. Examples include heating elements or units, cooling elements or devices (e.g., chillers) or structures (e.g., heat sinks or radiating fins) or systems (circulating fluids, radiators, etc.), electromagnetic devices (e.g., solenoids), photovoltaic devices, thermoelectric devices, and so on. Finally, the presence of electrical or thermal insulation (as barriers to energy flow) can provide valuable clues that energy flow is involved.

The *flow of fluids,* whether as liquids or gases, is also important for identifying or deducing role, purpose, and functionality during dissection during reverse engineering. Flow of liquid (e.g., water, ethylene glycol, oil, molten metals) for heating or cooling or actuation of a mechanism (e.g., hydraulic systems) by application of a force from pressure or gas (e.g., air, steam) for heating or cooling or actuation (e.g., pneumatic systems) by application of force from pressure can be valuable clues. Obviously, the presence of tubing, piping, channels, ducting, inlets or outlets, orifices, nozzles, valves, manifolds, and the like, identifies fluid flow. But there may be other, more subtle clues, such as a flow pattern evidenced by erosive wear, cavitation, deposition of fine particulate residue, or staining.

Table 7-3 summarizes the potential use of the flow of forces, energy, and/or fluids as either evidence of or clues to the role, purpose, and/or functionality of an item.

---

[9] Energy flow, while often part of energy field flow design, can actually also be described using a variant of the force flow diagram, as shown in Figure 5–9.

7-7 Using Functional Units or Subsystems from a Functional Model

Few would argue there is no more complex system than the human body. It is a remarkable assemblage of brain, organs, bones, arteries, veins and capillaries, muscles, ligaments and tendons, nerves, and skin.[10] To an engineer, however, the human body is not only a fascinating creation but is also valuable as a model for complex, nonliving technical systems (see Figure 5–3a). From an engineer's perspective—at least from the author's perspective as an engineer who teaches—the human body is made up of the following major functional units or subsystems that would constitute a functional model:

- Brain ⇒ central control system or subsystem
- Skeleton (i.e., bones and cartilage) or skeletal system ⇒ structural system or subsystem
- Circulatory (vascular) system ⇒ fuel system (for oxygenation and nutrition, as well as immunization and healing); heating and cooling system or subsystem
- Muscles, tendons, and ligaments ⇒ actuator system or subsystem; suspension system or subsystem
- Nervous system ⇒ sensory system or subsystem

Figure 7–5a through e show the human body, highlighting the organ system (a), skeletal system (b), muscle system (c), vascular system (d), and nerve system (e).

Missing from the preceding list are the organs of which, of course, there are a number, with a variety of principal functions. At first glance, there may appear to be no counterpart in nonliving technical systems. However, upon closer consideration, these organs perform major functions that tend to depend on either chemical processing and/or some form of energy conversion.

A short list (as there is really a rather long list, given at www.wikipedia.com under "Human Anatomy" in the section "Major Organ Systems") includes the following, with the essential technical function or functionality given for each:

- Brain ⇒ central control system; sensory signal processor; central processing unit
- Heart ⇒ pumping system for vascular system (i.e., pump and valves)
- Lungs ⇒ osmotic gas exchange system; gaseous filter system
- Stomach ⇒ chemical processing system or subsystem; energy conversion system or subsystem
- Liver ⇒ chemical processing system or subsystem; liquid (blood) filtration system
- Intestines ⇒ chemical processing system or subsystem; energy conversion system or subsystem; solid waste disposal (or exhaust) system or subsystem
- Kidneys ⇒ chemical processing system or subsystem; liquid (blood) filtration system or subsystem; liquid waste disposal system
- Pancreas ⇒ chemical processing system

From this standpoint, a complex technical system can be analyzed, during reverse engineering, for example, as the integration of numerous functional systems or subsystems, some number

---

[10] In fact, the skin is an organ itself—actually, the largest of all the organs in the human body.

Deducing or Inferring Role, Purpose, and Functionality during Reverse Engineering **131**

Human anatomy

(a)

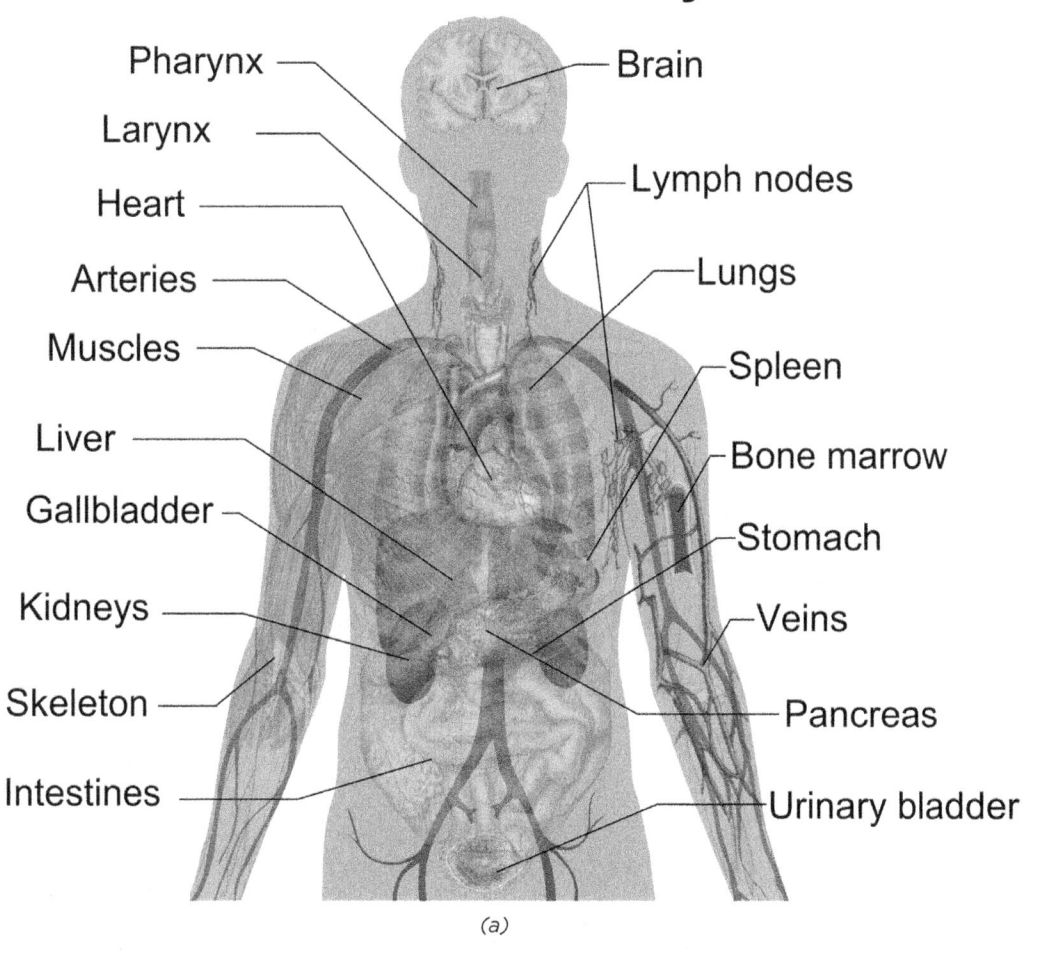

Figure 7-5 The human body, with various major systems highlighted, these being: the organ system (*a*), the skeletal system (*b*), the muscle system (*c*), the circulatory system (*d*), and the central nervous system (*e*). (*Sources:* Wikipedia Creative Commons, contributed by Michael Haggstrom on 17 December 2012 [*a*], by Bibi Saint-Poi on 3 June 2009 [*b*], by LadyofHats on 24 May 2012 [*c*], by KVDP on 16 January 2010 [*d*], and by The Emirr on 27 November 2011 [*e*].) ***Don't miss the color version of this figure, available at www.mhprofessional.com/ReverseEngineering.***

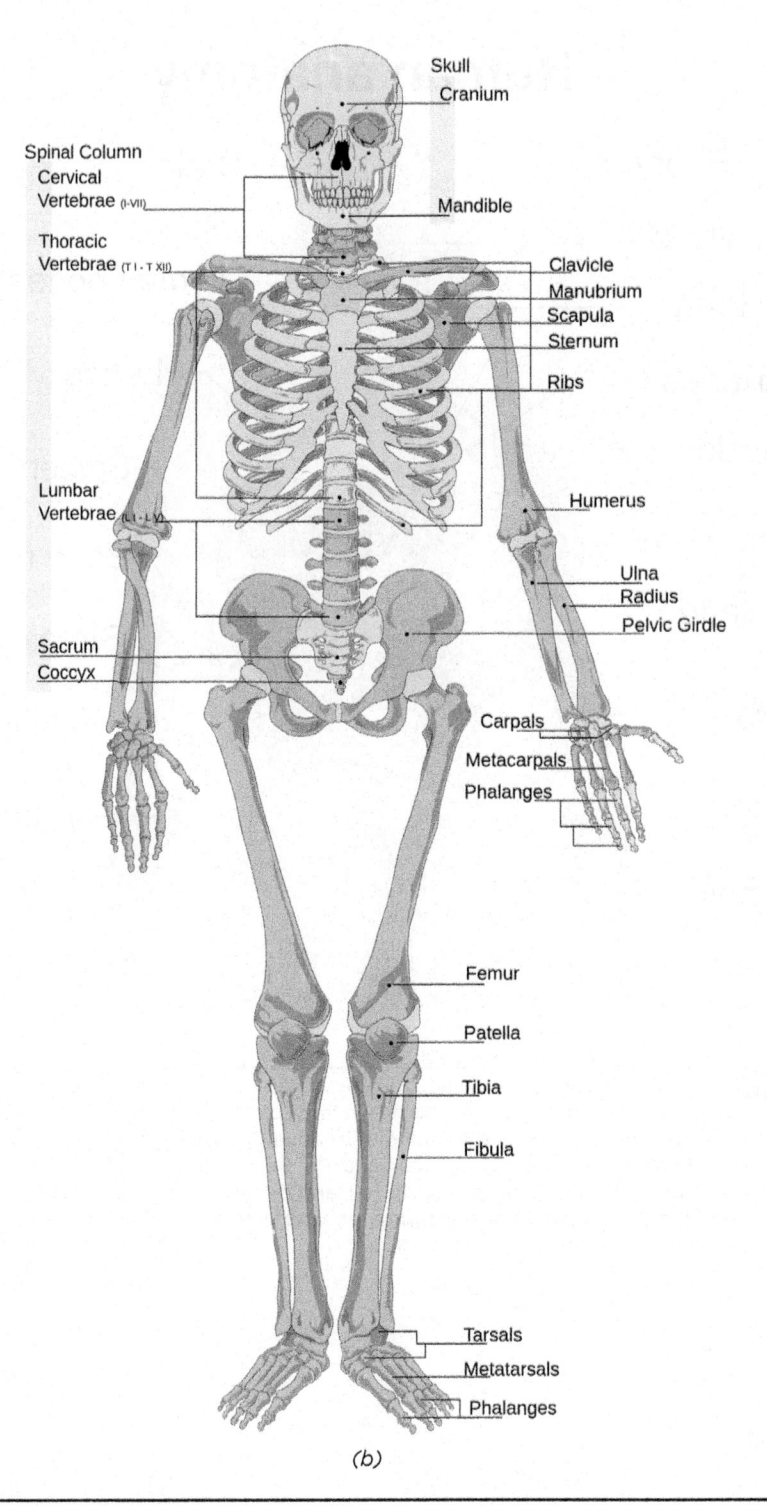

(b)

Figure 7-5 *(Continued)*

Skeletal muscles

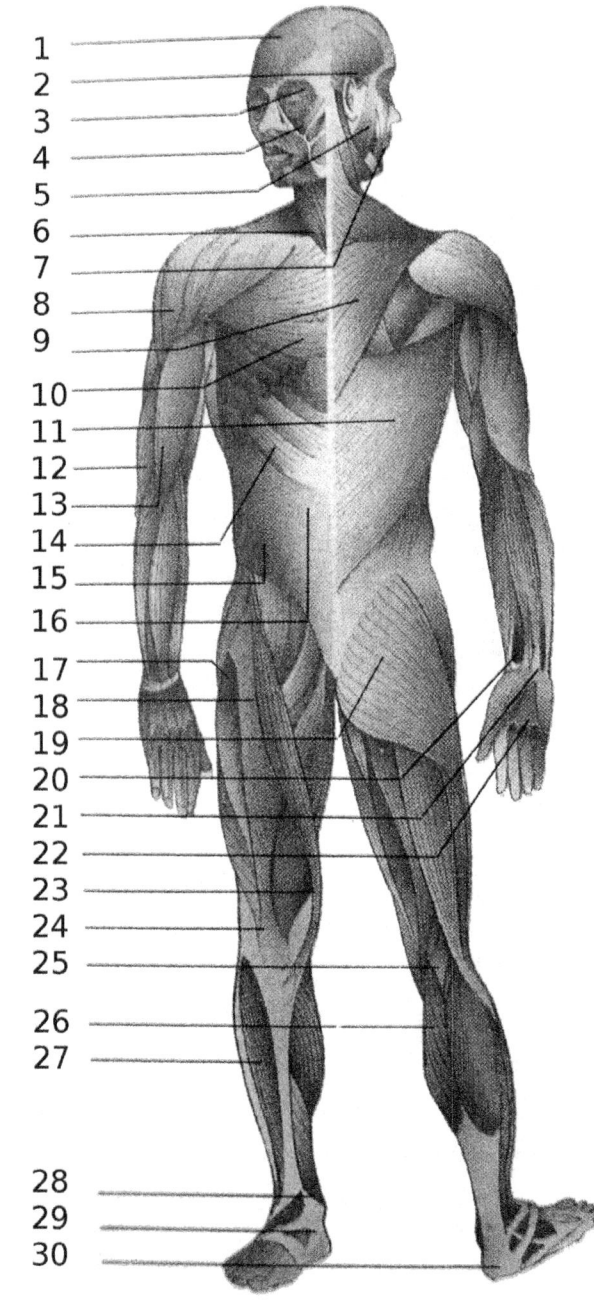

Musculus ...
1: occipitofrontalis
2: temporoparientalis
3: orbicularis oculi
4: levator labii superior
5: masticatorii
6: sternocleidomastoideus
7: orbicularis oris
8: deltoideus
9: trapezius
10: pectoralis major
11: latissimus dorsi
12: triceps brachii
13: biceps brachii
14: serratus anterior
15: rectus abdominis
16: obliquus externus abdominis
17: tensor fascia lata
18: rectus femoris
19: gluteus maximus
20: pronator quadratus
21: flexor retinaculum
22: flexor digitorum communis
23: sartorius
24: quadriceps femoris
25: ischiocrurale
26: gastrocnemius
27: tibialis anterior
28: soleus
29: extensor retinaculum
30: triceps surae

(c)

Figure 7-5 *(Continued)*

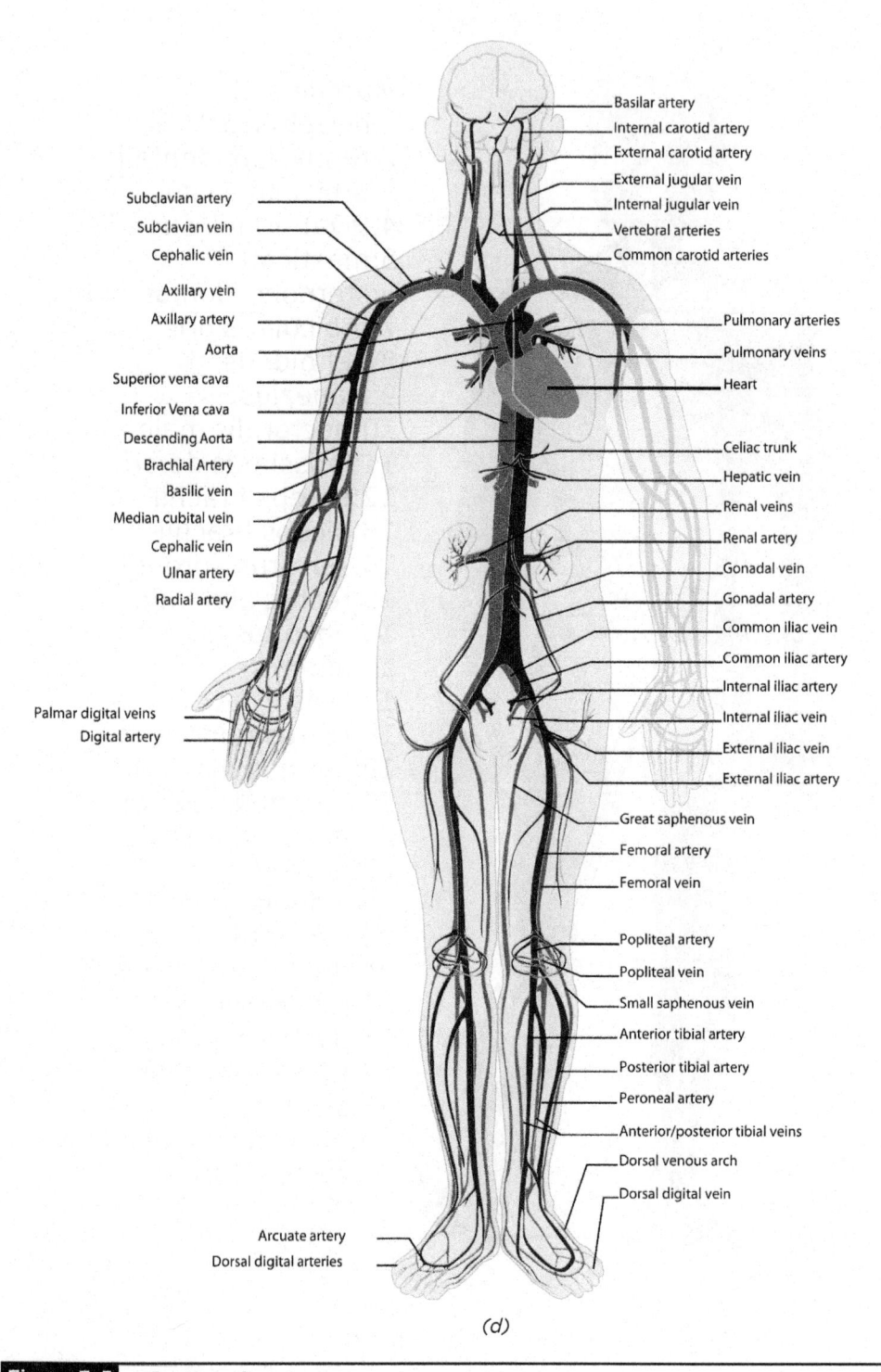

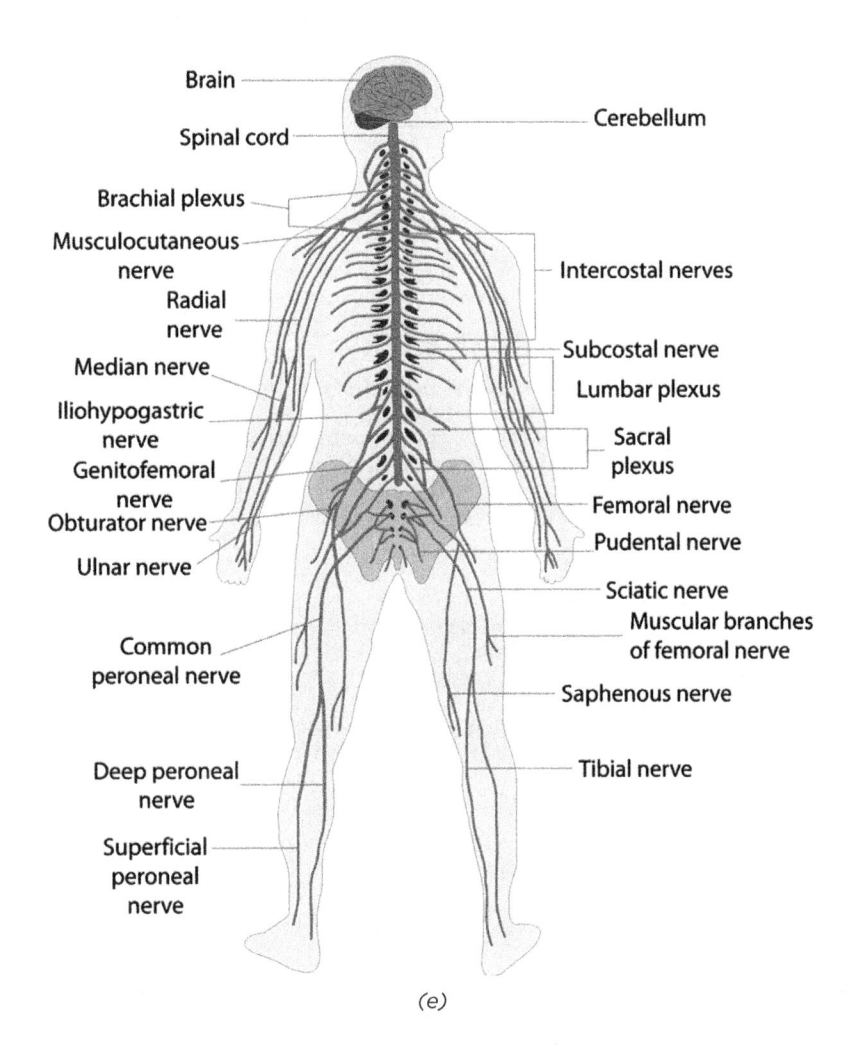

[Figure with labels: Brain, Cerebellum, Spinal cord, Brachial plexus, Musculocutaneous nerve, Intercostal nerves, Radial nerve, Subcostal nerve, Median nerve, Lumbar plexus, Iliohypogastric nerve, Sacral plexus, Genitofemoral nerve, Femoral nerve, Obturator nerve, Pudental nerve, Ulnar nerve, Sciatic nerve, Muscular branches of femoral nerve, Common peroneal nerve, Saphenous nerve, Deep peroneal nerve, Tibial nerve, Superficial peroneal nerve]

(e)

of which is, in their own right, an integration of components or parts. With this as a premise, Table 7–4 lists key generic functional systems or subsystems, along with key subsystems and/or major components about which an engineer should be vigilant during mechanical dissection during reverse engineering.

Let's work through the key categories in Table 7–4.

Power Sources

Devices, mechanisms, products, or systems that require or result in internal and/or external movement(s) require a power source (i.e., a source of energy over time). For the human body, that source is the combination of food, water, and air. For a technical system, the source can be electricity, natural or other hydrocarbon gas, a liquid hydrocarbon (gasoline, kerosene, diesel oil, etc.), steam, moving water, or moving wind, the latter two being kinetic energy sources.

TABLE 7-4 Using Functional Units in the Functional Model of a Technical System to Identify or Deduce Role, Purpose, and/or Functionality

Power Sources ⇒ *provide power*
- Electricity (external or internal; AC or DC) ⇒ quietness, cleanliness
- Natural gas ⇒ environmental considerations
- Gasoline or diesel oil ⇒ high horsepower, convenience
- Steam ⇒ high horsepower, long life
- Water ⇒ available natural moving water resource, cleanliness
- Wind ⇒ available prevailing wind, environmental convenience

Prime Movers ⇒ *serve to drive other components and cause motion*
- Engines ⇒ high horsepower
  - ✓ Internal combustion (e.g., gasoline, diesel, natural gas or propane, biofuel, hybrid, hydrogen)
  - ✓ External combustion (e.g., steam or gas turbine)
- Electric motors or drives ⇒ constant torque, speed control, small size
- Pneumatic motors or drives or piston actuators ⇒ sparkless
- Hydraulic systems (e.g., actuators) ⇒ limited motions, low weight
- Clockwork or windup motors ⇒ limited power, high reliability

Power or Motion Converters ⇒ *change power or energy from form to form or change the direction and/or type of motion*
- Coolers or chillers ⇒ remove heat
- Differentials ⇒ change direction of mechanical power
- Heaters/heating elements ⇒ produce heat
- Heat exchangers or dissipaters ⇒ extract heat
- Insulation ⇒ prevents heat flow
- Pistons and cylinders ⇒ convert expanding gases to motion
- Transmissions ⇒ allow changes in speed, power, direction

Structure ⇒ *carry, sustain, transfer loads, forces, or pressure*
- Primary support structure ⇒ major loads, often critical function
  - ✓ Arches ⇒ span a space, distribute loads outward to supports
  - ✓ Axles ⇒ allow rotation of parts as center supports
  - ✓ Beams ⇒ favor bending loads
  - ✓ Bulkheads ⇒ separate compartments, carry shear loads
  - ✓ Columns ⇒ intended for compressive loads
  - ✓ Decks ⇒ span supports, operate as horizontal beams
  - ✓ Domes ⇒ distribute loads outward to circular supports
  - ✓ Lugs ⇒ integral attachments for struts or other structures
  - ✓ Shells ⇒ thin domes or other shape enclosure, light weight
  - ✓ Struts ⇒ favor tension, tolerate light compression
  - ✓ Ties ⇒ transfer tensile loads (e.g., cables)
- Secondary structure ⇒ minor loads, noncritical function
  - ✓ Bearings ⇒ support axles in housings
  - ✓ Bushings ⇒ line holes as inserts for additional support
  - ✓ Ducts or ducting ⇒ contains and directs moving gases
  - ✓ Panels ⇒ partitions, enclosures
  - ✓ Piping ⇒ moves fluid (usually liquids)
  - ✓ Skins (e.g., bodies, cases, housings, enclosures, shrouds)
  - ✓ Sleeves ⇒ line outsides of shafts for additional support
  - ✓ Stiffeners ⇒ to increase resistance to buckling in 2D shapes
  - ✓ Suspension members ⇒ springs, shock absorbers

Actuators ⇒ *create a specific action requiring movement*
- Augers ⇒ transfer of material by screw
- Belts and pulleys ⇒ transfer power, rotating pulley to pulley
- Cams and cranks ⇒ control and create complex repeated motion
- Chains and sprockets ⇒ very positive power transfer as belts, etc.

Deducing or Inferring Role, Purpose, and Functionality during Reverse Engineering

- Gears ⇒ transfer rotational motion, change speed/power ratios
- Linkages ⇒ to transfer motion from component to component
  - ✓ Connecting rods ⇒ transfer motion from cam to piston
  - ✓ Hinges ⇒ allow a degree of rotational freedom
  - ✓ Links ⇒ transfer force and motion between components
  - ✓ Shafts ⇒ transfer rotational force or motion
- Pumps ⇒ force fluids
- Rack and pinion ⇒ converts rotational to linear motion
- Transducers ⇒ convert energy signal to motion

Fluid Movers ⇒ transport liquids or gases, possibly for force
- Channels ⇒ guiding troughs or pathways
- Fans ⇒ rotating blades for moving air or other gases
- Impellers ⇒ move liquids by rotation using blades or vanes
- Nozzles ⇒ constrict flow of fluids in a directed fashion
- Orifices ⇒ holes which allow passage of fluids
- Propellers ⇒ provide pushing or pulling forces for motion in fluids

Sensors ⇒ detect signals to cause an action
- Diaphragms ⇒ convert pressure waves to voltage signal
- Photocells ⇒ convert light to voltage
- Thermocouples ⇒ generate voltage in response to temperature
- Thermoelectric devices ⇒ convert heat to voltage, or vice versa
- Thermostats ⇒ convert temperature to movement
- Transducers ⇒ convert pressure or force into voltage, or vice versa

Control/controllers ⇒ direct actions
- Computers ⇒ access, process, and store data; interpret signals
- Dials/dial knobs ⇒ adjustable continuous change in signals/actions
- Governors ⇒ limit speed
- Gyroscopes ⇒ stabilize direction of motion
- Programmers ⇒ direct actions with preset values and limits
- Shift levers ⇒ change actions in fixed increments
- Slides adjustments ⇒ continuous adjustment for changes in actions
- Switches ⇒ discrete position changes in signals/actions
- Valves ⇒ regulate flow rate and/or pressure of fluids

Joints ⇒ unite primary and secondary structural components
- Adhesive bonds ⇒ chemical forces to spread loading, favor thin
- Brazes ⇒ >425°C/840°F primary bonding without melting base
- Integral attachments ⇒ design features to cause interference
  - ✓ Dovetails and grooves ⇒ rigid interlock for wood
  - ✓ Elastic snap-fits ⇒ retain parts with elastic catch and latch
  - ✓ Plastically deformed interlocks (e.g., stakes, crimps) ⇒ lock
  - ✓ Tongues and grooves ⇒ rigid interlocks
- Mechanical fasteners ⇒ supplements devices to cause interference
  - ✓ Bolts (and nuts) ⇒ provide security against tension
  - ✓ Machine screws ⇒ small bolts
  - ✓ Nails ⇒ resist shear between wood elements
  - ✓ Pins ⇒ allow rotation, fix position
  - ✓ Rivets ⇒ secure against shear via upset
  - ✓ Self-tapping screws ⇒ secure against shear making own hole
  - ✓ Spring clips ⇒ retain components with elastic compression
- Solder joints ⇒ <425°C/840°F primary bonding without melting base
- Welds ⇒ metallurgically bonded metal parts, permanent joint

Miscellaneous Devices
- Blades
- Cutters
- Shears
- Shredders

Prime Movers

A *prime mover* is a device that transforms energy from or to thermal or electrical energy or pressure to or from mechanical energy. Examples (for which, if uncertain, the reader should seek definitions on the Internet) include engines and motors.[11]

Power or Motion Converters

These devices or mechanisms change power or energy from one form to another form or from one direction or speed or torque level to another. Examples cited include heaters, coolers or chillers, heat exchangers, transmissions, and differentials. Also in this category is insulation.

Structures or Structure

Structures are intended to carry, resist, sustain, and/or transfer forces or loads. If the loads or forces are high, and failure of the structure would lead to catastrophic results, the structure is considered primary. If the loads or forces are low and failure of the structure did not lead to a catastrophic event, the structure is considered secondary. A variety of examples are given for each in Table 7–4.

Actuators

Actuators cause an action by transferred energy or, more often, motion from one subsystem or component to another. The motions transferred may be translational or rotational or both. Numerous examples are given, it being left to the reader to seek a definition for any unknown example.

Fluid Movers

In some technical systems, it is necessary to move a fluid or fluids, either gaseous or liquid or both. Examples of devices (e.g., fans and impellers) or features (e.g., channels and nozzles) for doing this are given in Table 7–4.

Sensors

Sensors detect what is taking place in the environment of a technical system, generate a signal as a response, and send that signal to either a control system (like the brain of a human being) for action to be taken or directly to an actuator. The input to the sensor can be electricity, light, heat, or force or pressure, analogous to the inputs to the sensory system of human beings. There are some modern technical systems that have sensors that detect smell/taste via chemical molecules (e.g., gas sniffers).

Controls or Controllers

Controls and/or *controllers* direct action(s). More correctly, *control* direct action, usually by employing input from one or more sensors to trigger a signal to an actuator to cause a particular action. *Controllers,* on the other hand, generally initiate the eventual action(s) to be

---

[11] The human body, as well as the bodies of other living animals, differs from most technical systems in that there is no central prime mover. The brain recognizes or creates a reason for movement, sends a signal to the appropriate part of the body, and action (in the form of movement) takes places there. Action is distributed versus centralized. An analog in a technical system is the use of either hydraulics or, in modern systems, fly-by-wire signals to cause actions (in the form of movements) to control surfaces (e.g., ailerons, flaps, elevators, rudders) of an airplane.

taken or provided by an actuator by processing an input signal from a sensor and generating a corresponding (generally, preprogrammed) signal to direct the actuator. Besides actuators, both controllers and controls may send a signal to a power source and/or a prime mover or a power or motion conversion device or system. The best example of a controller is a computer. Good examples of controls are governors and valves.

Joints

Complex technical systems (as well as the human body) consist of multiple parts. These parts need to be held together to create an assembly. *Joining processes* (e.g., mechanical joining, adhesive bonding, and welding) create *joints* that hold parts together, sometimes so the parts do not move relative to one another and other times so that one can still move relative to the other. Joints may be intended to be permanent (i.e., hold parts together so they cannot be separated or disassembled) or only temporary (i.e., to allow intentional disassembly).

While all of the joining processes listed in Table 7–4 are probably familiar, one form that may not be immediately recognized involving integral attachment will, most likely, be recognized once examples are cited. Integral attachments employ geometric features that are either designed into opposing detail parts to cause mechanical interlocking (e.g., a rigid dovetail-and-groove joint or an elastic snap-fit) *or* are created between mating parts to be joined by plastically deforming an interlocking feature at their interface (e.g., a stake or crimp). Readers interested in this very old (perhaps oldest!) method of joining are referred to a unique book written by the author (ref. Messler).

Miscellaneous Devices

The list of miscellaneous devices could be virtually endless, as it encompasses everything that cannot be placed into one of the other categories. The examples included in Table 7–4 are simply a few of the common devices that quickly came to the author's mind.

7-8 Summary

The procedure for conducting a reverse-engineering activity involves six steps. Step 1 (statement of purpose) is covered in Section 2–3; Step 2 (mechanical dissection) is covered in Chapter 4 as well as in Section 5–2. The remaining steps are covered here (Step 3, identifying or deducing role, purpose, and functionality of details) and in Chapter 9 (Step 4 to identify or deduce materials-of-construction), Chapter 10 (Step 5 to identify or deduce the method-of-manufacture), and Chapter 12 (Step 6 to assess design for suitability of purpose).

The differences between knowing something to be true with certainty from the outset, identifying something previously unknown to be true with certainty, and either deducing or inferring something to be likely were defined and discussed. The value of experience, as a tremendous asset during reverse engineering, was also described and discussed.

The use of available evidence, clues, and cues in reverse engineering was described and discussed, with the key both for finding evidence and clues being meticulous observation and for finding cues being experimental measurement. Specific use of geometry, flow of forces, energy, and/or fluids, and of functional units or subsystems as a functional model to assess the role, purpose, and functionality of details on the way to doing the same for the overall technical system were described and discussed, with tables summarizing each.

7-9 Cited References

King, Thomas, *UFOs That Crashed to Earth: Reverse Engineering of Alien Spacecraft, Mankind Creates the Atomic Bomb, UFO Enigma Solved,* AuthorHouse, Bloomington, IN, 2009.

Messler, Robert W., Jr., *Integral Mechanical Attachment: A Resurgence of the Oldest Method of Joining,* Butterworth-Heinemann/Elsevier, Burlington, MA, 2006.

7-10 Thought Questions and Problems

7-1 Step 1 in "The Procedure for Reverse Engineering" is "Identify and articulate the purpose or motivation for disassembly or mechanical dissection and analysis."
   a. Explain why this is so important to do first.
   b. The design process is typically divided into four major stages (see Section 1–4 and Table 1–1), which include Step 1—Problem Formulation, Step 2—Concept Generation or Conceptualization Stage, Step 3—Trade-off Studies or Embodiment Stage, and Step 4—Detail Design. A fifth Step 5 is often listed as Documentation Stage. The key point being made, however, is that Step 1 in the design process is essentially the same as Step 1 in "The Procedure for Reverse Engineering." Briefly but thoroughly explain why these are the initial steps in both processes.
   c. Too often, engineering students (if not also those new to engineering practice) are so anxious to solve an assigned problem that they fail to properly formulate the problem in language that is as simple and unambiguous as possible. A student-team in a senior design capstone course in mechanical engineering at the institution where the author worked and served as an advisor brought the mother of one of the students before the faculty as their solution to the stated problem of "Come up with a way to return different-colored and -sized balls to their proper shelves after they have been left in disarray by toddlers at a day-care center." What was really wanted by the faculty—but not unambiguously stated—was a "design for an electromechanical system to . . ."

   Prepare what you feel is the appropriate Step 1 statement for each of the following situations in reverse engineering:
   (1) You are assigned the job of dealing with an initial model of one of your company's products that is failing prematurely.
   (2) You are assigned the job of finding an engineering (versus marketing) solution to sales lost to a competitor's product like yours due to lower price for the same functionality.
   (3) You are assigned the job of finding a way for your company to technically support a product of a small company acquired out of bankruptcy receivership. Unfortunately, while the product has sold and continues to sell very well, there are no engineering drawings, CAD files, manufacturing specifications, etc., available.

7-2 a. Some things are known with certainty, some things can be identified with certainty, and some things are only suggested by clues. In your own words, define and differentiate the following four different levels of confidence associated with:
   - Knowing
   - Identifying

Deducing or Inferring Role, Purpose, and Functionality during Reverse Engineering 141

- Deducing
- Inferring

b. The most subtle difference is between *deducing* and *inferring* or *deduction* and *inference.* Briefly explain the differences so that they are clear to you and others who read your definitions.

7-3 Section 7–4 deals with "using available evidence, clues, and cues" during the process of reverse engineering. The three terms are defined rather briefly, but, hopefully, unambiguously.

Choose a problem related to science or engineering from your past or work or from the Internet that needed to be solved, and reflect on the evidence, clues, and cues that were used to arrive at a solution. Briefly describe the situation and experience in terms of how *observations, experiments* or *tests,* and *measurements* were or might have been used.

7-4 To accomplish the end goal of learning about and understanding an entity being examined for the first time, short of knowing or being able to identify with certainty, one has to deduce or infer what the role, purpose, and function or functionality were likely to have been.

For any *two* of the following entities, what has been presumed, if not known with certainty, about their role, purpose, and function or functionality:

- The Ark of the Covenant
- The Great Pyramids of Giza
- The Great Pyramids of the Aztecs
- Stonehenge
- The Lines of Nazca
- "Die Glocken" ("The Bell") of Nazi Germany in World War II

Be sure to describe how role, purpose, and function or functionality were determined or presumed to be.

7-5 A significant challenge faced by engineers who engage in reverse engineering is to use what is there to gain understanding and knowledge for their intended use. The three sources that represent what is there are: (1) geometry; (2) flow of forces, energy, or fluids; and (3) major functional elements or units.

Each of these sources is important in and of itself and *all* are important when they act in concert. Each is addressed and treated separately in Sections 7–5, 7–6, and 7–7, respectively. However, the book does not discuss how they can be—and should be—considered as they act in concert.

Choose any *two* of the three sources and consider how they act in concert, i.e., how what is found in one leads to or follows from the other. Prepare a thoughtful essay of less than two pages on your argument, logic, rationale, and conclusions. (This is very useful for you as you mature as an engineer and as one desiring to be expert in all that reverse engineering has to offer!)

CHAPTER 8

The Antikythera Mechanism

8-1 The Discovery

A severe October storm caused seas to rise as the winds increased and torrential rain reduced visibility to near zero. Captain Dimitrios Kondos had decided to seek shelter on the leeward side of Point Glyphadia on the tiny Greek island of Antikythera rather than risk his boat, his crew of sponge divers, and their valuable harvest of sponges (Figure 8–1). The next morning the skies had cleared and the sea had calmed so that the sun reflected brightly off the sea's mirrorlike surface. It was a perfect day for a final dive before heading home.

Elias Stadiatos, a deeply tanned 34-year-old Greek sponge diver among Captain Kondos's small crew, broke the surface of the blue Aegean Sea wearing his canvas diving suit and copper diver's helmet. Pulling off his helmet, he babbled excitedly about finding the wreck of an ancient Greek cargo ship 60 meters beneath the surface. Almost incoherently, it seemed, he told of seeing a heap of rotting corpses of naked women and the carcasses of badly decomposed horses littering the area around the wreck, and of rich treasure. Ancient pottery, statues, jewelry, coins, wine, and fine furnishings had the young diver so exuberant that his crewmates and captain thought he was suffering hallucinations from the effects of excess carbon dioxide. Curious, Captain Kondos dove to the bottom himself, whereupon he confirmed the excited sponge diver's claim as he returned to the surface with the arm of a large bronze statue.

Gathering as many of the relics as they could, the delighted crew set off for their home port. They were most excited by what they were sure were ancient Greek art treasures, gold jewelry, and ancient gold and silver coins from the first century BC. But what would prove to be their greatest find that bright October day in 1900 were the dark greenish-blue rocklike corroded lumps of bronze that ended up being one of the most remarkable discoveries of all times. Unbeknownst to them, young Elias Stadiatos had discovered an ancient mechanical clockwork analog computer that would not be duplicated until the fourteenth century AD in medieval Europe.

But what did it do? What was it for? Only reverse engineering could solve the mystery.

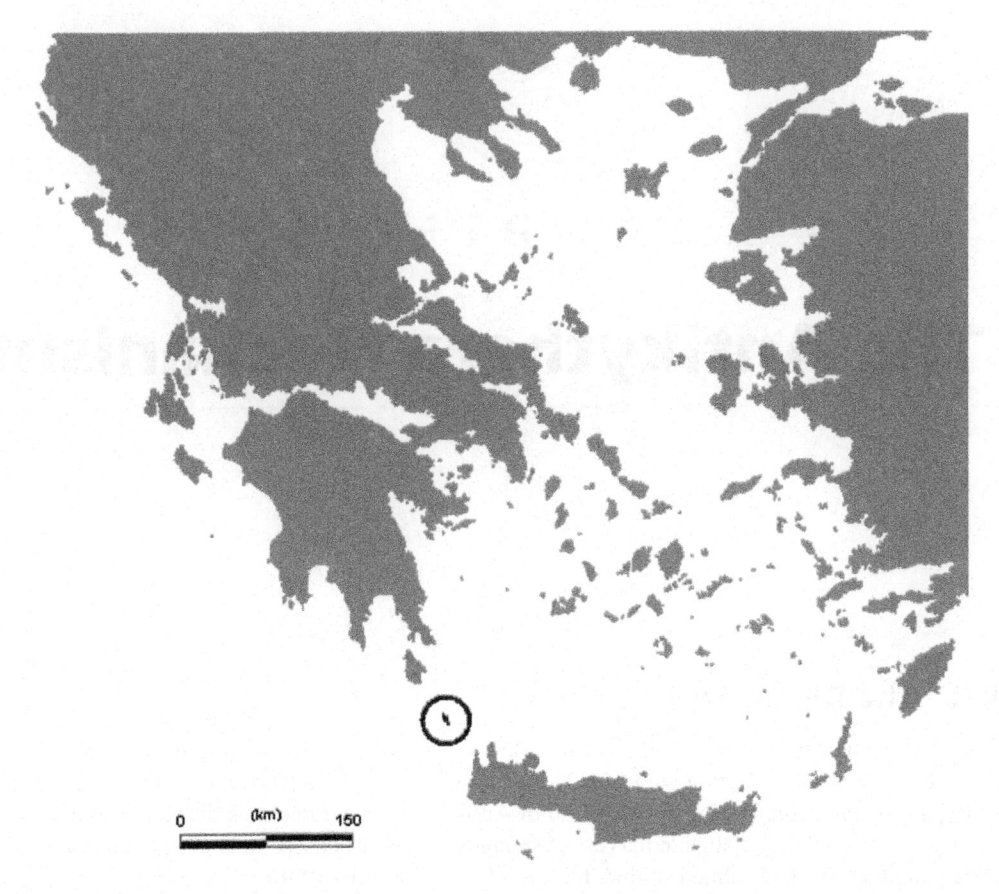

Figure 8-1 A map showing the location of the site of the discovery of the Antikythera mechanism among the wreckage of an ancient Greek ship from the first century BC off Point Glyphadia on the tiny Greek island of Antikythera in the Aegean Sea, northeast of Crete. (*Source:* Wikipedia Creative Commons, originally contributed by Lencer on 23 July 2008 and modified by Pitichinaccio on 18 December 2010.)

8-2 The Recovery

Captain Kondos and his crew of simple sponge divers had retrieved numerous ancient artifacts, including marble, porcelain, and bronze statues, decorative pottery, glassware, fine furnishings, jewelry, coins, and fragments of what was to become known as "the Antikythera mechanism," all of which were transferred to the National Museum of Archaeology in Athens for storage and analysis. The mechanism, however, went unnoticed for nearly two years, seeming to be an uninteresting collection of rocklike lumps of corroded bronze and waterlogged and rotting wood. The museum staff had far too many other far more interesting pieces to keep them occupied.

However, on May 17, 1902, former Minister of Education Spyridon Stais made a celebrated find when examining the recovered artifacts of the Antikythera wreck. He noticed that a severely corroded rocklike piece of material bore inscriptions in ancient Greek and had a gearwheel

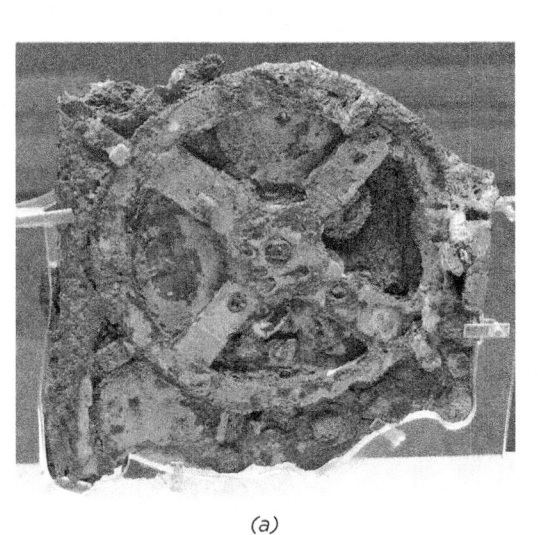

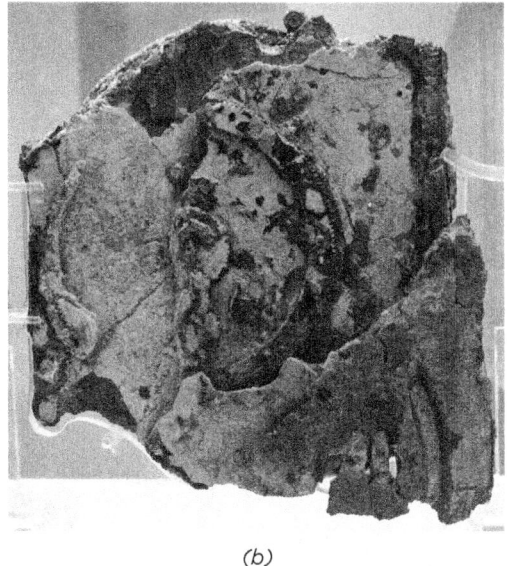

The Antikythera Mechanism **145**

(a)　　　　　　　　　　　　　　(b)

Figure 8-2 The largest fragment (Fragment A) from the Anitkythera shipwreck. The rocklike lump of severely corroded bronze contains an obvious embedded four-spoke gear, as seen from the front in (a), along with several visible smaller gears and inscriptions on the back (b). (*Source:* Wikipedia Creative Commons, both images contributed by Marsyas on 20 December 2005, with permission from the National Archaeological Museum in Athens.) **Don't miss the color version of this figure, available at www.mhprofessional.com/ReverseEngineering.**

embedded in it (Figure 8–2*a* and *b*). The object would become known as the *Antikythera mechanism*. Originally, Stais thought it to be one of the first forms of a mechanical clock known as an *astrolabe*. But it was far more complicated than others that had been found from the same time period. Study would eventually reveal it to be the oldest known analog computer, although, to be more technically correct, it was actually a mechanical calculator. His find led him to organize and lead further diving expeditions at the site with the Greek navy. While many other valuable artifacts, including a magnificent bronze bust statue, were recovered, too many decompression accidents and deaths caused dives to be ceased early in the twentieth century.

It wasn't until 1951 that Professor Derek J. de Solla Price became interested in the puzzling artifact and launched a detailed study. In 1971, Price and Greek nuclear physicist Charalampos Karakalos made x-ray and gamma-ray images of the 82 fragments that had been recovered (Figures 8–3 and 8–4). Price published an extensive 70-page paper on their findings in 1974 (ref. Price). Price's report set in motion a series of meticulous investigations using the principles and techniques of reverse engineering to unravel the mystery of the Antikythera mechanism.

The Antikythera shipwreck, as it became known, was formally dated to shortly after 85 BC by coins that were found by the great French oceanographer Jacques-Yves Cousteau (1910–1997) and a French naval officer in the 1970s. Inscriptions on the device itself indicate that it was in use for 15 to 20 years before that. The ship (itself a subject of great interest to archaeologists) also carried vases in a style found in the trading port of Rhodes, on the nearby island of Crete, around the

Figure 8-3 The 82 fragments of the *Antikythera mechanism* discovered in a first century BC wreck of an ancient Greek cargo ship. A total of 30 gears were eventually identified, including 27 in the largest fragment (Fragment A) at the upper left of the figure. Only seven pieces contained either gears or significant inscriptions in ancient Greek. (*Source:* © 2005 Antikythera Mechanism Research Project/National Archaeological Museum in Athens, with permission of Dr. Tony Freeth, Project Secretary, and the National Archaeological Museum in Athens.) **Don't miss the color version of this figure, available at www.mhprofessional.com/ReverseEngineering.**

same time (i.e., 85 BC). Suspicion arose among scholars that the device may have been related to astronomy as the great Greek astronomer Hipparchus (190–120 BC) was believed to have worked in Rhodes from 140 to 120 BC.[1] In fact, after his death, a school for astronomy was set up in his tradition. Further evidence the device may have come from Rhodes was the interest and capability artisans in that ancient city had in mechanical devices.

8-3 The Suspected Device

The remnants of the Antikythera mechanism are displayed at the National Archaeological Museum of Athens, along with a reconstruction made and donated by Professor Derek de Solla Price. Other reconstructions, by other researchers, are on display at other sites in Bozeman, Montana (U.S.A), New York City, Kassel, Germany, and Paris. All are the product of reverse engineering for deducing the role, purpose, and functionality of the device.

---

[1] Hipparchus was a Greek astronomer, geologist, and mathematician. He is considered the "Father of Trigonometry." In fact, he used his trigonometry to calculate the distance of the Sun and of the Moon from the Earth using what is known as *Hipparchus's construction* (see Wikipedia.com under "Hipparchus").

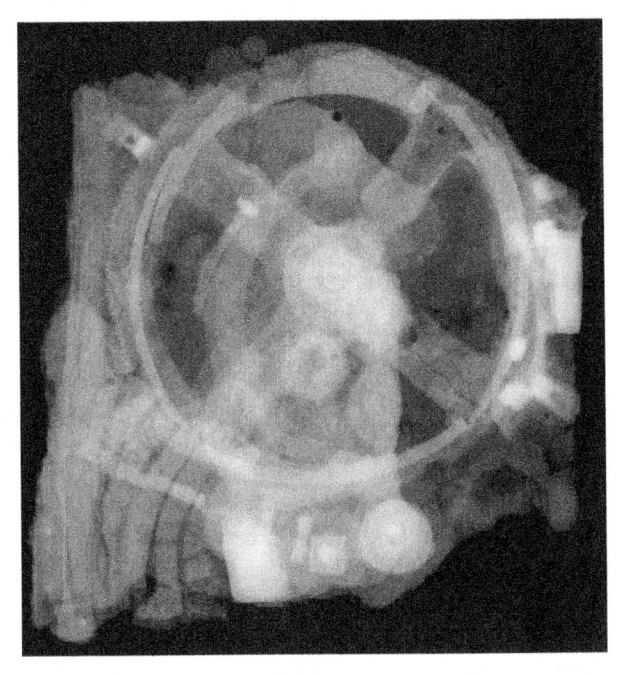

Figure 8-4 X-ray/gamma-ray images of the largest fragment (Fragment A) of the Antikythera mechanism. (*Source:* © Antikythera Mechanism Research Project, with permission from Dr. Tony Freeth, Project Secretary, and the National Archaeological Museum in Athens.)

The original mechanism was housed in a wooden box measuring approximately 340 × 180 × 90 millimeters (13½ × 7¼ × 3½ inches) in size and consisting of 30 bronze gears, although some might be missing.[2] The size of the box led scholars to unanimously conclude the device was intended to be portable, although some doubt it was intended for use on a ship, because they believe the bronze would corrode and interfere with use.[3]

The largest fragment (called "Fragment A," Figure 8–2 and upper-left corner of Figure 8–3) contains the largest gear, a four-spoke type, 140 millimeters (5⅝ inches) in diameter and having either 223 or 224 teeth. Of the 82 fragments, only 7 contain either gears or significant inscriptions.

Early in Price's work, the device was recognized as one of the oldest known complex gear mechanisms and was ultimately recognized as the first-known analog computer, although scholars suspect it may have had undiscovered predecessors, as its quality of manufacture was so high. (Recall the importance of using workmanship as a clue during reverse engineering.)

---

[2] The alloy is believed to be 95 wt.% copper/5 wt.% tin, although precise analysis has not been possible due to the severe extent of corrosion by saltwater, which is known to selectively leach out tin, thereby changing the composition.
[3] In fact, bronze is widely used by every navy in the world, and has been for centuries. While bronze alloys tarnish in salty air and saltwater, use out of the water allows long life, provided it is regularly cleaned, which might not have been practical for such a delicate mechanism.

8-4 Operation of the Mechanism

The original mechanism was operated by turning a small hand crank (either never discovered or lost after recovery) that was linked to the largest, four-spoked gear by a crown gear, setting the date on the front dial. In setting the desired date (past, present, or future), action of the crank simultaneously caused all interlocking gears within the mechanism to rotate. Different gear trains mechanically simulated astronomical time cycles that resulted in the "calculation" of the position of the Sun and the Moon, along with other astronomical information on phases of the Moon, solar and lunar eclipse cycles, and, theoretically (as the required gear trains were not found with the device, although there seems to be provisions for such gear trains; see Section 8–6) to calculate the positions of the five planets known to the ancient Greeks, Mercury, Venus, Mars, Jupiter, and Saturn. The purpose of the device was clearly to predict important astronomical events for a variety of reasons (see Section 8–8).

Key to the Antikythera mechanism is its gears and gearing.[4] The teeth of the gears were in the form of equilateral triangles with an average circular pitch of 1.6 millimeters, an average gear thickness of 1.4 millimeters, and an average gap between gear faces (not teeth!) of 1.2 millimeters. Reverse-engineering deductions suggest gears were probably fabricated (see Chapter 10) from a blank round of bronze using hand tools (dividers, scribers, saws, files, etc.), evidenced by slightly uneven sizes and spacings. All gears believed to be comprised by the device were found except a 63-toothed gear (labeled "r1" in schematics of the gear train arrangements) unaccounted for in Fragment A.

The mechanism is particularly remarkable for its degree of miniaturization, complexity of parts, and complexity and precision of arrangement. While the Antikythera mechanism contained at least 30 gears, scholars have suggested that the Greeks of the period were capable of creating systems with many more gears. In fact, some investigators of the device cite evidence of another set of gears on a complementary or peripheral device that would have attached to the front of the existing device to allow calculation of the positions of all of the planets known by the ancient Greeks—Mercury, Venus, Mars, Jupiter, and Saturn.

As important as having good, reasonably precise gears was to the mechanism, the real key to its ability to calculate astronomical events is the existence of a number of interconnected but discrete *gear schemes* or *gear trains*. Each of these, by using appropriate gear ratios within the train, allows the calculation of a particular cycle of importance in astronomy, including the following, to be described subsequently:

- The Sun gear
- The Moon train
- The Metonic train
- The Olympiad train
- The Saros train
- The Exeligmos train

Figure 8–5 is a schematic of the artifact's known mechanism, while Figure 8–6 shows the *gear chain diagram* for the known elements of the mechanism, as determined by evidence, deduction, or inference in reverse engineering the device. Hypothetical (presumed) gears in Figure 8–6 are labeled in italics.

---

[4] *Gearing* refers to the linkage of gears in a train to produce certain precise ratios of rotation to control output speed, torque, timing, and so on.

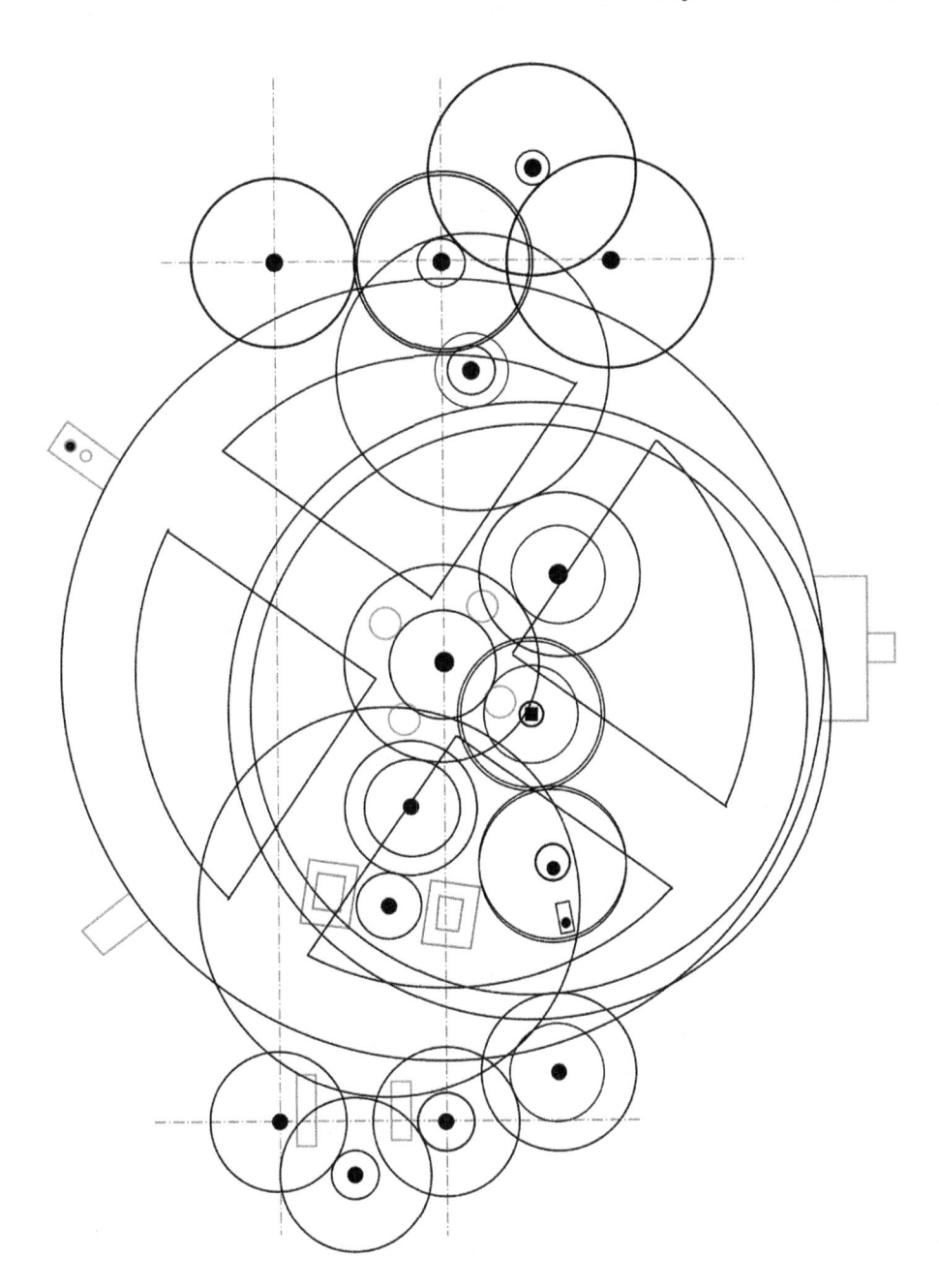

Figure 8-5 A schematic of the known interior of the Antikythera mechanism from x-ray/gamma-ray imaging and other investigations employing the technique of reverse engineering, albeit nondestructively and not using mechanical dissection. (*Source:* Wikipedia Creative Commons, originally contributed by Peryton*r on 6 February 2006 and modified by Lead Holder on 11 August 2012.)

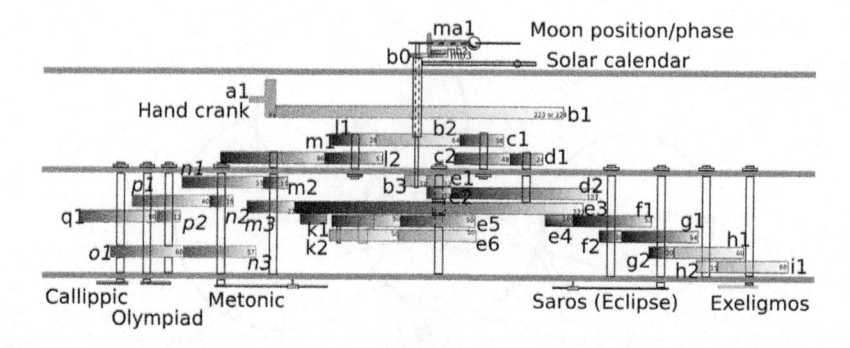

Figure 8-6 The gear train diagram from the known gears in the Antikythera mechanism, with unknown gears shown in *italics*. The location of the hand crank, solar calculator, Moon position/phase calculator, and various cycles of importance in astronomy are labeled. (*Source*: Wikipedia Creative Commons, contributed by Lead Holder on 21 July 2012.)

The Sun Gear

The *Sun Gear* is operated from a hand crank attached to gear a1. It drives the large, four-spoked mean Sun gear b1 and, in turn, drives the rest of the gear set, just as the Sun "drives" all of our solar system's planets and moons.[5] The Sun gear, itself, is b2, with 64 teeth.

The Moon Train

The Moon train is the most complex of the five gear trains that make up the mechanism. It connects to and follows the Sun train through gears b2/c1*c2/d1*d2/k2, transferring motion to the free-rotating e3 to e5. The system of e5/k1*k2/e6 (with 50 teeth each) is housed within the ring gear e4, lying on top of e3. Gears k1 and k2 rotate with it to serve as an epicyclic platform.[6] Gear e3, on the other hand, rotates at an angular velocity equal to the difference between the *sidereal months*[7] and the *anomalistic months*[8] with a ratio to the Sun of 0.1126:1. Gears e3 and e5 rotate in the same direction, with e5 rotating at the ratio of the sidereal month. Gears k1 and k2 are not coaxial but, rather, have their axes offset by about 1.1 millimeter. A pin that protrudes from k1 drives k2 via a slot. With their axes offset, k2 rotates at a varying angular velocity that depends on the position of k1's pin in k2's slot. Gear e6 is larger than e5 in order to merge with k2. The

---

[5] What is remarkable to the author is that the Antikythera mechanism seems to recognize the central role of the Sun in our solar system, even though it wasn't until Nicholas Copernicus (1473–1543), a Polish astronomer, published his book just before his death in 1543 on *heliocentrism*, i.e., the Sun as the center of the solar system, about which the planets orbit. Refusal to accept the Sun versus the Earth (in geocentric theory) as the center of the solar system was driven largely by the Roman Catholic Church until the Renaissance. After all, man was God's greatest creation, about which everything else revolved.

[6] *Epicyclic gearing* (or planet gearing) is a gear system that consists of one or more outer (i.e., planet) gears that revolve around a central (i.e., sun) gear. The planet gear(s) is(are) typically mounted on a moveable arm or carrier that may, itself, rotate around the sun gear.

[7] A *sidereal month* arises from the time it takes the Moon to return to a given position among the stars, equal to 27d7h43m11.5s (27.321661 days).

[8] An *anomalistic month* arises from the elliptical (versus circular) orbit of the Moon around the Earth that causes the apogee and perigee (i.e., apsises) to change over a repeating period of about nine years. These changes cause a lengthening of the time for the Moon to return to a particular apsis, with an average length for a month of 27d13h18m33.2s (27.554551 days).

unusual arrangement of these gears mimics the eccentricities of the Moon's orbit, as affected by the relative positions of the Earth and the Sun.

Motion then passes through to e1/b3 and through the centers of gears b2 and b1 and the Sun indicator shaft to the Moon spindle. The orbit of the Moon follows the rotational velocity of the Moon spindle in a ratio to the Sun gear of 13.368:1.

Finally, the system to indicate the Moon's phases employs further gear combinations, as gear b0 is attached to the Sun indicator shaft and, through the bevel gear train b0/mb3*mb2/ma1, the phase spindle imitates the *synodic month*[9] using a ratio to the Sun of 12.368:1.

As remarkable as this gear train is, there are others that are equally remarkable—especially considering the time frame of the device.

The Metonic Train

The *Metonic train* is driven from the Sun gear through b2/l1*l2/m1*m2/n1, with the total ratio for the train being 0.263. The Metonic cycle (proposed by Meton of Athens, an astronomer, mathematician, and geometer, in 432 BC) has a period of 19 years, which is remarkable for being nearly a common multiplier of the solar year and the lunar (or synodic) month, equal to 235 synodic months. In fact, $1/19$ of the 6940-day cycle gives a year of length $365 + 1/4 + 1/76$ days (about 365.263 days), very close to the modern length of a year (average about 365.243 days). It thus allows accurate measurement of a solar year.

The Callippic Train

The *Callippic train* follows from the *Metonic train*, using gears *n2/p1*p2/q1* with their total ratio to the Sun gear of 0.0132. The Callippic cycle (proposed by Greek astronomer and mathematician Callipus in 330 BC) has a period of 76 years, as an improvement on the Metonic cycle. It was likely used as another, more accurate measure of the length of the solar year when calculations were required over longer time spans.

The Saros Train

The *Saros train* is also driven by the Sun gear, following b2/l1*l2/m1**m3*/e3*e4/f1*f2/g1, with the total ratio for this train being 0.222. The Saros has a period of 223 synodic months, or 18 years 11 days (or 6585.3213 days). It is used to predict eclipses of the Sun and the Moon. One Saros after an eclipse, the Sun, Earth, and Moon return to the same relative geometry and a nearly identical eclipse occurs.

The Exeligmos Train

The *Exeligmos train* follows on from the *Saros train*, using gears g2/h1*h2/i1 with a total ratio to the Sun gear of 0.018. The Exeligmos cycle (which appeared around 100 BC) has a period of 54 years 33 days. It is used to predict successive eclipses that have similar locations and properties. Every exeligmos, a solar eclipse of similar characteristics will occur close to the location (in the sky) of the eclipse immediately before it.

---

[9] The *synodic month* is the average period of the Moon's revolution with respect to the Sun and Earth, and is the period of the Moon's phases. Phases occur because the Moon's appearance depends on the period of the Moon with respect to the Sun as seen from the Earth.

The Olympiad Train

The *Olympiad train* follows on from the Metonic train via gears *n3/o1*, which produce a total ratio to the Sun gear of 0.25. The Olympiad cycle of four solar years is associated with the Olympic Games of the Ancient Greeks. During the Hellenistic period,[10] beginning with Epherus (a Greek city), Olympiads were used as a calendar epoch. From the Olympiad cycle, the first Olympiad lasted from the summer of 776 BC to the summer of 772 BC.

All in all, the capabilities of the Antikythera mechanism as an astronomical calculator are staggering. The role, purpose, and functionality of the device were unraveled by not one but several reverse-engineering efforts during scholarly investigations.

8-5 Reverse-Engineering Investigations and Reconstructed Models

The dictionary defines an *enigma* as "one [or something] that is puzzling, ambiguous, or inexplicable." The Antikythera mechanism is an enigma, as for more than half a century from its discovery and recovery from the Aegean Sea in 1900, it was puzzling as to its role, purpose, and functionality.[11] When the puzzle began to be assembled (actually, reassembled) following several investigations (that often built on earlier findings), it was still inexplicable how such a remarkably sophisticated device could have been created, no less how it could have been conceived, 2000 years ago. What it was not, by any means, was ambiguous. It had a very specific purpose that it performed very well. It was one of the world's oldest known gear devices and its first analog astronomical computer or, more technically correct, mechanical calculator. It was not to be rivaled in complexity, capability, or precision for another 1500 years, if then.

With recognition on May 17, 1902, by former minister of education Spyridon Stais that an actual device of historical and scientific interest was encrusted within the rocklike lumps of corroded bronze retrieved from the seafloor 60 meters down and stored at the National Museum of Archaeology in Athens, decades of work were spent meticulously cleaning the device. This essential, tedious, and anonymous work enabled a series of investigators to advance the knowledge and understanding of the mechanism, including investigations by:

- Derek J. de Solla Price (1951– ca. 1974)
- Allan George Bromley (ca. 1984–ca. 2002)
- Michael Wright (ca. 1990–2006)
- The Antikythera Mechanism Research Project (2006–present)

Each of these investigations is described and discussed briefly in the following subsections. Table 8–1 lists the major fragments (and one minor fragment) of the Antikythera mechanism that were available for study.

Figure 8–7 shows computer-generated depictions of the front and back panels of the

---

[10] The *Hellenistic period*, or *Hellenistic civilization*, is the period of ancient Greek history from the death of Macedonian King Alexander the Great in 332 BC and the emergence of the ancient Roman Empire. Either of two dates is commonly considered: 146 BC, with the final conquest of the Greek heartland by Rome, or 31 BC, with the final defeat of the Ptolemaic Kingdom.

[11] Recall from Section 7–4, the following terms: *role* ⇒ the part something actually plays or played; *purpose* ⇒ the reason for which something is or was done or is or was created; *functionality* ⇒ the way (or how) something works or worked.

TABLE 8-1 List of the Recovered Fragments of the Antikythera Mechanism (7 major and 1 minor among 82 total)

Fragment	Size (mm)	Weight (g)	Gears	Inscriptions
A	180 × 150	369.1	27	Yes
B	125 × 60	99.4	1	Yes
C	120 × 110	63.8	1	Yes
D	45 × 35	15.0	1	—
E	60 × 35	22.1	—	Yes
F	90 × 80	86.2	—	Yes
G	125 × 110	31.7	—	Yes

Fragment A: The main fragment, containing the majority of the mechanism, most notably (in Figure 8-2a) is the huge four-spoked b1 gear. Behind b1 are parts of the l, m, c, and d trains visible as gears to the unaided eye. On the back (in Figure 8-2b) can be seen the rearmost e and k gears, including the pin-and-slot mechanism in the k gear train. Also contained are divisions of the upper-left quarter of the Saros spiral, with inscriptions; inscriptions for the Exeligmos dial, with remnants of the dial face on the back surface; and some backdoor inscriptions.

Fragment B: Contains approximately the bottom third of the Metonic spiral and inscriptions on both the spiral and the mechanism's back door. Also contained are 49 deciphered cells (of 235) for the Metonic scale and a single gear (o1) used in the Olympiad train.

Fragment C: Contains part of the upper right of the front dial face showing calendar and zodiac inscriptions. Also contained are the Moon indicator dial assembly, including the Moon phase sphere in its housing and a single bevel gear (ma1) used in the Moon phase indicator system.

Fragment D: Contains at least one unknown gear (M. T. Wright believes it contains two), the purpose of which has (have) not been deduced although it (they) lend to the argument for possible planetary displays on the face of the mechanism.

Fragment E: Contains six inscriptions from the uppermost of the Saros spiral (found in 1976).

Fragment F: Contains 16 inscriptions from the lower left of the Saros spiral, as well as remnants of the mechanism's wooden housing (found in 2005).

Fragment G: Is actually a combination of fragments taken from Fragment C during cleaning.

Fragment 19: A minor fragment containing significant backdoor inscriptions, including reference to "... 76 years ..." associated with the Callippic cycle.

Antikythera mechanism as it is believed to have originally appeared. Inscriptions on these panels were obtained (at least portions were obtained) from various recovered fragments (Table 8–1).

So let's look at the four investigations.

Derek J. de Solla Price

Derek J. de Solla Price (1922–1983), a British physicist, historian of science, and information scientist known as the "Father of Scientometrics" (Figure 8–8), was the first to undertake

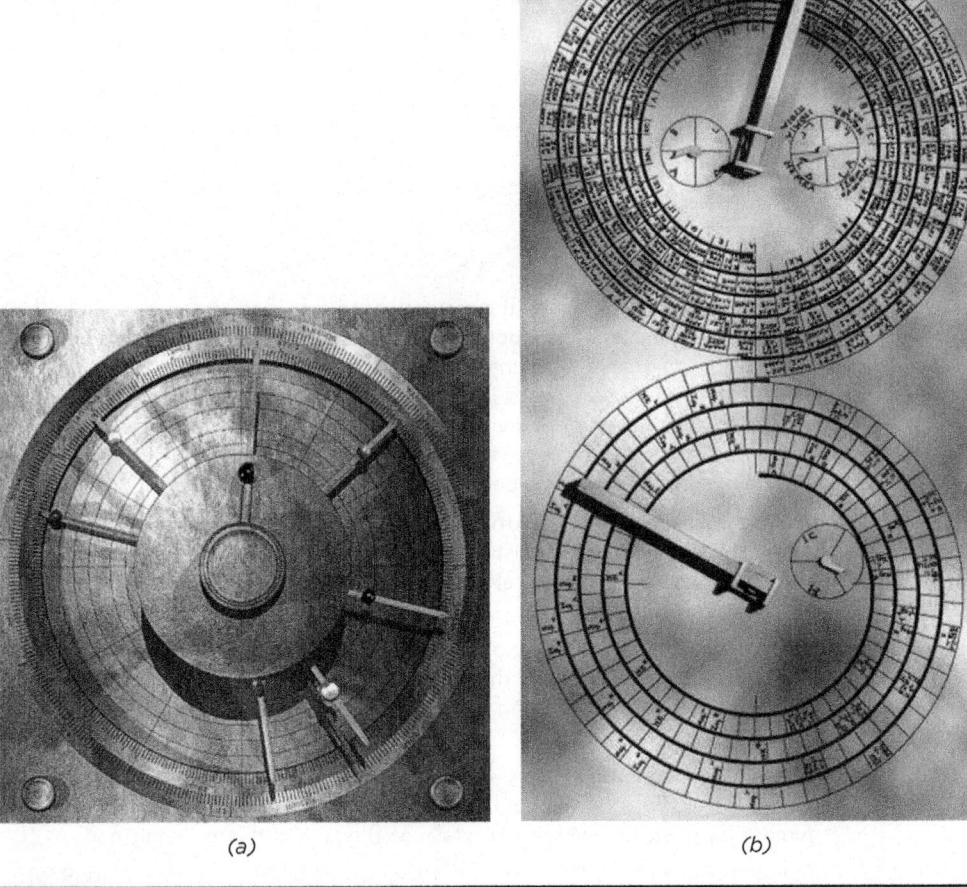

Figure 8-7 Computer-generated renderings of what the front (*a*) and rear (*b*) doors of the Antikythera mechanism probably looked like. (*Source:* Wikipedia Creative Commons, contributed by Skier Dude on 14 August 2012 [*a*], and by Americanplus on 15 July 2012 [*b*].)

systematic investigation of the mechanism in 1951. Unlike most efforts in reverse engineering (which is precisely what Professor Price was doing, i.e., working backward to identify, deduce, or infer the role, purpose, and functionality of an item), no mechanical dissection or teardown could take place. The relic, after all, was unique, irreplaceable, and would turn out to be priceless. Instead, Price used all the visual observations of evidence and clues he could.

By 1959, he advanced a theory that the Antikythera mechanism was a device for calculating "the motions of stars and planets," making it the first known analog computer.[12] In 1971, while

---

[12] Before Price's theory, the device's full purpose and functionality were unknown, although it had been correctly identified by Stais to be an astronomical device from inscriptions and was presumed to be an astronomical clock or astrolabe from its visible mechanism.

Figure 8-8 Photograph of Derek J. de Solla Price (1922–1983) and his early (1974) model of the Antikythera mechanism. (*Source:* Wikipedia Creative Commons, contributed by Richardfabi on 27 July 2007.)

serving as the first Avalon Professor of the History of Science at Yale University, Price teamed up with Charalampos Karakalos, a professor of nuclear physics at the Greek National Center of Scientific Research, to take the first x-ray and gamma-ray radiographs of the mechanism within the mineral-encrusted fragments and reveal critical information about the device's heretofore unknown interior configuration (see Figure 8–4).

In 1974, Price published *Gears from the Greeks: The Antikythera Mechanism—A Calculating Computer from ca. 80 BC,* and he presented to the archaeological, historical, and scientific community the first model of how the mechanism could have functioned (ref. Price). One of his more remarkable proposals was that the mechanism employed differential gears to enable the machine to add or subtract angular velocities and, thereby, compute the synodic lunar cycle by subtracting the effects of the Sun's movement from those of the sidereal lunar movement.

From the standpoint of reverse engineering, Price used his expertise as a physicist to decipher the motions by trains of gears and his expertise and experience in the history of science and information science[13] to begin to study the Antikythera mechanism by logical deduction (using visible gears and inscriptions) and inference (using apparently missing gears). The added expertise of Karakalos in nuclear physics and radiographic imaging allowed new observations to be made that further aided both identification and deduction. The enigma was beginning to unravel.

---

[13] *Information science* (or *information studies*) is an interdisciplinary field concerned with the collection, classification, analysis, manipulation, storage, retrieval, and dissemination of information.

Allan George Bromley

Allan George Bromley (1947–2002) was a computer scientist and professor at the University of Sydney (in Australia), as well as a historian of computing and one of the most avid private collectors of mechanical calculators. Working with retired Sydney engineer and clockmaker Frank Percival, the two picked up the mystery of the Antikythera mechanism where Price left off. In testing Price's theory of how the device worked by building a model of the main gear train using similar gears, they found the mechanism unworkable. Other advances with the Bromley-Percival solution (Figure 8–9) were:

- Improved mechanics of the device by altering the crank so that one complete turn represented one day (the most obvious of all astronomical phenomena)
- Conjecture that there were other missing gears where a gap appeared where there should have been a gear[14]
- Major discovery that one whole gear train that puzzled Professor Price, who assumed the 15 and 63 teeth counted by an assistant on mating gears had to be in error and should have been 16 and 64 to give a precise ratio of 4:1, was actually correct, and that the ratio of 4½:1 would yield an 18-year Saros cycle for predicting eclipses.

Bromley went on to make new, more accurate radiographic images in collaboration with Michael Wright before the former's early death.

From the standpoint of reverse engineering, Bromley and Percival were a remarkable team—the former an expert on computers and, particularly, mechanical calculators; the latter an expert in clockworks. Bromley himself said his "... collaboration with a mechanical-minded person" provided an advantage that Professor Price did not have. Again, the value of relevant experience shined through (see Section 7–3). Of course, there's also the benefit of playing off another's observations, ideas, theories, and so forth (in this case, those by Price), to test their validity and, if necessary, make refinements or take totally new directions. Finally, there was the use of experimentation and measurement via their model that added to earlier observations and provided new clues and cues for new actions or a new theory.

Michael Wright

Michael T. Wright (1948–), MA, MSc, FSA, was formerly the Curator of Mechanical Engineering at the London Science Museum and later joined the Imperial College in London, where he is Honorary Research Assistant at the Centre for the History of Science, Technology, and Medicine. Working with Allan George Bromley, he designed and built an apparatus for lineal tomography to allow the generation of 2D radiographic sections or slices to be made of the gear-containing fragments of the Antikythera mechanism. Resulting images showed that Professor Price's reconstruction was fundamentally flawed; however, they also supported Price's suggestion (in *Gears from the Greeks . . .*) that the mechanism could have served as a planetarium. Wright proposed that the Sun and the Moon could have moved in accordance with Hipparchus's theories and the two *inferior planets* (i.e., Mercury and Venus) and three *superior planets* (i.e., Mars, Jupiter, and Saturn) moved according to the simple epicyclic theory suggested by a theorem by another Greek of the period. To prove his suggestions were possible using the level of technology

---

[14]*Conjecture* (a verb) is defined as "formation of a conclusion from incomplete evidence, involving inference."

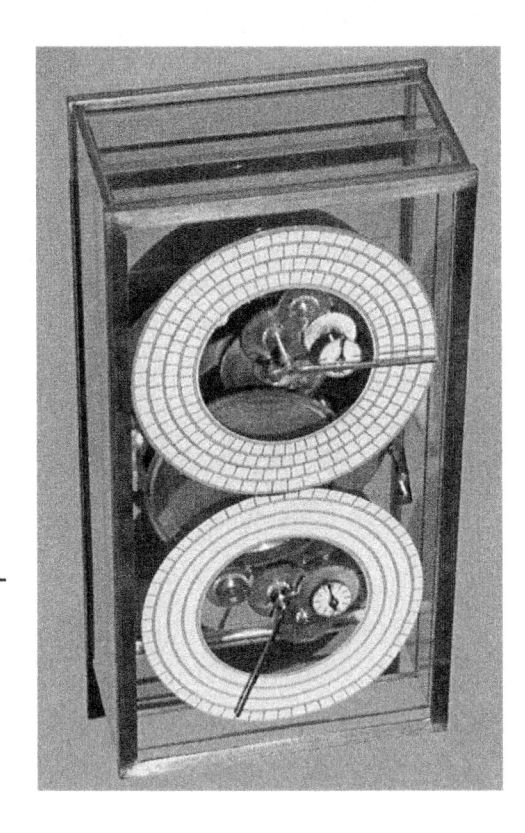

Figure 8-9 An elegant working model of the Antikythera mechanism built by Allan George Bromley and Frank Percival, in what is known as the Bromley-Percival solution. (*Source:* Wikipedia Creative Commons, contributed by Ezrdr on 17 September 2012.)

apparent in the Antikythera mechanism, Wright produced a working model of such a planetary device (ref. Wright).

Other advances by Wright included:

- Increase in Price's gear count from 27 to 31, including one in Fragment C (Table 8–1) eventually identified as part of the Moon phase display. (The technique that Wright suggested the inventor of the mechanism used preceded other known mechanisms for calculating Moon phases by 1500 years!)
- More accurate counts of teeth to allow a new gearing scheme to be advanced that included the 18-year Metonic cycle and the 76-year Callippic cycle.
- What is believed to be an almost exact replica of the Antikythera mechanism (Figure 8–10).

Somewhat arguable is Wright's contention that Price was in error when he suggested that the creator of the device used differential gears. Wright himself frequently speaks (in his paper) of the addition and subtraction of angular velocities by gears, which is precisely what differential gears do.

In 2006, Michael Wright joined the Antikythera Mechanism Research Project as a consultant.

From the standpoint of reverse engineering, Michael Wright offered exceptional expertise and experience, as a curator of mechanical engineering and on his website, a self-proclaimed "mechanician." His creation of physical models and a replica enabled experimentation to aid in

158 Chapter Eight

measurement and observation (Sections 4–3 and 4–4). His passion for antiquities was also an asset.

The Antikythera Mechanism Research Project

This project is a joint investigation between Cardiff University (Professor Michael Edmunds and Professor Tony Freeth), the National and Kapodistrian University at Athens (X. Moussas and Y. Bitsakis), the Aristotle University of Thessaloniki (J. H. Seiradakis), the National Archaeological Museum at Athens, X-Tek Systems (U.K.), and Hewlett-Packard (U.S.A.), funded by the Leverbalme Trust and supported by the Cultural Foundation of the National Bank of

Figure 8-10 Reconstruction of what most experts consider an exact replica of the Antikythera mechanism built by Michael T. Wright. (*Source:* Wikipedia Creative Commons, contributed by Marsyas on 20 December 2005.)

Greece. The effort is ongoing, with much of the measurement and experimentation to the device being restricted to the museum, due to the fragility of the artifact.

It was announced in October 2005 that new pieces of the Antikythera mechanism had been found (included among the 82 reported herein). A new imaging system has allowed many more of the Greek inscriptions to be viewed and translated, increasing the number of recognizable characters from about 1000 to 2160.

Additional papers relating to the Antikythera mechanism have appeared in the prestigious archival publication *Nature* in 2006 to date.

An appropriate ending to this section on investigations of this remarkable artifact is a partial November 30, 2006, quote from Professor Michael Edmunds, who led the 2006 study by the Antikythera Mechanism Research Project:

> This device is just extraordinary, the only one of its kind. The thing is beautiful, the astronomy is exactly right. The way the mechanisms are designed just makes your jaw drop. Whoever has done this has done it extremely carefully ... in terms of history and scientific value, I have to regard this mechanism as being more valuable than the *Mona Lisa.*

Surely, Leonardo da Vinci would agree!

In tribute to the Antikythera mechanism as perhaps the epitome of the application of reverse engineering. Look at the subtle image in the background, behind the cog, on the front cover of this book. Do you see the significance to the topic of the book?

8-6 Proposed Planet Indicator Schemes

The large space (gap) between the mean four-spoke Sun gear and the front of the case of the Antikythera device, along with the size of and mechanical features on the mean Sun gear, suggest to some scholar investigators that the mechanism contained additional gearing that was either lost in or subsequent to the shipwreck or was removed prior to the device being loaded onto the ship. The lack of solid *evidence,* and the nature of the front part of the mechanism, has led to several attempts to emulate (by deduction from inference of the existence of such a complementary mechanism) what the Greeks of the period might have done. Theories put forth for the existence of a planetary calculator or planetarium to show the positions of Mercury, Venus, Mars, Jupiter, and Saturn include those by:

- Michael T. Wright (2006–)
- Evans, Carmen, and Thorndike (2012)
- Freeth and Jones (2012)

Readers interested in this fascinating possibility are encouraged to seek additional information using a Google search by the investigator's name, followed by "Antikythera mechanism."

8-7 Similar Devices, Possible Predecessors, and the Possible Creator

Like ancient civilizations that preceded and succeeded them (e.g., the Ancient Egyptians, 3150–332 BC; the Babylonians, 1894–333 BC; and the Mayans, ca. 2000 BC–250 AD), the ancient

Greeks (in the Hellenistic period, see Footnote 10) not only were fascinated by astronomy but relied greatly on its cyclic events. Freedom from pollution in the atmosphere and by artificial light caused nights to be totally dark, making the always-intriguing twinkling heavenly bodies even more apparent than they are today (when we even take the time to look!). Phases of the Moon, positions of the stars and planets, eclipses involving the Sun, the Earth, and the Moon, and the appearance of comets[15] were all watched. Over centuries of fascinated observation, certain people learned about the periodicity and period of these events, often with astounding accuracy (even by modern standards). Astronomers and astrologers were revered—and often held in awe—for their ability to predict heavenly events, not just to know and tabulate tides, time the planting and harvesting of crops, and navigate long distances, but to foretell great events (e.g., the birth of Jesus of Nazareth, the Christ, for Christians).

Reference to machines and mechanisms that some modern scholars consider either astronomical calculators and/or some kind of planetaria to model and predict the movement of the Sun, the Moon, and the five planets known at the time began with Marcus Tullius Cicero (106–43 BC), or simply Cicero, a Roman philosopher and orator, in *De re publica,* a first century BC philosophical fictional dialogue taking place in 129 BC, in which he mentions two such machines. Not surprisingly, both were built by Archimedes (ca. 287–ca. 212 BC).

Archimedes was a brilliant Greek mathematician, physicist, engineer, inventor, and astronomer, whose work during his life was used but seldom credited to him. After mentioning the two devices, Cicero tells how the Roman general Marcus Claudius Marcellus brought them back to Rome following Archimedes's death in 212 BC at the siege of Syracuse. He tells of how the machines were constructed by Archimedes for use as aids in astronomy and were able to show the motion of the Sun, Moon, and five planets. The dialogue relates that Marcellus kept one of the devices as his only personal loot from Syracuse and donated the other to the Temple of Virtue in Rome. According to Cicero, Marcellus's mechanism was demonstrated to Lucius Furius Philus by Gaius Sulpicius Gallus. Cicero's dialogue states:

> I had often heard this celestial globe or sphere mentioned on account of the great fame of Archimedes. Its appearance, however, did not seem to me particularly striking. There is another, more elegant in form, and more generally known, molded by the same Archimedes, and deposited by the same Marcellus, in the Temple of Virtue at Rome. But as soon as Gallus had begun to explain, by his sublime science, the composition of this machine, I felt that the Sicilian geometrician must have possessed a genius superior to any thing we usually conceive to belong to our nature. Gallus assured us, that the solid and compact globe, was a very ancient invention, and that the first model of it had been presented by Thales of Miletus (ca. 624–ca. 546 BC). That afterwards Eudoxus of Cnidus (410 or 408–355 or 347 BC), a disciple of Plato (427–347 BC), had traced on its surface the stars that appear in the sky, and that many years subsequent, borrowing from Eudoxus this beautiful design and representation, Aratus had illustrated them in his verses, not by any science of astronomy, but the ornament of poetic description. He added, that the figure of the sphere, which displayed the motions of the Sun and Moon, and the five planets, or wandering stars, could not be represented by the primitive solid globe. And that

---

[15] The name *comet* comes from the Greek for "long-haired star," first used by Aristotle when he saw heavenly bodies he described as "stars with hair."

in this, the invention of Archimedes was admirable, because he had calculated how a single revolution should maintain unequal and diversified progressions in dissimilar motions.

When Gallus moved this globe it showed the relationship of the Moon with the Sun, and there were exactly the same number of turns on the bronze device as the number of days in the real globe of the sky. Thus it showed the same eclipse of the Sun as in the globe [of the sky], as well as showing the Moon entering the area of the Earth's shadow when the Sun is in line . . . [missing text].
[i.e., it showed both solar and lunar eclipses].

Much about the Antikythera mechanism evokes images of the mechanical genius of Archimedes, but whether the device spoken of in Cicero's fictitious dialogue is the artifact recovered from the wreck off Antikythera is uncertain. What is very likely, however, is that the Antikythera mechanism was created based on earlier work, since, as stated earlier, its complexity, accuracy, and elegance were simply too great for it to be a prototype.

8-8 Speculation on Role, Purpose, and Functionality

The deduced purpose (i.e., intended use) of the Antikythera mechanism, supported by considerable hard evidence (e.g., instruction-like inscriptions), is almost certainly to allow calculation of key astronomical events, including positions of the Sun and the Moon, phases of the Moon, the occurrence of lunar and solar eclipses, and measurement of time (e.g., days, months, and years). Prediction of lunar and solar eclipses was clearly based on Babylonian arithmetic-progression cycles. Inscriptions on two doors suggest the device was also intended to display planetary positions, even though no remnants of any additional (or peripheral) mechanism have been found.

The functionality (i.e., actual operation), based on presumed, deduced, and/or inferred purpose, is quite well identified and understood from several investigations that reverse engineered the mechanism. In fact, a reconstruction by Michael Wright is believed to be an exact replica of the original.

The role (i.e., actual use) of the Antikythera mechanism is the most obscure. Professor Price suggested it might have been for public display, perhaps as part of the extensive display of mechanical engineering devices at Rhodes (where machines and mechanisms were a specialty). However, the small size of the device, and especially the front and back dials, seems to discount this possibility, as the inscriptions would not be readable by observers. Furthermore, 2160 characters on two doorplates resemble detailed instruction manuals for users, not for observers.

The role was probably not for maritime use, however, for two reasons. First, some data (e.g., eclipse prediction) are not needed for navigation and would unnecessarily complicate a device intended for navigation. Second, damp, salt-laden sea air would seemingly quickly render delicate bronze gears immoveable and the mechanism useless.

A very readable account of the Antikythera mechanism is contained in a book by Jo Marchant (ref. Marchant), while a novel involving the Antikythera mechanism conjectures that there were sister devices (ref. Barbosa).

Whatever the role, the Antikythera mechanism is one of the most remarkable scientific and technological creations of all times!

8-9 Summary

Sponge diver Elias Stadiatos discovered and recovered, among obvious valuable ancient Greek artifacts, a rocklike corroded lump of bronze and rotten wood that once was an astoundingly accurate mechanical calculator of astronomical events, including position of the Sun and Moon in the sky, phases of the Moon, eclipses involving the Sun, Earth, and Moon, and, perhaps even the five known planets of the period. But unraveling the puzzle took more than 60 years using the principles, although not the use of mechanical dissection, of reverse engineering.

Among 82 recovered fragments of the Antikythera mechanism (named for the tiny Greek island off which it was discovered), 8 contain the majority of the evidence and/or clues to the object's role, purpose, and functionality. The largest (Fragment A) contains 27 of 31 gears that operate in five different trains to calculate five astronomical cycles of key importance, as well as a large, four-spoked Sun gear that drives the entire mechanism just as our Sun "drives" our own solar system.

The progressive investigations of Price, Bromley and Percival, Wright and Bromley, Wright, and the Antikythera Mechanism Research Project founded in 2006 are described and discussed for their relevance to reverse engineering. Evidence and/or clues for the existence of a planetary calculator complementary or peripheral yet-undiscovered mechanism are presented, along with discussion of predecessors and the possible creator of this remarkable historical, scientific, and technological marvel.

8-10 Cited References

Barbosa, Charles, *The Shipwreck at Antikythera,* CreateSpace Independent Publishing Platform, 2010.

Marchant, Jo, *Decoding the Heavens: 2000-Year-Old Computer—and the Century-Long Search to Discover Its Secrets,* DaCapo Press, Cambridge, MA, 2010.

Price, Derek de Solla, *Gears from the Greeks: The Antikythera Mechanism, a Calendar Computer from Ca. 80 BC,* American Philosophical Society, Philadelphia, PA, 1974.

Wright, Michael T., "Antikythera Mechanism Working Model.mov," YouTube.

Wright, Michael T., and M. Vicentini, "Virtual Reconstruction of the Antikythera Mechanism," YouTube.

8-11 Thought Questions and Problems

8-1 The *Antikythera mechanism* was discovered and recovered in a routine dive by sponge divers, in October 1900, 60 meters beneath the surface of the Aegean Sea. On May 17, 1902, Minister of Education Spyridon Stais, going through the recovered artifacts stored at the National Museum of Archaeology at Athens, noticed a gear wheel embedded in a corrosion-encrusted lump of bronze and immediately thought it to be an *astrolabe*.
  a. What is an astrolabe? Provide a brief but complete explanation and description of such a device.
  b. Use the Internet to find examples of astrolabes (other than the Antikythera mechanism)

from ancient times. Explain and describe how these are different from the Antikythera mechanism.

8-2 In 1951, Professor Derek J. de Solla Price became interested in the puzzling artifact at the National Museum of Archaeology at Athens and immediately engaged in a detailed study. Along with his seminal 70-page paper with Greek nuclear physicist Charalampos Karakalos, *Gears from the Greeks: The Antikythera Mechanism, a Calendar Computer from Ca. 80 BC*, Price built a reconstruction of what he was convinced was "the first analog computer."
   a. Explain what is meant by an "analog computer."
   b. Explain how an analog computer is different from a digital computer.
   c. Many analog computers in more modern times (i.e., the mid-1900s) were electronic devices. Explain how they performed mathematical calculations involving each of addition or subtraction, multiplication or division, and differentiation or integration.
   d. Explain how a mechanical analog computer works.

8-3 Michael T. Wright, an accomplished and renowned mechanical engineer, was convinced there was a significant portion of the Antikythera artifact that was never found in the ancient shipwreck in the Aegean Sea. He believed this missing portion contained additional gears that allowed the device to also calculate the positions of the five planets known to the ancient Greeks: Mercury, Venus, Mars, Jupiter, and Saturn.

Describe what led Wright to this conclusion, and describe what he created as a supposed "re-creation" of the complete device.

8-4 The brilliant ancient Greek mathematician, physicist, engineer, inventor, and astronomer Archimedes (ca. 287–212 BC) has been suggested to be the creator of the *Antikythera mechanism*. He certainly had the right mix of skills!

Write a cogent essay of less than two pages that supports or refutes (as you choose) this theory. Be sure to defend whichever position you take with logical argument.

8-5 There have been many other artifacts created by ancient or old civilizations that have baffled modern scientists and engineers, beyond just archaeologists and historians. In many cases, archaeologists have either totally dismissed or underestimated the significance of what they often discovered, leaving it to other scientists and/or engineers to suggest something significant. Reverse engineering was often the tool used to attempt to decipher the mystery.

Four examples are:
- The "Baghdad Battery"
- The Rosetta Stone
- The Dresden Codex or Codex Dredensis (a Mayan codex)
- Runestones (e.g., the Kensington Runestone in Minnesota and the Heavener Stone in Oklahoma)

Choose *one* of these four and write an essay of less than two pages on how reverse engineering helped unravel the mystery of the artifact(s).

CHAPTER 9

Identifying Materials-of-Construction

9-1 The Role of Materials in Engineering

With the exception of software engineering, engineers design, manufacture, test, operate, and maintain hardware, and all hardware is composed of material(s). As described and discussed in Section 9–4, there are three fundamental types or classes of materials: metals, ceramics, and polymers. Within and between (i.e., at the intersections of) these three there are a number and variety of subtypes, including inorganic glasses, semiconductors, and various natural and synthetic composites. Since our physical world is made up of materials and since engineers are charged with manipulating our physical world to make our lives safer and better, it is essential for every engineer, regardless of discipline, to understand the basics or essence of materials (ref. Messler).

In creating a new or improved design, an engineer needs to consider four aspects of the design: functionality; manufacturability; aesthetics, and cost. Despite what some marketing people will say, the first and foremost of these is *always* functionality—or should be. A design must work as it was intended to work. If it doesn't, what good is it that it was easy to manufacture, looked good, or was inexpensive? This is not to say these other aspects or factors are not important. They are. But functionality is essential!

As used here, at least, *functionality* means what something is capable of doing relative to the intended function of the design or by the designer.

In almost every case, the second most important factor (behind functionality) is manufacturability. Again, what good would a great idea be in terms of designed-in functionality if that idea could not be built?[1] *Manufacturability* includes ease of fabrication of detail parts, components, or structural elements, and ease of assembly of devices, products, systems, or structures. The relative ease of either fabrication or assembly can be measured by the labor

---

[1] *Built* includes manufactured for designs executed indoors (e.g., airplanes, automobiles, and smartphones) and constructed for designs executed, of necessity, outdoors (e.g., bridges, buildings, ships).

intensity time/cost required, skill level required for either fabrication or assembly (including the use of sophisticated and expensive automation), throughput (as units per time period), and yield (as fraction or percentage of acceptable units out of the total number of units produced).

The relative importance of *cost* or *aesthetics* is less straightforward, as different situations either call for or dictate different priority or value of one or the other of these factors. Some would say either cost or aesthetics could be the most important factor during design, but this really isn't true. Cost or aesthetics (for visual appeal) could be a major constraint on or even a driver for a design, but neither is more important than functionality and manufacturability.

There are, without question, situations where *cost* is a major consideration—for example, for competitiveness or acceptability in the marketplace. But achieving a limiting cost (e.g., fixed by market research) simply constrains design options for achieving functionality and, secondarily, may force some compromises to be made. Likewise, there are situations where *aesthetics* is a major consideration. Examples might include, to name but two, fine Danish modern-style furniture or a Rolls-Royce. But ask either the furniture designer or Rolls-Royce and they will say: "Our product is aesthetically beautiful, but, more important, it is highly functional." When Rolls-Royce advertised that "the loudest sound you hear when you ride in a Rolls-Royce is the clock," they didn't mean because the engine doesn't work!

As one conducts a product teardown or engages in an effort to reverse engineer an existing design, there is much to be learned from the material(s) used to create the entity (i.e., the material-of-construction). In many instances, the material provides major clues as to the intended functionality of the part, component, structural element, device, product, structure, or system. This will become more apparent in the rest of this chapter, especially Sections 9–2 and 9–3.

9–2 The Structure-Property-Processing-Performance Interrelationship

Convention seems to hold that there are three fundamental sciences—physics, chemistry, and biology—these three being quantifiable with mathematics.[2] Normally taught in secondary school in the sequence biology, chemistry, physics, it has been argued (by a group of Nobel prize recipients) that the order should be reversed: physics, chemistry, biology. The reason is that physics underlies chemistry, and both chemistry and physics underlie biology, the most complex of the three. In fact, the author contends there is only one science, that being physics. The other two are just applications of physics.

Here's the author's point:

Physics involves the interaction and/or relationships between matter, force, and energy in or through space and time.[3] Chemistry involves the interaction (as reactions) between matter and

---

[2] In at least the author's humble opinion, mathematics is purely a creation, an invention, of Man and not of Nature! It is a system of counting, accounting, and manipulating symbolic (abstract) representations of real things and, so, proves useful for quantifying the other three sciences. The fact that certain things in Nature (e.g., physics, chemistry, or biology) seem to follow a particular mathematical function is not because Nature created those things using mathematics and that equation, but because the equation was invented to fit certain types of behaviors. Two apples are a reality, without mathematics needing to give them a number, that is, 2. And the fact that $1 + 1 = 2$ reflects the reality of one apple next to another with an abstract, albeit convenient, convention.

[3] Force is the derivative of energy, so it is related to energy, and energy and matter are related by Einstein's $E = mc^2$. What else is there?

energy, which makes it physics. Biology involves the physics and chemistry of living things, with chemistry (including organic chemistry) being physics. So biology is also physics.

Well, enough of this! What about *materials science*?

As opposed to being a new science, *materials science* is simply science applied to materials, with materials being those elements, compounds, mixtures, and substances that constitute the physical components of the universe. The underlying principle of materials science is: *Structure determines properties.* The way a material is put together (or built up) directly affects (or determines) how it will respond to an external stimulus, the latter being the definition of a property (see Section 9–3).

Without being consciously aware of it, most people know this principle to be true. Everyone knows—or learns from experience—that wood is stronger and tolerates bending better when loaded along (or parallel to) the direction of its grain than across (or perpendicular to) its grain.[4] This shows up in a wood baseball bat. Another simple demonstration to make the point is this: Remove a complete sheet (i.e., a double page) from a newspaper and carefully attempt to tear 2-inch (5-centimeter) -wide strips from the top to the bottom of the sheet. You should be successful at tearing strip after strip, each with parallel long edges. Now attempt to do the same thing with strips torn horizontally across the sheet from left to right or right to left. Did you succeed? This usually proves quite difficult, with a ragged tear often resulting, since the low-grade paper used for newspaper stock is composed of coarse fibers that align in one direction (during the process in which they are laid down and rolled), top to bottom, of the newspaper page as it is cut from a very long roll. The structure (fiber orientation) determines properties (tear resistance).

The principle that structure determines properties applies across all solid materials, irrespective of type (i.e., metal, ceramic, or polymer) and at all scales, from the atomic level ($\sim 10^{-10}$ meters) through the microscopic level ($\sim 10^{-6}$ meters or 1 millimeter) to the macroscopic level (e.g., a plank of wood or a hot-rolled structural steel I-beam).

Since engineers use materials (usually, but not only, solid materials) to manufacture or construct things, they need to be aware of two other factors beyond structure and properties. Materials engineers (as those engineers who deal specifically with materials, as experts) are concerned with the interrelationship among *structure-properties-processing,* since the act of processing a material into a form (i.e., shape and/or condition) different from its initial form (i.e., shape and/or condition) affects the material's structure and, in turn, its properties.

Here's a simple demonstration to try for yourself: Take a common metal (usually steel) wire paper clip and carefully straighten it out. Notice that the wire is fairly pliable (i.e., ductile and easy to deform). Now, slowly and repeatedly bend the wire 90 degrees and, then, straighten it out again. You should notice two—maybe three—things. First, after several reverse-bends, the wire becomes progressively more difficult to bend (if you do it slowly!). It gets stronger and stronger. Second, after some number of reverse-bends (which should be fairly repeatable from one paper clip to the next, of the same type) it will break. It became brittle instead of remaining ductile, as it was at the outset. You just discovered work hardening, that is, the effect that cold (here, room-temperature) plastic deformation (processing) has on strength and ductility (properties) due to the effect of such

---

[4] Wood is actually a mixture of two different materials, cellulose, which is tough but soft and weak, and lignin, which is hard and strong but brittle. The lignin strengthens the cellulose, while the cellulose toughens the lignin in the composite material we call "wood." Grain results from the effect of the relative proportions of lignin and cellulose and air cells, as well as some natural coloring agents, on the appearance of the wood.

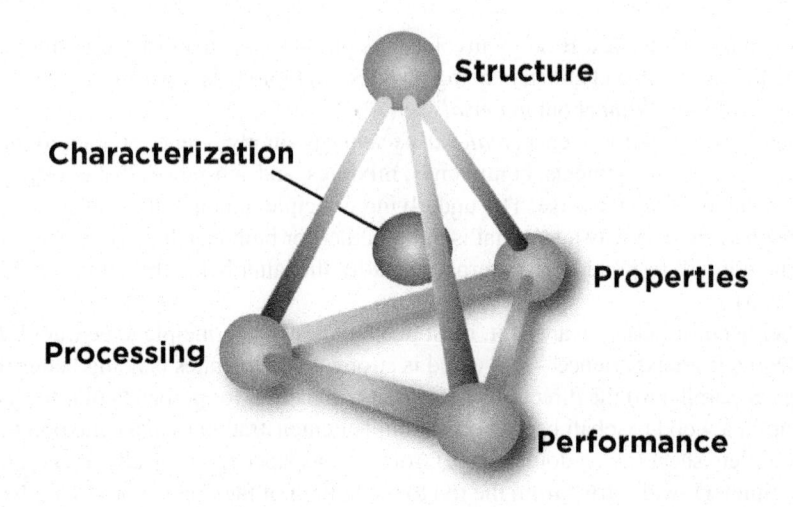

Figure 9-1 Representation of the interrelationship among the structure-properties-processing-performance of materials as a tetrahedron. The base or foundation of the tetrahedron is the structure-property-processing interrelationship within the material and the apex of the tetrahedron is the performance the material brings to a design.

deformation on the metal's atomic-level crystal structure (structure). The third thing you might have noticed is that the metal became warmer as some of the work was converted into heat at the bend. The faster you create reverse bends, the greater the sensible heat produced.

Each and every process (bending, rolling, forging, machining, molding, heat treating, welding, etc.) has some effect on the structure of a material, typically, at both the atomic level and its microscopic level, which, in turn, has some effect on the various properties of that material (strength, ductility, electrical conductivity, etc.).[5] In fact, the interrelationship runs six ways, with structure, property, and processing being, one each, at an apex of an equilateral triangle, and arrows pointing both directions along each of the three sides of the triangle. In other words (or *in* words): Processing affects structure (i.e., processing ⇒ structure), but so, too, does structure affect processing (i.e., structure ⇒ processing), so that, in fact, the relationship is structure ⇔ processing.

For example, something about the structure of a fired clay ceramic brick affects the way it can be processed into a different shape compared to how a like-sized and like-shaped brick of melted and cast gold could be processed into a different shape. Likewise, for each pair of factors, the relationship runs both ways: structure ⇔ processing; processing ⇔ properties; structure ⇔ properties. *Everything affects everything in material!*

Finally, for all engineers who use materials, regardless of their discipline, the materials are a means to an end—that end being performance in a design. Thus, the full interrelationship that exists for materials is *structure ⇔ properties ⇔ processing ⇔ performance,* graphically portrayed as a three-dimensional tetrahedron, as shown in Figure 9–1.

---

[5] For the steel wire paper clip, the strength increases, the ductility decreases, and, if you measured the wire's electrical resistance, you'd find a loss of electrical conductivity indicated by an increase in electrical resistivity (which, along with length and cross-sectional area, gives rise to electrical resistance).

The structure ⇔ property ⇔ processing interrelationship is represented as the equilateral triangular base of the tetrahedron, while performance, as the end goal of design, is shown at the apex of the tetrahedron.

Keep in mind that the interactions go both ways along edges of the tetrahedron and there is no escaping an interaction. This interrelationship is a two-edged sword for engineers, being at once a powerful way to manipulate a material to achieve needed performance for certain properties via processing and reality to be reckoned with when processing can adversely affect properties (through unwanted and unintended structural changes) to degrade performance.

9-3 Material Properties and Performance

As stated in Section 9–1, the most important aspect or goal of design is functionality, which includes performance.[6] Because of this, and as a direct consequence of the fact that physical entities are made of materials, materials have a direct and significant effect on the performance of a designed entity.[7] It is thus important to spend a little time considering the relationship between material properties and material performance, as performance at the material level affects performance in the parts, components, or structural elements that make up a device, mechanism, product, system, or structure. The place to begin is with a definition of a property.

A property for a material is the response of that material to a particular external stimulus (i.e., a stimulus originating outside the material itself).[8] For this reason, generic material properties tend to be divided or classified based on the particular generic stimulus as follows for primary material properties:

- *Mechanical properties* are the response to an externally applied force or attempted deformation.
- *Electrical properties* are the response to an externally applied electric or electromagnetic field.
- *Thermal properties* are the response to externally applied heat.
- *Optical properties* are the response to externally applied light.
- *Magnetic properties* are the response to an externally applied magnetic field.
- *Chemical properties* are the response to externally applied chemical agents.

There are also what are known as *physical properties.* These are, actually, quite small in number, but they share the common feature of being independent of and unchanged by the method of measurement.[9] Three examples are:

---

[6] As used here, *functionality* refers to *what* something (such as a design) does, while performance refers to how well (or to what level) it does it.

[7] We could equally well be talking about the performance of something from Nature, like an eagle, whose flight characteristics and performance are directly affected by the materials that constitute the eagle's wings, including muscles, bones, and feathers.

[8] In fact, properties, in general, are a response to a stimulus. Even we humans respond in particular, albeit slightly different, ways or degrees to external stimuli, as our response to a threat as flight for some and fight for others. The lion that threatens us, in turn, responds in his or her own way from the stimulus that we provide. It will run, maybe after us or maybe away from us. It depends on the "property"—or trait—of the particular lion.

[9] This is not a true mechanical property—*hardness,* for example—as different testing methods (MOH scratch test, indentation test, Shore rebound test, etc.) give different results. In addition, the hardness of the material is changed in the locale of the site where the hardness test was applied to make a reading by introducing a scratch or an indentation from an indenter or from a hardened steel ball that struck the surface.

- *State* (or state of matter), as solid, liquid, gas, or plasma
- *Density,* as mass per unit volume (e.g., g/cm^3)
- *Specific volume,* as volume per unit mass (e.g., cm^3/g)

Some reference sources include luminescence, specularity, color, and the like, among physical properties, although these are all really optical properties, as they are responses to applied light. Others include smell and taste, which the author contends are chemical properties.

There are some other properties that are, without question, responses to external stimuli but, one could argue, are probably the result of some combination of the previously listed primary properties. These include:

- *Acoustic (or acoustical) properties,* as the response a material has to applied sound, actually arising from the material's modulus of elasticity and density and, perhaps, some other properties
- *Radiation (or radiological) properties,* as the response a material has to applied radiation or a material having inherent radioactivity, actually arising from the electron structure as well as the nucleus of the atoms constituting the material
- *Biological properties,* as the response a material has to the environment found in living organisms, actually arising from chemical properties, for the most part

Finally, there are some complex properties that have particular meaning to manufacturing and/or construction but are actually the result of often ill-defined or unknown combinations of two or more specific properties within one or more particular primary properties. Examples include, but are not limited to:

- *Machinability,* as the degree of difficulty a material poses to shaping by machining to remove material
- *Formability,* as the degree of difficulty a material poses to reshaping by permanent plastic deformation (performed hot or cold)
- *Castability,* as the degree of difficulty a crystalline material poses to being made into a shape by melting and resolidification in a mold or die
- *Moldability,* as the degree of difficulty an amorphous glass or amorphous or semicrystalline polymer poses to being made into a shape by heating and softening and application of pressure
- *Weldability,* as the degree of difficulty a metal or alloy poses to joining by welding, especially involving fusion and resolidification

Each of the properties presented herein affects or, more correctly, determines a particular aspect of the performance of the physical entity (e.g., part, component, or structural element) comprising it. Likewise, properties of these details (from their material-of-construction) interact to give rise to the performance of the device, assembly, structure, or system they integrate to become.

In short, physical things perform in a way that manifests how the materials that make up the particular thing perform in response to similar stimulus or similar stimuli.

Table 9–1 summarizes how material properties give rise to performance in the things they constitute. Specific properties (covered in Section 9–4) are given as examples in each property group considered here.

Identifying Materials-of-Construction

TABLE 9-1 Summary of How Materials Give Rise to Performance in Designs

Property	Performance in Design
Physical Properties:	
■ Density (g/cm³)	■ Minimize weight (low)
	■ Maximize momentum or kinetic energy (high)
■ Specific volume (cm³/g)	■ Minimize/save space (low)
	■ Fill space (e.g., prevent choke hazard)(high)
Mechanical Properties:	
■ Strength (σ_y or σ_u)	■ Load carrying at minimum weight to resist yielding (σ_y) or fracture (σ_u)
■ Stiffness (E)	■ Resist deflection (in columns, beams, shafts)
	■ Absorb/store elastic energy
■ Ductility (% elong./% RA)	■ Facilitate deformation processing
	■ Provide robustness
■ Hardness	■ Resist wear or ballistic penetration
■ Fracture toughness (K_{IC})	■ Resist sudden flaw growth
	■ Provide robustness
■ Impact toughness (Charpy)	■ Tolerate impact or shock loads
Electrical Properties:	
■ Electrical conductivity (σ_{el})	■ Facilitate charge transport/current carrying
■ Electrical resistivity (ρ_{el})	■ Facilitate I^2R (joule) heating
■ Dielectric constant (κ)	■ Provide electrical insulation
	■ Store charge in capacitor
Thermal Properties:	
■ Melting point (MP or T_{MP})	■ Prevent unwanted melting
	■ Raise safe operating temperature
	■ Improve creep resistance
■ Glass transition temp. (T_g)	■ Provide polymer pliability (above T_g)
	■ Provide polymer stiffness (below T_g)
■ Thermal conductivity (K)	■ Allow heat transport (if high)
	■ Prevent heat transport or loss (if low)
■ Coefficient of thermal expansion (α_l)	■ Provide dimensional stability (if low)
Optical Properties:	
■ Transparency	■ Allow passage of light
■ Absorption	■ Prevent reflection or loss of light
■ Index of refraction	■ Bend light path
■ Specularity/reflectivity	■ Provide reflection and/or brightness
■ Color	■ Aesthetics
	■ Camouflage
Magnetic Properties:	
■ Ferromagnetism (magnetic)	■ Provide or allow magnetic attraction
■ Paramagnetism (nonmagnetic)	■ Prevent magnetic attraction or induced force
■ Magnetization force	■ Allow magnetization (if low), prevent (if high)

(continues)

TABLE 9-1 *Continued*

Property	Performance in Design
Chemical Properties:	
■ Corrosion resistance	■ Provide chemical stability to environment
■ Reactivity	■ Allow or facilitate reaction (if high)
	■ Remain inert (if low)
Acoustical Properties:	
■ High loss coefficient	■ Block sound (as insulation)
	■ Damp out ringing
■ Low loss coefficient	■ Allow ringing (e.g., bells)
Radiological Properties:	
■ Radioactivity	■ Spontaneously emit radiation (α, β, γ)
■ High absorption	■ Block (insulate against) radiation
■ Low absorption	■ Allow passage of radiation (as "window")
Biological Properties:	
■ Biocompatibility	■ Remain inert in body; prevent rejection
■ Reabsorption	■ Dissolve in body over time
Machinability:	
■ High	■ Facilitate metal removal
	■ Allow complex shapes, intricate details
■ Low	■ Pose challenges to metal removal
	■ Limit shape complexity and details
	■ Poor surface finish
Formability:	
■ High	■ Facilitate plastic deformation without problems
■ Low	■ Allow shape complexity and intricacy of details
	■ Limit shape complexity and intricate details
Castability/Moldability:	
■ High	■ Facilitate complex shapes and intricate details
■ Low	■ Limit shape complexity and intricacy of details
Weldability:	
■ High	■ Facilitate joining by welding without defects
■ Low	■ Complicate (if not preclude) fusion welding
	■ Force use of other joining methods

*Moldability refers to the degree of difficulty posed by a polymer (typically a thermoplastic polymer) to shaping using a mold, usually with some pressure assist.

9-4 A Primer on Materials

While some of what engineers do involves gases (e.g., air or steam) or liquids (e.g., water, hydraulic fluids, liquid-metal coolants, lubricants), most of the time they employ solid materials. Solid materials allow physical structures to be created and can carry and/or transfer forces or loads without having to be constrained (as fluids do).

Every material that occurs naturally on Earth (or in the universe, for that matter), as well as all of those that are synthesized, is made up of atoms of the elements, shown organized by their atomic number arranged in vertical groups and horizontal series in the *Standard Periodic Table* (Figure 9–2).[10] The elements in the first vertical column or group (Group IA in Figure 9–2), at the left of the table, all have low densities, are all metals, all react violently with water (to strip the oxygen atoms from H_2O molecules to oxidize the metal), and all produce strong alkaline solutions with water (hence their name, *alkali metals*). The elements in the seventh vertical group (Group VIIA, second from the right end of the table), on the other hand, are all nonmetals, all have a pungent odor and are toxic, and all produce strong inorganic acids with water. They are known as *halogens.* Similar situations, albeit built around different common chemical or physical characteristics and behaviors, occur for each of the other major vertical columns (i.e., Groups IIA, IIIA, IVA, VA, and VIA). An eighth group (Group VIII) contains the inert gases discovered well after Mendeleev conceived his periodic table[11] but still reflect periodicity of behavior in that all are chemically inert and rare.

In time, particularly with the development of the quantum theory of atoms during the first two decades of the twentieth century, it was recognized that the group number of the Standard Periodic Table (IA through VIIA) reflected the number of electrons in excess of (i.e., beyond) the number needed to exactly fill the next-lower-energy shell for certain inherently *stable electron configurations* to result. This stable electron configuration caused an atom to seek to neither lose nor gain any other electron(s), as the atom was fully satisfied with its overall electron energy state. On the other hand, atoms of elements that appeared in vertical columns or groups other than Group VIII needed to either give up (i.e., lose) or add (i.e., gain) one, two, three, or four valence electrons to achieve a stable configuration like that of the inert gas immediately before or after it (by atomic number).

Group IA elements, for example, like Na^{11} (atomic number 11, with 11 protons in its nucleus and 11 electrons orbiting that nucleus), with an electron configuration of $1s^2 2s^2 2p^6 3s^1$ in quantum notation, need to give up one $3s^1$ outermost valence electron to achieve an electron configuration of $1s^2 2s^2 2p^6$ like the inert gas Ne^{10} immediately preceding it. The net deficit of one negative-charged electron (i.e., 10– electrons versus 11+ protons) results in a sodium ion with a net +1 charge: Na^{1+}. In other words, sodium "prefers" (i.e., is more stable and at a lower energy state) when it exists as Na^{1+} ions than as Na atoms. Hence, no atomic or nascent sodium is found on Earth (or elsewhere in the universe), unless it is synthesized and properly stored). Instead, sodium is found as 1+ ions in compounds (e.g., NaCl in seawater and deposits, Na_2O in soda ash deposits, etc.).

All of the elements in Group IA tend to lose their one valence electron to form 1+ ions. Similarly, all of the elements in Group IIA (e.g., Mg^{12}, with a $1s^2 2s^2 2p^6 3s^2$ electron configuration) tend to lose their two $3s^2$ valence electrons to form 2+ ions, and so forth, for Group IIIA. Group VIIA elements, on the other hand, find it easier (probabilistically) to gain one extra valence electron than to lose seven, in the process creating 1– ions, as F^9 with a $1s^2 2s^2 2p^5$ electron configuration forms F^{1-} with a $1s^2 2s^2 2p^6$ electron configuration, like Ne^{10} immediately following it. Likewise, for Group VIA elements add two valence electrons to create 2– ions and Group VA elements add three valence electrons to create 3– ions.

Here's where the first division of elements occurs. Elements that tend to lose electrons to form positive ions are said to be *electropositive* and are defined as *metals*. On the other hand, elements

---

[10] The periodic arrangement of the elements, suspected and attempted by others as triads or octets, was brought to fruition by Dimitri Ivanovich Mendeleev (1834–1907) around 1863. The Russian chemist noted that the atoms of the 56 elements known at the time exhibited similar chemical characteristics and behaviors (e.g., reactivity), as well as similar physical properties (e.g., density) approximately every eight elements, based on atomic number, the atomic number reflecting the number of protons found in the nucleus of the element's atoms. The same basic arrangement is still used today with great efficacy.

[11] The inert gases were discovered, usually by progressive evaporation of liquefied air, between 1898 and around 1902.

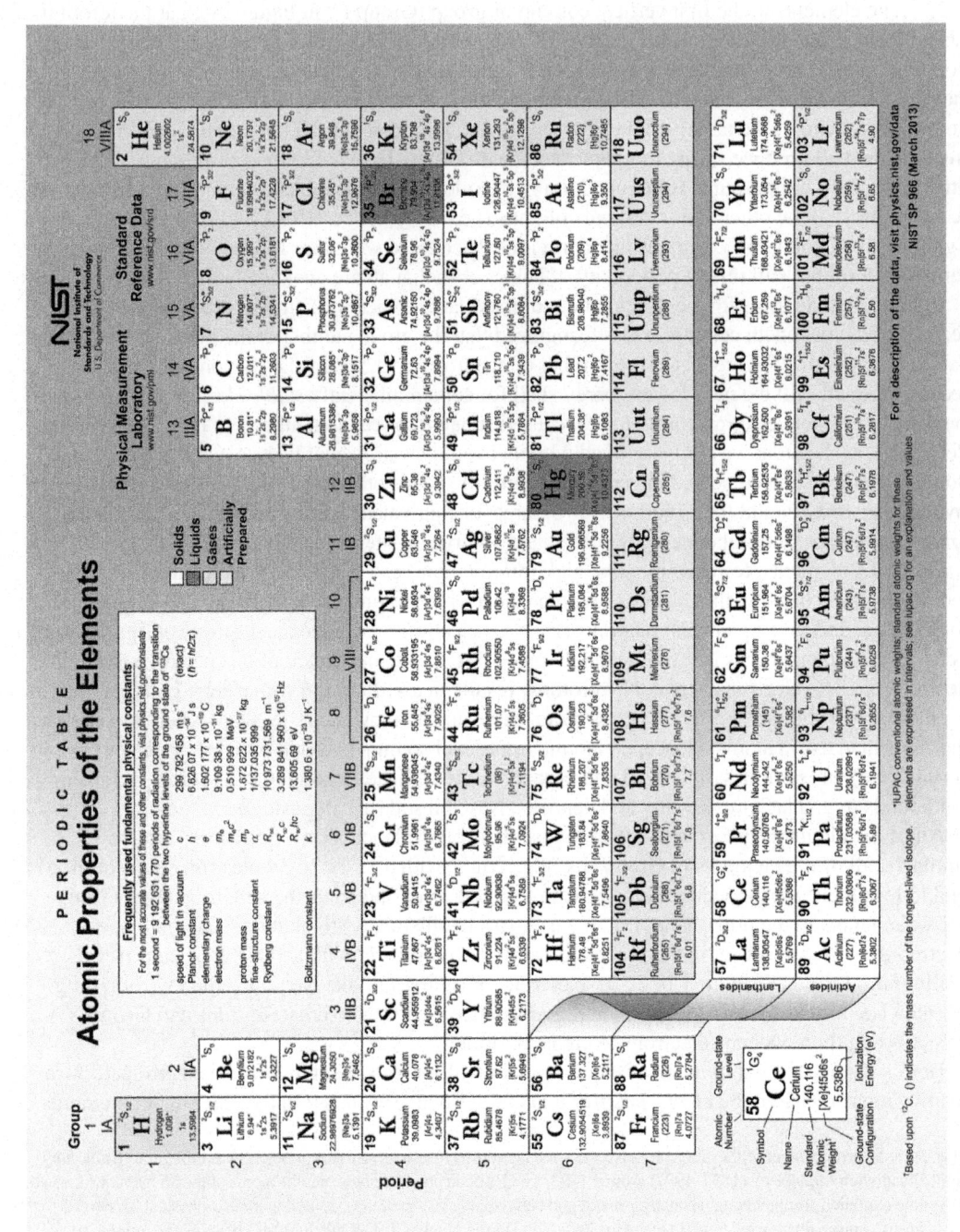

Figure 9–2 The Standard Periodic Table of Elements, the concept for which was developed by Dmitri Ivanovich Mendeleev in 1863. (*Source:* National Institute of Standards and Technology [NIST].)

that tend to add electrons to form negative ions are said to be *electronegative* and are defined as *nonmetals*.

Metals are found to always be electrical conductors, while nonmetals are found to always be nonconductors or electrical insulators. This particular behavior arises from the structure (actually, the "band structure") of the outermost valence electron and next-highest energy, empty conduction states (or bands). It shouldn't come as much of a surprise that elements found in Group IVA, which can either lose four or gain four electrons, depending on the opportunity presented, to attain a stable electron configuration, are neither metals nor nonmetals, and neither electrical conductors nor insulators. In fact, they are *electrical semiconductors* and are called either *semimetals* or *metalloids*.

While atoms of the elements other than the inert gases in Group VIII try to lose or add electrons to their valence shells to achieve stable electron configurations and, in the process, to create + or − ions, respectively, they can only do so if they have someplace to move or obtain the surplus or needed electron(s). This can happen as the atoms of elements form atomic-level bonds with other atoms in what is known as *atomic bonding*. In the process of bonding, the atoms create aggregates we know as solid materials. Such bonding can occur in one of three primary ways:[12]

- *Ionic bonding,* in which one atomic species gives up excess electron(s) to another atomic species, resulting in a net + charge on the former and a net − charge on the latter, and a strong electrostatic force of attraction between the two (or more) (e.g., NaCl, CsCl, MgO).
- *Covalent bonding,* in which one atomic species shares some number of electrons with another atomic species (or, sometimes, another atom like itself, as in O_2, Cl_2, etc.) in an intimate fashion to form a covalent bonded molecule (e.g., H_2O, CH_4, CO_2) or solid (e.g., C diamond).
- *Metallic bonding,* in which like metal atoms (or atoms of different metals) share their valence electrons in a communal fashion in what is known as *delocalized bonding,* but more clearly as *extended covalent bonding,* to create a periodic array of positive metal ions held together by a "sea" or "cloud" of permeating free or conduction electrons (e.g., in solid Cu, solid Al, or solid Fe).

The reader seeking more details on atomic bonding, and its ramifications, is referred to any of several basic textbooks on materials (ref. Callister, Messler).

Depending on an element's location (or position) in the Standard Periodic Table, different types of primary atomic bonds form to create different types of solid materials. In brief:

- *Metals* appear to the left on the table, from about one-third over from the right. They all involve metallic bonding (although some, most to the far right, exhibit mixed metallic-covalent bonding) and all are electrical, as well as thermal, conductors.
- *Nonmetals* appear at the upper right. There are only about 17 to 20, depending on how one counts, i.e., He, Ne, Ar, Kr, Xe, Rn, F, Cl, Br, I, O, S, (Se), N, P, As, C, Si, (Ge), (B). Those in parentheses ( ) are semimetals. All true nonmetals bond to themselves (e.g., O_2) or other nonmetals (e.g., SO_2) using covalent bonding. All true nonmetals are electrical insulators.[13]

---

[12] Such bonding is said to be *primary*, as the resulting strength of the bond created is high (i.e., strong) versus low (i.e., weak). The strength of the bonds holding a solid material together directly result in mechanical strength as well as melting point. This is so because it takes a lot of either mechanical energy or, alternatively, thermal energy to break the bonds and cause the atoms and material to come apart.

[13] Some nonmetallic solids created by very strong and stiff covalent bonds (e.g., C in its diamond form, AlN) are electrical insulators but exhibit very high thermal conductivity. This occurs because heat is conducted by atom vibrations on the crystal lattice (i.e., as phonons) rather than relying on transfer by free-moving conduction electrons (of which there are none).

TABLE 9-2 Key Characteristics of Different Types of Bonding in Materials

Primary Bond Type	Material(s)	Characteristics
Ionic	Many ceramics (oxides)	■ Water-soluble salts (e.g., halides) ■ Can be hard ■ All tend to be inherently brittle ■ Tend to be refractory ■ Conduct electricity dissolved in water or fused ■ Can have high heat capacity and latent heat of fusion ■ Generally thermal/electrical insulators as solids ■ Many are nonreactive; environmentally stable ■ Low CTEs compared to metals and polymers
Covalent	Some high-perf. ceramics; within polymer chains (especially Group III–V)	■ Usually not water soluble ■ Soluble in some organic solvents ■ All tend to be inherently brittle ■ Among the hardest materials ■ Many are very refractory ■ Can be nonreactive; environmentally stable ■ Generally electrical insulators ■ Some are very good thermal conductors (e.g., AlN) ■ Low CTEs compared to metals and polymers
Metallic	All metals and alloys	■ All are electrical and thermal conductors ■ Melting points vary over wide range ■ Hardness varies over wide range ■ Corrosion resistance varies over wide range ■ Most exhibit ductile behavior
Mixed Ionic-Covalent	Some ceramics (toward upper-right)	■ Like ionic and covalent ceramics; more like ionic
Mixed Metallic-Covalent	Increasingly covalent with progression to right	
Secondary (van der Waals) Bonding	Between polymer chains	■ Weak, except hydrogen bonding ■ Limits melting, softening temperature of solids ■ May promote low coefficient of friction

- *Semimetals, metalloids, or semiconductors* appear in Group IVA and include $C_{diamond}$, Si, and Ge, although there are also compound Group III-V semiconductors (e.g., GaAs).

Some additional behaviors relative to the location of elements in the Standard Periodic Table are:

- Metals to the left, and especially the lower left, form ionic bonds with nonmetals to the right, and especially the upper right, to create *ceramics*.
- Some ceramics (often referred to as *high-performance ceramics* for their exceptional properties), which are always compounds of metals and nonmetals, exhibit covalent or mixed ionic-covalent bonding, especially among Group IIIA, IVA, and VA elements (e.g. SiC, Si_3N_4, B_4C, BN).
- *Polymers* are molecular (versus atomic) materials, consisting of very large, long-chain molecules based on a backbone of C or Si atoms. Bonding within the chain is covalent. Bonding between chains is weak van der Waals secondary bonds based on dipoles.[14]

The different types of bonding (which determine the solid material's structure) determine the solid material's properties. Table 9–2 summarizes key characteristics of the different types of bonding found in solid materials, while Section 9–5, which follows, identifies and describes the major properties of materials of value to engineers.

Appendix A contains a list of all major material classes and subtypes, as well as an extensive list of the most significant members of each.

9-5 A Primer on Material Properties

In Section 9–3, the variety of *primary properties* (e.g., mechanical, electrical, thermal, optical, magnetic, and chemical), as well as somewhat *secondary properties* (e.g., acoustical, radiological, and biological) and complex *combination properties* (e.g., manufacturability, formability, castability, moldability, and weldability) were presented and discussed. Each was correctly defined to be the response of a solid material to a particular external stimulus. It is now time, after the primer on materials (intended solely as a refresher for degreed engineers, as all engineering students take a basic course in materials) in Section 9–4, to identify and define those specific primary properties that are most often used by engineers.

The list of specific primary properties of particular importance to practicing engineers, by category, are:

- Mechanical properties:
  - ✓ Yield strength
  - ✓ (Ultimate) tensile strength
  - ✓ Modulus of elasticity
  - ✓ Ductility (as percent elongation or percent reduction in area)
  - ✓ Hardness
  - ✓ Fatigue limit or endurance strength

---

[14] Another major contributor to the strength and elasticity of polymers is entanglement among the long-chain molecules due to entropy, which tries to promote disorder over order.

- ✓ Creep strength
- ✓ Impact strength and ductile-brittle transition temperature
- ✓ Fracture toughness
- Electrical properties:
  - ✓ Electrical conductivity
  - ✓ Electrical resistivity
  - ✓ Dielectric constant
- Thermal properties:
  - ✓ Melting point (T_{MP}) for crystalline materials
  - ✓ Glass transition temperature (T_g) for amorphous materials
  - ✓ Thermal conductivity
  - ✓ Specific heat
  - ✓ Coefficient of thermal expansion/contraction
  - ✓ Thermal shock resistance

For other applications, optical, magnetic, chemical, acoustical, radiological, or biological properties are important, but these are not covered in this overview or primer.

Figure 9–3 shows typical engineering stress–engineering strain diagrams for both (*a*) a generic nonferrous metal or alloy (other than iron or iron-based) and (*b*) ferrous (iron-carbon or steel) alloy.

Following are the definitions of the key mechanical properties shown on one or the other or both of the engineering stress-strain diagrams in Figure 9–3.[15]

Mechanical Properties

- *Yield stress/strength* (σ_y)[16] is the stress level (determined with 0.2% strain offset) at which a metal or alloy is defined to stop exhibiting elastic behavior, at which point strain under load is not recovered upon release of the load.[17] This property is important for engineers to know when a structural member will change shape permanently (i.e., yield), including use for deformation processing (rolling, forging, bending, etc.). For iron-carbon steels, an *upper* and a *lower yield point* appear from interstitial carbon atmosphere effects on dislocation blocking in slip deformation.
- *Ultimate tensile strength* (σ_u) is the stress level at which a structural member loaded in tension exhibits localized deformation in the form of a reduced cross-section neck. From this point, a

---

[15] An engineering stress–engineering strain diagram plots load or force/initial cross-sectional area (i.e., engineering stress) versus length change/initial length (i.e., engineering strain). On this plot, an apparent maximum stress/strength appears at the peak or ultimate tensile stress. A similar curve results for loading in shear, as well as for compressive loading, although there is no ultimate strength (due to the absence of localized necking) under compression. It is also possible, and sometimes useful, to plot true stress versus true strain, for which the effect of localized necking in tension is accounted for by using actual cross-sectional area and extended length.

[16] The *stress* in a material is the applied load or force divided by the load-bearing cross-sectional area, while the *strength* is the material's response to resist applied stress. *Strain* in a material is the change in length per unit length, while *elongation* is the change in length as a percentage of the original length or other dimension.

[17] Elastic behavior actually ends at what is known as a material's *elastic limit*. Practical difficulties for determining this limit led to an internationally accepted definition of offset yield stress or strength for 0.2%, or 0.002, strain. In the dislocation theory of plastic deformation by slip, the yield stress is essentially the stress level at which dislocations begin to move to cause slip.

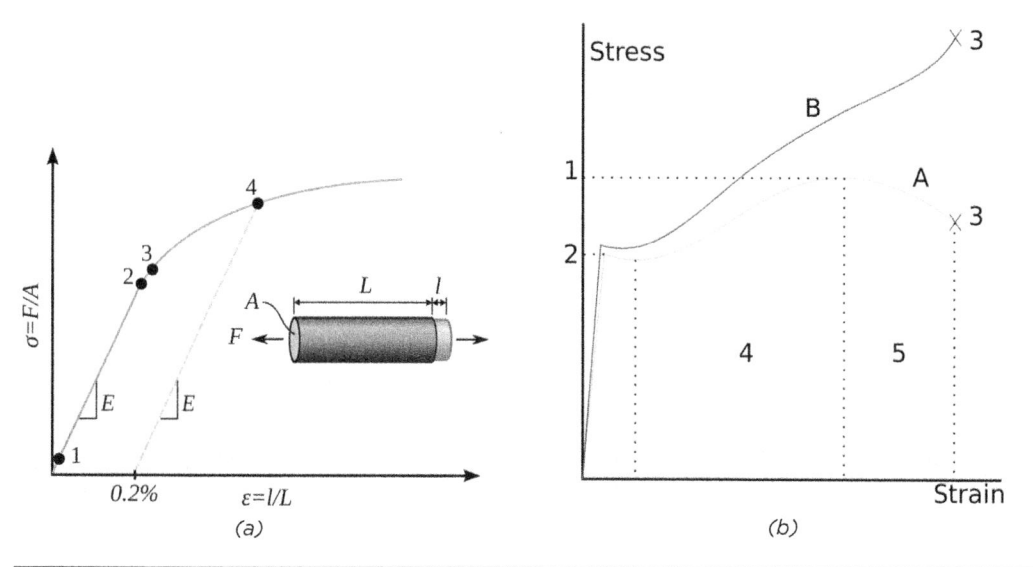

Figure 9-3 Schematic plots of engineering stress versus engineering strain for (a) a generic nonferrous metal or alloy exhibiting ductility (i.e., reasonable strain-to-failure) and (b) a generic carbon-strengthened ferrous alloy (i.e., steel) exhibiting a yield point phenomenon. In (a), Point 1 represents a low stress within the elastic region in which it is proportional to the strain by the factor of the modulus of elasticity E (in Hooke's law). Point 2 represents the proportional limit, beyond which stress and strain are no longer linearly related, while Point 3 represents a point where stress and strain no longer follow a linear relationship, but where behavior is still elastic. Point 4 represents the defined offset yield stress at 0.2% strain. In (b), Curve A represents apparent stress (i.e., engineering stress), while Curve B represents true stress. Point 1 indicates the ultimate tensile stress or strength in Steel B, where localized necking begins. Points 2 represent upper yield points for two different steels, while Points 3 represent points at which fracture occurs. Region 4 is the region of strain hardening, while Region 5 represents the toughness of the steel. (*Source:* Wikipedia Creative Commons, originally contributed by Sigmund on 4 September 2007 and modified by Wizard191 on 17 May 2011 [a], and by Slashme on 25 February 2009 [b].)

structure proceeds to rapid failure by fracture unless the applied load or force can be removed, so it must be safely avoided (by some factor) in design.

- *(Young's) Modulus of elasticity* (E) is the rate of increase in stress (σ) for a given incremental increase in strain (ε), from $\sigma = E\varepsilon$ (Hooke's law). It is a measure of the stiffness of the material (independent of cross-sectional shape, as reflected in the moment of inertia I). For metals and ceramics, this elastic portion of the stress-strain diagram is linear. This is usually not the case for polymers. Modulus is important in designs to resist buckling in compression-loaded columns or panels, deflection in beams under bending, and twist in shafts under torsion. It is also useful for storing elastic energy that is to be recovered (e.g., highway guardrails and vaulting poles).
- *Ductility* measures the ability of a material to tolerate plastic strain or deformation without fracturing. It can be measured as the percentage elongation to failure or the percentage reduction in cross-sectional area during rolling, drawing, etc., without cracking. It is especially useful in fabrication, and is important to impart robustness to a structure.

- *Toughness* is the ability of a material to tolerate impact energy without fracturing. It represents the area under a stress-strain curve to the point of fracture. A common measure is Charpy impact energy, in Newton-meters (N-m) or foot-pounds (ft-lb). *Resilience* is the ability of a material to absorb impact energy elastically and is represented by the area under the stress-strain curve to the point of yielding. In body-centered cubic metals and alloys, a transition from high energy absorption to low energy absorption occurs at a particular temperature (or narrow temperature range) known as the ductile-brittle transition temperature (DBTT).
- *Hardness* is a measure of the resistance of a material to scratching, indentation, or penetration using MOH, Brinell or Rockwell (for example), or ballistic penetration tests. It is especially useful in design to resist wear, including retaining a sharp edge.
- *Fatigue limit or endurance limit* is a measure of the ability of a material to tolerate repeated, cyclic loading without fracturing. A high value (as a stress) indicates high resistance to fatigue. *Fatigue limit* is the stress that can be safely tolerated (with some factor) for some number of cycles of loading (i.e., life), while *endurance limit* (found in steels) is supposedly the stress below which failure will never occur.
- *Creep strength* is a measure of the ability of a metal or alloy (or ceramic) to resist continued elongation (or strain) under a sustained load when the material is above some fraction of its absolute melting point (MP)—about $0.4\ T_{MP}$ for metals/alloys, $0.5\ T_{MP}$ for ceramics, and about $0.3\ T_g$ for polymers.

Electrical Properties

- *Electrical conductivity* (σ_{el}) is a measure of how well a material conducts electricity using electrons in metals and electrons and holes in semiconductors. It is useful in designs where an electrical signal is to be transmitted or where some device is to be shielded from a static charge or electric or electromagnetic field (i.e., using a Faraday shield). High values of σ_{el} tend to be favored.
- *Electrical resistivity* (ρ_{el}) is the reciprocal of electrical conductivity, as it is a measure of the impediment a material presents to the conduction of electrons as electric current. It is usually favored (as a high value) for electrical insulation (see also *dielectric constant*) or for I^2R (joule) heating elements or devices.[18] When I^2R losses must be kept to a minimum, resistivity must be low.
- *Dielectric constant* (κ) is a measure of the ability of a material to prevent the passage or leakage of electric charge Q across a gap (i.e., between plates in a capacitor or through insulation over a conducting wire core). The value of the dielectric constant reflects the effectiveness to prevent charge passage compared to a perfect vacuum, as a number greater than 1. A high value of κ is favored for insulation. The value of capacitance is $C = Q/V = \kappa A/d$, where V = voltage, A = plate area, and d = plate separation distance.

Thermal Properties

- *Melting point* (T_{MP}) for crystalline materials (e.g., metals, alloys, ceramics, or crystalline or semicrystalline polymers) represents the discrete temperature at which a pure solid transforms from its solid to its liquid state. It is useful in design to set an absolute upper use temperature,

---

[18] Electrical resistivity (r_{el}) is the material contribution to electrical resistance (R) measured using an ohmmeter. Electrical resistance also includes the effect of the length (L) and cross-sectional area (A) of a conductor, so that $R = r_{el}\ L/A$.

but it also determines more practical limits as softening temperatures (in service), working temperatures (in processing), and recrystallization temperatures in cold-worked/strain-hardened metals or alloys. Designers often think in terms of the *homologous temperature*, defined as T/T_{MP} as a fraction on an absolute temperature scale.

- *Glass transition temperature* (T_g) is the temperature at which an amorphous or semicrystalline polymer changes from rigid/glasslike to soft/fluidlike behavior under applied stress. It is used to ensure safe use of polymers in structural elements or structures (by operating below T_g) or, alternatively, to allow shape processing (by operating above T_g).[19]
- *Thermal conductivity* (K) is a measure of the ease (or difficulty) for heat to flow in a solid material. A high value is favored to move or remove heat, while a low value is favored to block the passage of heat (i.e., for insulation). K is the mobility term in $Q = K\, dT/dx$ (Fourier's equation).
- *Specific heat* (C_p) is a measure of how much heat a solid material can absorb before its temperature rises by 1 degree (typically K or °C, but possibly °F). High values are favored for heat storage.
- *Coefficient of thermal expansion* (α_l), or CTE, measures the linear dimensional stability of a solid material relative to temperature changes or excursions (ΔT). High values (in mm/mm/°C or in/in/°F) give rise to large dimensional changes. Designers need to consider CTE differences between joined or adjoining materials of more than about 15%, as such differences give rise to thermally induced stresses from $\sigma_{th} = \alpha_l \Delta T \Delta L$, in which L is the length dimension of the member of concern. Thermal stresses also arise in a solid material in which a severe temperature gradient dT/dL prevails.
- *Thermal shock resistance* (TSR) is an index which indicates how well (or how poorly) a solid material tolerates a rapid change in temperature (especially, but not only, during cooling or quenching). TSR includes effects from the material's fracture strength (σ_f), linear coefficient of thermal expansion (α_l) thermal conductivity (K), and modulus of elasticity (E) as TSR ~ $\sigma_f K / \alpha_l E$. High values of TSR tolerate thermal shock better.

The preceding should serve most engineers for most situations encountered during reverse-engineering dissections and analysis. Once again, as a reminder, Table 9–1 summarizes how materials give rise to performance in engineering design. Table 9–3 summarizes the specific properties found under each major classification based on stimulus.

9-6 Relationships for Material Properties in Material Selection Charts

Michael F. Ashby (1935–), Emeritus Professor of Materials Science and Engineering at Cambridge University, popularized the use of plots of pairs of certain properties of materials over the full range of values those properties could exhibit in solid materials (ref. Ashby). Known as Material Selection Charts, these plots allow materials that might seem radically different to be shown to be equivalent or comparable for certain design situations using mathematically derived relationships for certain material combinations as Material Performance Index or Indices that are

---

[19] Glasses, which are fully amorphous materials (often with formulations consisting of mixed ceramic oxides), use working temperatures, softening temperatures, and annealing temperatures that relate to the viscosity of the glass.

TABLE 9-3 Specific Properties within the Major Classifications of Materials Properties (by stimulus)

Mechanical Properties
Tensile strength
Compressive strength
Shear strength
Bearing strength
Proportional limit
Yield strength
Ultimate tensile strength
Modulus of rupture
Young's (tensile) modulus
Shear modulus
Bulk modulus
Poisson's ratio
Ductility
Hardness
Impact toughness
Resilience
Fatigue limit strength/endurance strength
Creep strength
Plane strain fracture toughness

Electrical Properties
Electrical conductivity
Electrical resistivity
Permittivity
Dielectric constant
Dielectric (breakdown) strength
Piezoelectric constant
Seebeck coefficient

Chemical Properties
pH
Hygroscopy
Surface energy
Surface tension
Reactivity
Corrosion resistance
Passivity

Biological Properties
Toxicity
Biocompatibility

Environmental Properties
Embodied energy
Embodied water

Magnetic Properties
Permeability
Hysteresis
Curie point or temperature

Optical Properties
Absorptivity
Reflectivity
Refractive index
Color
Photosensitivity
Transmissivity
Luminosity

Thermal Properties
Thermal conductivity
Thermal diffusivity
Coefficient of thermal expansion
Emissivity
Specific heat
Melting point
Glass transition temperature
Boiling point
Flash point
Triple point
Heat of vaporization
Heat of fusion
Pyrophoricity
Autoignition temperature
Vapor pressure

Acoustical Properties
Acoustic absorption
Speed of sound

Radiological Properties
Neutron capture cross section
Specific activity
Half-life

Physical Properties*
Density

*There is no stimulus needed for a physical property.

used as design guidelines. Materials that lie on or above the same design guideline are comparable or superior, respectively, to one another for the given property combination.

A good and familiar example of seemingly quite different materials offering an important comparable property set is found in the frames of bicycles. Usually made from hollow tubular members, the trapezoidal frame of a bicycle, while obviously needing to be strong enough not to either yield or fracture under the weight of the rider while pedaling, is actually designed to limit the amount of elastic deflection. Resistance to deflection for a given shape (e.g., a hollow circular or elliptical tube) and moment of inertia comes from the modulus of elasticity of the tube material. But the weight of the frame is also important and is generally kept as low as possible. Weight of structural members depends on the density of the material-of-construction. Hence, any materials that exhibit comparable ratios of modulus to density (often with the modulus E taken to some power, such as 1, ½, or ⅔) would be suitable for fabrication into a frame. For this reason, bicycles with Al alloy, Ti alloy, steel or stainless steel, fiberglass, graphite-epoxy, and even bamboo all offer comparable stiffness-to-weight performance.

Figures 9–4 and 9–5 give two examples, the first for strength versus density and the second for Young's modulus versus density. The former would allow materials having comparable (or superior) ratios of strength to density to be found, while the latter would allow materials having comparable (or superior) ratios of Young's modulus to density to be found. These combinations are important in design because they represent strength to weight and stiffness to weight, respectively, which has value for minimizing the weight of a structure whose performance is limited by either strength (e.g., tensile yielding or overload fracture) or stiffness (e.g., compressive buckling, deflection in bending, or twist in torsion).

In fact, these particular ratios are so important, design engineers frequently consider what are known as *specific strength* (as strength/weight) and/or *specific modulus* (as Young's modulus/weight or stiffness/weight). A wide variety of charts are available plotting strength, Young's modulus, fracture toughness, and density, as well as other properties, against one another. Readers interested in these extremely clever and valuable charts are referred to any of several of Michael Ashby's fine books (see Recommended Readings).

Before moving on, however, it is worth spending a moment considering what one of these charts (as an example) portrays about materials. Let's consider the strength-versus-density plot in Figure 9–4.

First, each property (i.e., tensile strength/compression strength for all materials except ceramics, for which only compressive strength is used, in MPa, and density in kg/m^3, which is comparable, dividing by 1000, to g/cm^3) is plotted for the full range or spectrum over which materials exhibit values. For example, the most dense materials are metals and alloys (as they tend to have close-packed crystal structures and more massive atoms), with the densest engineering metals/alloys being tungsten and its alloys at around 19,300 kg/m^3 or 19.3 g/cm^3, and the least dense engineering metals/alloys being magnesium and its alloys at around 1740 kg/m^3 or 1.74 g/cm^3. These values fix the range of the envelope (or "bubble") that contains all engineering metals and alloys. Likewise, the upper and lower limits on strength tend to be around 2800 MPa and 5 MPa (e.g., maraging steel at 2700 and pure indium, used in solders, at under 10 MPa).

As all metals have similar structures (involving metallic bonding), they all exhibit properties that cluster in "bubbles." Likewise, for other materials (e.g., engineering ceramics, glasses, porous ceramics, composites, polymers, rubbers, wood and wood products, and foams), structures within groups are similar so properties cluster in "bubbles" for the groups.

184 Chapter Nine

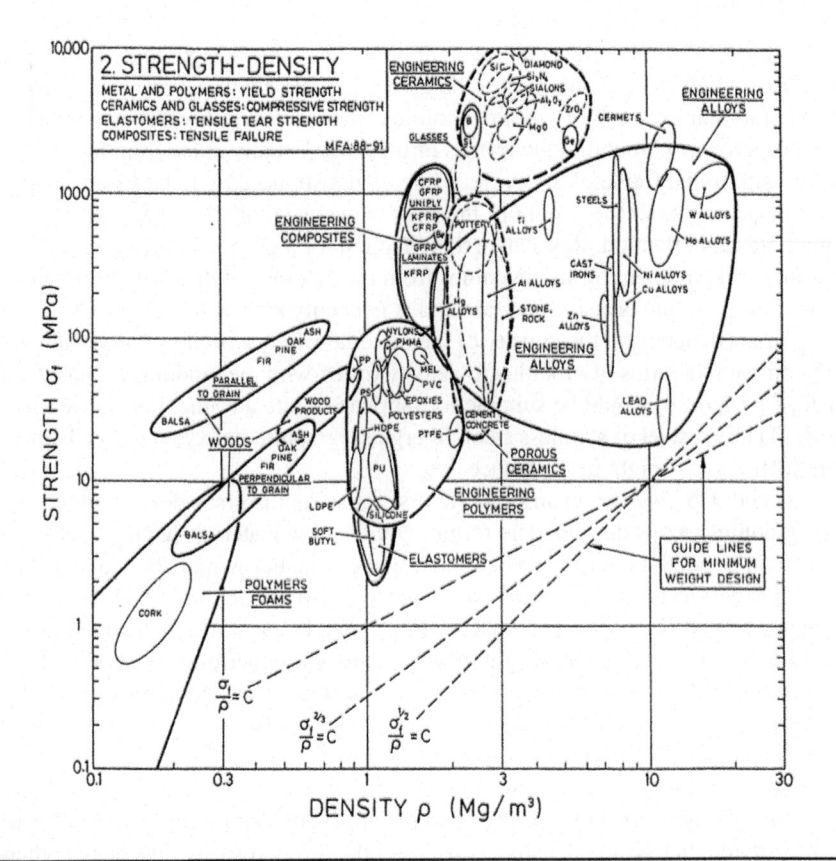

Figure 9-4 Material Selection Chart for strength versus density showing clustering of these properties for each material type or subtype. (*Source:* Michael F. Ashby, *Materials Selection in Mechanical Design,* 3rd edition, Butterworth-Heinemann/Elsevier, Burlington, MA, 2007, page 54, Figure 9-7.)

Second, the relative positions of the "bubbles" for the other material groups make sense based on their structures compared to metals. As but one example, ceramics are compounds of metals and nonmetals, with the nonmetals having much lower atomic weights than most metals, and packing, while often close, that results in overall lower densities than exhibited by metals. Strengths for ceramics, on the other hand, are higher (in compression) since ionic bonding tends to be stronger than metallic bonding, hence the higher melting points of ceramics compared to metals also.

Obviously, foams (e.g., foamed polymers, such as foam rubber or Styrofoam) are much lower in density than other materials (without porosity), as air weighs virtually nothing and foams are 60 to 80 percent or more air.

The point being made is this: The position of various materials on Material Selection Charts makes sense in terms of structure-property relationships group to group.

Ashby presents 18 different Material Selection Charts in his book (ref. Ashby) that cover the most widely needed combinations of one property plotted against another, as follows:

Identifying Materials-of-Construction 185

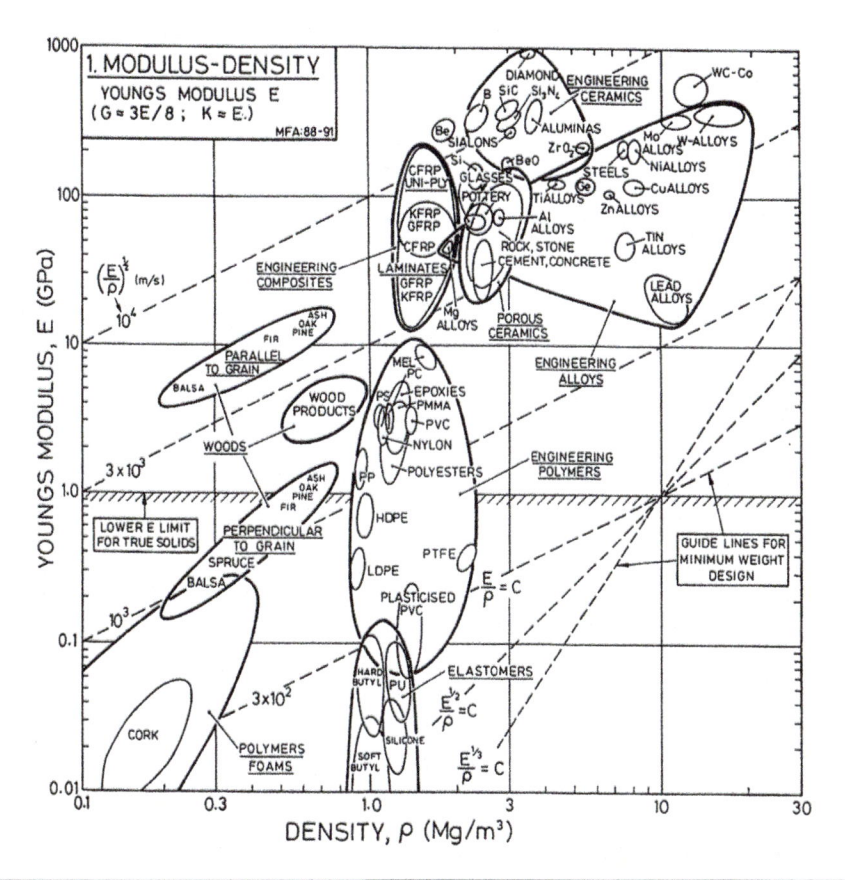

Figure 9-5 Material Selection Chart for Young's modulus versus density showing clustering of these properties for each material type or subtype. (*Source:* Michael F. Ashby, *Materials Selection in Mechanical Design,* 3rd edition, Butterworth-Heinemann/Elsevier, Burlington, MA, 2007, page 50, Figure 4.3.)

Chart 1   Young's Modulus /Density
Chart 2   Strength/Density
Chart 3   Fracture Toughness/Density
Chart 4   Young's Modulus/Strength
Chart 5   Specific Modulus/Specific Strength
Chart 6   Fracture Toughness/Modulus
Chart 7   Fracture Toughness/Strength
Chart 8   Loss Coefficient/Young's Modulus
Chart 9   Thermal Conductivity/Thermal Diffusivity
Chart 10 Thermal Expansion/Thermal Conductivity
Chart 11 Thermal Expansion/Young's Modulus
Chart 12 Normalized Strength/Thermal Expansion
Chart 13 Strength/Temperature
Chart 14 Young's Modulus/Relative Cost

Chart 15   Strength/Relative Cost
Chart 16   Normalized Wear Rate/Bearing Pressure
Chart 17   Young's Modulus/Energy Content
Chart 18   Strength/Energy Content

9-7 Identifying Materials by Observation Only

Figure 9–6 summarizes materials using a Venn diagram to depict the three fundamental types of classes: metals, ceramics, and polymers. Subtypes within types (or groups within classes) and either hybrids possessing characteristics of two types (e.g., semiconductors or metalloids between metals and ceramics) or composites of two or all three types are shown at intersections of the fundamental types. When conducting a reverse-engineering dissection, an early step (either just before or right after identifying the known, suspected, or deduced role, purpose, and functionality of each detail, as covered in Chapter 7) is to identify the material-of-construction. In fact, knowing the material-of-construction may provide clues as to what a particular part might do based on its properties.

The process begins with identifying the generic type (or class) of material as an engineering metal or alloy, an engineering or porous ceramic, an engineering polymer or elastomer, a glass, or a composite. Once this is done, an attempt should be made to further identify the specific metal or alloy-base, ceramic, polymer, or composite. This is easier for metals and alloys than for ceramics and polymers or most composites.[20]

So how to begin using only observation versus laboratory analysis?

The keys to distinguishing among the three fundamental types using observation only rely on clues to the senses—appearance (using sight), feel (using touch), and sound (using hearing).

Bare metals or alloys (without any tarnish layer, coat of paint, or covering of polymer, etc.) are most recognizable by their *luster,* by which is meant "appearance of the surface of a material [typically a mineral or metal] that exhibits brilliance and ability to reflect light." As used here, luster, however, is subtly different from just smoothness and shininess, like a freshly waxed floor. (We're talking about "shiny and reflective like a metal," which is meaningful only if one has seen lustrous metals.) Metals and alloys are always opaque and exhibit color (e.g., yellow gold, reddish-orange copper, silvery-white aluminum). To check for luster, a part or detail should be scratched (e.g., with a knife blade or file or other sharp object) to reveal the material itself (versus any natural or artificial coating). If the underlying material exhibits luster, the material is a metal.

Engineering metals and alloys[21] all tend to be more dense than polymeric materials (e.g., polymers and polymer-matrix composites), and, except for Al and Mg alloys, are more dense than most ceramics.[22] Density can be assessed by *heft,* which is the judged weight for a given volume versus some familiar material for comparison. For metals, a good, well-known standard for comparison is steel, with a density of about 7.9 g/cm^3. Al, Mg, and Ti alloys feel lighter, while

---

[20] Composite materials can easily be identified as polymer-matrix, metal-matrix, ceramic-matrix, or carbon-matrix (using observable clues). Beyond this, they can usually be identified as having continuous aligned versus random chopped fibers, random particles, or laminates as reinforcements. Identifying specific types beyond this is not usually possible without laboratory analysis.

[21] *Engineering metals and alloys* refer to those metals and alloys commonly used in design for their functionally specific properties, as opposed to the more exotic metallic elements.

[22] There are some engineering ceramics (also known as *advanced ceramics* or *high-performance ceramics*) that are more dense than Ti alloys (at about 4.3 to 4.5 g/cm^3), but not many. Porous ceramics, like cement, are less dense (at about 3.1 to 3.2 g/cm^3).

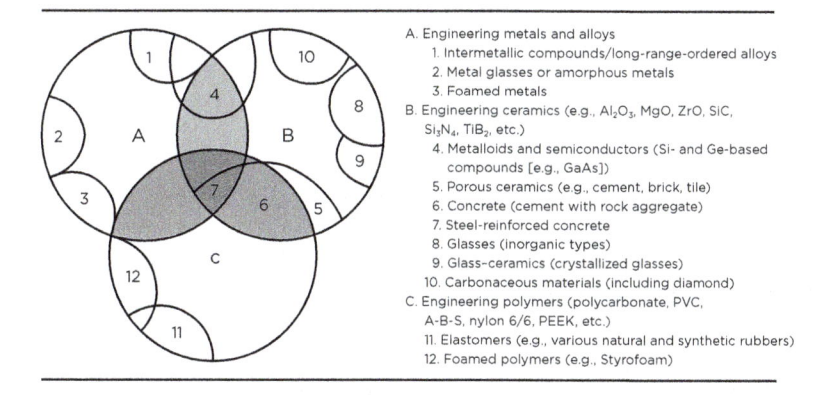

Figure 9-6 Venn diagram schematically depicting the fundamental types and subtypes of materials.

A. Engineering metals and alloys
  1. Intermetallic compounds/long-range-ordered alloys
  2. Metal glasses or amorphous metals
  3. Foamed metals
B. Engineering ceramics (e.g., Al_2O_3, MgO, ZrO, SiC, Si_3N_4, TiB_2, etc.)
  4. Metalloids and semiconductors (Si- and Ge-based compounds [e.g., GaAs])
  5. Porous ceramics (e.g., cement, brick, tile)
  6. Concrete (cement with rock aggregate)
  7. Steel-reinforced concrete
  8. Glasses (inorganic types)
  9. Glass-ceramics (crystallized glasses)
  10. Carbonaceous materials (including diamond)
C. Engineering polymers (polycarbonate, PVC, A-B-S, nylon 6/6, PEEK, etc.)
  11. Elastomers (e.g., various natural and synthetic rubbers)
  12. Foamed polymers (e.g., Styrofoam)

Cu and Ni alloys feel slightly heavier than a comparable volume of steel. Experience in handling metals helps greatly!

Another clue from touch is apparent *coolness* compared to other objects in the same surroundings or environment as the object of interest, so temperature has equilibrated in them all. Since metals generally have higher values of thermal conductivity than most ceramics, all polymers and polymer-matrix composites (e.g., fiberglass, graphite-epoxy, and Kevlar-epoxy), and wood, they will feel cooler, even though they are not actually at a lower temperature. The sense of *coolness* comes from the rate at which heat is drawn from one's hand by the material. Higher thermal conductivity metals, like Al, feel cooler than, say, steel, and both feel cooler than polymers or wood. The degree to which one metal, in particular, feels cooler than another (e.g., steel) is directly proportional to its higher thermal conductivity.

Finally, metals tend to have a certain sound when they are tapped or struck with a metal object or a stone, for example. Commonly referred to as a *metallic sound*, it is the result of how much faster—and easier—sound waves travel through metals than through other materials. The speed of sound is related to the bulk modulus and density of a material, as the square root of the ratio of the modulus to the density.[23] Polymers tend to "thud" when tapped, more so the lower their modulus, as the densities of all polymers are about the same (i.e., 1.0 to 1.3 g/cm^3). Woods tend to exhibit a "knock," while ceramics exhibit a sharp sound, but definitely different from metals.

One can also use the modulus of material to assess flex (in bending) versus steel of about the same shape and thickness. Higher-modulus materials flex less for a given applied bending force.

Another observation from touch is hardness, which can be assessed by attempting to scratch the surface of the unknown material with a sharp metallic object (knife blade, screwdriver tip, common nail, car key, etc.). Polymers are all softer than most engineering metals and alloys and all ceramics, so they will scratch easily (often producing a "shaving" and scar). Ceramics are generally harder—much harder—than most engineering metals and alloys, and will rarely be scratched by a metal.

---

[23] *Bulk modulus K* is related to Young's modulus *E* through Poisson's ratio n as $K = E/3(1-2\nu)$. Values for ν for metals typically lie between 0.3 and 0.4, with 0.35 being a good average value.

188 Chapter Nine

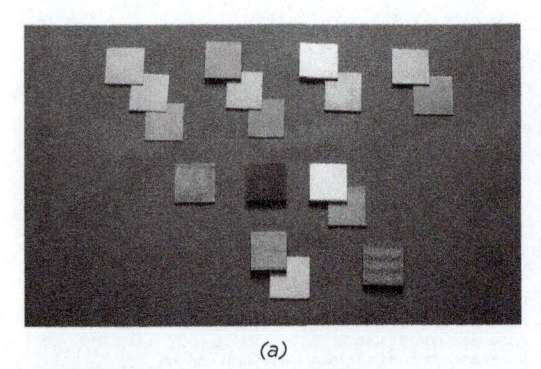

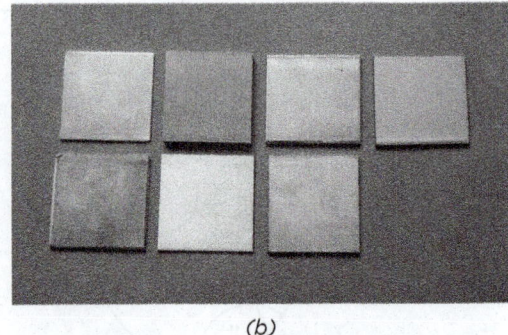

Figure 9-7 Photographs showing the colors of various engineering metals and alloys. From left to right in the top row, the samples are: 2024 Al alloy with Cu, O annealed, and T4 and T6 aged conditions (top to bottom within a set); 7075 Al alloy with Zn and Mg, O annealed, and T6 and T651 aged conditions; 6061 Al alloy with Mg+Si, O annealed, and T6 aged conditions; 5083 Al alloy with Mg, O annealed, and T6 aged conditions. From left to right in the middle row, the samples are: plain carbon 1045 steel in annealed condition; 4.8% C gray cast iron; pure Cu and 70Cu-30Zn yellow brass (top to bottom in the set), both in annealed condition. From left to right in the bottom row, the samples are: Ti alloy with 6Al and 4V (i.e., Ti-6-4) in annealed condition and 304 austenitic stainless steel (nominally 19Cr-9Ni) and a red bronze with around 10% Sn in Cu, both in annealed condition (a). From left to right in the top row, the samples are: 5083 Al alloy with Mg in O annealed condition; 5083 Al alloy with Mg in T6 aged condition; Ti-6Al-4V in annealed condition; Ni alloy Inconel 625 in annealed condition. From left to right in the bottom row, the samples are: 0.45% C 1045 plain carbon steel in annealed condition; 304 austenitic stainless steel in annealed condition; and commercially pure (CP) Ti in annealed condition (b). (Source: Photograph by Kris Qua Photography for Robert W. Messler, Jr.; property of Robert W. Messler, Jr.) **Don't miss the color version of this figure, available at www.mhprofessional.com/ReverseEngineering.**

So you know a part or detail is a metal (versus being a ceramic or a polymer), but what metal? You need to use more observable clues from your senses, specifically:

- Color (silvery like clean Ag, yellow like gold and Cu-Zn brasses with more than 30wt.% Zn, reddish-orange like copper or Cu-rich brasses, greenish or greenish-yellow like Cu-Sn bronzes, "white" or silvery-white like Al, light to medium gray like steel, dark gray like cast iron, etc.) (see Figure 9–7 as a full-color print or online photograph).
- Heft (relative to steel) from density
- Coolness (relative to steel) from thermal conductivity
- Metallic ring (relative to steel) from modulus of elasticity
- Hardness (relative to annealed steel) using a scratch test with a sharp steel object
- "Flex" (relative to like-shaped and -thickness steel) from modulus of elasticity

and, of course,

- Attraction to a magnet; strongest for carbon and low-alloy steels, moderate for Co and Co alloys, and light for Ni and Ni alloys. Other metals are nonmagnetic.

Comparative qualities or identifying characteristics of major engineering metals and alloys to help with observations are given in Table 9–4. Similar information is provided for nonmetallic materials

TABLE 9-4 Identifying Characteristics for Metals and Alloys

Metal/Alloy	Density (g/cm³)	(vs. steel)	Thermal Conductivity (W/m-K)	(vs. steel)	Stiffness (GPa)	(vs. steel)	Color	Other
Ag	10.49	1.33	429	8.6	83	0.4	silvery-white	soft
Al	2.70	0.34	237	4.7	69–70	0.33	silver ⇒ light gray	very soft
Al alloys	2.63–2.83	0.35	160–190	3.5	70–75	0.35	light to dark gray	soft ⇒ hard
Au	19.30	2.44	318	6.4	79	0.38	rich yellow	very soft
Cu	8.96	1.13	401	8.0	110–128	0.56	red-orange	soft
Brasses	8.2–8.4	1.05	100–120	2.2	100–125	0.54	yellow ⇒ red	medium
Bronzes	8.8–8.9	1.12	50–65	1.1	95–120	0.51	greenish-yellow	med. ⇒ hard
Monels	8.9	1.13	20–25	0.5	180	0.86	white ⇒ yellow	med. ⇒ hard
Fe	7.874	1.0	80.4	1.6	211	1.00	silver gray	soft; magnetic
Steel	7.85–7.95	1	35–60	1	210	1	gray ⇒ dark gray	hard; magnetic
Cast iron	7.15	0.91	42–70	1.1	110–160	0.64	dark gray	hard; brittle; mag.
Stainless	8.0–8.1	1.02	12–15	0.3	190–200	0.93	silver-gray	med.; nonmag.
Mg	1.738	0.22	156	3.1	45	0.21	shiny gray	very soft
Mg alloys	1.8–2.3	0.26	60–80	1.5	50–60	0.26	shiny gray	soft ⇒ hard
Ni	8.909	1.14	90.9	1.8	200	0.95	silver, gold cast	soft; slight mag.
Ni alloys	8.4–8.8	1.09	12–30	0.4	200–205	0.96	lt. gray, gold cast	med. ⇒ hard
Ti	4.506	0.57	21.9	0.4	103	0.49	greenish cast	med. ⇒ hard
Ti alloys	4.3–4.6	0.56	7–10	0.2	105–117	0.53	greenish cast	med. ⇒ hard
W	19.25	2.44	173	3.5	411	1.96	grayish-white	hard
Zn	7.134	0.90	116	2.3	108	0.51	silver-gray	soft ⇒ medium
Zn alloys	6.6–6.9	0.85	110–150	2.6	85–87	0.41	silver-gray	medium

TABLE 9-5 Identifying Characteristics for Nonmetallic Materials

Material	Density (g/cm³)	Th. Cond. (W/m-K)	Stiffness (GPa)	T_{glass} (°C)	Other Properties
Polymers					
Polycarbonate	1.20-1.22	0.19-0.22	2.0-2.4	+147	Hardness M70/R118
Polyethylene (HD)	0.93-0.97	0.49	0.8	-130/-80	Hardness Shore D55-70
Polymethylmethacrylate	1.18	0.167-0.25	1.8-3.1	+85/165	Hardness R92
Polypropylene	0.86-0.95	0.1-0.22	1.5-2	-20	Hardness R95
Polystyrene (dense)	1.06-1.22	0.2-0.25	3-3.5	+100	Hardness R95
Polystyrene (foamed)	0.015-0.03	0.033	1.8-2.0	+100	n/a
Polytetrafluoroethylene	2.2	0.25	0.5	+115	Hardness Shore D50-65
Polyvinylchloride	1.1-1.45	0.19	2.5-4.5	+82	Hardness Shore D85
Nylon 6,6	1.14	0.25	2-3.6	+45/50	Hardness R88
ABS	1.02	0.17	2.3	+90	Hardness R110
Ceramics					
Alumina	3.95-4.1	20-30	300-400	n/a	VHN 1550/KHN 2100; white
Silicon carbide	3.21	120	450-500	n/a	KHN2480; black color
Silicon nitride	3.2	30	300-400	n/a	VHN 1580; gray/black color
Zirconia	5.68	2-23	205-300	n/a	VHN 900-1600; white
Composites					
E-Glass-epoxy	1.3-1.4	0.04	17	n/a	Commonly "fiberglass"
Graphite-epoxy	1.6-1.7	150-250	150-300+	n/a	Various fiber types
Kevlar-epoxy	1.1-1.25	<0.10	65-75	n/a	Known as Aramid
Wood, soft (pine)	0.37-0.5	0.14-0.16	8-10	n/a	Parallel to the grain
Wood, hard (oak)	0.65-0.8	0.15-0.20	10-12	n/a	Parallel to the grain

TABLE 9-6 Typical Applications of Some Major Engineering Polymers

	Key Properties	Key Applications
Polycarbonate (PC) (comes as Lexan™)	Strong; durable; high impact resistance; low scratch resistance; optical clarity	CDs/DVDs; construction materials (e.g., sound walls); bottles; electronic components; optical reflectors; eyeglass lenses
Polyethylene, high-density (HDPE)	Inert; thermally stable; tough; high tensile strength	Arena boards; backpack frames; chemical-resistant tanks; piping; plastic lumber
Polymethyl-methacrylate (PMMA) (comes as Plexiglass™)	Extremely strong; low impact resistance	Orthopedic bone adhesives; optical lenses; optical fibers; transparent glass substitute
Polypropylene (PP)	Resistance to acids and bases; high tensile strength	Auto parts; industrial fibers; food containers; "living" hinges; tubing
Polystyrene (PS), dense or foamed	Thermal insulator; when dense, strong and tough; when foamed very light and shock absorbing	CD cases; impact-resistant cases; plastic utensils; foamed packaging
Polytetrafluoro-ethylene (PTFE)/Teflon™	Very low coefficient of friction; excellent dielectric properties; chemically inert	Armor-piercing bullets; bearings; slide plates, nonstick cookware; coating against chemical attack; sterile artery bypasses; thread sealing tapes; tubing
Polyvinylchloride (PVC)	Insulator; chemically inert	Architectural trim; cable and wire insulation; home siding; pipes; window frames; fencing; toys
Nylon 6,6	High strength; good rigidity; tough	Gears; ropes; strings; tubes
Acrylonitrile-butadiene-styrene (ABS)	Superb impact resistance; good strength	Automobile bumpers and trim; enclosures for electronics; protective headgear; LEGO blocks

in Table 9–5, although it must be said that identifying a specific ceramic or polymer is much tougher and, generally, impossible. For polymers, a better approach may be to consider the specific application compared to common applications of specific polymers (Table 9–6).

The point is this: Do the best you can to identify the fundamental material type (by class), and then do the best you can to try to narrow down the specific subtype (Al alloy, common steel, hard thermoplastic, perhaps polycarbonate, etc.). Positive identification can be made using laboratory tests described in Section 9–8. However, during reverse engineering, a general sense of what material was used to fabricate a particular part or detail helps greatly in deducing role, purpose, and functionality.

9-8 Laboratory Identification Methods

A sure way to identify a material, much more reliably and specifically than using observations, is to use laboratory techniques, of which there are many (ref. "Characterization (Materials Science)," Wikipedia.com). These run the gamut from simple-to-use chemical spot tests to highly sophisticated techniques that rely on expensive apparatus and highly skilled technicians.

A very simple test, which can even be applied in the field, no less in a general engineering laboratory, uses *metal test kits* and so-called chemical spot tests. Available from a number of suppliers, these all use chemical reagents that fluoresce into brilliant colors in the visible or UV spectrum. Specific tests are used to identify specific base metals, including kits to distinguish metals from one another (e.g., Al, Mg, Fe, Ni, Cu, Ti, Zn) as well as to identify specific generic alloys within a base system (Al alloys by classification, stainless steel types from other types of steel, etc.).[24]

The two most popular analysis techniques used to identify metals and alloys are (1) x-ray fluorescence (XRF) and (2) optical emission spectroscopy (OES).

In XRF, the emission of characteristic "secondary" (or fluorescent) x-rays from a material excited by bombardment with high-energy x-rays or gamma rays is used for identification. The technique allows for simple elemental analysis (i.e., makeup) or chemical analysis (i.e., composition) of metals, glasses, and ceramics. A typical energy dispersion XRF spectrum is shown in Figure 9–8.

In OES, a small (5- to 10-mg) sample of a material to be analyzed is vaporized using a laser beam or an electric arc, which excites electrons in the various atoms that make up the sample to emit light as the excited electrons fall back to their ground state. The wavelengths of the various light signals emitted provide a unique signature of the element. By analyzing the spectrum, including wavelengths and intensities, it is possible to calculate the range of elements present and their composition.

Before any chemical analysis is conducted (if one ever is conducted!), a materials engineer (and, especially, a metallurgist) would examine the microstructure of the metal or alloy part or detail. Such metallographic examination using optical microscopy gives most of the information a good metallurgist would need to identify most metals and alloys. Readers interested in metallographic techniques are encouraged to seek references (e.g., ref. vander Voort).

For polymer materials, a variety of other techniques are used to identify one type from another, where observable clues alone help little. Without going into details here, the list of techniques used to identify a polymer includes the following:

- Elemental analysis (using Lassaigne's test)
- Solubility tests
- Infrared (IR) analysis
- Flame test/melting test
- Specific gravity determination
- Dilatometry tests (to determine T_g)

---

[24] One supplier is Koslow Scientific Testing Instruments, 172 Walkers Lane, Englewood, NJ 07631.

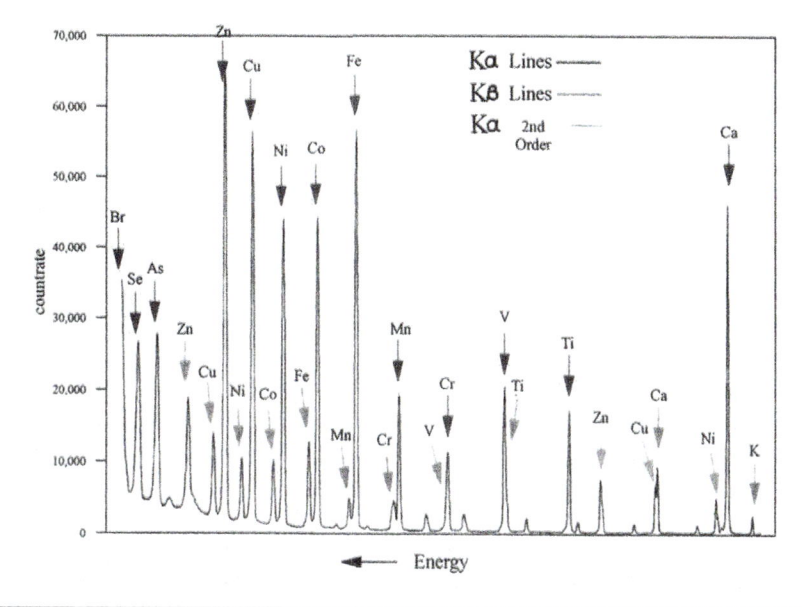

Figure 9-8 Example of the dispersive energy spectrum obtained during x-ray fluorescence analysis. (*Source:* Wikipedia Creative Commons, contributed by Magnus Manske on 14 March 2009.)

A few particularly common useful techniques for identifying and characterizing polymers are the following:

- *Differential scanning calorimetry* (DSC): Used to identify the glass transition (T_g) and melting temperature (T_{MP}) of polymers. It is also useful to identify whether a part contains multiple types of polymers; however, this would work only if the different polymers have glass transition temperatures of more than about 5 degrees and if the polymers are immiscible.
- *Wide-angle x-ray diffraction/scattering* (WAXD, WAXS, or XRD): Suitable only for semicrystalline polymers and could provide the percentage of crystallinity in addition to lattice information.
- *Fourier transform infrared spectroscopy* (FTIR) *or Raman spectroscopy*: FTIR is commonly used to identify chemical (or radical) groups on a polymer chain. It is a relatively easy technique, but the raw data can be quite extensive to analyze. Newer machines are computer controlled, and most of them include a database to identify and match peaks to chemical groups.
- *Nuclear magnetic resonance* (NMR): Different types are available, but C^{13}-NMR is extremely powerful and not only could provide detailed information about the chemical structure of the polymer chain but any defects along the chains can also be deduced. The experimental data is not easy to analyze.

Readers interested in identifying polymers are encouraged to seek additional information from online or other references by technique. The amount of effort that is put into identifying materials during reverse-engineering dissection depends on the goals of the procedure.

9-9 Summary

There is an inextricable interrelationship among the structure, properties, processing, and performance of solid materials, with performance at the material level directly influencing performance in the part or detail comprising that material. The structure of a solid material arises at the lowest level from the nature of the atoms and the type of bonding that make up the material. There are additional contributions at the nanoscale (10^{-9} to 10^{-8} m), microstructural scale (10^{-6} to 10^{-4} m), and the macroscopic scales (10^{-2} and up). The properties of a material are its response to a particular external stimulus or set of external stimuli, including mechanical, electrical, thermal, optical, magnetic, and chemical stimuli and corresponding properties. More complex properties of interest in manufacturing, in particular, are ill-defined or unknown combinations of these basic properties.

Because material properties are the basis for its selection for a particular detail of a design, identifying the material-of-construction during reverse-engineering dissection often provides valuable clues about the role, purpose, and/or functionality of the parts, details, and overall entity. Since reverse engineering relies so heavily on observation, one needs to use observations as the major source (or, at least, first-level source) for identification. Relying on the senses, clues can be obtained from color, opacity versus transparency, luster, heft (as a reflection of density), coolness (as a reflection of thermal conductivity), stiffness against flexural bending (as an indicator of modulus), sound when tapped (as a reflection of modulus and density as a ratio), and hardness (using a simple scratch test).

Positive identification of a material-of-construction is possible using any of a number of laboratory analysis techniques, including metallography or materialography.

9-10 Cited References

Ashby, Michael F., *Materials Selection in Mechanical Design,* 4th edition, Butterworth-Heinemann/Elsevier, Burlington, MA, 2010.

Callister, William D., Jr., and David G. Rethwisch, *Materials Science and Engineering: An Introduction,* John Wiley & Sons, 2010.

Messler, Robert W., Jr., *The Essence of Materials for Engineers,* Jones & Bartlett Learning, Burlington, MA, 2011.

vander Voort, George F., *Metallography: Principles and Practice,* ASM International, Materials Park, OH, 1999.

"Characterization (Materials Science)," Wikipedia.com (provides an extensive list of techniques for identifying materials and gives reference to many specific *highlighted* techniques).

9-11 Recommended Readings

Ashby, Michael F., *Materials and the Environment,* 2nd edition, Butterworth-Heinemann/Elsevier, Burlington, MA, 2012.

Ashby, Michael F., and Kara Johnson, *Materials and Design: The Art and Science of Material Selection in Product Design,* Butterworth-Heinemann/Elsevier, Burlington, MA, 2002.

9-12 Thought Questions and Problems

9-1 The fundamental premise—and fact—underlying the study of materials (i.e., *material science*) is that *structure determines properties*. Without consciously being aware of it, you have known this since you first handled a wood baseball bat or observed how flagstones were strong in the plane of the stone but easily separated through its thickness. This effect begins at the atomic scale (10^{-10} m) for mechanical properties (at the even smaller electron level for electrical, magnetic, thermal, optical, and radiation properties) and reappears and/or is further developed at nano- (10^{-9} to 10^{-8} m), micro- (10^{-7} to 10^{-4} m), and macro- (10^{-3} to 10^{-1} m) scales.

   a. Use the Internet or a materials science textbook (Callister, Messler, Shackelford, Askland et al., Smith, etc.) to find *two* examples of this relationship for each of the following, giving a specific value of some appropriate property to support your choice in each case:
      (1) a hardwood versus a softwood (i.e., natural composites) parallel *and* perpendicular to the grain
      (2) a naturally occurring soft versus hard mineral (i.e., a ceramic or ceramic composite)
      (3) a pure soft versus pure hard elemental metal
      (4) synthetic (i.e., man-made) versus natural fiber (i.e., both polymers) (This is more challenging.)
      (5) human cancellous bone versus hydroxyapatite mineral (i.e., a natural ceramic composite versus a ceramic) (This is more challenging.)

Since engineers manufacture or construct things from materials, it is usually necessary to subject these materials to one or more processes to create a needed shape or develop a needed property. Hence, engineers in general, and materials engineers in particular, must deal with the more elaborate, but equally inextricable, interrelationship between structure ⇔ property ⇔ processing.

   b. Use the Internet or a materials science and engineering textbook (especially, Messler) to find *two* examples of this interrelationship for each of the following, giving a specific value for some exemplary property affected by the processing to support your choice of each:
      (1) cold rolling a metal or alloy, before and after cold working
      (2) heat treating a steel by quenching, before and after quench treatment
      (3) drawing a polymer fiber versus producing the same polymer as a bulk form (This is more challenging.)
      (4) tempering a soda-lime window glass versus the same glass without tempering (i.e., untempered) (This is more challenging.)
      (5) the region immediately adjacent to a fusion arc weld made in a low-alloy steel (e.g., AISI 4130), in the heat-affected zone and in the unaffected base metal (This is more challenging.)

9-2 When it comes to selecting a material to meet the requirements of a design, it is all about the properties that give rise to the required performance (i.e., the structure ⇔ property ⇔ processing ⇔ performance interrelationship).

   a. Use the Internet or a materials science and engineering textbook (Callister, Messler, Shackelford, Askland et al., Smith, etc.) to find *two* examples of a design situation or application for which each of the following is important for ultimate performance in the design:

(1) mechanical strength: *one* for yield and *one* for ultimate strength
(2) mechanical toughness: *one* requiring high toughness and *one* requiring very low toughness (This is more challenging.)
(3) high electrical conductivity (for one example) versus high electrical resistivity (for the other example)
(4) high thermal conductivity (for one example) versus very low thermal conductivity or high insulation (for the other example)
(5) need for two different optical properties for different applications
(6) very good resistance to corrosion (for one example) versus very low resistance or high reactivity (for the other example) (This is more challenging.)
(7) very low density (for one example) versus very high density (for the other example) (This is more challenging.)

b. Because of the engineer's need to create devices, structures, etc., more complex combination properties (e.g., machinability) are sometimes very important.

Use the Internet or a materials science and engineering textbook to find a good example where each of the following complex combination properties would be desirable, if not required:
(1) good to excellent machinability
(2) good cold formability
(3) good castability
(4) good moldability
(5) good weldability
(6) good solderability

9-3 Every engineer needs to have a basic understanding of materials science since all engineers (except software engineers) design, build, test, operate, and maintain *things* made from materials. Toward this end, as a refresher for earlier exposure to materials in an introductory course or to reinforce what is covered in Section 9–4, respond to the following:

a. Why are all *metals* and *alloys* electrically conductive?
b. Why are most *metals* and *alloys* ductile, while virtually all *ceramics* are brittle?
c. Where do *polymers* come from? Are they on the periodic table? Are they all synthesized (i.e., man-made), or do any occur naturally?
d. What is it that causes some *metals* to be high melting and some to be low melting?
e. Why are *ceramic* materials almost always higher melting than most *metals* and *alloys*?
f. Why are *ceramic* materials resistant to corrosion?
g. What is it—from their structure—that causes *polymers* to be called "plastics"?
h. Are there any ceramics that conduct electricity? If so, name one. If not, why not?

9-4 A powerful tool for selecting materials in design is the use of *Material Selection Charts* (Section 9–6). Figures 9–4 and 9–5 give two examples.

Using Figure 9–5 for Young's modulus (i.e., stiffness) versus density (i.e., a key contributor to the weight of structure), respond to the following:

a. Why do like materials (e.g., all engineering metals and alloys or all foamed polymers) cluster in "bubbles" on the chart?
b. Explain the relative position or location of the "bubble" for engineering ceramics versus engineering metals and alloys, and then of engineering polymers versus engineering metals and alloys.

c. Use the Internet to find the most dense engineering material, and also the least dense of all solid materials. Give the density of each in grams per cubic centimeter (g/cm³).
d. Materials that lie on the same *design guideline* (i.e., sloping dashed lines on the chart) are equivalent for *the plotted properties*. Give *two* examples of structures, products, or devices where at least three different materials (often from different "bubbles") have been used for their stiffness-to-weight relationship.
e. Why can one *not* simply use any of the materials that lie on the same dashed design guideline for an application for which they seem equivalent or comparable for the plotted properties? Explain your answer.

9-5 Having the ability to distinguish one material from another and even one general metal or alloy group from another is a valuable skill for engineers, especially during reverse engineering.
  a. Explain what you would try to *observe* about two materials that both appear dull gray but you suspect are different in that one is a *metal* or *alloy* and the other is a *ceramic*.
  b. You are presented with six parts, each made from a different bare (unpainted, not plated, not coated, etc.) metal or alloy. You have reason to believe they are the following, but you do not know which is which:
  - Pure aluminum (AA1100, annealed)
  - Fully age-hardened Al alloy (e.g., AA7075-T6)
  - A titanium alloy in the annealed condition (e.g., Ti-6-4)
  - A plain carbon steel in the annealed condition (e.g., AISI-SAE 1018)
  - A low-alloy steel in the quenched-and-tempered condition (e.g., AISI-SAE 4130)
  - An austenitic stainless steel in the annealed condition (e.g., AISI-SAE 304)

  All you have available, besides your own senses, are a ruler, a magnet, a magnifying glass, and a small pocketknife.

  Briefly, but completely, describe how you would identify each of the six parts by its material-of-construction.
  c. You are presented with three identical-appearing materials, in the form of a 5-centimeter-long by 0.5-centimeter-diameter solid rod. You are told one is "a metal," one is "a ceramic," and one is "an elemental semiconductor." You have a beaker of water and ice, a beaker of boiling water, a voltmeter, and a nail file. You can use any, all, or none of these—only!

  Briefly, but completely, describe how you would tell which is which using one test involving measurement of one property.
  d. Why is it that it is *so* difficult to tell polymers apart? How has this made recycling of "plastics" more difficult?

CHAPTER 10

Inferring Methods-of-Manufacture or -Construction

10-1 Interaction among Function, Material, Shape, and Process

In Section 9–2, the inextricable interrelationship among structure, properties, processing, and performance was presented and discussed. In Section 9–6, a process for selecting materials to meet critical design requirements (first and foremost, functionality) using Material Selection Charts was introduced, with full treatment left to a seminal work by Michael F. Ashby (ref. Ashby). Before embarking on consideration of the methods used for manufacture or construction, it is important to focus once more on the materials-of-construction.

At the heart of the process for selecting materials in design lies another inextricable interrelationship or interaction: the *interaction among function, material, shape, and process.* As used here, and originally by Ashby, the role of each of these factors is:

- *Function* drives (if not "dictates") the choice of material in design.
- *Shape* is chosen to perform the required function(s) using the selected material.
- *Process* is strongly influenced by the complex material properties of formability, machinability, castability, moldability, weldability, heat-treatability, and, sometimes, others.[1]

Recall that material is chosen based on the required or desired responses to stimuli acting on or arising from within the design. Once again, the interactions between pairs of these three factors are two-way.[2] For example, function (e.g., to resist or limit deflection) dictates shape (e.g., a cross

---

[1] The reverse of the order of *formability* and *machinability* here versus earlier (Section 9–3) is to reflect the taxonomy of manufacturing methods presented in Section 10–3.
[2] See Section 9–2 for the two-way interactions among structure ⇔ properties ⇔ processing ⇔ performance.

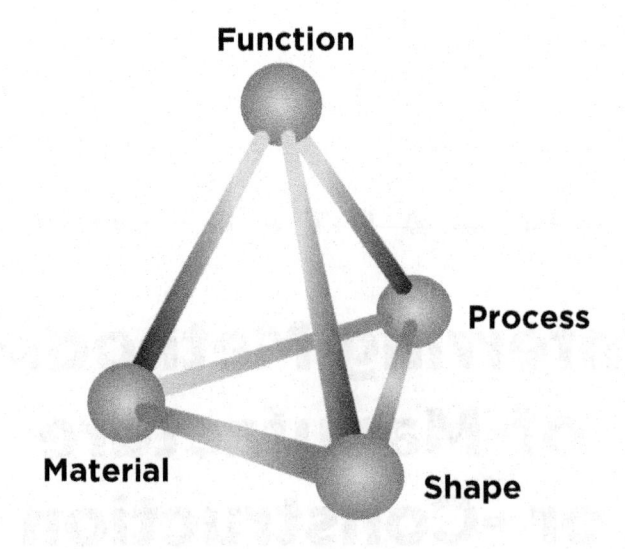

Function

Material **Shape**

Process

Figure 10-1 Schematic depiction of the interaction between and among function-material-shape-process, with all-important function at the apex of the a regular tetrahedron.

section with a high moment of inertia I) but so, too, does the shape (e.g., offering stiffness against bending) dictate function (i.e., to resist or limit deflection). Ashby correctly points out: The more demanding (or "sophisticated") the design, the more restrictive (or "tighter") the specifications and the greater the interaction. He uses the wonderfully vivid analog: "It is like designing a wine: for cooking wine, almost any grape and fermentation process will do. For champagne, both grape and process are tightly constrained." This American author (versus the aforementioned wonderful European author) would have chosen an analogy of blended whiskey and 12+-year-aged, single-barrel Tennessee or Kentucky bourbon. But to each his or her own.

Figure 10–1 schematically depicts the interaction among *function* ⇔ *material* ⇔ *shape* ⇔ *process,* with function at the apex of a regular tetrahedron having an equilateral triangle base of material ⇔ shape ⇔ process.[3]

The basis for the inextricable interactions among function, material, shape, and process is that function requires both material and shape, but to achieve shape, the material must be subjected to processes, which, taken together, constitute manufacturing or construction.[4]

So let's look at the role of manufacturing or construction.

---

[3] Recall the paramount role of function (or functionality) in design (Section 9–1), hence the positioning of *function* at the apex, analogous to *performance* being the apex of the structure-property-processing-performance tetrahedron in Figure 9–1.
[4] *Shape* is used herein to represent two scales. First, external shape together with size (as geometry), constitute what is most correctly the *macroshape* of a physical entity. Second, physical entities can possess a *microshape* that is internal to a specific component or an entity. Three examples are: (1) honeycomb cells in lightweight Al alloy structural sandwich panel cores used in aerospace, (2) air cells in foamed polymers or metals or human cancellous bones, and (3) grain structure in wood or directionally rolled metals or alloys.

10-2 The Role of Manufacturing or Construction

The King James Version of the *Holy Bible* opens with Genesis I, verse 1: "In the beginning God created the heavens and the earth." With all due respect, in this statement in Judeo-Christian belief, God (Yahweh), as Creator, was/is architect, engineer, and fabricator. He or She designed and made (i.e., fabricated) everything for both functionality and aesthetics.

As God's ultimate creation, we human beings, as an intelligent and curious species, seek to understand how our universe, our planet, and we ourselves came to be. Physics has come along to tell us what the universe comprises and how it works (ref. Hawking and Mlodinow). In the so-called big bang theory, physics proposes how the universe was manufactured or constructed, as it were. And, as part of that universe, physics has deduced how our own Planet Earth was "manufactured." But, despite its efforts, physics has not yet resolved the mystery of how the original matter was manufactured. It cops out by saying: Before there was time, there was nothing. At the moment time began, all matter was concentrated in an infinitely dense sphere which exploded from its own repulsive energy (and forces) to create all that there is in the universe. Uh-huh! But where did "all the matter" come from at time $t = 0$?

The process by which physics has come to understand (or think it understands!) what it does, is, in fact, a long and elaborate process of reverse engineering. Is it not? Rather than intending to re-create or improve upon the original, the motivation for physicists is solely (as opposed to simply!) to understand it: its role, purpose, and functionality. The process used has, very much, used dissection—both physical dissection (e.g., by geologist and high-energy-particle physicists) and intellectual dissection (e.g., by theoretical physicists, cosmologists, astrophysicists, and astronomers) based on deduction and inference. Deduction, as always, has been based on observable, measured, or experimentally derived clues, while inference, as always, has been based on a reality (as an effect) that suggests a cause. We are here, so we must have been created.[5]

A design—no matter how meticulously detailed—is only a concept (i.e., is abstract) until it is built (i.e., to become a reality). As used throughout this book, there are two ways by which designs get built: (1) they are manufactured, if the work is done or could have been done indoors, or (2) they are constructed, if the work, of necessity, is done outdoors. Airplanes are manufactured in airplane factories or aircraft plants, as are automobiles manufactured in automobile factories or plants. But dams are constructed where they are needed. In modern bridge building, most of the structural elements, as well as subassemblies, are manufactured indoors, while the bridge itself is constructed outdoors, on-site by erecting the prefabricated details.

Because of the interaction among function ⇔ material ⇔ shape ⇔ process (Section 10-1), a great deal can be learned from the way something was fabricated and assembled. If, for example, steel (material) I-beams are used for the columns (i.e., vertical compression-loaded structural members) of a building, it is clear the I-shaped cross section (shape) was chosen for greater structural stiffness against buckling (function). Closer examination (observation) of the I-beams will reveal whether they were produced in one piece (i.e., were monolithic) or from multiple pieces (i.e., were built up). In the former case, they would have been manufactured by hot-rolling (in a

---

[5] Recall Socrates's (Greek philosopher, 499–369 BC) inference: "If there is a statue, there must be a sculptor." Also, take another look at the quote from the great rocket engineer, Dr. Wernher von Braun, at the front of this book.

steel plant), while in the latter case, they may have been manufactured (in a plant) or constructed (on-site) by welding, bolting, or riveting. How they were manufactured (or constructed), in turn, reveals a great deal about the engineering (versus material) structure. Let's see why.

One-piece hot-rolled steel I-beams are reasonably structurally efficient (although there may be more material in some features, e.g., horizontal caps or vertical web, than is needed)[6] and are, without question, cost effective, since they are (1) prefabricated by experts as (2) commodity product in (3) standardized shapes and size at (4) high speeds. But there are limits on how large (in area) their cross section can be and, thus, on how much load they can safely support. Being monolithic (i.e., one piece), they have somewhat limited damage tolerance (beyond what can be derived from selection of a steel with suitably high critical fracture toughness K_{IC}). A crack anywhere in the beam's cross section is free to propagate across the entire section to cause complete fracture. Hence, the alternative of built-up I-beams arises.

A built-up I-beam typically includes: (1) horizontal cap members; (2) additional narrower cap members for use as "doublers," "triplers," and so forth; (3) vertical web members; (4) additional narrower web members as doublers, and so forth; and (5) perhaps, right angles to allow joining of the horizontal cap and vertical web members. By building up the beam from smaller details, the cross section can be optimally designed and precisely fabricated to provide exactly the moment of inertia desired. There is no wasted material (cost) and no excess weight. The drawback is that the various details of the I-beam-to-be must be joined.

There are two fundamental joining options, with two suboptions for one of these. First, the details could all be fusion welded (e.g., using shielded metal-arc or flux-cored arc welding either in a prefab shop or on-site). This approach would keep weight to a minimum (i.e., no added material for fastener heads and feet) and would preclude most water entrapment and subsequent crevice corrosion. On the other hand, in creating a monolithic structure, this approach would sacrifice damage tolerance, as a crack anywhere in the beam's cross section would be free to propagate through the section unabated by any physical interface, leaving damage tolerance solely to the steel's K_{IC}.

The second option is to build up the details using mechanical fasteners, of which there are two suboptions: (1) high-strength bolts with nuts or (2) hot-set steel rivets. Either option adds some weight (for protruding heads on bolts and for extending bolt shanks and nuts or for rivet heads and upset feet) compared to welding (which adds only modest fillet material). The particular advantage added by this approach is improved structural damage tolerance, as every interface between details serves to arrest any cracks growing within any structural element. An example of where this is very important is in the massive vertical columns and horizontal beams or arches used in underground subway systems or in overhead, elevated rail systems. Load-carrying requirements frequently require cross sections that are too massive to hot roll.[7] In addition, severe low- and

---

[6] Structural steel I-beams are manufactured to specifications developed under codes by organizations responsible for public safety (e.g., the American Society of Civil Engineers). Certain standard designs are available for which the relationships between key dimensions are fixed; e.g., horizontal cap width-to-vertical web height, cap thickness-to-cap width, web thickness-to-web height, and cap thickness-to-web thickness. These relationships represent a consensus response to consensus agreement on loading and structural requirements. The need for customized dimensions sometimes exists, however.

[7] One reality an engineer must contend with in manufacturing is that there is a limit to how large an entity can be produced by any particular manufacturing method. One can only pour so much molten metal to produce a casting, only deform so much material by rolling or extrusion or drawing, only forge so much material, only machine something so big. Size and shape (complexity) limits exist from every manufacturing method.

high-frequency vibrations (from heavy trains) lead to potentially severe fatigue loading, with a great asset being crack arrestment at interfaces.

So manufacturing (including construction, where appropriate) plays a major role in an executed design. It is not surprising, therefore, that there is information to be gleaned during reverse-engineering dissection on the role, purpose, and/or functionality of a component, subassembly, or entire structure or system. Before considering how such information can be obtained, however, let's take a look at the organization of manufacturing processes into a logical taxonomy (Section 10–3),[8] as well as an overview of key manufacturing methods (Section 10–4).

10-3 The Taxonomy of Manufacturing Processes

Processing materials has three principal purposes, as aims. These three involve the achievement of one or more of the following: (1) geometry (i.e., shape and dimensions), (2) properties, and (3) finish. In turn, these three aims are achieved using one or more of various options or methods. *Shape and dimensions* (as these create geometry) can be achieved by three basic methods: (1) flow (or rheological) processes, (2) machining, and (3) assembly of premade parts by joining. *Properties* are achieved in materials selected for their potential property(ies) when brought to the right condition (via microstructure) using heat treatment.[9] *Finish,* which includes engineering tolerances, surface quality, and surface protection and appearance, is achieved on a part by either machining or a variety of chemical, mechanical, or thermal methods or treatments.

After looking at a logical way of arranging manufacturing processes and/or methods into a taxonomy, it will become apparent that these can also be divided into processes or methods that (1) move material around without removing or adding any material, (2) removing material, or (3) adding material. The latter two methods are known as *subtractive* and *additive processes,* respectively, while the former are known as *flow processes.* In the case of heat treatment processes, which neither add nor subtract material, the "flow" involves the movement of atoms around, generally by diffusion mechanisms but, possibly, by a massive shear mechanism.

A scheme that divides manufacturing processes into flow, subtractive, and additive types proves useful during reverse-engineering dissection, as it potentially provides clues to the specific method-of-manufacture.

Figure 10–2 presents a taxonomy of manufacturing processes that, while generic, was popularized by Ashby (ref. Ashby) and found highly useful for discussion of processes.

The elegance of the taxonomy shown in Figure 10–2 is that it shows, in the form of a flowchart, how a raw material can be taken through various steps of manufacturing to end up with a detail part ready to use for itself or be integrated into a device, mechanism, product, assembly, structure, or system. Nine classes of process are shown. Those classes in the first long horizontal row include all of the processes used to accomplish primary shaping. In some cases, within each class, the

---

[8] *Taxonomy* is defined as "[a system of scheme that results in a] division into ordered groups or categories. The most familiar taxonomy is that used in biology to divide living things into kingdom (e.g., animal), phylum (e.g., cordata, having a spine), class (e.g., carnivora, meat-eater), family (e.g., canidae, dog-like), genus (e.g., carnu, dog), and species (e.g., familiaris, domesticated).

[9] *Composite materials* are unique among engineering materials in that they allow the tailoring of desired properties in the composite by mechanically combining two or more specific materials from one or more classes of materials. In doing this, they take advantage of what is known as the "principle of combined action" (ref. Messler).

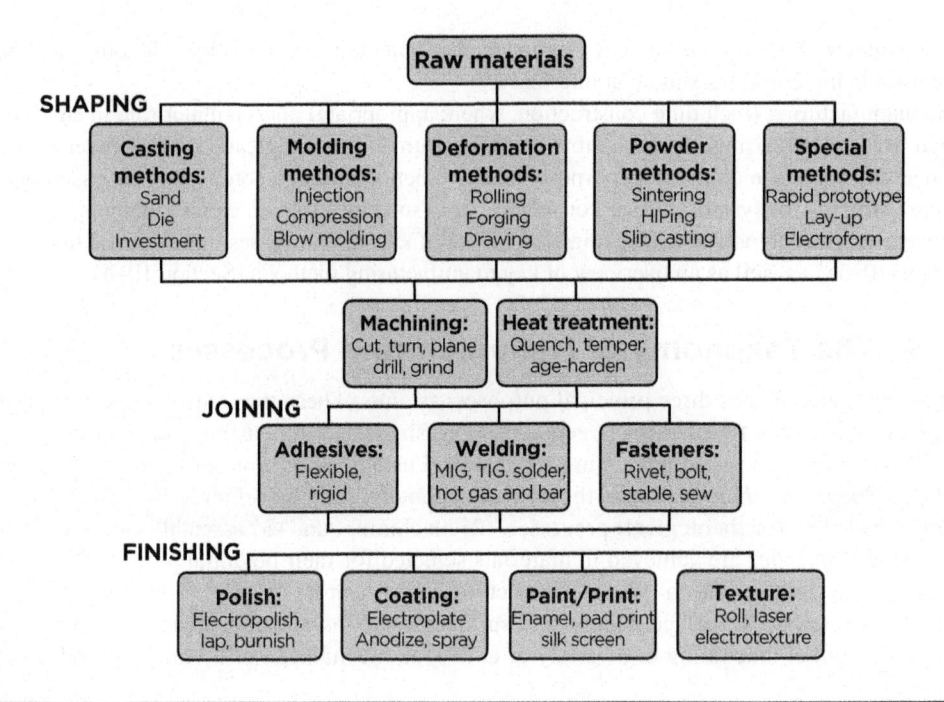

```
                          Raw materials
SHAPING ┌──────────┬──────────┬──────────┬──────────┐
    Casting      Molding    Deformation   Powder     Special
    methods:     methods:   methods:      methods:   methods:
    Sand         Injection  Rolling       Sintering  Rapid prototype
    Die          Compression Forging      HIPing     Lay-up
    Investment   Blow molding Drawing     Slip casting Electroform

              Machining:        Heat treatment:
              Cut, turn, plane  Quench, temper,
              drill, grind      age-harden

JOINING ┌──────────┬──────────┐
    Adhesives:   Welding:       Fasteners:
    Flexible,    MIG, TIG. solder, Rivet, bolt,
    rigid        hot gas and bar   stable, sew

FINISHING ┌──────────┬──────────┬──────────┐
    Polish:      Coating:       Paint/Print:   Texture:
    Electropolish, Electroplate  Enamel, pad print Roll, laser
    lap, burnish  Anodize, spray silk screen    electrotexture
```

Figure 10-2 Taxonomy of manufacturing processes showing nine classes of processes: five primary shaping processes in the top horizontal row and four secondary processes below in a vertical format. (*Source:* Michael F. Ashby,, *Materials Selection in Mechanical Design,* 1st edition, Pergamon Press, Oxford, UK, 1992, page 169, Figure 9.3; used with permission.)

resulting shape is no more than what is known as a *product form* (e.g., a cast ingot or billet, a bulk molded polymer, a rolled plate or I-beam, a block forging, a draw wire, an extruded shape, a powder preform, or a general composite lay-up or preform). The *primary shaping processes* create shape but often only as a starting point for further, secondary, processing to refine the shape, add details, and/or achieve desired dimensions. In other cases, on the other hand, primary processes are capable of and used to produce either *near-net shape* or *net-shape parts*.

Near-net-shape parts have nearly the required final shape but with some detail still needing to be created. Moreover, they seldom have the required final dimensions, so a great deal of additional processing may be required to achieve these dimensions. A couple of examples are: (1) a cast Al-alloy automobile engine block (which requires considerable machining and finishing, as well as joining to be made complete) and (2) a forged steel truck wheel (which requires finish machining and application of finish). *Net-shape parts* are, for all intents and purposes, ready to use. Little or no additional shaping is required, although some machining might be required for some details to be brought into tolerance, and some finish may need to be applied. A couple of examples are: (1) a precision cast Ti-alloy hydraulic value body for aerospace application (which might require light finish machining of key surfaces or interfaces and/or machining of any required internal threads) and (2) powder injection-molded stainless steel orthodontic appliances (which require only heat treatment to drive off volatile binder and allow solid-state sintering of powder particles to eliminate porosity and form a dense metal part).

Included among the primary shaping processes are:

- *Casting methods,* in which a crystalline metallic or, occasionally, ceramic material is melted and poured into a sacrificial mold or permanent, reusable die. Then, either relying on gravity or, occasionally, employing some pressure to force the molten material into all areas of the mold or die cavity, the filled mold or die is allowed to cool and produce a solid replica of the cavity as a casting that is removed by either destroying the sacrificial mold or opening the permanent mold or die along its parting plane(s). Casting allows repeated production of complex shapes and is capable of intricate details and good control of some (but not all) dimensions using more expensive permanent dies and, often, pressure during casting. Casting is a flow process.
- *Pressure molding,* in which an amorphous glass or amorphous or semicrystalline polymer is made soft by heating sufficiently above its glass transition temperature (T_g) to allow flow into all areas of a mold or die cavity under an applied pressure, and the filled mold is allowed or, more often, forced to cool until the part becomes suitably rigid below the material's T_g that it will retain shape upon ejection using ejector pins. Pressure molding allows complex shapes, intricate details, and precise dimensional control. Pressure molding is a flow process.
- *Deformation processing,* in which a metal or alloy is forced to change shape by employing sufficient pressure or stress to cause plastic deformation by crystalline slip. Such pressure or stress may be applied slowly (as in closed-die isothermal or creep forging), progressively in steps or stages (as in rolling, drawing, and some open- or closed-die forging), continuously (as in extrusion), or as a single large or repeated small blows (in many open-die forging processes). Deformation can be performed hot, suitably above the metal or alloy's recrystallization temperature that work/strain hardening never occurs due to dynamic recrystallization and rapid and significant reductions in cross-sectional area occur, or cold, well below the metal or alloy's recrystallization temperature to allow work/strain hardening to occur with slower and more limited reductions in cross-sectional area, or in between, that is, warm. Cold deformation requires higher power (more work) and takes place more slowly but results in more precisely controlled dimensions and surface finish. Deformation processing is a flow process.
- *Powder methods or processing,* in which metals, ceramics, polymers, or mixtures of these materials are used in particulate (i.e., powdered) form, often with an added volatile binder (to be removed later), compacted under pressure in precision permanent, reusable dies, and subsequently sintered (using solid-state diffusion) to cause particles to grow together at points of contact (i.e., "necks") to remove porosity and yield either a dense part or, if desired, a part with controlled porosity level (e.g., a porous bronze bearing to allow lubrication with entrained oil). The compacting stage can be done cold or hot, with dynamic sintering occurring with the latter. The method is uniquely capable of creating special microstructures (mixed materials, functionally graded materials, hard-phase or soft-phase impregnated materials, controlled porosity materials, etc.). Geometric complexity requires elaborate and expensive dies but is achievable. Replication of mold or die details is very faithful. Compacted parts are ejected from molds or dies using ejector pins. The general method should *not* be used if other primary processes or machining is capable of producing the needed part. Powder processing is a flow process.

- *Special methods* is a catchall category or class for processes that do not logically fit in any of the other four primary shaping classes. Many of the specific manufacturing processes or methods found in this class are employed with composite materials and involve lay-up methods (for tapes, fibers, filaments, broad-goods, and laminates) or filament winding or braiding or weaving, and so on. (Many of these processes or methods were taken from the textile industry and were either adopted by or adapted to composite materials.) These composite processing methods are all additive processes. Other special methods also include additive processes, albeit not intended for composites. Examples include electroforming (shape creation via electrodeposition), weld-forming (shape creation via weld metal deposition), chemical and/or physical vapor deposition (shape creation via deposition by a chemical or evaporation process).

Beneath the primary shaping process classes are *secondary processes* for shaping, conditioning, joining, and/or finishing. The first and most extensively used secondary process is *machining*. Machining comprises a host of processes and/or methods, but all share the common feature that they remove material to create the desired shape and dimensions, often of a final part. *Machining* includes subclassifications that involve cutting (e.g., sawing, planing, milling, threading), turning (e.g., using lathes), drilling (including boring and reaming), and grinding (i.e., removing either hard material or very small quantities of material to achieve a precise dimension and/or surface finish, often using a ceramic abrasive). In order to machine a part, there must be relative motion between the cutting tool or element and the workpiece. One or the other can move, but not both. Many machining processes require machines that provide 1 to 5 degrees of freedom among three axes of translation ($x, y,$ and z) and three axes of rotation (around $x, y,$ or z). Machining processes leave telltale marks (e.g., kerf marks or machining marks), which, depending on the particular method and final surface finish requirement, can be very noticeable or very subtle. But they are always there and indicate the direction of material removal.

Machining is often required following a primary shaping process, in order add required geometric features not achievable via the employed primary shaping process, create intricate details, achieve precise dimensions (i.e., close tolerances), produce internal or external threads, and so on.

Heat treat (or *heat treating* or *heat treatment*) is unique among the nine classes of processes, as it neither adds, subtracts, nor moves material by flow to create shape and dimension. Rather, *heat treatment* is used to create the needed or desired microstructure in the material constituting a part, structure, or assembly. The function of the heat is almost always to accelerate diffusion of atoms in the solid state.[10] Pure metals may be heat treated to remove locked-in, residual stress (i.e., stress-relief heat treatment), remove unwanted work/strain-hardening (i.e., recrystallization anneal heat treatment), or refine grain size (i.e., grain refinement heat treatment). Alloys, depending on the method by which they are strengthened (ref. Messler) may be heat treated to quench harden them (e.g., carbon, low-, medium-, and high-alloy steels), soften them (e.g., annealing heat treatment), reduce brittleness in quenched steel (i.e., temper heat treatment), increase strength/hardness (i.e., solution-and-age heat treatment), level composition (e.g., solution annealing or homogenization annealing heat treatment), or refine grain size in steels (e.g., normalization annealing heat

---

[10] A useful rule of thumb is that the rate of diffusion—or any temperature-dependent process that follows an exponential Arrhenius relationship—doubles with for every 30°C/50°F increase in temperature, and halves for every 30°C/50°F decrease in temperature.

treatment). Heat treatment can be done before machining or joining, or after, depending on the circumstances and desired outcome.

Joining allows small detail parts to be used to make larger assemblies or structures. Three primary categories of joining are: (1) mechanical joining, using only the physical interference of parts to cause interlocking (e.g., mechanical fastening and integral mechanical attachment); (2) adhesive bonding, using chemical forces at surfaces with the aid of a wetting chemical agent known as an "adhesive" (e.g., gluing, bonding, cementing); and (3) welding, using the natural attraction between atoms to form permanent atomic bonds (e.g., fusion welding, nonfusion welding, brazing, soldering). Joining is an additive process that allows, among many other things, the achievement of complex geometry and/or large size (ref. Messler2).

Finish (or *finishing*) refers to processes intended to modify the surface of parts with the proper shape, dimensions, and condition (via heat treatment). Specific processes for achieving the required surface roughness include grinding, lapping, honing, and polishing, all of which remove a thin layer of material. Processes to improve the fatigue resistance of a part by modifying the surface include burnishing, peening (e.g., using a hammer or shot), and certain laser surface-modification techniques. Wear resistance can be improved by hard plating, weld hard-facing, or surface heat treatment (induction hardening, case carburizing, nitriding, anodizing Al alloys, etc.). Corrosion protection can be achieved using anodizing or zinc-chromate treating Al alloys, galvanizing steel with Zn, and painting). Decorative finishing techniques include polishing, burnishing, plating, painting, dying, and polymer coating). Finishing for wear protection, corrosion protection, and decoration are additive processes, while processes to improve resistance to fatigue tend to be flow processes.

A fairly comprehensive list of specific processes in each primary and secondary processing class is given in Appendix B. Readers seeking information on specific processes are encouraged to search of information online or in any of several good books on manufacturing processes (ref. Kalpakjian and Schmid; Thompson).

10-4 Process Attributes

Just as materials are characterized by their set of properties, manufacturing processes are characterized by their *attributes*. The *attributes* of a process describe the things the process can do. Process attributes include:

- The materials the process can handle or for which the process offers particular benefit
- The size range of the things a process can produce (i.e., how big and how small?)
- The shape a process favors (e.g., turning on lathes is suited only to shape having rotational symmetry)
- The geometric complexity (including suitability to detailed features)
- The dimensional accuracy the process is capable of imparting
- The surface finish (as roughness) the process is capable of producing
- The speed with which the process produces parts

These attributes, and a couple of others, are listed in Table 10–1, along with definitions of each.

By observing features of parts or details of parts during reverse-engineering dissection, visual evidence or clues may be found that help identify the method-of-manufacture from the likely method's attributes. For example, a metal part with tremendous geometric complexity and intricate

TABLE 10-1 Attributes of Manufacturing Processes

Attribute	Definition
Material class	Materials to which the process is amenable based on hardness and, in some cases, T_{MP} or T_g
Size	Minimum and maximum size suitability, based on either volume or weight
Shape	Symmetry; aspect ratio; thickness to depth; surface-to-volume ratio
Complexity	Amount of dimensional and surface finish data necessary to fully characterize the shape; symmetry versus asymmetry
Tolerance	Accuracy or precision attainable in dimensions
Roughness	Surface finish as measured by rms roughness
Surface detail	Smallest radius of curvature; intricacy of details
Minimum batch size	Minimum number of units practical
Production rate	Cycle time; time to produce one unit
Cost	Cost per unit, including capital cost and labor

details would lend itself to certain casting processes, powder processing, or machining. Clues on the surface of the part might help differentiate machining (where there would be evidence of kerf marks) from casting or powder processing. These latter two methods might be distinguishable for the precision of dimensions in more than one or two directions, as it is difficult to obtain precision in all three orthogonal directions with casting due to the shrinkage accompanying the solidification of most metals and alloys and all engineering metals and alloys.

At one time, shortly after he began to popularize Material Selection Charts, Michael Ashby attempted to employ comparable charts for aid in the selection of a processing method to allow the creation of a part. Like their materials' counterparts, Process Selection Charts plotted full-range data for various process attributes against one another. These charts never caught on, as there was far too much overlap among not only specific methods within a process class but, worse, between and among process classes (Figure 10–3). As a consequence, plotting a point on a chart that represented the attributes of the designer's part of interest revealed too many suitable options. With too many choices, an engineer not familiar with specific processes and processing, in general, was no better off than before. Process selection comes back to experience.

Ashby created five different Process Selection Charts, the first of which consisted of three levels of detail, as follows:

Chart P1	Surface Area/Minimum Thickness
Chart P1(a)	Surface Area/Minimum Section [Thickness] (detail)
Chart P1(b)	Surface Area/Minimum Section [Thickness] (detail)
Chart P2	[Geometric] Complexity/Size
Chart P3	Size/Melting Temperature
Chart P4	Hardness/Melting Temperature
Chart P5	[Dimensional] Tolerance/Surface Roughness

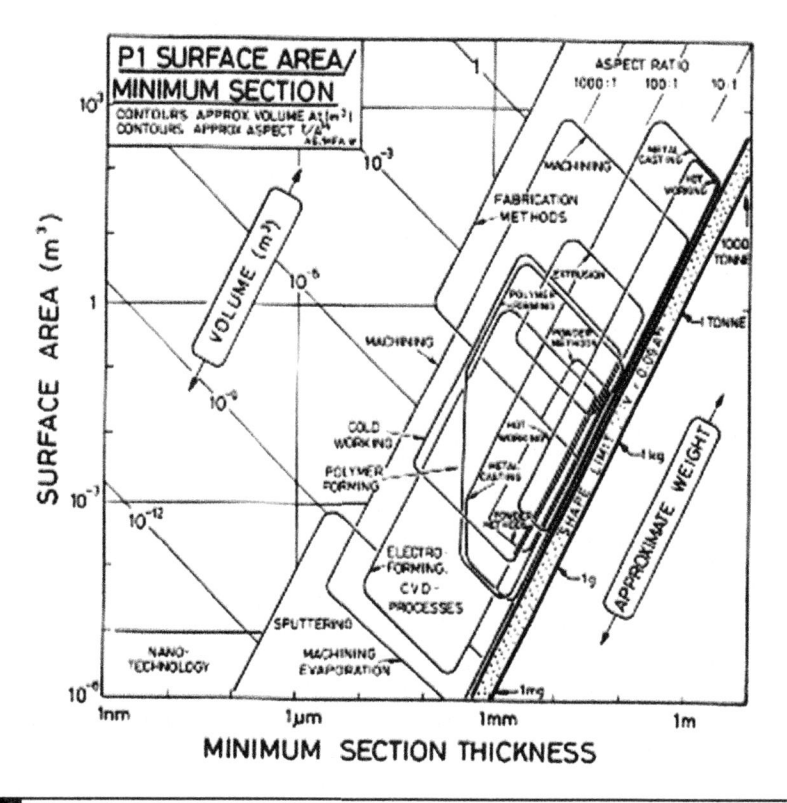

Figure 10-3 An example of one of Michael Ashby's (ref. Ashby) Process Selection Charts, here Chart P1 for Surface Area/Thickness for fabricated parts. (*Source:* M. F. Ashby, *Material Selection in Mechanical Design,* Material and Process Selection Charts [supplement], Pergamon Press, Oxford, UK, 1987, page 44; used with permission of the copyright owner, Elsevier.)

10-5 Inferring Method-of-Manufacture or -Construction from Observations

If it exists as a part, component, structural element, feature, detail, assembly, or structure, it must have been manufactured or constructed.[11] Knowing this, another important set of information to be extracted from an entity being subjected to reverse-engineering dissection is the identification by solid evidence or deduction from clues of the certain or likely *method-of-manufacture* for each part, component, feature, detail, structural element, mechanism, device, structure, or assembly. Besides helping one understand how an entity of interest was created, such information (like

---

[11] For the remainder of this chapter, at least, *manufacture, manufactured,* and *manufacturing* will be used to include *construct, constructed,* and *construction.* While there are, without question, some processes and methods that are used exclusively in construction (e.g., grading and/or excavation of a building site), there are, absolutely, close counterparts to manufacturing processes (e.g., pouring cement or concrete is a "casting process," cutting and polishing stone masonry is "sawing," a "machining process," and "grinding" is a "finishing process."

"design signature") can reveal much about the creator,[12] not the least of which is capability of and concern for quality workmanship.

The goal of observations at this stage of dissection, that is, after assessing role, purpose, and functionality and identifying or deducing materials-of-construction, is to gather evidence and/or clues for identifying or deducing method-of-manufacture. While experience with manufacturing helps tremendously, the way to proceed is to consider each attribute from Table 10–1 for each part, component, or structural element. Each of these can greatly help narrow down the candidates for likely method-of-manufacture to at least a *class* and, by looking closely for more subtle evidence or clues, to some specific *method(s)* within a class.

So let's consider how each attribute can help, with more subtle evidence or clues indicated for each.

Using Material Class

Based on conclusions for the material-of-construction for each part, component, or structural element, the options for the primary shaping class can be narrowed down. Metallic material generally suggests either casting or deformation processing and, very occasionally, may suggest powder processing. There is a slim chance that a special process (e.g., electroforming or weld-forming) may have been used, but this is rare.

Cast metal parts will be indicated by five key characteristics or features: (1) 3D (bulk) versus 2D (flat) form; (2) geometric complexity and intricacy of details; (3) generally rough, textured (e.g., mottled) surface finish in areas not machine-finished (except for specially finished mold or die cavities for more precision casting); (4) evidence of a parting line (where mold or die opens to allow the casting to be removed) on external surfaces; and, on occasion, (5) evidence of marks on cast part surfaces from ejector pins used to assist with removal of the casting while still hot and soft but solid or, more commonly, of special raised "bosses" against which ejector pins can push without causing damage. However, keep two things in mind: First, casting is used far less often than deformation processing (as final mechanical properties are inferior) and, second, castings will only infrequently be used for primary load bearing and virtually never for safety-critical parts, components, or structural elements.[13] The most commonly cast metals/alloys are, in descending order: Al and Al alloys, bronzes, cast irons, stainless steels, Ni alloys, Mg alloys, Zn alloys, and Ti and Ti alloys. Secondary machining on castings will be largely limited to mating surfaces, as most casting processes (except sand casting) create near-net shapes.

Figure 10–4*a* shows *cope* (top) and *drag* (bottom) die halves for a sand-casting process with *cores* in place in the drag to produce hollow portions of the casting. The mold halves shown are divided by a *parting plane*. Figure 10–4*b* shows bronze (left) and aluminum-alloy (right) parts

---

[12] By "design signature" is meant some uniquely identifying aspect, feature, or quality of an individual's or an organization's design. While sometimes hard to explain or quantify, many manufacturers, especially of high-quality products, structures, or systems, have a certain look to their design. Nowhere is this more apparent than in automobiles. One doesn't need to see a logo to recognize a certain manufacturer's style. BMW designers are indoctrinated to "be sure the cars look like they are moving even when they are standing still" and "to be sure an observer can tell the front from the back looking from the side."

[13] As a rule, the tensile yield strength (and, to a lesser extent, the tensile ultimate strength) of cast metals and alloys is about 70 percent that of a deformation-processed (i.e., wrought) counterpart. Ductility is even less than 70 percent, typically 50 percent, that of a wrought product. This is due to the dendritic structure, compositional inhomogenieties, and susceptibility to porosity and other voidlike defects in castings.

Inferring Methods-of-Manufacture or -Construction **211**

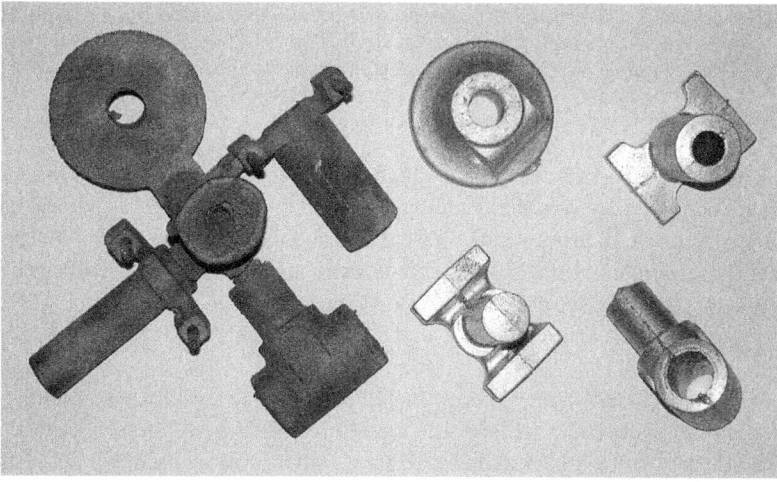

(a)

(b)

Figure 10-4 Upper (*cope*) and lower (*drag*) halves of a mold for making sand castings are shown, with *cores* inserted in the drag for creating hollow regions in the cast parts (*a*). The mold halves are divided by the *parting plane*, which can leave evidence in the form of *parting lines* on some castings. Bronze alloy (*left*) and Al-alloy parts from the mold in (*a*) are shown. Clear evidence of parting lines can be seen on three of the Al-alloy castings, while none are apparent in the photograph for the fourth Al-alloy part or any of the bronze parts. The surfaces of the Al-alloy castings show the rough, mottled finish characteristic of most sand casting (*b*). (*Source:* Wikipedia Creative Commons, photographs taken by Glenn McKechnie on April 6, 2005 and posted by Graibeard on 15 April 2005.)

Figure 10-5 A wonderful example of the geometric complexity and intricacy of detail achievable with precision casting; here, die casting of an Al-alloy six-cylinder engine block for a BMW. (*Source:* Wikipedia Creative Commons, originally contributed by 160SX on 26 September 2009 and modified by Wizard 191 on 6 August 2010.)

from the mold shown in (*a*). There is clear evidence of a parting line on three of the Al-alloy parts, but no such evidence is apparent in the photograph for the bronze parts. Also visible on the Al-alloy castings is the rough, mottled surface left by the sand used to make the mold. Figure 10–5 shows a wonderful example of the geometric complexity and intricacy that can be achieved by casting for a die-cast Al-alloy six-cylinder engine block.

An informative video on casting can be found on YouTube at "Mold making—Mass Casting Complex Parts."

For *deformation processed metallic parts,* forged parts will almost always require some secondary machining, while heavy, rolled billets and thick plate are often heavily machined, usually, to add details. Drawn and extruded product is only occasionally machined. Forgings will tend to be 3D shapes with simple to moderately complex geometry. Machined billet, plate, and large-diameter rounds will tend to be heavily machined. Thinner, 2D product, notably sheet, will tend to be hot-, warm-, or cold-formed but may also be stamped to produce formed shapes which have been blanked (to create a peripheral shape) and/or punched (to create interior cutouts) and/or formed to create bends or curved contours. Deformation processing is most common for the following metals/alloys, in descending order: carbon steel, low-alloy steel, stainless steels, Al and Al alloys, Cu and brass, and Ni and Ti alloys.

Forged parts will often provide clues of flashing (where metal flowed outward, under forging pressure, into the seam between closed mating dies or in the gap between upper and lower open dies) (Figure 10–6*a* and *b*). Heavy section parts are much more likely to have been forged (if they were not machined from heavy billets or plates) than cast. Also, recognize that both cast and

Inferring Methods-of-Manufacture or -Construction   **213**

(a)

(b)

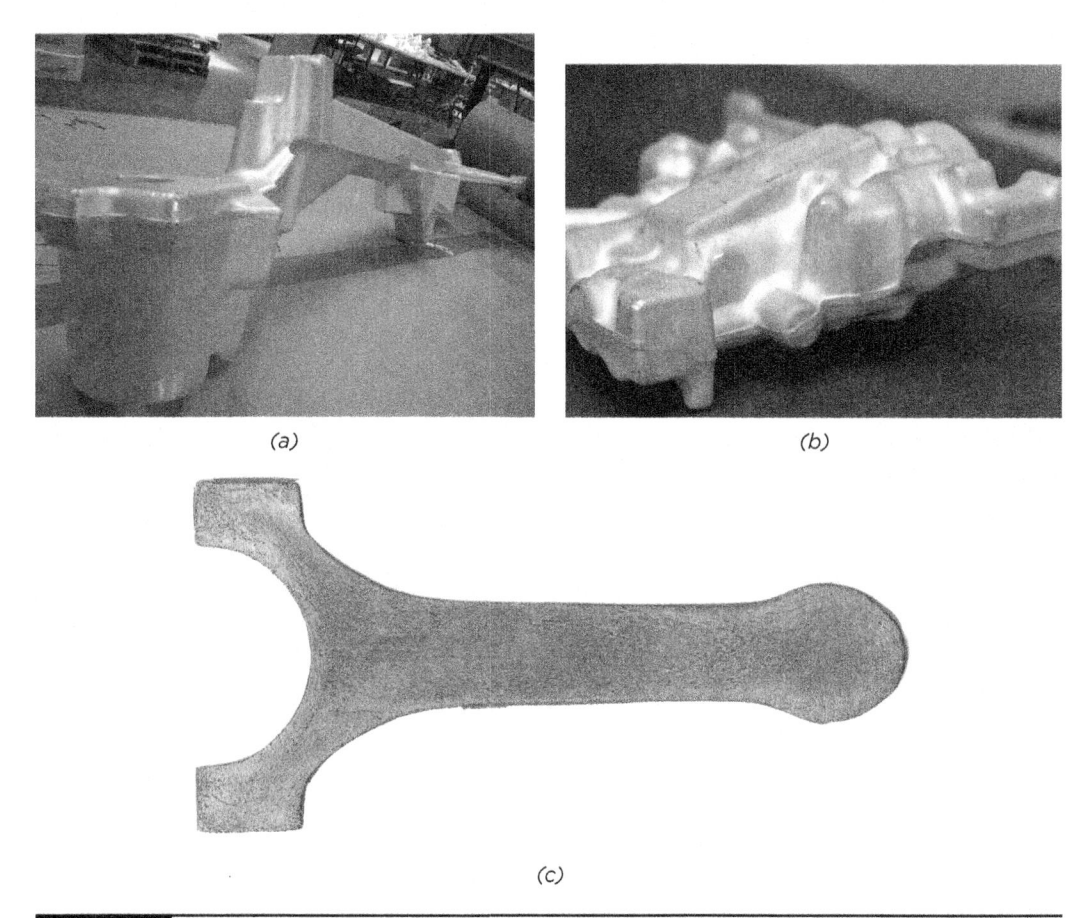

(c)

Figure 10-6 Two examples of forged Al-alloy parts showing evidence of the *flashing* along planes running through the part's thickness directions that facilitate part extraction from dies (*a* and *b*). The grain structure of a forged connecting rod that was sectioned, polished, and etched reflects the plastic flow of hot, but solid, metal during forging (*c*). Such *flow lines* tend to enhance the strength and toughness of forged parts. (*Sources:* Photographs from Continental Forge Company, Compton, California, producer of precision Al-alloy forgings, used with their kind permission [*a* and *b*]; and from Wikipedia Creative Commons, contributed by Graibeard on 8 September 2005 [*c*].)

forged parts have to be extracted from molds or dies, so there will usually be some symmetry in the through-the-thickness direction, as well as evidence of shallow draft angles on surfaces perpendicular to mold or die closure direction(s). Marks from ejection pins are sometimes seen. Especially telling is evidence of plastic flow of hot solid metal during forging to produce *flow lines* (Figure 10–6*c*). These tend to enhance strength and toughness of forged parts as compared to cast parts.

Powder processed metallic parts will be rare, and are difficult to distinguish from castings, as they will generally both have complex shapes. But, as opposed to many castings, powder-processed parts will tend to be small and exhibit intricate surface details. The as-processed, nonmachined surface finish of powder-processed parts is generally better than that for castings. Pure iron, steel, bronze, and Ti and Ti alloys are the most commonly powder-processed metals/alloys.

Polymer parts will almost always have been molded, usually by either compression molding (using open or closed molds or dies), or, most often, by injection molding. Injection molding almost always leaves ejector-pin marks on internal surfaces (Figure 10–7a) as still-hot soft material beneath the stiffened cooler surface layer is deformed (as production rates for injection-molded polymer parts tends to be high) and may, if not done properly, show evidence of flow lines (Figure 10–7b) and/or *knit lines,* where viscous polymer flows together from different directions (Figure 10–7c). Polymer parts will rarely be machined.

Ceramic parts will have been cold cast (using a slurry of powdered ceramic in a binder) or powder processed by compacting and sintering (or firing, in the case of glassy ceramics). Ceramic parts will not be machined but may exhibit ground interfaces or close-fitting mating surfaces.

Composites will always have been the product of special processing, with overall shape being a major clue as to the specific method. Flat or curved, essentially 2D parts will have been made by lay-up of broad goods or tapes. Hollow 3D parts will likely have been filament wound.

Using Size

Unless they have been joined, the largest metallic parts will have been either cast or forged, with forging being favored for most heavily loaded structures, while casting will be favored for the most complex geometric shapes or for corrosion-resistant alloys. Large polymer parts will have been compression molded or, possibly, reaction molded. Small size (e.g. smaller than a golf ball) metal parts are usually precision cast or powder processed but could be precision machined. Extremely small parts are likely to have been produced by a special process (e.g., electro-, chemical-, or vapor-deposition).

Using Shape

Metallic parts having three-dimensionality and other than small size will have been cast or forged. Small, intricate parts could have been powder processed but may, alternatively, have been precision cast. Parting lines will help distinguish castings from powder parts in many cases. Flat or contoured but essentially 2D sheet-stock parts will have been deformation processed by forming (including stamping and forming) or stamp-forming).

Using Complexity

The most complex parts will have been cast, unless they were completely machined from heavy-section billets, rounds, or plate. A tip-off is quantity produced, with high quantities favoring casting (as die making labor cost is amortized over many units versus recurring with each part). Very small, geometrically complex, and intricate-detail metallic or ceramic parts may have been powder processed.

Using Tolerance

Close tolerances usually demand secondary machining, unless costly precision casting or powder processing was used. Machining is always evidenced by machining or marks or kerf marks on the

Inferring Methods-of-Manufacture or -Construction **215**

(a)

(b)

(c)

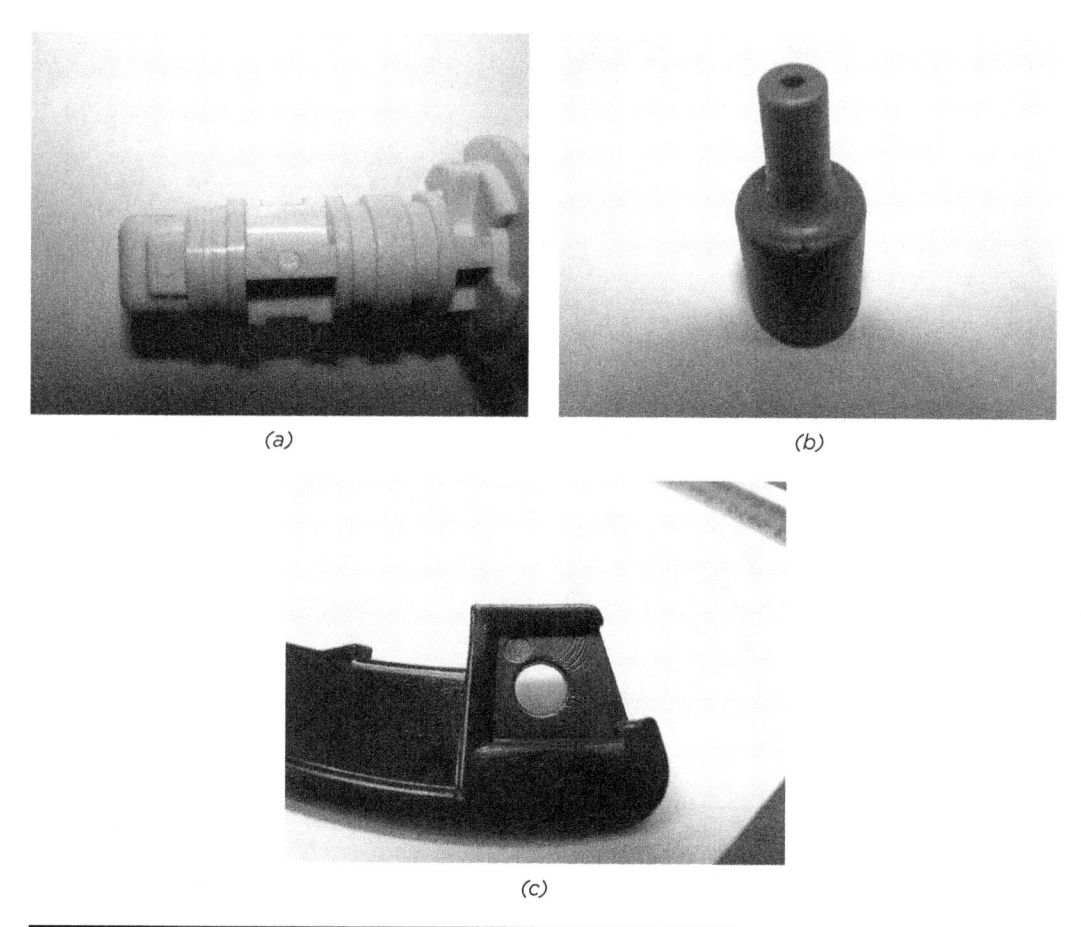

Figure 10-7 Marks from ejector pins used to aid in the removal (usually automatic ejection) of still-soft polymer parts after molding (a), as well as flow lines created when molding parameters (e.g., temperature and/or flow rate from pressure) are not optimum (b) and/or *knit lines* where viscous polymer flows together from different directions (c). (*Source:* Photographs courtesy of Rebling Power Connectors, 170 Franklin Drive, Washington, PA, used with permission from Nate Bower, president.)

surface (Figure 10–8 *a, b,* and *c*), with the appearance differing for the different types of milling cutters (e.g., end mills, surface mills, edge mills; for example, see Figure 10–9).

Using Roughness

A rough surface indicates either inexpensive casting (e.g., sand casting) or drop-hammer, or press forging (open- versus closed-die precision forging). High-quality surfaces indicate machining, with the smoothest surfaces indicating grinding. A high-quality surface finish should have an apparent purpose, such as allowing parts to fit tightly together.

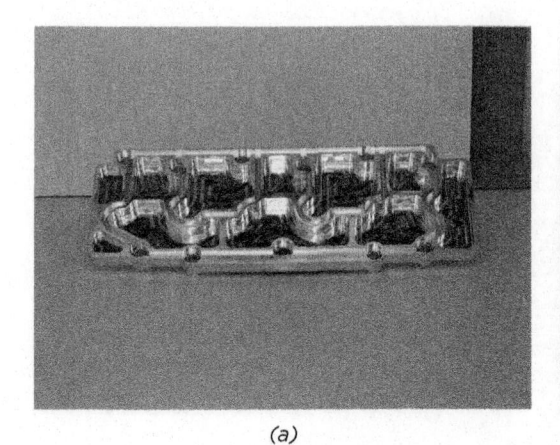

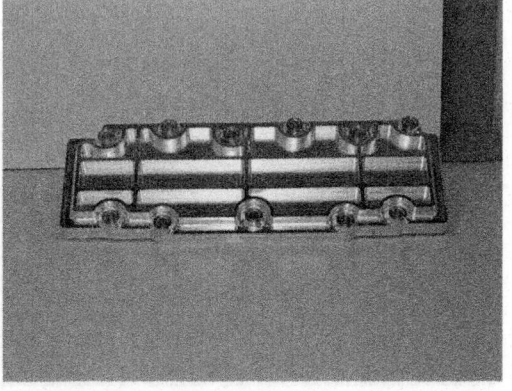

(a)

(b)

(c)

Figure 10-8 Examples of varying severity and appearance of marks left by machining (i.e., *machining marks* or kerf marks), from coarse contours left in rough-machined pockets (*a*) to less severe but still easily noticeable marks also left from milling pockets (*b*) to virtually indistinguishable marks left after precision machining (*c*). (*Source:* Hargett Precision Products, Vista, California, with permission from Mark Hargett.)

Using Surface Detail

Intricate surface detail could be achieved with casting or powder processing, unless it was produced by machining (evidenced by machining marks). Another possibility is that intricate surface details were embossed on by forging or by special processes (e.g., laser engraving).

Using Batch Size and/or Production Rate

Very small batch sizes (down to one of a kind) favor machining, as mold or die costs would be prohibitive for one-off production.[14] Large batch sizes tend to put machining at a disadvantage, with casting, forging, stamping, and/or forming being favored.

---

[14] One-off production is now facilitated by *rapid prototyping processes,* of which there are many. When used to produce actual parts, the process is called *rapid manufacturing.*

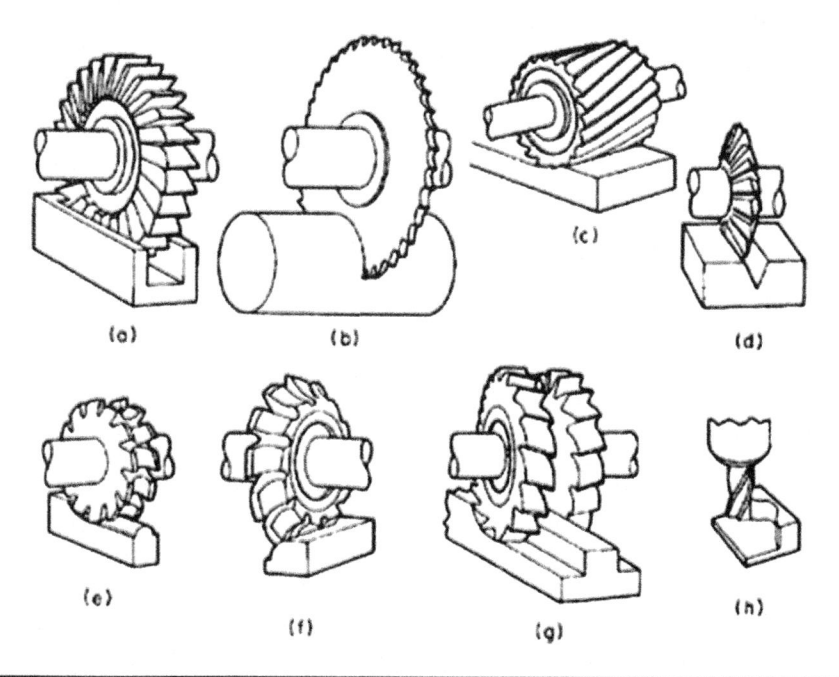

Inferring Methods-of-Manufacture or -Construction **217**

Figure 10-9 Schematic illustrations of several different types of milling cutters used in machining. Each type leaves a distinctive machining or kerf mark. (*Source:* Niagara Cutter Inc., used with permission.)

Cost

More costly manufacturing processes must be justified by demanding performance and/or requirements for exceptional reliability. Higher-cost processes include precision (die) casting, precision (closed-die isothermal) forging, specialized machining, and most powder processing. Less costly, lower-end manufacturing processes should also be justified by the intended market and use for the manufactured item. Remember, though, workmanship says a lot about a manufacturer, and even low-end processes can be done well or badly.

Beyond using the attributes of manufacturing processes as a means of identifying or deducing the method-of-manufacture of parts during reverse-engineering dissection, an engineer should look at how the hold parts were joined into assemblies or structures, and also how parts were finished.

Using Joining

There are three fundamental process options for joining materials (or parts comprised by materials), each relying on one of three possible forces as to how things are joined together. *Mechanical joining,* which includes the use of fasteners (in *mechanical fastening*) or design- and fabricated-in features (in *integral mechanical attachment*), uses strictly mechanical forces derived from and giving rise to physical interference. *Adhesive bonding* uses chemical agents (known as adhesives) to create adhesion between surfaces using chemical forces arising from atomic-level secondary bonding. *Welding* uses the physical forces resulting from the natural tendency of atoms to form bonds with other atoms in the solid state, with brazing and soldering being subtypes of welding in which a low-melting filler alloy facilitates joining. Much can be learned about the role, purpose, and functionality of parts in a structure or assembly, as well as about the structure or the assembly, by the type of joining process used (ref. Messler2).

Here are some clues:

- Designed- and molded-in features (e.g., various catches and latches) create snap-fits used to join polymer parts, as these greatly facilitate automated assembly and overcome the many logistical issues and labor intensity associated with fasteners. Joining security is not as great for snap-fits as for fasteners, but snap-fits eliminate the potential choking hazard associated with loosened fasteners in children's toys, for example.
- Nonstandard fasteners (e.g., special head designs beyond slotted- and Phillip's head types, wire-securing, thread-locking adhesives) indicate an attempt by the manufacturer to prevent tampering by the user.
- Welding suggests a need for permanence (with no chance for disassembly, including for maintenance or service), sealing against fluid leaks or intrusion, or high load bearing.
- Brazing or soldering suggests a need to limit exposure of parts (and/or surrounding parts in an assembly) to elevated temperature and/or a need to make joints en masse (e.g., furnace or dip brazing or soldering).

Special finishes, other than bare base material, suggest a purpose or need. Examples of need include improved resistance to wear, improved resistance to corrosion, friction reduction, friction increase or grip, or decorative appearance.

What is presented here is intended to guide observations during reverse-engineering dissection. Once again, the challenge in reverse engineering is "to put all the pieces of the puzzle together to decipher the puzzle." Observed *evidence* tells one what something is and allows positive identification. Observed *clues* point one in a direction toward what something may be and allow only a probability of being correct. Reading clues better comes only with experience. The more clues an engineer has seen, and interpreted, the more likely he or she will get it right in the future!

Table 10–2 summarizes clues provided by manufacturing process attributes.

10-6 A Word on Heat Treatment

All metals and alloys can be made more resistant to complications during machining as well as more resistant to fatigue if they are *stress relieved* to remove the worst locked-in stresses from earlier processing (by forming, welding, etc.). But some alloys can be made stronger by heat treatment, depending on the mechanism by which the alloy achieves strengthening beyond solid-solution strengthening. The two predominant mechanisms for additional strengthening relying on heat treatment are quench hardening of steels and age-hardening of certain wrought Al alloys with either Cu (i.e., 2000-series), Mg+Si (i.e., 6000-series), Zn (i.e., 7000-series), or Li (e.g., 8000-series) as the primary solute addition. For both mechanisms, heat treatment creates a single-phase solid solution at elevated temperature and then develops either a meta-stable shear-transformed hard, string, but brittle, martensite in susceptible steels or second-phase intermetallic precipitates

TABLE 10-2 List of Characteristics and Clues (*) for Indicating Manufacturing Processes

Casting (metals; some ceramics):
- 3D more common than 2D.
- Favors complex geometry.
- Seldom used for primary structure; low stress compatibility.
- Details and close-fitting interfaces are machined.
- Suits the most corrosion-resistant alloys.
- Favors high production volumes; low to moderate rates.
- Cost rises rapidly with complexity and precision.
- Parting lines and ejection marks are common.*

Molding (polymers; glasses):
- Favors housings and shells versus bulk.
- Favors complex geometry and intricate details.
- Facilitates snap-fit attachment.
- Low stress capability.
- Rarely, if ever, machined.
- Favors high production volumes; medium to high rates.
- Low cost, in general.
- Ejection marks are very common; flow lines suggest quality problem; may be evidence of sprue or runner break-off.*

Deformation processing (metals):
- Simple geometry from rolled products.
- Forging favors complex geometry.
- Extrusion allows complex cross sections in long lengths.
- Forming favors flat, bent, or contoured thick sections.
- High stress capability due to wrought microstructure.
- Precision and surface finish better for cold than hot working.
- Machining is common for heavy-section roll product and other than net-shape forgings and extrusions.
- Flash remnants common with some forgings.*

Powder processing (metals and ceramics or composites thereof):
- Favors very complex geometry and intricate details.
- Favors special materials and/or microstructures (e.g., controlled porosity).
- Mechanical properties match wrought product.
- Favors low production rates and volumes in most cases.
- Cost tends to be high, so must be justified.
- Should not be used if casting or machining would suffice.
- Ejector marks are possible.*

Special processing (polymer-matrix composites):
- Costs tend to be high and production rates slow.
- Restricted to exotic needs beyond composites.

Machining (metals; limited to grinding for ceramics):
- Allows complex geometry and complicated details.
- Capable of very high precision (highest of all processes).
- Allows high-quality surface finish in terms of roughness.
- Labor cost is recurring; tooling costs are usually modest.
- Machining marks or kerf identify method.

(continued)

TABLE 10-2 *Continued*

Heat treat (metals):
- Allows development of mechanical properties via microstructure.
- Relieves potential detrimental residual stresses.
- Adds cost.
- Requires protective atmospheres.

Joining (all materials):
- Mechanical joining uniquely allows motion and intentional disassembly.
- Welding is for permanence or fluid tightness; ripples show speed.
- Adhesive bonding spreads loading; suited to shear only.
- Soldering is for electrical/electronic connectivity.

Finishing (all material):
- If not for wear or corrosion protection, for aesthetics.

in Al alloys. The way to tell whether a part made from either steel or an Al alloy has been heat-treat strengthened is to test its hardness, either by a simple scratch test or using a relatively nondestructive hardness test (e.g., portable Rockwell).

Readers interested in the details of heat treatment are referred to a good textbook on basic materials (ref. Callister, Messler).

Table 10-3 lists the hardness of several of the most important engineering alloys in both un-heat-treated or annealed (i.e., full soft) conditions and heat-treat hardened condition. Hardness, obtained from simple nondestructive tests, can help identify a material-of-construction.

TABLE 10-3 Hardness Values for Non-Heat-Treatable or Fully Annealed and Hardened Conditions for Major Engineering Metals and Alloys

Metal or Alloy (Condition)	Brinell Hardness (BHN)	Rockwell Hardness (RHN-Scale)
Pure 1100 Al (annealed)	~20–25	~10–15B
Al alloys (non. ht. trt.) (annealed)	~75–90	~38–54B
Al alloys (ht. trt.) (annealed)	~30–60	~20–35B
Al alloys (ht. trt.) (fully aged)	~95–150	~60–90B
Pure ETP Cu (annealed)	~80–95	~40–60B
Pure Cu (cold worked)	~115–155	~68–82B
Brasses (annealed)	~140–160	~77–84B
Brasses (cold worked)	~160–210	~84–95B
Bronzes (annealed)	~80–220	~40–95B
Pure Fe (annealed)	~95–105	~55–70B

Carbon steels		
1018 (annealed)	~125–130	~75–80B
1018 (fully hardened)	~325–380	~35–40C
1030 (annealed)	~150–190	~79–92B
1030 (fully hardened)	~450–510	~48–52C
1045 (annealed)	~180–230	~90–98B
1045 (fully hardened)	~575–650	~56–60C
1060 (annealed)	~200–250	~92–100B
1060 (fully hardened)	~690–740	~62–65C
1095 (annealed)	~285–330	~30–35C
1095 (fully hardened)	~900–1000	>65C
Low-alloy steels (e.g., 4330–4340)		
Annealed	~270–330	~27–35C
Fully hardened	~450–615	~48–58C
Austenitic stainless steel		
Annealed	~200–210	~92–94B
Cold worked	~285–375	~30–40C
Martensitic stainless steel		
Annealed	~190–230	~90–98B
Fully hardened	~575–650	~56–60C
Mg alloys		
Annealed	~110–160	~70–80B
Fully aged	~220–280	~95–105B
Ni alloys		
Non-heat-treatable	~200–250	~92–100B
Heat-treatable (annealed)	~185–235	~90–98B
Heat-treatable (fully aged)	~300–360	~30–37C
Pure CP Ti (annealed)	~140–160	~75–82B
Ti alloys		
Annealed	~270–310	~25–31C
Fully aged	~350–385	~35–39C
Zn alloys	~80–115	~40–65B

10-7 Summary

How an item or entity was manufactured (if done in a factory or a plant) or constructed (if done, of necessity, outdoors) can reveal a great deal about not only the role, purpose, and/or functionality but also its creator. Recall that the quality of workmanship is usually a key indicator of the quality of a manufacturer or construction firm. There are a wide variety and large number of manufacturing processes or methods, and these are conveniently organized in a taxonomy of primary shaping processes (e.g., casting, molding, deformation processing, powder processing, and special processing) and secondary processes of machining (to refine shape), heat treatment (to adjust property condition), joining (to assemble), and finishing (to achieve final surface finish).

    Just as materials exhibit characteristic properties, processes possess certain attributes, which include material preference, size range, shape capability, complexity, tolerance, surface roughness, batch size and production rate suitability, and cost. These attributes are useful for seeking evidence or clues as to what specific method-of-manufacture was, or probably was, used. More subtle clues, such as parting lines in castings, flash on forgings, ejector-pin marks for either casting or forging (or, occasionally, powder processing), and machining marks, help narrow down choices to a specific method.

10-8 Cited References

Ashby, M. F., *Materials Selection in Mechanical Design,* 1st edition, Pergamon Press, Oxford, UK, 1992.

Ashby, Michael F., *Materials Selection in Mechanical Design,* 4th edition, Butterworth-Heinemann/Elsevier, Burlington, MA, 2010.

Callister, William D., Jr., and David G. Rethwisch, *Materials Science and Engineering: An Introduction,* John Wiley & Sons, Hoboken, NJ, 2010.

Hawking, Stephen, and Leonard Mlodinow, *The Grand Design,* Bantam, New York, 2010.

Kalpakjian, Serge, and Steven Schmid, *Manufacturing Processes for Engineering Materials,* 5th edition, Prentice-Hall/Pearson, Upper Saddle River, NJ, 2007.

Messler, Robert W., Jr., *The Essence of Materials for Engineers,* Jones & Bartlett Learning, Burlington, MA, 2011.

Messler2, Robert W., Jr., *Joining of Materials and Structures: From Pragmatic Process to Enabling Technology,* Butterworth-Heinemann/Elsevier, Burlington, MA, 2004.

Thompson, Rob, *Manufacturing Processes for Design Professionals,* Thames & Hudson, London, 2007.

10-9 Recommended Reading

Charles, J. A., F. A. A. Crane, and J. A. G. Furness, *Selection and Use of Engineering Materials,* 3rd edition, Butterworth-Heinemann, Burlington, MA, 1997.

Duvall, J. Barry, and David R. Hillis, *Manufacturing Processes: Materials, Productivity, and Lean Strategies,* 3rd edition, Goodheart-Wilcox, Tinley Park, IL, 2011.

Groover, Mikeil, *Fundamentals of Modern Manufacturing Materials, Processes, and Systems,* 4th edition, John Wiley & Sons, Hoboken, NJ, 2010.

Lennox, John C., *God and Stephen Hawking: Whose Design Is It Anyway?,* Lion UK, Oxford, UK, 2011.

10-10 Thought Questions and Problems

10-1 At the heart of selecting materials to meet the property requirements of designs lies an inextricable interrelationship or interaction besides structure ⇔ properties ⇔ processing ⇔

performance. This interrelationship or interaction involves *function* ⇔ *material* ⇔ *shape* ⇔ *process*.
  a. Briefly, in less than one page, explain this interrelationship in your own words. Also, indicate how it is really not so different from the other interrelationship.
  b. Engineers involved in the design of structures (i.e., structural engineers, principally civil and mechanical engineers) use specific terms to refer to structural elements that provide a particular function and tend to have a particular shape. These terms are:
     (1) columns
     (2) beams
     (3) shafts (for some applications called "axles")
     (4) ties (or, occasionally, "struts")
     (5) shells
     Use the Internet to aid you in providing a clear and concise definition of each term. Then, for each, give a brief explanation of how *function* (e.g., intended, preferred loading) and *shape* are related within each type.
  c. Attaining the needed shape based on intended function obviously requires processing. But the choice of process depends on two interrelated factors: shape (as geometric complexity) and material.
     Briefly explain how and why each of the following arises:
     (1) The more complex the shape, the more expensive the process(ing).
     (2) Complex shapes are much more difficult to process into ceramics than metals or polymers.

10-2 Figure 10–2 presents a taxonomy for manufacturing processes that is a hierarchical ordering of things somehow related in order to show similarities, differences, and familiar relationships (e.g., parent and child, parent and grandchild, cousin). At the highest level are so-called *primary processes* used to transform materials into basic product forms (e.g., ingots; billets; thick, flat plates; long, thin strips; rough-shaped 3D forms) or, sometimes, *near-net shapes* or even *net shapes*.
  a. In your own words, using the Internet as a resource, define the following, giving one or two examples of each as used by companies involved in manufacturing or construction:
     (1) product form
     (2) near-net shape
     (3) net shape
  b. Using the Internet or some reference on manufacturing processes (ref. Duvall and Hillis, Groover, Kalpakjian and Schmid, Thompson) give *one* example for the use of each of the following primary processing methods for each of a product form, a near-net shape, and a net shape. Give the example as the object and a specific process (e.g., "precision pressure die casting") by which the object is made. The primary processes are:
     (1) casting
     (2) molding
     (3) rolling (within deformation processing)
     (4) extrusion (within deformation processing)
     (5) forging (within deformation processing)

224 Chapter Ten

10-3 Just as materials are characterized by their set of properties, manufacturing processes are characterized by their *attributes,* which describe the things processes can do. *Attributes* for manufacturing include the process's capability relating to:
- material suitability
- size range suitability
- shape suitability
- geometric complexity capability
- dimensional accuracy capability
- surface finish capability
- speed of production capability

a. For each of the following manufacturing processes, provide the requested response relating to an *attribute,* defending your answer in each instance:
   (1) the best suited to *cold casting, slurry casting,* or *slip casting*
   (2) the *two* generic materials best suited to *blow molding*
   (3) the size range(s) as very large, large, intermediate, small, or very small) to which *fusion welding* is best suited
   (4) the size range(s), as in (3) to which *metal* or *ceramic powder injection molding* is best suited
   (5) the shape to which *lathe turning* is best suited
   (6) the shape complexity to which *hot extrusion* of metals or alloys is best suited
   (7) the level(s) of dimensional accuracy (as very high, high, moderate, low, or very low) to which *block forging* is best suited
   (8) the level(s) of dimensional accuracy, as in (7) to which *electrical discharge machining* is best suited
   (9) the surface finish(es) (as very smooth, smooth, intermediate, rough, and very rough) to which *lapping* is best suited
   (10) the speed capability (as very high, high, moderate, slow, and very slow) to which *cold heading* of fasteners (e.g., bolts) is best suited

b. Experienced manufacturing engineers tend to know the particular attribute(s) of each particular manufacturing process. Give what you find on the Internet or, if you can't find it on the Internet (which is doubtful!), give what you suspect are the *two* most significant *attributes* of each of the following manufacturing processes:
   (1) *cold wire drawing*
   (2) *permanent die casting*
   (3) *wave soldering*
   (4) *electroslag welding*
   (5) *abrasive grinding*
   (6) *abrasive water-jet cutting*

10-4 The manufacturing processes by which materials can be shaped all tend to leave more or less subtle evidence or clues to their identity. Such evidence or clues, not surprisingly, almost always appear on the exterior (surface) of parts and, to a lesser extent, for some processes, in the bulk (interior) structure or microstructure of parts. Table 10–2 summarizes some of the more important of these clues.

a. Suggest the general primary or secondary process most likely used to manufacture

a part with the following visible evidence or clues on its exterior or in its internal microstructure, briefly supporting your choice in each case:
   (1) The overall geometry of a 3D metal part the size of a lunch box is extremely complex, with many intricate details on its surface.
   (2) A metal part about a meter long and around 10 centimeters across with rotational symmetry exhibits fine, very closely spaced circumferential marks that seem to spiral in a small-pitch helix.
   (3) A fairly simple metal part the size of a full-size automobile tire has considerable bulk, various recesses in its cross section, and apparent symmetry about a plane through its midthickness evidenced by a very fine line around its entire perimeter. However, there is no evidence of any circular marks anywhere on its surface.
   (4) A hollow, thin-walled polymer part with a nearly round cross section and closed at one end, reveals subtle lines running from top to bottom down each side, diametrically opposite one another.
   (5) 8- to 12-foot-long metal parts with rather complex open and closed cross sections have a distinct line on their entire surface running lengthwise.
   (6) A thin sheet of metal reveals grains in a metallographic cross section (looking in from the long edge of the sheet) that are very elongated with indistinct grain boundaries.
b. Other clues found on manufactured parts and related to certain process attributes help indicate the likely process of manufacture. Suggest the likely process of manufacture from each of the following, defending your choice in each case:
   (1) Steel support columns with a general H- or I-shaped cross section seem far too big in cross-sectional area to have been produced by hot rolling. (This could be determined by calculating how much rolling pressure would be required to cause plastic deformation—at rolling temperatures—for the area and looking into modern rolling mill capacities.)
   (2) The hull of a silvery white to very light metallic gray canoe seems to be seamless, except at the forward and aft ends, which were riveted.
   (3) Hundreds of thousands of small, precision steel machine screws (not self-tapping screws!).
   (4) The intricately detailed, close-tolerance stainless steel hardware (i.e., brackets) used in orthodontics.
   (5) Tens of thousands of metal manhole covers bearing the name of the city in their design.

10-5 Performance in a design is critically linked to having selected the right material processed to have the proper geometry but, especially, to have the right properties. A key secondary process for developing the mechanical properties of strength, ductility, and toughness in metals and alloys is *heat treatment*. There are several extremely important heat treatment processes which have very specific purposes with which all engineers, not just materials engineers, should have familiarity.

Using the Internet or a good textbook on materials science and engineering (Callister, Messler, Shackelford, Askland, Smith, etc.), provide a brief description of the purpose *and* procedure (i.e., time-temperature-cooling rate) for each of the following:

a. Annealing of carbon or low-alloy steels
b. Quench hardening of carbon or low-alloy steels
c. Tempering of carbon or low-alloy steels
d. Stress relief of carbon or low-alloy steels
e. Stress relief of Cr-Ni austenitic stainless steels for use in corrosion-resistant applications (This is more challenging.)
f. Age-hardening a precipitation-strengthening Al alloy
g. Recrystallization or recrystallization annealing a cold-worked Cu-Zn yellow brass alloy sheet

CHAPTER 11

Construction of Khufu's Pyramid: Humankind's Greatest Engineering Creation

11-1 Herodotus Reveals the Pyramids to the World

Its base is square, each side eight hundred feet long, and its height is the same; the whole is of stone polished and most exactly fitted; there is no block of less than thirty feet in length. This pyramid was made like stairs, which some call steps and others, tiers. When this, its first form, was completed, the workmen used short wooden logs as levers to raise the rest of the stones; they heaved up the blocks from the ground onto the first tier of steps; when the block was raised, it was set on another lever that stood on the first tier, and the lever was again used to lift it from this tier to the next. It may be that there was a new lever on each tier of steps, or perhaps there was only one lever, quite portable, which they carried up to each tier in turn; I leave this uncertain, as both possibilities were mentioned. But this is certain, that the upper part of the pyramid was finished off first [with polished casing stones], then the next below it, and last of all the base and the lowest part.

So goes the first telling of the existence of the Great Pyramid at Giza outside of Egypt by Herodotus, a Greek historian from Ionia. Born in Halicarnassus, Caria, a Greek city in southwest Asia Minor (near modern-day Bodrum in Turkey), he lived in the fifth century BC (ca. 484–425 BC). He is most notable for his writing of *The Histories,* the first six books of which deal with the growth of the Persian Empire under the rulers Croesus (595–547 BC) and later Cyrus the Great (580–529 BC). The second of these six books (known as *Euterpe*) is largely concerned with Egypt and the annexing of it by Cyrus's son and successor, Cambyses II (?–522 BC). It is here that Herodotus speaks of the Great Pyramids at Giza, revealing for the first time their existence to the Hellenic world, and, for many centuries later, the rest of the world (Figure 11–1).

228 Chapter Eleven

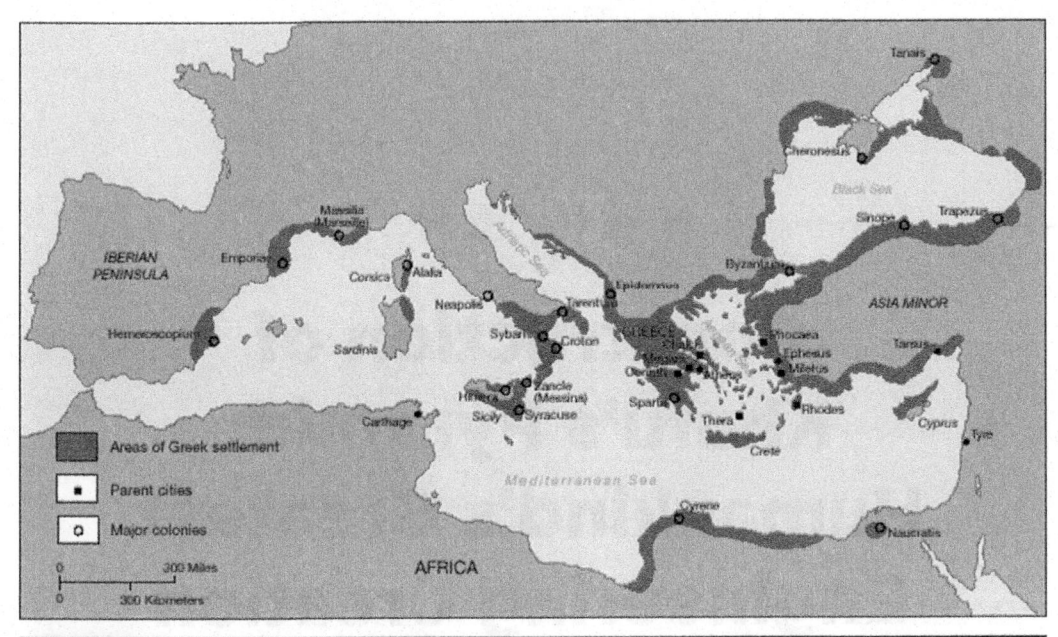

Figure 11-1 A map showing the Hellenic world highlighted. (*Source:* Wikipedia Creative Commons, created and originally contributed by Regaliorum on 15 October 2010 and modified by Athens2004 on 30 April 2012.)

A great traveler, Herodotus visited Egypt in 430 BC for a couple of years before he died in 425 BC. He even claimed he traveled up the Nile River from Alexandria, where he alleged he visited the pyramids, although many modern scholars doubt he actually did.

While Herodotus was the first to use the word *history* in his works, and became known as the "Father of History" for this reason, his method of gathering his information by speaking with people in an interview raises serious questions about the veracity of his facts. The problem is, not only are recollections of people prone to fade with time, but they are prone to become more and more exaggerated, if not fanciful, with each retelling. Beyond this reality, Herodotus himself is said to have had "a tendency to report fanciful information," according to some scholars. For this reason, he also picked up the less flattering title of "Father of Lies."[1] Even his description at the head of this chapter raises questions, including: (1) the height (or altitude) of the Great Pyramid of Khufu (to be discussed in the rest of this chapter) is 481 feet (146.5 meters), while each side of the square base is 765 feet (230.4 meters), so the height is not "the same" as the sides; (2) "there is no block of less than thirty feet in length" is easily found to be untrue simply by looking at the pyramids, as most stones are about 6 feet (2 meters) on an edge; (3) the questions of why and how one would use "short lengths of wood logs" as levers, when effective leverage for heavy stones would come from long levers, not short ones,[2] should cause pause; and (4) the suggestion

---

[1] An article entitled "Father of History or Father of Lies: The Reputation of Herodotus," by J. A. S. Evans, in *The Classical Journal*, October 1968, pp. 11–17, deals with this troubling possibility.

[2] Of course, it was the great Greek mathematician, scientist, and engineer Archimedes (287–212 BC) who was the first to have said, "Give me a *long enough lever* and a place to stand, and I will move the earth," 300 years later! Hence, poor Herodotus couldn't have known one needed a long log for a lever.

Construction of Khufu's Pyramid: Humankind's Greatest Engineering Creation

Figure 11-2 A collage of the Seven Wonders of the Ancient World chosen by Herodotus and depicted by the sixteenth-century Dutch artist Maarten van Heemskerck. The seven are, top left to right and working from the top to the bottom: (1) Great Pyramids of Giza, (2) Hanging Gardens of Babylon, (3) Temple of Artemis at Ephesus, (4) Statue of Zeus at Olympia, (5) Mausoleum of Halicanassus, (6) Colossus of Rhodes, and (7) Lighthouse at Alexandria. (*Source:* Wikipedia Creative Commons, originally contributed by Magnus Manske on 25 November 2007.) **Don't miss the color version of this figure, available at www.mhprofessional.com/ReverseEngineering.**

that the "lever" be moved was absurd, or the job would have taken even longer than the 20 years Herodotus claimed it took later in his *Euterpe*.

Veracity aside, Herodotus got it right that the ancient Egyptians built something very remarkable, however they did it. He, himself, was so impressed, that he ranked the Great Pyramids of Egypt as the greatest of his *seven wonders of the ancient world,* by which he meant the Hellenic world (Figure 11–2).

Subsequent reports of the Great Pyramids of Giza (as well as of the enigmatic Sphinx) by Greek scholars, Arabian scholars, and European scholars, as well as in ancient Egyptian documents found later in history (Table 11–1), all agree about one thing: the Great Pyramids at Giza are the greatest engineering creation of humankind ever!

11-2 The Great Pyramid of Khufu

Rising strikingly out of the flat desert sands of the Giza Plateau 5 miles (9 kilometers) west of the north-flowing Nile River near what is now Al-Jizah (Giza), Egypt, 15 miles (25 kilometers) southwest of Cairo, is the oldest and only remaining of Herodotus's Seven Wonders of the Ancient World (Figure 11–3*a* and *b*). The Great Pyramid of Giza, also known as the Pyramid of Khufu (known as Cheops in ancient Greek), is the largest and oldest of the three great pyramids in what has come to be known by archeologists as the Giza Necropolis ("City of the Dead"). It is also the only one that remains largely intact (Figure 11–4*a* and *b*). A masterpiece of engineering, the Great Pyramid still poses mysteries as to its true purpose for being built, its seemingly thought-filled location, its startling embedded mathematical symbolism, and its remarkable construction over 4600 years after its creation—mysteries that can only begin to be unraveled by reverse engineering.

TABLE 11-1 Ancient to Modern Historical Accounts of the Great Pyramids

Ancient Egyptian Sources
- The "Inventory Stella"
- Middle Kingdom Papyrus at Leiden
- The "Dream Stella" (1420 BC)

Ancient Greek Sources
- Herodotus (ca. 450–435 BC)
- Manetho of Sebennytos (ca. 280–270 BC)
- Josephus (AD 37–ca. 100)
- Diodorus Siculus (ca. 56–60 BC)
- Strabos's account (24 BC)
- Pliny's account (AD 20)
- Solinus (AD 250)
- Dionysius of Temahre (AD 818–845)

Early Arabian Sources
- Al-Manuns's entry (AD 820)
- Masoudi (died AD 956)
- Ibn al-Nadim (?–AD 995/998)
- Abdallah Muhammed bin Abd ar-Rahim al Kaisi
- Edresi (AD 1236–1245)

Early and More Modern European Sources
- Sir John Mandeville (AD 1340)
- Professor John Greaves (AD 1637–1638)
- Benoit de Maillet (AD 1735)
- Nathaniel Davison (AD 1763)
- Napolean's account (AD 1798–1799 AD)
- Captain Giovanni Battista Caviglia (1,817)
- Giovanni Batista Belzoni (AD 1818)
- Colonel Howard Vyse and J. Perring (AD 1837)
- John Taylor (AD 1859)
- Charles Piazzi Smyth (AD 1865)
- Waynman Dixon and D. R. Grant (AD 1872)
- Sir William Mathew Flinders Petrie (AD 1880)

But before beginning our quest to unravel some of the mysteries, let's first look at some data.

Egyptologists believe that the Great Pyramid was built as a tomb for the fourth-dynasty Egyptian Pharaoh Khufu over a period of 14 to 20 years, concluding around 2560 BC.[3] The original pyramid, at construction, was 280 Egyptian cubits tall 480.6 feet (146.5 meters), capped by either a solid gold or gilded pyramidion. In the absence of its pyramidion (believed to have

---

[3] Khufu was the son and successor of Pharaoh Sneferu, who reigned 24, 30, or 48 years ca. 2600 BC. Pharaoh Khufu reigned 23 years, according to most modern historians, from ca. 2589 to 2566 BC. He was succeeded by his son Djedefre, who reigned 10 to 14 years. Djedefre, in turn, was succeeded by his brother, Khufu's other son, Khafra, who reigned for 26 years. Sneferu's pyramid, known as the "Bent Pyramid," was 332 feet (101.1 meters) tall and had a square base with 619-foot (188.6-meter) sides. The lower portion had a 54°27' slope, which was changed midconstruction to 43°22', giving the pyramid its unique shape and name. Djedefre's and Khafra's pyramids, also true pyramids like Khufu's, stand near the Pyramid of Khufu. (See Figure 11–3*b*.)

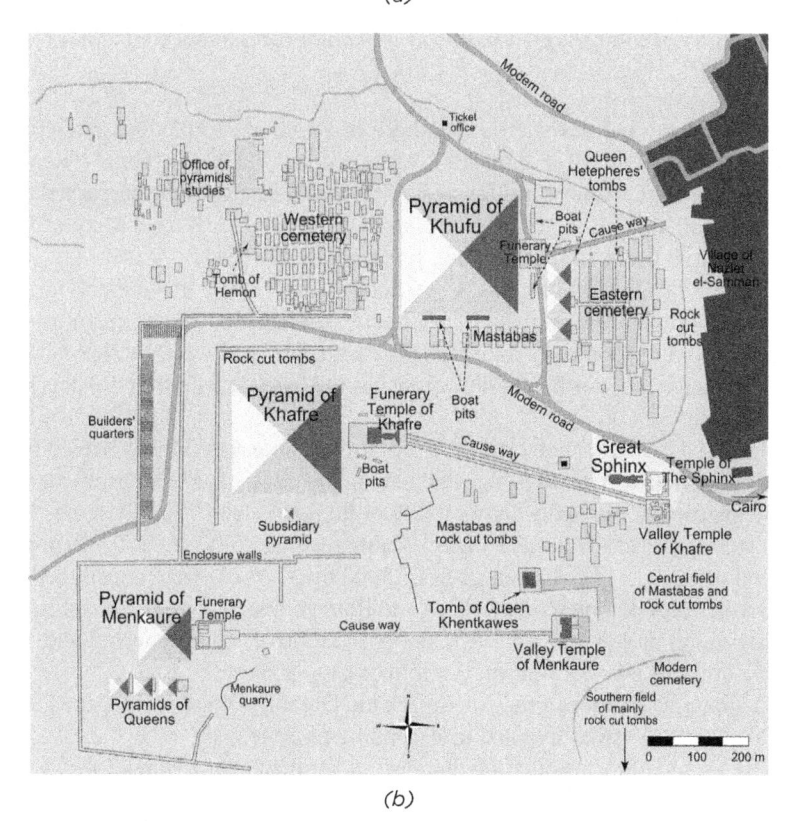

Construction of Khufu's Pyramid: Humankind's Greatest Engineering Creation 231

(a)

(b)

Figure 11-3 A satellite image of the Pyramids of Giza (*a*) and a site map to show details (*b*). Khafra was the son of Khufu, and Menkaure was the son of Khafra and grandson of Khufu. (*Sources:* The satellite image [*a*] is in the public domain, while the map [*b*] is from Wikipedia Creative Commons, originally contributed by MesserWoland on 10 August 2006 and modified by Jeff Dahl on 14 November 2007.)

232 Chapter Eleven

Figure 11-4 Photographs of the three Pyramids of Giza (a), with the Great Pyramid (of Khufu) at the right of the photograph, and shown alone at the left of the figure (b). (*Source:* Wikipedia Creative Commons, originally contributed by Ricklib on 17 June 2007 and modified by Ikiwarer on 19 July 2007 [a], and originally contributed by Minto on 20 August 2005 and modified by A. Parrot on 7 May 2010 [b].) **Don't miss the color version of this figure, available at www.mhprofessional.com/ReverseEngineering.**

been stolen within a thousand years, if not hundreds of years, of completion) and with erosion, the present height is 455.4 feet (138.8 meters). At its original height, the Great Pyramid was the tallest man-made structure in the world for over 3800 years, finally being surpassed by the 160-meter (~525-foot) -tall spire of the Cathedral Church of the Blessed Virgin Mary of Lincoln (or, commonly, the Lincoln Cathedral) in Lincoln, England, in AD 1311.[4]

Each base side of the Great Pyramid was 440 cubits (755.9 feet/230.4 meters) long. The base is horizontal and level to within ±0.6 inch (15 millimeters). The sides of the square base are closely aligned to the four cardinal compass points (north, east, south, and west), within 3 minutes of arc, based on true (not magnetic) north, and the corners of the base deviate from 90 degrees by a mean error of only 12 minutes of arc.[5]

The Great Pyramid consists of an estimated 2.3 million limestone blocks, weighing an average of 5500 pounds (2.5 metric tons), most of which are believed to have been hand-cut from and hewn at nearby quarries, for a total weight of 12.6 billion pounds (5.75 million metric tons). The largest granite stone lintels used in the King's Chamber weigh 55,000 to 176,000 pounds (25 to 80 metric tons), and more than 17.6 million pounds (8000 metric tons) were transported from Aswan, more than 500 miles (800 kilometers) away. In addition, another *1.1 billion pounds* (*500,000 metric tons*) of gypsum mortar was used. The entire pyramid was once covered with polished white Tura limestone cut into slant-faced, flat-topped casing stones.

Figure 11–5 shows a schematic cross section of the Great Pyramid of Khufu, which was unique in that it contained both an ascending and a descending passage.[6]

Table 11.2 lists some of the astounding dimensions for the Great Pyramid.

---

[4] The Lincoln Cathedral remained the tallest man-made structure for 238 years, from AD 1311 to 1549.
[5] There are 60 minutes (′) in 1 degree (°) and 60 seconds (″) in 1 minute of arc.
[6] The unique Descending Passage goes down at an angle of 26°31′23″ through the masonry of the pyramid and into the bedrock beneath it, becoming level after 345.2 feet (105.23 meters) and ending at a Lower Chamber. Some Egyptologists suggest the Lower Chamber was intend to be Khufu's original burial chamber, but Khufu himself changed his mind during construction, wanting to be buried higher up in the pyramid.

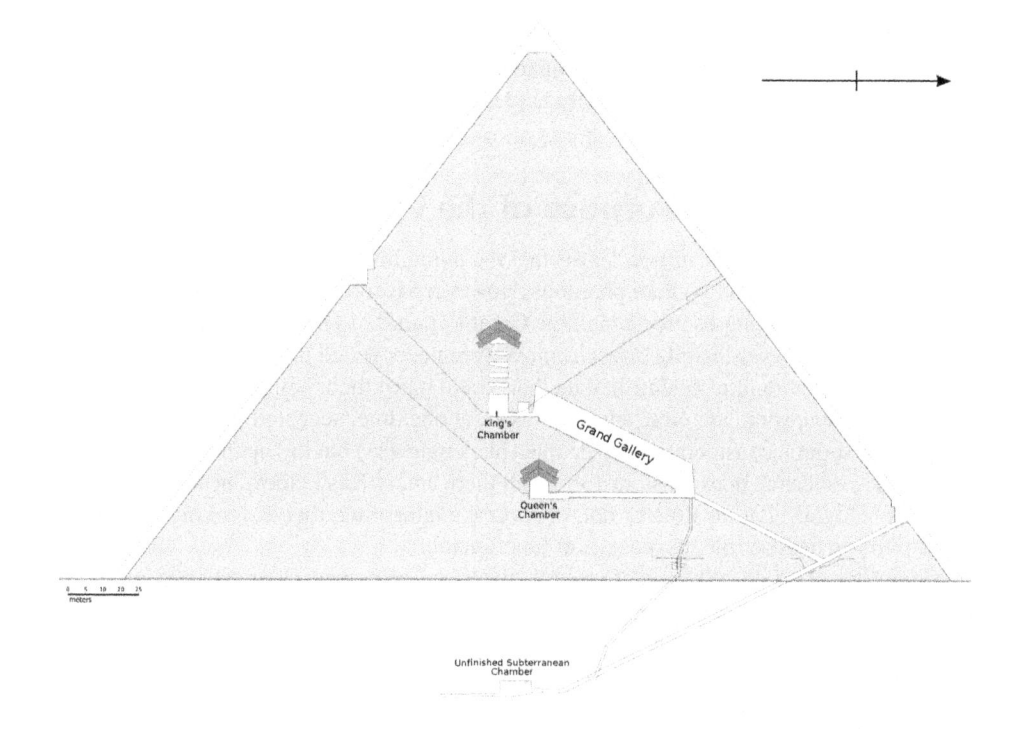

Construction of Khufu's Pyramid: Humankind's Greatest Engineering Creation 233

Figure 11-5 A simple schematic of the cross section through the Great Pyramid of Khufu, showing key features. (*Source:* Wikipedia Creative Commons, originally contributed by Jeff Dahl on 14 November 2007 and modified by Hardwigg on 22 July 2012.)

TABLE 11-2 Key Dimensions of the Great Pyramid of Khufu*

Height (including capstone)	439 ft/146.55 m
Courses of stones (with a mean height of 2.274 ft/0.693 m)	
West	755.76 ft (230.42 m)
North	755.41 ft (230.32 m)
East	755.87 ft (230.45 m)
South	756.08 ft (230.51 m)
Perimeter	3023.22 ft (921.70 m)
Angle of corners	
Northwest	89°59'58"
Northeast	90°3'02"
Southwest	89°56'02"
Southeast	90°3'02"
Slope (of north face)	51°50'40"
Area of base	43,614 ft^2 (4050 m^2)
	13.6 acres (0.405 hectares)

*Data were selectively taken from measurements recorded by Peter Lemesurier in his 1977 book, *The Great Pyramid Decoded*, reprinted by Elements Books Ltd., Rockport, MA, 1998.

Especially astounding is the accuracy of linear and angular dimensions. The 756-foot (230.6-meter) lengths of all four sides of the square base are accurate to ±0.69 foot/±8.28 inches (0.26 meters), or 0.044 percent, while the right angles at the four corners are accurate to ±0.048° or 0.054 percent.

11-3 Theories on the Purpose of the Pyramids

Every schoolchild, including the author, learns that the pyramids were built as tombs for the ancient Egyptians' beloved and godlike pharaohs. However, over the centuries, many theories have been proposed for the true purpose for which the Great Pyramid of Giza was built (ref. Bonwick; Tompkins). More *hypotheses* than theories, as most were never tested using scientific method and are not backed by the amount of scientific data that would merit them being considered theories, some are quite unusual, some are quite trite (but were, at one time, accepted by many people), and a couple bear serious consideration by people (like engineers) having open minds. For those theories for which evidence does exist, or for which there are, at least, clues, these have been the result, whether recognized at the time or not, of reverse engineering, that is, looking at what one can observe to try to deduce role, purpose, and functionality.[7]

What follows are short descriptions of several "theories," in no particular order of implied merit, excluding outrageous suggestions, with brief description.

The Tomb of the King (fourth century BC)

Hypothesis: The theory that the Great Pyramid, like other earlier and later of the 138 pyramids built by the ancient Egyptians over a period of nearly 2000 years (from around 2600 BC), was a tomb to contain the earthly remains of their beloved king, Pharaoh Khufu, began with Herodotus in the fourth century BC. Herodotus wrote: "Cheops [the Greek name for Khufu] ordered Philitis to prepare him a tomb." The idea that the Great Pyramid was a tomb was repeated by a Syrian writer in the ninth century AD when he wrote: "They are not granaries of Joseph as some say [see "Joseph's Granaries"], but mausoleums erected upon the tombs of ancient kings." Author Auguste Mariette-Bey (1821–1880) was adamant about the tomb theory, stating: " . . . with regard to the use of which the pyramids were destined, it is to do violence to all that we know of Egypt, to all that archeology teaches us of the monumental customs of that country, to see them as any other thing than tombs." Most modern books on Egyptology state that all pyramids, including the Great Pyramid of Khufu, were built as tombs.

Evidence: There is, in fact, disagreement about the Great Pyramid being built as a tomb. Other pyramids were absolutely built as tombs, as they contained the mummified remains of kings and queens and other nobles. While it may have been started as a planned tomb for Khufu, no mummies or any other human remains were ever found in the Great Pyramid, and it is believed unlikely that any such remains were removed by tomb robbers. The Great Pyramid is the only

---

[7] A book that espouses a theory that the Great Pyramid was a power plant which drew its energy from "acoustical resonance in harmony with the Earth" and is promoted (by its publisher!) as "a brilliant piece of reverse engineering," otherwise the author of the current book wouldn't even mention it, is *The Giza Power Plant: Technologies of Ancient Egypt*, by Christopher Dunn, Bear & Company, Rochester, VT, 1998. Without having read the book, I am not trying to throw stones; however, the author, Christopher Dunn, is hailed as a "craftsman" but is *not* an engineer. Use this fact as a data point if you choose to read the book. And, most of all, act like a scientifically educated person, skeptical of but open-minded to new possibilities supported by data.

pyramid built with an ascending system of passages, while it, like all other pyramids, has a system of descending passages. In all other pyramids, the pharaoh is buried in a chamber at the end of the descending passage. Furthermore, the Great Pyramid is unique in containing a Grand Gallery, for which no consensus opinion of purpose exists. Most striking of all, there are no hieroglyphics, paintings, inscriptions, or the like, in or on the Great Pyramid, while all other pyramids, monuments, and the like, are covered with inscriptions. It is considered unlikely the king would have been buried without any inscriptions or paintings. So why was the Great Pyramid built?

Display of Royal Despotism (from ca. 359 BC to nineteenth century)

Hypotheses: Aristotle (Greek, 384–322 BC) thought the priests of ancient Egypt convinced the Pharaoh Khufu to undertake construction of the Great Pyramid to find employment for the idle as a way of diverting them from rebellions. Pliny the Elder (Roman, AD 23–79) thought it was built so the Pharaoh could keep his captives busy. Rev. E. B. Zincke (nineteenth century AD) suggested that Egypt was so fertile, so much excess food had accumulated from in-kind taxation of subjects by the pharaoh, and people's wants were so few that surplus labor was available. Hence, the Great Pyramid was built to employ workers who had no job and to use up excess money in the treasury and food in the warehouses.

Evidence: Ancient Egypt during the fourth dynasty *was* rich, and there *was* plenty of labor available. Also, scholars no longer believe that either slave labor or forced labor of captives was used, but, rather, regular Egyptian citizens. The idea is not outrageous, by any means, but there was likely a greater and loftier motivation for such a monumental undertaking, most logically, devotion to their pharaoh.

An Astronomical Observatory (ca. AD 200)

Hypothesis: British astronomer Richard Proctor found reference in the works of the Roman philosopher Tiberius Claudius Severus Proculus (AD 163–218) that the Great Pyramid was used as an observatory before its completion. He went into great detail as to how this was done.

Evidence: Many modern investigators have found what appear to be more-than-coincidental alignments of the Great Pyramid's base, openings on the north face, and positioning of the two subsequent pyramids at Giza to the constellation of Orion. There are many constructions by ancient people (e.g., Stonehenge, Mayan temples) that relate to astronomy and probably allowed astronomical observations for both practical (e.g., seasonal changes) and religious reasons. Concern for careful alignment of the Great Pyramid, and all of the Pyramids of Giza, by the ancient Egyptians would also make sense based on their own concept of an afterlife.

Preservation of Learning from the Expected Deluge (AD 992)

Hypothesis: Many early Arabian authors (most notably Murtadi in AD 992) tell of a great flood that was foretold by astrologers, and contend that the Great Pyramid was built to preserve the memory of the then-existing learning, as well as, according to some authors, to also preserve medicines, magic, and talismans. To quote a portion of a 1672 translation of Murtadi's story:

> The priests having thus spoken, the king [Pharaoh] commanded them [the astrologers] to take the height of the stars, and to consider what accident they portended. Whereupon they [the astrologers] declared that they promised first the Deluge and, after that, fire. Then he [Pharaoh] commanded pyramids should be built, that they might remove and secure in them what was of

most esteem in their treasures, with the bodies of the kings, and their wealth, and aromatic roots which served them, and that they should write their wisdom in them, that the violence of the water might not destroy it.

Evidence: To date, nothing of the sort (medicines known as "alakakirs"; written science of astrology, arithmetic, geometry, and physics; etc.) has been found. On the other hand, many scholars believe mathematic knowledge is embedded in the design of the Great Pyramid (see "Standard of Weights and Measures"), and knowledge of astrology and astronomy may also be embedded in the site selection and positioning of the three great Giza pyramids (see the previous section "An Astronomical Observatory" as well as Section 11–4).

Joseph's Granaries (Middle Ages)

Hypothesis: In the Middle Ages (fifth to fifteenth century in Europe), Benjamin of Toledo proposed that the pharaoh had the Great Pyramid built for use in storing wheat in times of famine.

Evidence: Contrary to Benjamin's belief, the Great Pyramid has a solid core and is not hollow. Therefore, there is not room to store a very large quantity of grain.

Imitation of Noah's Ark or the Tower of Babel (1833)

Hypothesis: In 1833, Thomas Yeates came up with the odd opinion that the Great Pyramid modeled the Tower of Babel, which he alleged took its dimensions from Noah's Ark. He said:

> The Great Pyramid soon followed the Tower of Babel [3100 BC], and had the same common origin. Whether it was not a copy of the original Tower of Babel . . . and, moreover, whether the dimensions of these structures were not originally taken from the Ark of Noah? The measures of the Great Pyramid at the base do so approximate to the measures of the Ark of Noah in ancient cubit measure, that I cannot scruple, however the idea, to draw some comparison.

Evidence: In fact, the Book of Genesis describes Noah's Ark as having a length of 300 cubits, which equates to ~443 feet (135 meters), while the base of the Great Pyramid has side lengths of 755.9 feet (230.4 meters). Not close!

Barriers against Shifting Desert Sands (1845)

Hypothesis: In 1845, M. Fialin de Persigny expressed the opinion that the purpose of the pyramids was to act as barriers against the wind-driven sands of the desert in Egypt and Nubia.

Evidence: There is credible evidence that this didn't work, if that was the intent, as Thutmosis IV allegedly found the Sphinx buried to its neck in sand and excavated it to reveal its full glory, erected the inscribed "Dream Stele" between its paws, and painted it in bright colors in 1400 BC.

Filtering Reservoirs (mid-1800s)

Hypothesis: A Swedish philosopher of the 1800s suggested that the pyramids were simply contrivances for purifying the muddy water of the Nile River by having it pass through the passages through the core.

Evidence: There is no evidence of any canals for diverting water from the Nile to the pyramids (even though the river has almost certainly changed its path over the ages. Furthermore, the passages are simply not numerous enough or long enough to provide adequate filtering.

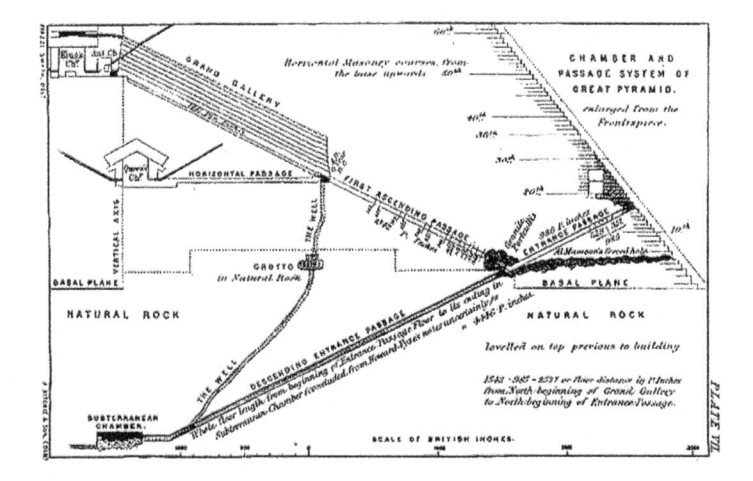

Construction of Khufu's Pyramid: Humankind's Greatest Engineering Creation 237

Figure 11-6 A diagram from Charles Piazzi Smyth's *Our Inheritance in the Great Pyramid* (1877) shows some of his measurements and chronological determinations made from them. (*Source:* Wikipedia Creative Commons, contributed by WolfgangRieger on 25 October 2009.)

Standard of Weights and Measures (1877)

Hypothesis: Edinburgh Professor Charles Piazzi Smyth (1819–1900), Astronomer Royal for Scotland (1846–1888), is well known for many innovations in astronomy and for his meticulous study of the Great Pyramid of Giza. Although similar ideas had been suggested by many others before him, Piazzi Smyth was the one who developed a formal theory with mathematical skill and attracted public attention for its appeal. It was his books that brought popularity to the Great Pyramid, but it was his statement that intellect shown by the ancient Egyptians was a gift from God.

Evidence: A few of Smyth's findings and proposals follow:

- The sacred cubit used by the builders of the Great Pyramid (at 25.025 English inches) was the same length as the one used by Moses to construct the tabernacle and by Noah to build the Ark. Furthermore, because the twenty-fifth part of this cubit is within a thousandth of a part of being the same as the English inch, Smyth concluded that the British had inherited this sacred unit of length down through the ages.
- Measurements of the coffin in the King's Chamber led to Smyth's conclusion that it was a standard of linear and cubic measurement.
- Smyth confirmed the value of 2ρ being built into the Great Pyramid if one divides the perimeter of the base by the height (or altitude) of the tetrahedron, to astounding accuracy, thus: (2)(3.14159)(480.69 feet) = 3020.26 feet versus 3023.22 feet, an error of only 0.098 percent, or, working backward, a value of ρ of 3.145 (versus 3.142, to four significant figures).[8]

---

[8] What is mysterious about this finding is this: The perimeter of the base of the Great Pyramid is virtually the same as the circumference of a circle with a radius equal to the altitude of the structure. For what other reason would the ancient Egyptians have this happen than to embed their knowledge of π into the Great Pyramid?

- Smyth also determined that the perimeter of the Great Pyramid was 36524.2 inches, which he believes was meant to correspond to a year of 365.2 days (versus a modern value of 365.25 days).

Smyth summed up his findings with this statement: "The linear measurement of this colossal monument, viewed in the light of philosophical connection between time and space, has yielded a standard measure of length which is more admirably and learnedly earth-commensurate than anything which has ever yet entered into the mind of man to conceive."

Figure 11-6 shows some of Smyth's interesting measurements.

Two other interesting and intriguing demonstrations of the ancient Egyptians' knowledge of mathematics are these: First, the ratio of the linear dimensions of the diagonal of the end face to the length to the diagonal of the volume of the King's Chamber in the Great Pyramid is 3:4:5, the most basic of right triangles that demonstrate the Pythagorean theorem. Second, dividing the length of the face (i.e., the apothem) by half the length of the base gives the Golden Ratio to astounding accuracy, thus: 610 feet/377.9 feet = 1.614 (versus 1.618), for an error of less than 0.25 percent.[9]

The most likely purpose of the Great Pyramid, like other Egyptian pyramids, was to serve as a tomb for the king, specifically for the Great Pyramid, Pharaoh Khufu. Whether he was actually ever buried in his pyramid is questionable, but that seemed to be the intent (as he directed the movement of his burial place from the traditional location at the end of the descending passage to higher up in the King's Chamber. The encoding of mathematics in the Great Pyramid seems to be supported by data, but is not, of itself, inconsistent with the Great Pyramid still being intended primarily as a tomb.

Other fascinating findings from reverse engineering the Great Pyramid's location, alignment, and orientation make much more sense when one considers the importance of and interpretation of an afterlife to the ancient Egyptians.

11-4 Theories on the Location of the Great Pyramid

If the last section taught us (you, as well as the author) nothing else, it taught us that the line between fact and fiction, reality and fantasy can easily be crossed. So, too, can the line between real *science* and *pseudoscience*.[10] While not the only subject to fall victim to pseudoscience, the subject of pyramids in general, and the Great Pyramid of Giza in particular, certainly has—so much so that this particular pseudoscience has been given a name: *pyramidology*. *Pyramidology* is a term, often used disparagingly, to refer to pseudoscientific speculation regarding pyramids, most often the Giza Necropolis and the Great Pyramid of Khufu in Egypt.[11]

The author assures the reader, there is absolutely no intent for what is presented in this

---

[9] Other fascinating and intriguing mathematics embedded in the Great Pyramid are discussed on http://greatpyramidmath.weebly.com/32.html.

[10] *Pseudoscience* is "a discipline or approach that pretends to be or has close resemblance to science [but lacks the required systematic approach and rigor]".

[11] With the appearance of pyramids outside of Egypt, particularly in Mexico and Central and South America (i.e., Mezoamerica), *pyramidologists* extend their interest and sometimes wild speculation to these as well, including a belief that the common geometric form must, somehow, either be mystical or guided by knowledge from a common race, such as extraterrestrials. The more logical reason is that a pyramid is easier to build and have remain stable than a rectangular prism.

chapter to drift toward, no less lie within, the area of pseudoscience. Rather, the intent is to use the Great Pyramid of Khufu as a familiar, wonderful, and rich example of how reverse engineering can be—and, in many cases, has been—used to gain knowledge and understanding about something constructed by others than ourselves, in this case, by a marvelous civilization in very ancient times.

With the author's true intentions in mind, this section attempts to summarize the findings, as well as speculations, relating to theories on the location of the Pyramids of Giza and, among these, the Great Pyramid of Khufu. By *location* (here) is meant three details: (1) geographic location on our and the ancient Egyptians' world, (2) alignment to the cardinal points of the compass, and (3) orientation relative to one another.

So let's look at each of these, in turn, with an effort to separate fact from fiction or, at least, to identify fact from speculation.

Geographic Location

The three Pyramids of Giza are located in the Giza Necropolis on the Giza Plateau in the southeastern outskirts of the sprawling modern city of Cairo in Egypt. Figure 11–7*a* is a satellite image showing the location (in the small black square), at the southernmost tip of the fertile and verdant delta of the north-flowing Nile River. Higher-resolution insets in the photographic image show the position of the three great pyramids constituting the Pyramids of Giza. The Great Pyramid of Khufu is at the upper right of the highest-resolution inset (nearest the upper right of the figure).[12]

There are considerable data to show that the Great Pyramid of Khufu is located almost precisely at the center of the landmass of the Earth, or "the Navel of the World" (Figure 11–7*b*), using an equal surface projection. Its east-west axis corresponds to the longest land parallel across the Earth (passing through Africa, Asia, and North America) and also the longest land meridian on Earth (passing through Asia, Europe, Africa, and Antarctica), and passes right through the Great Pyramid. Some interpret this (along with data on latitude and angle variations from the tilt of the Earth's rotational axis as it results in changes with its revolution around the Sun) as indication that the ancient Egyptians had exceptional knowledge of the Earth's size. Still others believe the Great Pyramid models the Earth itself in both size and mass (ref. Fix; Mendelssohn). Much of this is, by any engineer's measure, speculation!

The author believes the Pyramids of Giza are located where they are because that's where the ancient Egyptians lived, that is, in the fertile delta of the Nile River. Location close to the Nile was logical, as it was a major transportation "highway" for travelers and for barging the heavy cut-stone blocks from limestone quarries also located in the valley of the great river. And, to survive the long period of construction by ten of thousands of workers, there would need to be a readily available source of water, but also near the desert for preservation of the structure.

---

[12] At the time the Great Pyramid of Khufu was built, it is believed (and supported by geological evidence) that the site was much closer to the west bank of the Nile River, perhaps so close the river could easily be seen. In fact, the Nile River, like all major rivers of the world, changed paths over time, the original site being much closer than today (at about 5 miles /9 kilometers), based on this quote from Herodotus: "For ten years, the people wore themselves out building the road over which the stones were dragged, work which was, in my opinion, not much lighter at all than the building of the pyramid (for the road was nearly a mile long and twenty yards wide, and elevated at its highest to a height of sixteen yards, and it is all of stone, polished and carved with figures."

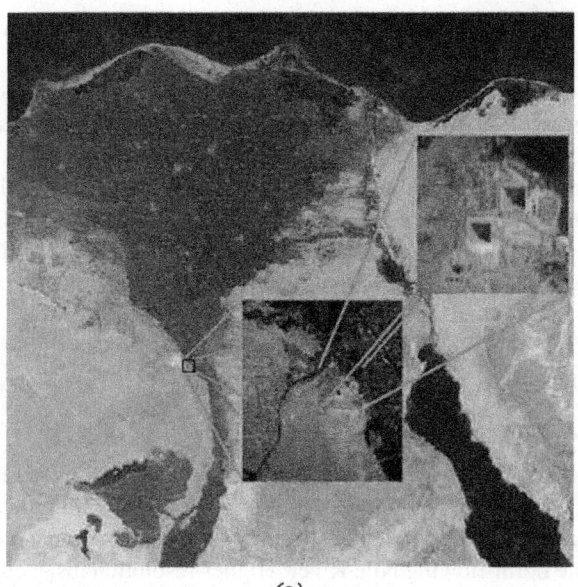

240 Chapter Eleven

(a)

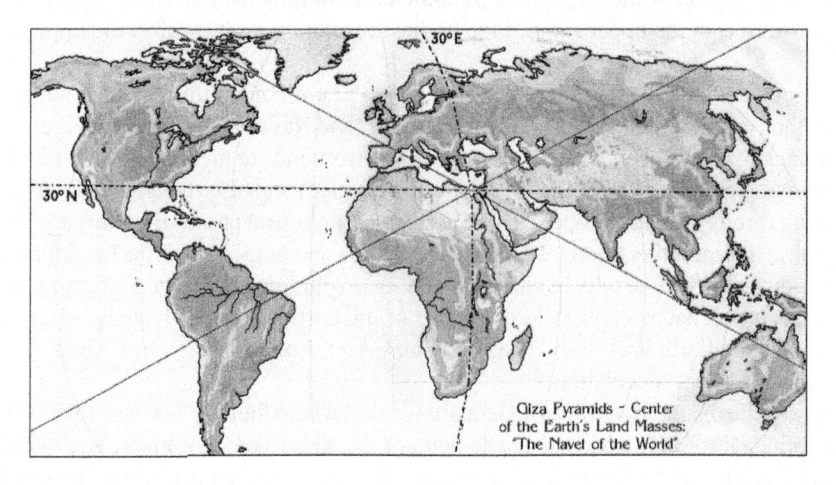

(b)

Figure 11-7 A satellite image of the fertile, verdant delta of the north-flowing Nile River indicating (by a small black square) the location of the Pyramids of Giza (at 29°59'N/31°09'E), along with two progressively higher resolution images (as insets) showing the site on the Giza Necropolis and arrangement of the three great pyramids (*a*). **Don't miss the color version of this figure, available at www.mhprofessional.com/ReverseEngineering.** The Great Pyramid of Khufu is at the upper right of the highest-resolution inset. A schematic illustration showing the location of the Great Pyramid at what is alleged to be the precise center of the landmass of the earth, that is, "the Navel of the World" (at the bottom of the figure) (*b*). (*Sources:* The satellite images are in the public domain [*a*]; the schematic illustrations are from www.pillar-of-enoch.com, courtesy of Helena Lehman, with her permission [*b*].)

Alignment to the Cardinal Points of the Compass

The four sides of the square base of the Great Pyramid, or, more precisely, its north-south axis, is nearly perfectly aligned to true (versus magnetic) north. The deviation is an astounding 2'28" of arc (about $\frac{1}{30}$ of a degree!). That alignment is to true north and not magnetic north is not surprising, since magnetic compasses were not invented until the Qin Dynasty (221–206 BC) in China. The much more obvious approach for a people who got to stare up into a perfectly dark sky at night, undimmed by artificial light or atmospheric pollution, was to align to the stars. But to what star? And why?

The obvious choice of using Polaris, the North Star, was not an option, as it had not yet taken up its current position at the end of the handle of the Little Dipper (Ursa Minor) at the time of the ancient Egyptians (Figure 11–8). Instead, according to a theory by Dr. Kate Spence, university lecturer in archaeology at Cambridge University, in a paper in *Nature,* the ancient Egyptians used beta–Ursae Minoris (also known as Kochab) in the Little Dipper and epsilon–Ursae Majoris (also known as Mizar) in the Big Dipper to align the Great Pyramid.[13] But there's a problem! These two stars would have been aligned on a plumb bob suspended between them to indicated true north in 2467 BC, 100 years after Pharaoh Khufu died, and far too late to have helped align the Great Pyramid to the accuracy with which it is aligned. In fact, Dr. Spence uses her theory to date the construction of the Great Pyramid as 2485–2475 BC, which doesn't even fit with her own calculation of 2467 BC. This would explain why Khufu's remains were not found in the Great Pyramid, but it ignores a great deal of other evidence for the pyramid's construction, most obviously the subsequent construction of the pyramids of Khufu's son Khafra and grandson Menkaure.

A rather simple solution to the problem of precisely aligning the Great Pyramid to true north is presented in the book *On the Orientation of Pyramids,* by Otto Neugebauer (Munksgaard, Copenhagen, 1980). Neugebauer has shown that orientation of the square base of the Great Pyramid required no sophisticated astronomical observation, nor is there any evidence that the Egyptians ever performed any such observations. A level pyramid, large or small, may serve as a *gnomon* or sundial, and the shadow cast by the Sun will indicate true north just as any sundial. The length of the shadow upon the ground will be at its shortest precisely at noon, when the Sun is due south, with the shadow itself pointing due north. Neugebauer goes on to make the point that the practical difficulties of maintaining orientation during construction posed a far greater challenge than the mathematics of it. We will look at this problem, along with others relating to construction of the Great Pyramid, in Section 11–5.

But why such concern for precise alignment to the cardinal points of the compass? The reason goes back to the religion of the ancient Egyptians and their belief in and concept of an afterlife. According to the ancient *Book of the Dead,* the soul of the beloved pharaoh would exit a door at the west of his "resting place" to enter the afterlife over the horizon.[14] In testimony to this fact is the ancient Egyptian name for the Great Pyramid: in translation, "Khufu's Horizon."

So one can speculate in the hope of finding some mystical reason for such precise positioning and alignment of the Great Pyramid, but the real reason is probably much simpler, as stated by

---

[13] Kate Spence, "Ancient Egyptian Chronology and the Astronomical Orientation of Pyramids," *Nature,* Vol. 408, November 16, 2000, pp. 320–324.

[14] The *Book of the Dead* is an ancient Egyptian funerary text first used at the beginning of the New Kingdom (around 1550 BC until 50 BC).

242 Chapter Eleven

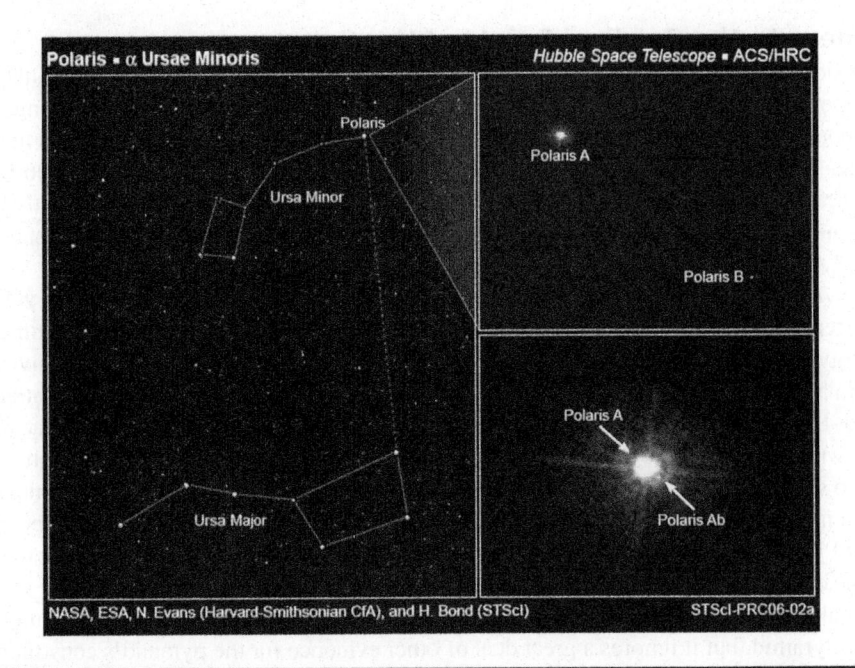

Figure 11-8 Sky-map (*left*) and telescopic photographs (*right, top and bottom*) showing the location of Polaris, the North Star, in the modern sky. Polaris was not at this location in Ursa Minor at the time the ancient Egyptians built the Great Pyramid. (*Source:* Wikipedia Creative Commons, contributed by Kristaga on 22 November 2006.)

Occam's razor: "Among competing hypotheses, the one that makes the fewest assumptions should be selected," as simplest is best![15]

Orientation of the Three Pyramids of Giza to One Another

Once pyramidologists, even if not archaeological scholars and/or serious archaeologists, began to speculate about why the Great Pyramid was located where it is and aligned how it is, it shouldn't come as much of a surprise that they also speculated there is meaning behind why the three great Pyramids of Giza are oriented as they are to one another. A favored "theory," also attributing great importance of astronomical observations to the ancient Egyptians (for which, at least, Neugebauer is convinced there is neither need nor evidence) is the so-called Orion Correlation Theory. First popularized by Robert Bauval (ref. Bauval), the theory proposes that the Pyramids of Giza are aligned relative to one another to correlate with the stars in the belt of the constellation of Orion (Figure 11–9*a* and *b*), that is, from largest pyramid to smallest, Al Nitak, Al Rai, and Mintaka. The "theory" has been pushed even further to suggest that the three great pyramids built by the Aztec civilization also correlate with the belt stars in Orion (Figure 11–9*c*).[16]

---

[15] *Occam's razor* (or *Ockham's razor*) is attributed to the fourteenth-century logician and Franciscan friar William of Ockham, Ockham, England.
[16] The Mayan civilization began about 3000 years ago (i.e., about 1000 BC) in Central America, in what is now Guatemala and southern Mexico, on the Yucatan Peninsula. Like the ancient Egyptians, of whom the Maya probably knew nothing, they constructed pyramids as temples to their gods (Figure 11–10). In fact, a fascinating video can be found on YouTube entitled "World's Largest Pyramid Discovered, Lost Mayan City of Mirador in Guatemala."

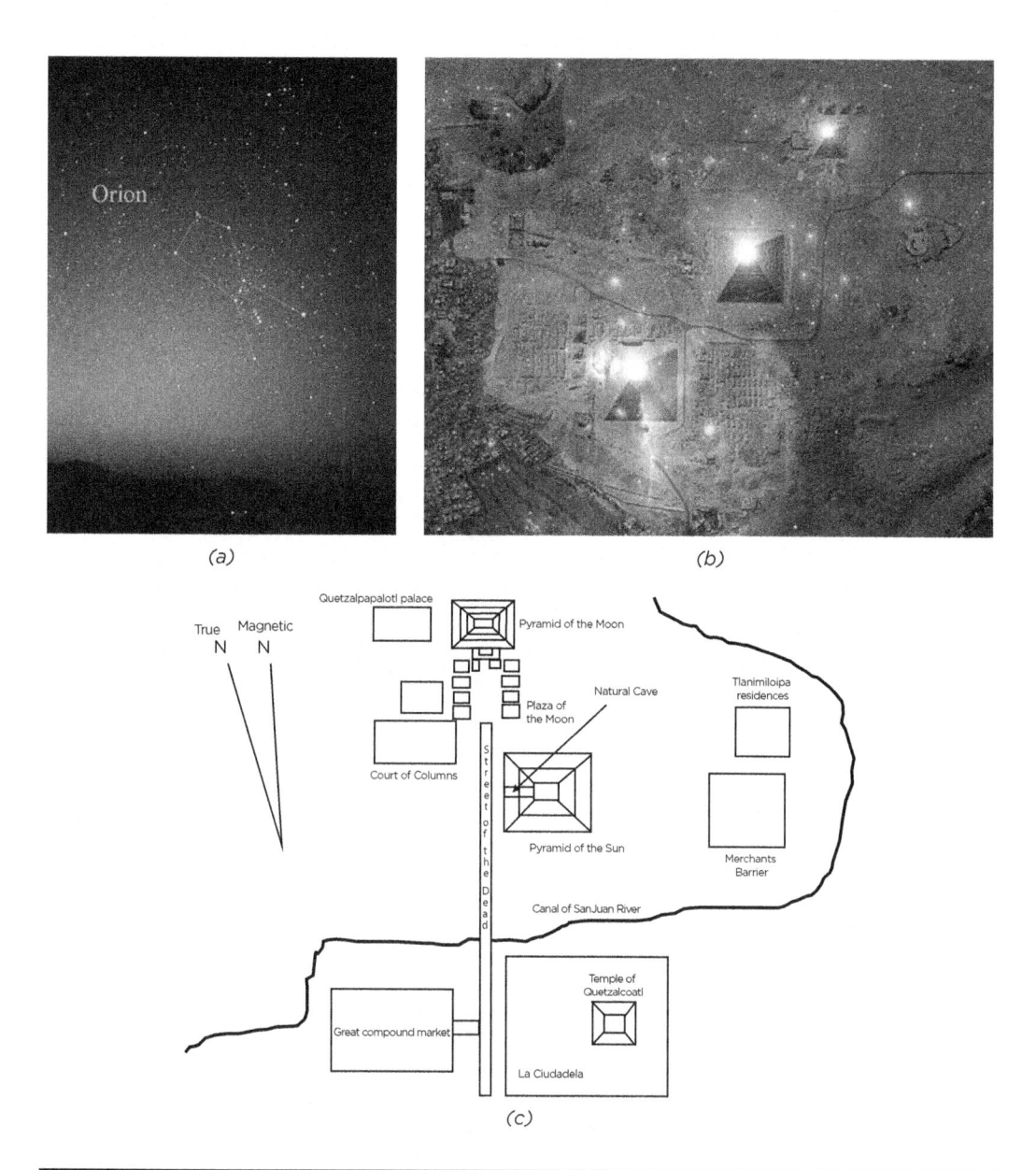

Figure 11-9 Lines tracing the constellation of Orion (a) show the three stars in Orion's belt to which the relative positions of the three great Pyramids of Giza seem to correlate. An overlay of the three stars in Orion's belt over the three great Pyramids of Giza, with Khufu's Pyramid at the upper right, shows the apparent correlation (b). **Don't miss the color version of this figure, available at www.mhprofessional.com/ReverseEngineering.** In (c), a schematic map shows the location of the "three great pyramids of the Aztec" at Teotihuacan. The oldest and largest at the site (and one of the largest pyramids in Mezoamerica) is the Pyramid of the Sun (center) built around 100 BC. The other two pyramids are the Pyramid of the Moon (top) and Temple of Quetzalcoatl. These three appear to also align with the three stars in Orion's belt. (*Sources:* Wikipedia Creative Commons, contributed by Till Credner on 26 June 2012 [a]; from www.spacecollective.org, used with permission [b]; from Wikipedia Creative Commons, contributed by Maunus on 21 August 2006 [c].)

Figure 11-10 The huge La Tigre complex in modern Guatemala contains some of the largest pyramids (by volume) ever built, including La Danta. In the photograph shown, the heavy tropical jungle covers the ruins of La Tigre. These were the creation of the ancient Maya that populated the region for millennia. (*Source:* Wikipedia Creative Commons, contributed by Authenticmaya on 20 January 2007.)

Peter Tompkins, in his book *Secrets of the Great Pyramid: Two Thousand Years of Adventures and Discoveries Surrounding the Mysteries of the Great Pyramid of Cheops* (ref. Tompkins), described a theory proposed by Russian mathematicians that maintained the three pyramids were arranged so that lines drawn between their centers, along sides, and to true north and other compass points created a series of right triangles that encoded the ancient Egyptians' understanding of what much later became known as the Pythagorean theorem.

11-5 Theories on the Construction of the Great Pyramid

No aspect of the creation of the Pyramids of Giza—and Khufu's Pyramid as the first and largest, in particular—is more intriguing than their actual construction. Historians, archaeologists, Egyptologists, and pyramidologists have their theories, but how these huge edifices were constructed is—or should be—most intriguing to engineers, who build great structures. How did the ancient Egyptians build the Great Pyramid more than 4600 years ago? Theories abound, but the answer, perhaps still to be discovered or, perhaps, just to be formally confirmed from among the many existing theories, will come only from systematic reverse engineering and *not* from casual (if not wild) speculation or, even worse, delusion.

The challenges posed by the construction of such huge structures to such precision, using materials, tools, and technologies available at the time, are many and involve:

Construction of Khufu's Pyramid: Humankind's Greatest Engineering Creation

- Site selection
- Site preparation
- Design layout and measurement
- Checking of progress (for adherence to plan)
- Preparing materials-of-construction
- Transporting massive stones to the site
- Placing massive core stones in position
- Adding casing stones
- Completing internal details

What follows is but a brief summary of what is known and/or logically deduced and proposed as answers or possible answers to the preceding list.

Let's begin.

Site Selection

For an engineer concerned about structural integrity (as opposed to an architect, who is much more concerned with aesthetics), *site selection* is about providing stability, not placement at the center of the Earth's landmass (to "focus acoustical energy") or arrangement to correlate with stars in the "belt" of an imagined mythological hunter in the sky or placement on a site to "be in harmony with Nature." Alignment to true north, on the other hand, is not unreasonable if it serves a purpose for achieving some design goal (e.g., orienting entrance and exit doors to east and west to comply with religious beliefs in life and the afterlife).

In the case of the Great Pyramid of Khufu, the site was selected to have the massive structure, estimated at more than 13.2 billion pounds, or 6,000,000 metric tons, (for comparison, the Twin Towers of the former 110-story, ~1368-foot World Trade Center in New York City was 3 billion pounds, or 1.36 million metric tons), rest on a solid bedrock base. The pyramid builders carefully chose the building plot, having seen what would happen if they did not. The Khufu pyramid lies on the best ground existing on the Giza Plateau, being on top of a bedrock core that extends at least 26 feet (7.9 meters) beneath the pyramid. Major problems had developed at earlier pyramid building sites when the strength of the underlying ground was overestimated. Most notable, the ground beneath the Pyramid of Sneferu (Khufu's father) at Dahshur yielded under the huge weight when construction was about two-thirds complete (based on volume) and led to collapse of the outer casing stones near the base. As a result, another layer of casing stones had to be added to reduce the angle of inclination from 60 to 54°27' and, still later, the angle of the core for the upper portion of the pyramid had to be further reduced to 43°22'. The result of this in-process reverse engineering saved Sneferu's pyramid, to become known as the "Bent Pyramid" (Figure 11–11).

Site Preparation

A key requirement for a building site, especially for very large structures, is that it be level. Remarkably, the bedrock foundation on which the 755.9-foot/230.4-meter 13.6-acre/0.405-hectare) square base was set was horizontal and flat to within 0.6 inch (±15 millimeters), a feat that would challenge the best modern excavating and grading techniques, surveying technologies (e.g., laser leveling), and firms. The measurement technique for accomplishing this is nicely described (along with many other interesting facts relating to the Great Pyramid's construction) in a website at www.cheops-pyramide.ch (ref. Löhner). A *square level,* in the shape of a large letter

246 Chapter Eleven

Figure 11-11 The "Bent Pyramid" of Pharaoh Sneferu (ca. 2600 BC) is unique in having its angle of inclination change from 54°27' for the lower portion to a shallower 43°22' for the upper portion. The angle was reduced to avert collapse when signs of instability appeared during early construction due to an overestimate of the strength of the underlying ground. (*Source:* Wikipedia Creative Commons, contributed by Magnus Manske on 30 January 2010.) **Don't miss the color version of this figure, available at www.mhprofessional.com/ReverseEngineering.**

"A," was constructed from wood beams. A plumb line suspended from the apex is used to check level against a mark in the middle of a crossbeam. Field-testing of a replica device showed that an accuracy of 1 centimeter over a distance of 90 meters could be maintained. To actually level the ground, it is most likely that the ancient Egyptians flooded the area being leveled with a thin layer of water, adding or removing material as necessary to get rid of pools or islands, respectively.

Design Layout and Measurement

Most construction experts, if not Egyptologists, believe the builders of the Great Pyramid transferred dimensions from plans drawn to scale on papyrus (as remnants have been found!), and likely from scale models, to the actual ground in full size and then actually placed the first tier of stones on these lines.[17] It is known that the ancient Egyptians employed land surveyors known as *harpedonaptai* or *harpedonapts*. These were rope stretchers using knotted ropes, with 11 equally spaced knots (or painted marks) dividing the rope into 12 equal parts. These ropes could then be

---

[17] This same basic technique was and is used by wood-boat builders (and was used by manufacturers of aluminum-alloy airplanes from the 1930s through 1950s). The technique is known as *lofting* (ref. Messler), with the taking of measurements from scale models known as *lines taking*.

used to form a right triangle with sides in a ratio of 3:4:5 (later to be called a *Pythagorean triple*), about which the ancient Egyptians clearly knew (recall the dimensions of the King's Chamber mentioned earlier). For right angles, the ancient Egyptians employed a simple construction using a rope more than half the length between two points on a straight line to scribe arcs from each endpoint of the line. By drawing a straight line through the intercepts of these two long scribed arcs, a new line perpendicular to the first could be made (ref. Löhner).

Most engineers familiar with manufacturing and/or construction recognize that in order to achieve the phenomenal accuracy of the lengths and corner angles of the large square base, construction would have had to begin with placement of the outermost stones to form the perimeter to the desired accuracy, with other stones then being added to fill the interior core.[18] With this approach, it was a matter of regularly checking accuracy against the plans until the pyramid was complete. We will see next that this requirement helps eliminate some theories about how the huge stones used to build the pyramid were put in place.

Checking Progress (for Adherence to Plan)

Every engineer knows, or should know: You plan your work and work your plan. For the builders of the Great Pyramid, this meant having to maintain a continuous direct line of sight with the pyramid as it grew in size. Only by sighting on the edges of the outermost stones as they were placed (including, especially, the polished tapered casing stones) could the head builder be sure dimensions were being held. This means that several of the theories proposed by other than engineers (e.g., archaeologists and Egyptologists) could not have been used, as they obstructed direct observation of the structure. Examples include the use of ramps and certain suggested lifting techniques requiring scaffolding and, perhaps, either levers or rope lifting devices on the inclined faces.

Preparing Materials-of-Construction

The Great Pyramid contains an estimated 2.3 million limestone blocks weighing an average of 5500 pounds (2.5 metric tons) and averaging 31½ inches (0.693 meters) square by about 70 inches (2 meters) long. Most of these were transported from nearby quarries. White Tura limestone used for casing stones was quarried across the Nile River from the construction site. Huge granite stones (used as lintels for the roof or ceiling of the King's Chamber, for example) weighed 55,000 to 176,000 pounds (25 to 80 metric tons) and were transported downriver (north) on the Nile from Aswan more than 500 miles (800 kilometers) away. Another 1.1 billion pounds (500,000 metric tons) of crushed gypsum mortar (made by roasting crushed limestone) was used to fill gaps between hand-hewn blocks that often were only 1/50 inch (0.5 millimeters) wide.

Since the ancient Egyptians had not yet discovered or used iron, it is believed by archaeologists that they cut the fairly soft limestone blocks by scoring a line with a hard point (perhaps bronze or a harder, sharpened stone) and hammering wooden wedges into the rock, which they then soaked with water to cause swelling, and fracturing the rock along the scored line. Finishing of the blocks to achieve dimensions and smoothness was believed by archaeologists to have been done using copper chisels, although *none* have ever been found at either quarries or at the construction site, where it is unimaginable that none would have been lost.

---

[18] Since casing stones would be added once the core was completed, compensation for the dimensions of these stones would have to be taken into account so that the overall dimensions of the true, smooth pyramid would be correct.

Upon completion of the stepped or tiered core, the Great Pyramid was allegedly surfaced with white Tura limestone casing stones having slanted faces and flat tops, bottoms, and sides. These were carefully cut to the required face angle with a *seked* (horizontal run for an ancient Egyptian cubit of height) of 5½ palms.[19] Just before placement on the face of the pyramid, casing stones were given any required final adjustment of dimensions to create the precise angle of inclination for the faces.

Transporting Massive Stones to the Site

Removal of cut-stone blocks from quarries, transport to the construction site, and, eventually, placement into position on the ever-growing pyramid to create the core and, finally, polished surface, was hindered by the lack of lifting machines, pulleys, or wheels, which had not yet been invented in ancient Egypt. Large granite blocks, as well as all of the limestone core and casing stones, were transported northward or across the Nile River by huge, specially built barges. Once near the site, the blocks were moved from the barges along a specially constructed roadway, which Herodotus said took as much effort to build as the pyramid itself. The workers probably used sleds that slid on their own runners or on water-lubricated wooden rails, as using log rollers would dig into the compacted desert sand. Another suggestion is a system of four quarter-round cages that were used to encase the stone as the cage was then rolled over and over along the roadway.

Several investigators have, often based on tests using archaeology students and blocks of cast concrete that simulate the size and weight of the average core stones, estimated the number of workers needed to pull such sleds carrying a 2.5-metric-ton stone across horizontal compacted sand or an inclined ramp of 5 to 10 degrees, also of compacted sand. Estimates range from 8 to 10 men for a horizontal move to 20 to 40 men for a 5-degree or 10-degree ramp, respectively. Obviously, larger, heavier stones would require larger, heavier sleds and many more men.

Placing Massive Core Stones in Position

Nowhere does the mystery of how the ancient Egyptians become more baffling—and controversial—than with how so many massive blocks of stone were put into place by the ancient Egyptians to create the huge pyramid.[20] Theories (if that is what some of the wildest ones deserve to be called!) abound, with the greatest number involving ramps of some kind.[21] Until quite recently, these were all external to the pyramid itself, some being straight toward the structure, some zigzagging up the pyramid from tier to tier, some spiraling around the pyramid along the

---

[19] In ancient Egypt, 1 cubit (*Meh Nesut*) = 7 palms (*Shesup* or *Shep*) = 28 fingers (*Yeba* or *Zebo*) = ~21 inches (52.4 centimeters). 1 palm = about 3 inches (7.48 centimeters), and 1 finger = about ¾ inch (1.87 centimeters). A *seked* of 5½ palms gave a slope of about 54, determined by the builders to be optimum for a stable pyramid.

[20] Recall that the pulley and block and tackle, with their offering of mechanical advantage, were not in existence at the time of the ancient Egyptians as they built the pyramids. The Greek Archimedes (287–212 BC) is credited with inventing the pulley, and Hero of Alexander (AD 10–70) is the first to describe the use of a block and tackle, both of which offer considerable mechanical advantage for lifting heavy objects.

[21] Outrageously wild "theories" that the author refuses to lend any credibility to by discussing them here include the use of levitation (by concentrating the "acoustical energy of the Earth" to be "in harmony" and "resonance" with the pyramid, advanced knowledge provided by the superior race of people of the Lost Continent of Atlantis described by Plato [424/423–348/347 BC], and assistance by extraterrestrials [Erich von Däniken].

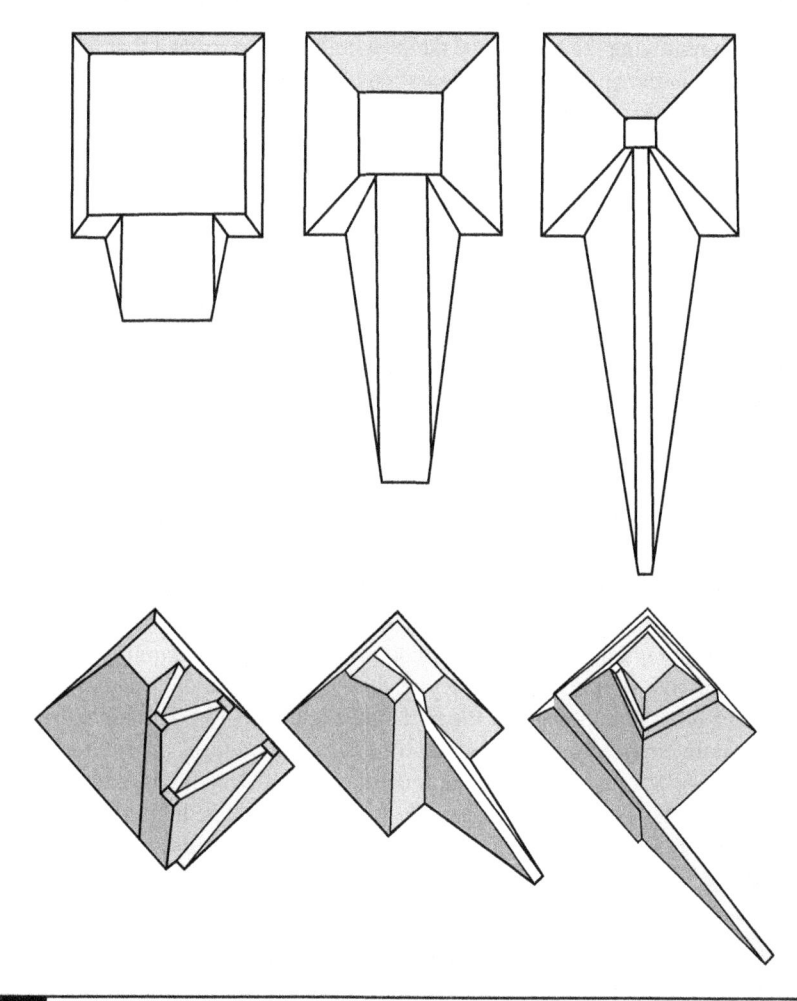

Construction of Khufu's Pyramid: Humankind's Greatest Engineering Creation

Figure 11-12 Examples of external ramps that might have been used to allow massive blocks of stone to be moved to allow setting into position as the pyramid grew ever larger include straight, zigzag, and spiral designs, some of which become more elaborate. Even more elaborate designs can be found online, but, as Occam's razor states in paraphrase: Most obvious is most likely.

tiers, and others having odd or outrageous designs.[22] Some of the more reasonable ramp designs are shown schematically in Figure 11–12.

From the standpoint of reverse engineering, many, if not most, of the external ramp designs suggested are untenable for one or both of two reasons. First, the labor associated with

---

[22] The so-called combination ramp is said by the author to be outrageous because the sidewalls of the many ramps are far too steep for the likely materials-of-construction (i.e., compacted sand and mud, with added rock and wood slats at the very top "road" surface). This is easy to prove from reverse engineering based on the *angle of repose* of compacted materials.

constructing and then removing the temporary ramps would rival or exceed that required to build the pyramid itself. This is shown by simple calculation using an incline of 5 or 10 degrees and a reasonable angle of repose for the sidewalls (ref. Messler, Exercise 4–1). For a 5-degree incline, for which a much more manageable team of 12 to 16 (versus 30 to 40) haulers would be required, a straight ramp would extend more than a 1.5 kilometers, or a mile, with the volume of material needed being more than that used in the Great Pyramid itself! Second, many external ramps interfere with the direct observation of the growing structure required to have linear and angular dimensions comply with the plan.

While many (including Herodotus) claim the builders used levers to incrementally lift stones from tier to tier, first lifting one side and inserting a shim and then lifting the other side and inserting a shim, and so on, the number of such levers required would pose a problem, as they would cover the face of the pyramid with wood structures, again obstructing direct observation of the growing pyramid. In fact, a simple calculation that should be used early in any reverse engineering effort relating to construction of the Great Pyramid reveals that a structure composed of 2.3 million blocks of limestone would require a block to be placed in its final position in about 1 minute and 50 seconds working 10 to 12 hours a day, 6 days a week for almost 20 years (ref. Messler, pp. 177–183)! Try it yourself.

So, while experiments by archaeologists and others have shown levers are capable of lifting a stone from one tier to the next in about 2 minutes, the simple logistics of placing stones everywhere around the pyramid at that rate would demand an inordinate number of lever devices.

Franz Löhner proposed a simple *rope roll* device that uses a system of counterweights (i.e., other stones) to move blocks up the face of the pyramid (Figure 11–13). The basis for his device is this: Why build separate ramps when the pyramid has four inclined faces as an integral part of its structure? Absolutely correct, and very much in accordance with Occam's razor (see Section 11–4 and Footnote 14). One other key point in favor of Löhner's proposal is this: Finished casing stones would have to be put in place as the pyramid grew (and not after the entire core was completed) to provide the inclined face. In fact, this would help the builders, as the finished face could be checked in progress and not at the end, when any errors could be too late to overcome.

So there are ways to lift stones into position, some more practical than others. But little or no evidence has been found by archaeologists at the Giza site or in any hieroglyphics (which the ancient Egyptians used to record much of their history) for ramps. However, there has been lots of evidence to support a novel theory for the use of an internal ramp, described in Section 11–6.

One final theory is that many, although not all, of the blocks composing the Great Pyramid were cast in place using a formulation for a "geopolymer cement," which the ancient Egyptians are alleged to have fallen upon. This theory was first proposed in the late 1980s by Dr. Joseph Davidovits, director of the Geopolymer Institute in San Quentin, France. He proposed that many of the blocks, especially farther up on the pyramid, are a "synthetic stone" made from a mixture of finely crushed limestone, nanoparticle kaolin clay, lime, and water, making a very early form of concrete. The idea was picked up and promoted by Distinguished Professor Michael Barsoum (of Egyptian heritage) at the Department of Materials Science and Engineering at Drexel University.

In this author's opinion, as a materials engineer himself, the idea may be intriguing but is difficult to prove, in the first place, and not supported by evidence, in the second place. Limestone is a sedimentary rock formed from lime, sand (silicon dioxide), clay, and water as a natural cementitious process. Thus, it would be easy to be fooled into thinking blocks hewn from a natural deposit in a quarry were cast, as the microstructures would be very similar to cast blocks. Claims

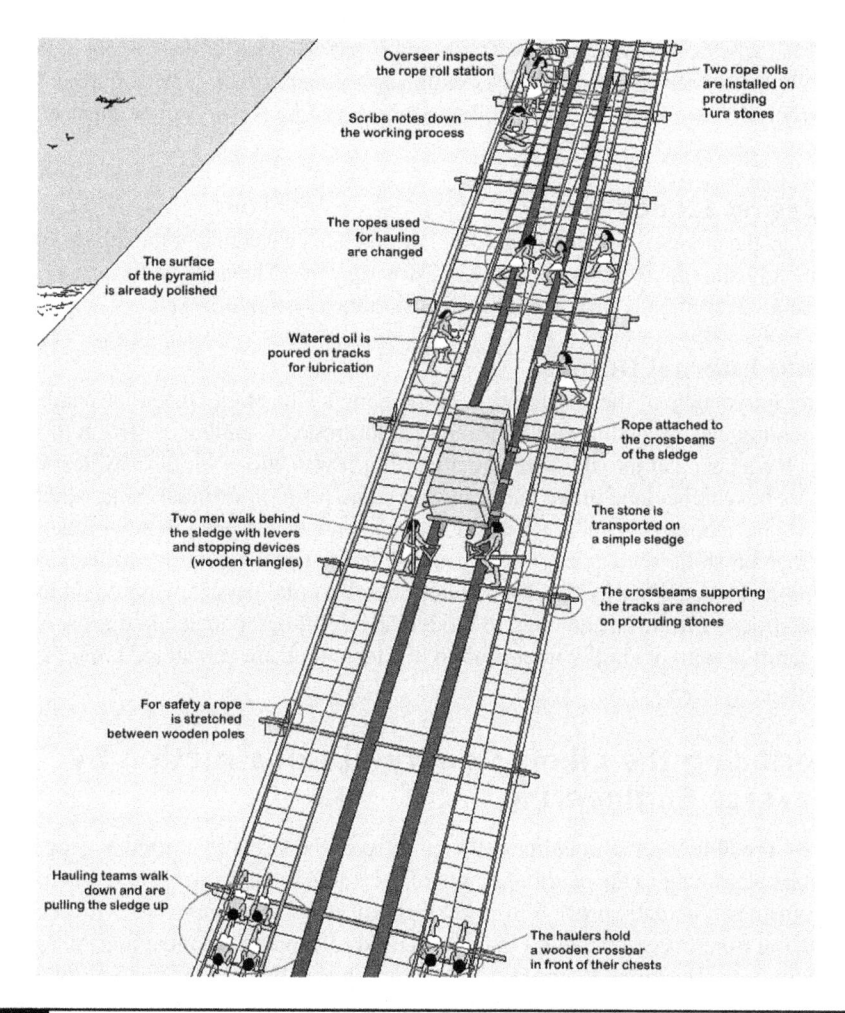

Construction of Khufu's Pyramid: Humankind's Greatest Engineering Creation 251

Figure 11-13 Franz Löhner's proposed use of a *rope roll*. The rope roll consists of a small wooden stand constructed from planks and a round axle supported by lubricated copper bearings around which a long rope is wound. Heavy limestone blocks are lifted up the inclined face using a system of counterweights. (*Source:* From www.cheops-pyramide.ch/khufu-pyramid/rope-rol.html, "Building the Great Pyramid," used with permission of Franz Löhner and Teresa Zuberbühler.)

about the proportions of "amorphous" versus crystalline material seem unsupportable. Exposure to weather—drought and rain—for 4600 years could easily change blocks of limestone that were once part of a massive deposit and now have much more exposed surface, perhaps altering the microstructure from atmospheric carbon dioxide and water.

Of greater significance is the fact—easily observed in the Great Pyramid—that the dimensions of the blocks vary significantly from one to the next, with no evidence of a standard block. One would expect blocks cast from "geopolymer concrete" to be made using reusable molds, which would reproduce block after block having the same dimensions and surface finish. This is simply not the case, as Professor Barsoum freely admits in his papers. Also, if the ancient builders could cast blocks, why stop with such small blocks requiring so many castings to be made? What not cast much larger blocks or even the bulk of the pyramid, just as great concrete dams are cast today?

Adding Casing Stones

The key with casing stones is to get the face angle right. This was done at the quarry, with any necessary fine adjustments being made at the construction site. The method for achieving the desired face angle was already discussed, that is, using the concept of *seked*.

Completing Internal Details

Without going into details of the construction of the King's Chamber, Queen's Chamber, Grand Gallery, descending and ascending passages, and Underground Chamber, suffice it to say here, they need not have been cut into the core structure after it was built. Nor did any decoration of these chambers have to be done in the dark or using some form of artificial light, which some argue could not have been oil-fired torches since there is no soot deposit in any of the chambers, so this must mean that the ancient Egyptians had an electric lightbulb! If one reflects on the Great Pyramid being constructed tier by tier, the various chambers and passageways could have been built in as the structure rose. The analogy to modern stereolithographic techniques is obvious. Readers not familiar with or simply interested in this technique are encouraged to seek information on the Internet.

11-6 Deducing the Likely Reality of Construction by Reverse Engineering

All of the most credible theories relating to the purpose of the Great Pyramid, its geographical location, and its alignment to the cardinal points of the compass have, in reality, whether realized or not, come from a systematic process of reverse engineering. Demonstrable evidence and logical deduction from observable clues have been used to come to the most logical and generally simplest conclusions. The Great Pyramid of Khufu was located where the pharaoh ruled, where the laborers and materials were the most readily available, and where access to the waters of the sacred Nile River were at hand for helping level the building site, for transporting materials by barge, for helping level the building site by flooding, for lubricating skid-way rails, and for cooling the bodies and quenching the thirst of the tens of thousands of workers. The pyramid was aligned to true north in accordance with the ancient Egyptians' belief in and concept of an afterlife, that is, to allow the pharaoh's soul to depart his body in its earthly resting place and cross over the horizon in the west. Less clear is why the other two great pyramids of Khufu's son Khafra and grandson Menkaure are arranged relative to the Great Pyramid as they are, but more likely to correlate to the stars in the belt of the constellation of Orion the Hunter, perhaps because that is where the bedrock provided the most stable foundation for immensely heavy structures.

In each case—geographic location, alignment, and, perhaps, orientation to other pyramids—the hypothesis requiring the fewest assumptions prevailed as the friar William of Ockham suggested.

The solution to how the Great Pyramid was constructed, it would seem, should also be the simplest, if not the most obvious. No labor-intensive temporary external ramps for which there is no solid evidence, no synthetic cast-in-place blocks that don't match, and, surely, no levitation by resonance with the Earth's "natural rhythms" was used. But what method, then, was used?

In 1986, a French team of investigators led by Gilles Dormion and Jean-Yves Verd'hurt, working on the mystery of the Great Pyramid of Khufu under the aegis of the Fondation EDF made two startling findings that hinted at an answer to the question of how the great structure was built, even though they didn't, at the time, fully comprehend the significance of their findings. First, the team had been shown ancient plans (drawn on papyrus) by Professor Huy Duong Bui. On the plans was what they all first believed was a construction anomaly for which earlier hypotheses of pyramid construction involving external ramps could not account.[23] The anomaly, dubbed the "spiral structure," looked exactly like a ramp built inside the pyramid, hidden from view by the outermost blocks. The spiral ramp suggested a major role in the pyramid's construction, but the team failed to immediately follow up on the anomaly. Second, some members of the team had fortuitously spotted a desert fox in a hole next to a notch 280 feet (90 meters) up on the pyramid near one edge between abutting faces, not once, but on two separate occasions (Figure 11-14). Short of scaling the impossibly steep face and high-stepped tiers, how else could the fox have gotten there other than by navigating some kind of internal passageway[24]?

Segue forward to 1999. Henri Houdin, a French civil engineer, had become fascinated by the Great Pyramid while watching a television documentary on its construction. The theories espoused, largely employing external ramps, simply didn't make sense to him. He soon enlisted the help of his son, Jean-Pierre Houdin, an established architect in Paris with expertise in 3D computer graphics, to develop his idea, spawned by practicality, that the Great Pyramid had been built from the inside out, using an internal ramp that spiraled around the structure with a gradual incline from tier to tier. During 2000, father and son met with the leaders of the 1986 team simply to gather some additional data on measurements the team had made. To the Houdins' amazement and delight, they learned of the "spiral structure" and the mysterious, reappearing desert fox.

In 2003, Henri created the Association of the Construction of the Great Pyramid (ACGP), in order to promote his research. The ACGP enabled Henri and Jean-Pierre to meet a number of experts, as their hobby morphed into an all-consuming obsession, and the project moved ahead.

In 2005, Mehdi Tayoubi and Richard Breitner of Dassault Systemes invited Jean-Pierre, who had taken over the project from his failing father, to join a new sponsorship program called Passion for Innovation. Together, they decided to examine the theory with the aid of Dassault Systemes' industrial and scientific 3D solutions. Using software applications like CATIA to reconstitute the site of the gigantic construction in three dimensions allowed them to test in real time whether such an approach was plausible. The result was the spectacular *Khufu Reborn,* "a 3-D Experience" that has played in La Geode in Paris, one of the finest virtual-reality theaters in the world. Fortunately, for those of us not able to get to Paris to go to the theater, the video can be seen on the Internet at www.3ds.com/khufu. Readers are strongly encouraged—for both education and sheer amazement—to view this remarkable and beautifully produced documentary, updated in 2011 from the original 2007 version.

---

[23] Open-minded scientists and engineers should not dismiss evidence, even while remaining skeptical, because that evidence does not fit with preconceived notions.

[24] Returning to the Great Pyramid again in 1998, members of the 1986 team found a notch leading to an internal space, perhaps part of the internal spiral ramp (Figure 11-14).

254 Chapter Eleven

(a)   *(b)*

Figure 11-14 Investigator and author Bob Brier contemplating the notch (*a*) about 280 feet (90 meters) up from the base on the northeast edge of the Great Pyramid (*b*), which led to a chamber beyond, behind the outer face of the pyramid, and may be part of the internal spiraling ramp proposed by Henri and Jean-Pierre Houdin, upon a return visit to the site. (*Source:* From an article by Bob Brier and Jean-Pierre Houdin, "Return to the Great Pyramid," *Archaeology,* Vol. 62, No. 4 [July/August 2009] [Gedeon Programme/Dassault Systemes]; used with the kind permission of Bob Brier.) **Don't miss the color version of this figure, available at www.mhprofessional.com/ReverseEngineering.**

Figure 11–15*a* shows a computer 3D rendering of the Great Pyramid and the internal spiraling ramp proposed by Henri and Jean-Pierre Houdin, while Figure 11–15*b* shows a detail of how a two-level arrangement of the internal ramp allowed teams of haulers returning from bringing a block to its final location in the structure to pass incoming teams of haulers without causing obstructions in the passageway. In fact, returning teams could assist with turning the heavy blocks at corners. Figure 11–16 shows how a reasonable-size external ramp could have been used to bring blocks to build the greatest mass and volume of the pyramid to a certain level, with ever-diminishing numbers of blocks being hauled up the internal spiraling ramp to create smaller and smaller volumes of higher and higher tiers.[25] Both thermal (IR) images and gravimetric (density)

---

[25] The bulk of the volume of a pyramid lies near the bottom. At half its final height, two-thirds of the eventual volume has been reached. In the remaining half height, another half (i.e., to reach the three-quarter height) would account for two-thirds of the remaining third of the final volume, for a total of eight-ninths of the volume (i.e., about 89 percent). In fact, less than 3 to 4 percent of a pyramid's volume is created by the last 10 to 15 percent of its height or altitude. This fact, by the way, is a major reason for the selection of a pyramidal shape for a large structure. Surely the shape of the Great Pyramid symbolizes a link between the heavens and the Earth, with the apex pointing to the heavens and the base being on the Earth. The shape also evokes the image of the rays of the Sun, so important in the religion of the Ancient Egyptians (as Ra), shining on the Earth. But, as Occam's razor states (to paraphrase): The simplest explanation is the most likely.

Construction of Khufu's Pyramid: Humankind's Greatest Engineering Creation 255

(a)

(b)

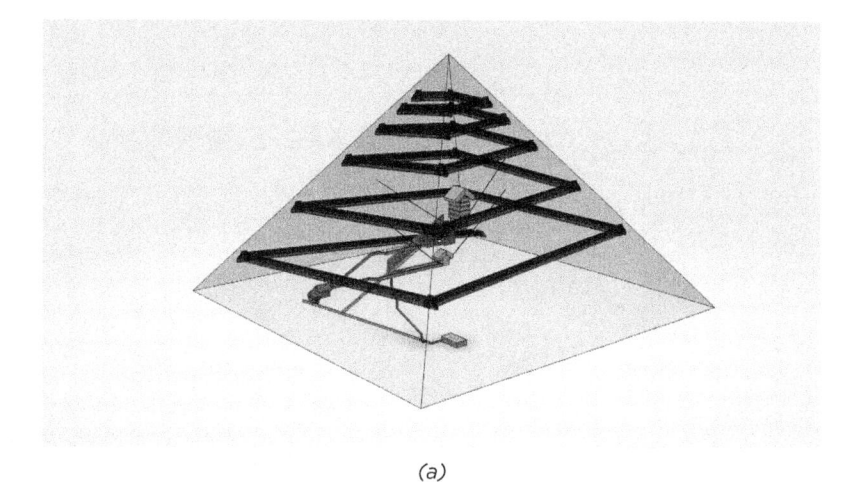

Figure 11-15 A 3D computer rendering of the internal ramp that spirals within the Great Pyramid, as proposed by Henri and Jean-Pierre Houdon (a) along with a detailed rendering to show a two-level arrangement in the passageway that allowed outgoing teams to pass incoming teams of haulers without obstruction, but, also, to allow two teams to turn the heavy blocks and sleds at corners (b). (*Source:* Images from Figures CT-2 and CT-3 in Robert W. Messler, Jr., *Engineering Problem-Solving 101: Time-Tested and Timeless Techniques,* McGraw-Hill, New York, 2013, page 265, were provided by Dassault Systemes' Passion for Innovation Program at the DVD *Khufu Reborn,* with the kind permission of Richard Breiner, director.)

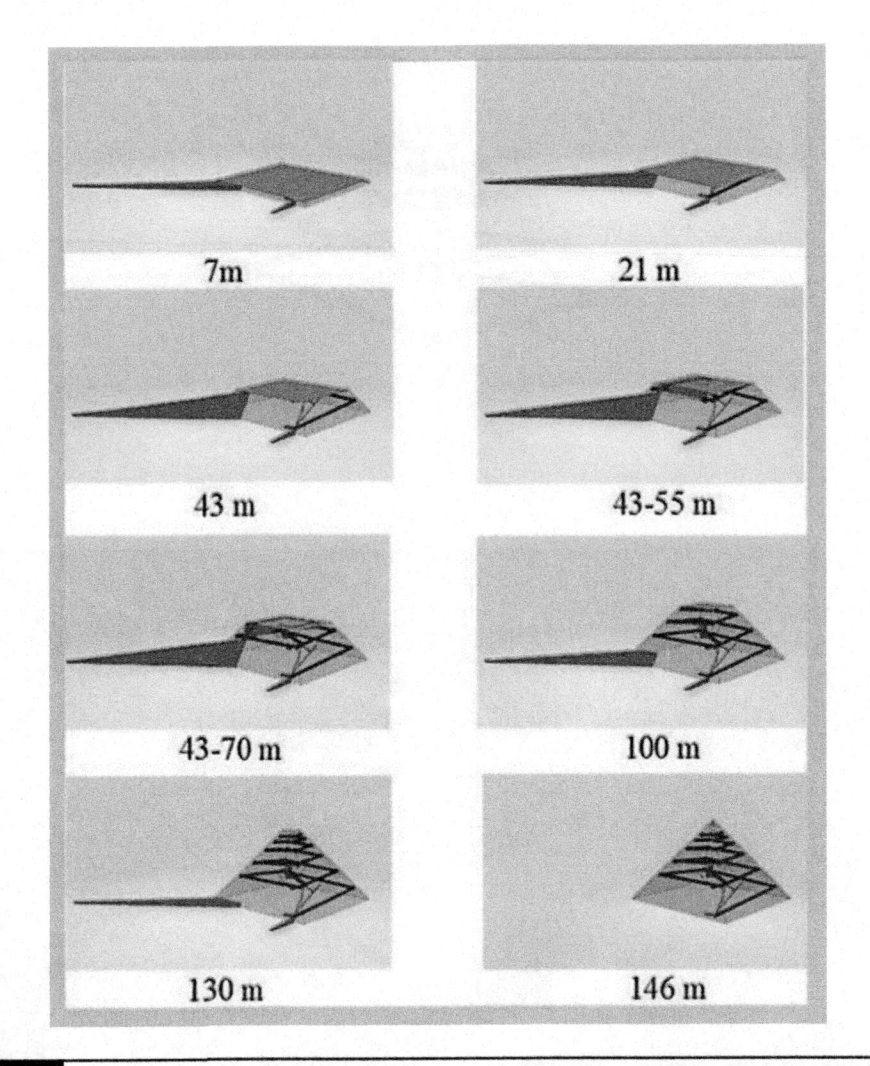

256 Chapter Eleven

7m	21 m
43 m	43-55 m
43-70 m	100 m
130 m	146 m

Figure 11-16 Schematic illustration showing how a much-smaller-volume straight external ramp could be used for moving limestone blocks into position for the lower tier of the Great Pyramid, while the internal spiral ramp of Henri and Jean-Pierre Houdin could be used to move blocks to the upper tiers. (*Source:* Courtesy of Dassault Systemes' Passion for Innovation Program, courtesy of Richard Breitner, director.)

images of the Great Pyramid have lent impressive evidence for the existence of an internal ramp system that spirals upward to allow construction. Figure 11–17 shows a thermal (IR) image, while gravimetric scans made more than two decades ago can be found online with a Google image search. Newer techniques have allowed even more detailed, and impressive, gravimetric imaging.

The internal ramp theory has great appeal, as it adds no additional labor (to construct and then remove huge temporary ramps), it actually eliminates the need for some core blocks, and it overcomes the problem of obstructing direct observation to regularly check construction against design plans.

Construction of Khufu's Pyramid: Humankind's Greatest Engineering Creation **257**

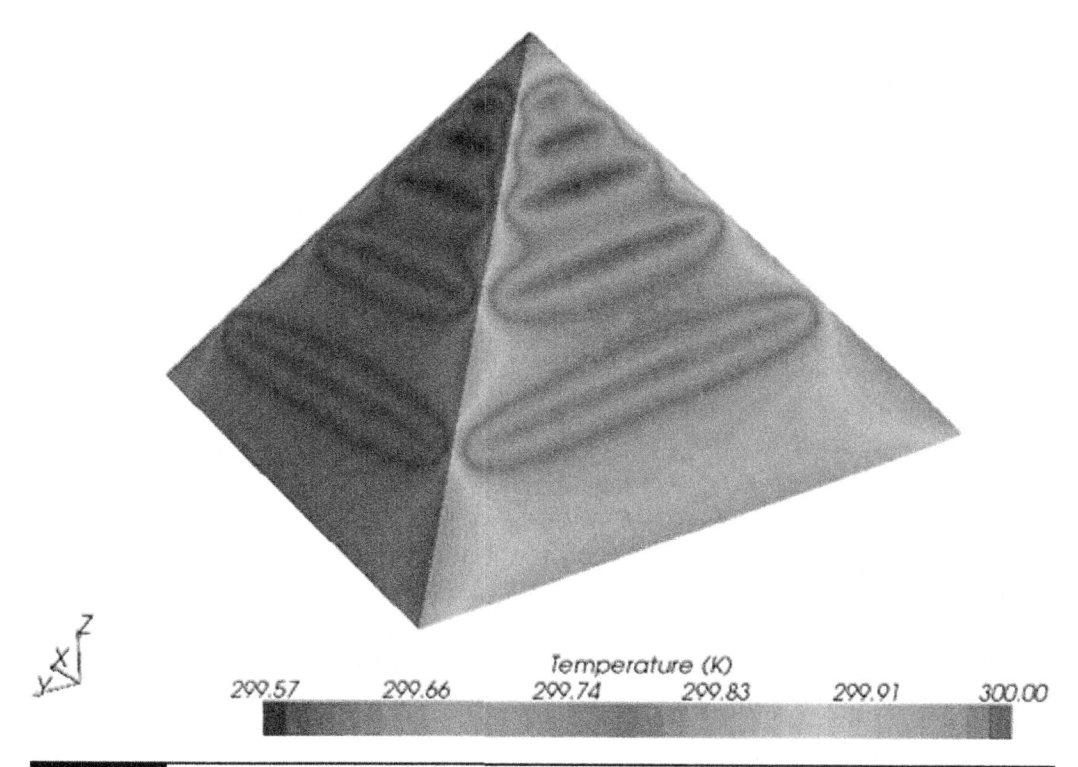

Figure 11-17 Infrared (IR) thermal imaging of Cheops's (or Khufu's) pyramid, shows evidence of the internal spiraling ramp proposed by Henri and Jean-Pierre Houdin. (*Source:* From Dassault Systemes' *Khufu Reborn,* used with permission from Richard Breitner, director.) **Don't miss the color version of this figure, available at www.mhprofessional.com/ReverseEngineering.**

11-7 Summary

The Great Pyramid of Khufu at Giza is not only the oldest and sole-surviving member of Herodotus's Seven Wonders of the Ancient World, it is, almost without question, the most impressive and majestic creation of humankind in all history. For an engineer, it is a marvel for its size and precision, with many mysteries surrounding its Earth-central geographic location, precise alignment to true north, and, most of all, its method-of-construction.

Dr. Craig Smith, a member of a team from the construction management firm of Daniel, Mann, Johnson & Mendenhall, along with Egyptologists including Mark Lehner, estimated that an average workforce of 14,600 people would be required, with a peak workforce of 40,000, and could complete the Great Pyramid with the available tools and technology in about 10 years minimum. Dr. Smith was extremely impressed, however, saying:

> The logistics of construction at the Giza site are staggering, when you think that the Ancient Egyptians had no pulleys, no wheels, and no iron tools. Yet, the dimensions of the pyramid are extremely accurate and the site was leveled within a fraction of an inch over the entire

13.6-acre base. This is comparable to the accuracy possible with modern construction methods and laser leveling. That's astounding. With their rudimentary tools, the pyramid builders of Ancient Egypt were about as accurate as we are today with 20th [21st] century technology.

Reverse engineering is the key to understanding the Great Pyramid, and the Great Pyramid is a wonderful way to bring reverse engineering alive!

11-8 Cited References

Bauval, Robert, and Andrew Gilbert, *The Orion Mystery: Unlocking the Secrets of the Pyramids,* Broadway Books, New York, 1995.

Bonwick, James, *The Great Pyramid of Giza: History and Speculation,* Courier Dover Publications, Mineola, NY, 2003.

Fix, William R., *Pyramid Odyssey,* Smithmark Publications, New York, 1978.

Löhner, Franz, "The Great Pyramid of Khufu (Cheops): Alignment of the Pyramids and Controlling the Shape of the Pyramid," www.cheops-pyramide.ch, 2006.

Mendelssohn, Kurt, *The Riddle of the Pyramids,* W. W. Norton & Co., New York, 1986.

Messler, Robert W., Jr., *Engineering Problem-Solving 101: Time-Tested and Timeless Techniques,* McGraw-Hill, New York, 2013.

Tompkins, Peter, *Secrets of the Great Pyramid: Two Thousand Years of Adventures and Discoveries Surrounding the Mysteries of the Great Pyramid of Cheops,* BBS Publishing Corporation, Edison, NJ, 1997.

11-9 Recommended Readings

Capt, E. Raymond, *The Great Pyramid Decoded: God's Stone Witness,* Artisan Books, Muskogee, OK, 1978.

Jackson, Kevin, and Jonathan Stamp, *Building the Great Pyramid,* Firefly Books, Richmond Hill, ON, 2003.

Lehner, Mark, *The Complete Pyramids: Solving the Ancient Mysteries* (The Complete Series), Thames & Hudson, London, 2008.

Romer, John, *The Great Pyramid: Ancient Egypt Revisited,* Cambridge University Press, Cambridge, UK, 2007.

11-10 Thought Questions and Problems

11-1 Every schoolchild knows about the *pyramids* of ancient Egypt, especially the Great Pyramids of Giza. But, pyramids are not found only in Egypt, nor were they built only by ancient Egyptians. They are quite literally found around the world. Details of the geometry vary slightly, but the pyramidal form is common to all.
    a. Go to www.historvius.com (or research "pyramids around the world") to help you prepare an essay of less than two pages on "Pyramids Around the World." Besides giving

Construction of Khufu's Pyramid: Humankind's Greatest Engineering Creation 259

or listing the wide variety of locations, builders (or cultures), and construction periods (i.e., age), provide some logical argument for the common form (pyramidal). (Do *not* fall victim to wild speculation relating to migration of Ancient Egyptian technology, *unless* you can provide convincing evidence or, worse, influence of extraterrestrials.)

b. The modern world has its share of pyramids, too—probably far more than you think! The most famous may be the Louvre Pyramid, located at the Louvre Museum in Paris, France, and designed by the world-renowned architect I. M. Pei.

Use the Internet to find as many modern examples as you can, but not less than six to eight. Prepare a list and describe the purpose of each example you find.

c. As an engineer familiar with mathematics and physics, what is there about a pyramidal shape that makes it an extremely logical shape for a large structure, and, from an architectural standpoint, what additional appeal is offered by this shape?

11-2 There are a striking number of facts and/or features concerning the Great Pyramid of Khufu that continue to confound modern scientists, engineers, archaeologists, and other rational people. Respond to each of the following with a brief but thoughtful essay of less than one page each:

a. The sheer magnitude of the effort to construct the Great Pyramid of Khufu is startling in terms of the demands it placed on construction logistics. Given that Khufu's Pyramid contains more than 2.3 million blocks of limestone and was constructed in what most reasonable estimates place at less than 20 years and some say was 14 years, perform the following:

(1) Calculate the rate (in time per block) required for block placement by laborers working 12 hours per day, 6 days per week, for 20 years, and then for 14 years

(2) Estimate the number of haulers required if each block required around 12 to 20 men. The estimate is for how many haulers are needed on an ongoing basis, based on your estimate for *placement time* in (1), assuming it took the team two hours from start to finish per block.

(3) Refine your estimate of how long a team would be engaged in moving a block that weighed an average of 2.5 metric tons from information you find online for estimates of the speed with which such a weight could be moved. If you can't find a value, run the estimate for a rate of movement of 5 meters per minute, and use the estimated length of an external ramp.

b. The accuracy of linear and angular dimensions for Khufu's Pyramid allegedly (in a variety of references) challenges modern construction capability (although this author doesn't think so!). Using the Internet, find the typical tolerance for linear and angular dimensions in modern on-site (versus prefab) construction of large concrete structures and then compare the accuracy of the Great Pyramid to this capability. Briefly comment on what you find.

c. There are many serious investigators (as opposed to publicity seekers, sensationalists, conspiracy theorists, etc.) who believe the ancient Egyptians intentionally embedded their advanced knowledge of mathematics in the geometry of the Great Pyramid. Prepare an essay of less than two pages on this premise, giving as many examples as you can find on the Internet or other *credible* references.

11-3 Science is based on logical argument, supported by irrefutable data. Speculation (or worse!) is based on bias, unsupported theories, or intent to defraud others if not one's self. The geographic location, alignment to the compass, and orientation relative to one another of the Pyramids of Giza cause serious investigators to ponder and wild speculators to run rampant.

Use the Internet to look into these three features relating to the Great Pyramids of Giza. After gathering what you feel are credible facts, prepare an essay of less than two pages on location, alignment, and relative orientation that appeals to you.

11-4 Theories as to how the Great Pyramids were constructed abound. For engineers, who deal with practicality beyond feasibility, very few "traditional" theories proposed by archaeologists, historians, or Egyptologists make sense when subjected to scrutiny.

Use the Internet to look into the most serious theories of pyramid construction relying on external ramps, internal ramps, or cast-in-place blocks. Make a logical and rational argument for *one* theory among these three. Be sure to fully support your argument with facts!

11-5 Prepare a brief but thoughtful essay of less than two pages that describes the essential role of *reverse engineering* in deducing role, purpose, and function or functionality of confounding ancient constructions, such as (but not limited to) the Great Pyramid of Khufu. (If you prefer, focus your essay on one of the following instead of Khufu's Pyramid: (1) Stonehenge, (2) Tunnel of Samos, (3) "Nazca lines," or (4) stone monoliths of Easter Island.)

End your essay with your thoughts on whether there is any engineering structure created after the start of the twentieth century that will cause archaeologists, scientists, and engineers to ponder its role, purpose, and function or functionality, as well as marvel at our capability.

CHAPTER 12

Assessing Design Suitability

12-1 Different Designs, Different Role, Purpose, and Functionality

Imagine it's the twenty-sixth century. The population of the Earth has swelled to 15 billion people, while the number of human beings has swelled to 16 billion, even though stringent population control policies and practices have been in place since the 2200s. Colonization of the Moon and Mars by adventurous humans began more than 200 years ago and has grown to nearly 100 million on the Moon and almost 1 billion on Mars. Fortunately, technology has continued to grow almost exponentially since the nineteenth century, with the dawn of the Industrial Age, or, as referred to in some history books, the "Industrial Revolution." Agriculture long ago became a distant secondary source of nutrition to hydroculture, ocean farming, and chemistry.[1]

Virtually every major city of the ancient world (New York, Chicago, Los Angeles, Toronto, Montreal, Paris, Berlin, London, Rome, Johannesburg, Nairobi, Lagos, Durban, Khartoum, Seoul, Tokyo, Beijing, Rio de Janeiro, Sao Paulo, Caracas, and others) grew into a megalopolis several hundred years ago, with new cities of more than 20 million people having been founded in what were once rural areas doing the same. As a result, the planet is, quite literally, one urban sprawl, and, of necessity, personal transportation became a thing of legend. The masses move in high-speed pods propelled through an intertwining network of tubes by counter gravity and other repulsive-drive systems. Travel to distant points on the Earth, if 3D holographic video and virtual reality meetings or sightseeing do not suffice, is accomplished using teleportation operated by an agency of the Common Government, which oversees all human needs, everywhere.[2]

---

[1] *Hydroculture* is the growing of plants in a soilless medium or an aquatic-based environment, with required nutrients for the plants being provided via additives to the water. *Ocean farming* allows the growing of nutrient-rich seaweed, as well as protein-rich krill, fish, crustaceans, and mollusks. Chemistry creates an entire gamut of "processed foods," kind of like Cheeze Whiz gone wild!

[2] *Teleportation* is the transfer of matter from one point to another without traversing the physical space between them.

Keep imagining! Laser excavation of a site for a new subcity within the megalopolis of Tuscan in what was once northern Italy (part of the ancient continent of Europa) has been halted by agents of the Common Government's Ministry of Ancient Civilization Relics. Experts in cultural and technological archaeology (if what the ancients of the twentieth through the twenty-second centuries had could rightfully be called "technology") impounded three nearly perfectly preserved "automobiles," an ancient mode of transportation that once, long ago, pervaded the planet and led to the near destruction of the habitable environment before atmosphere synthesizers were developed in the 2300s.[3]

At the Ministry's Northeastern Global Quadrant laboratory complex on the relatively sparsely populated Greenland Island-Continent at Nuuk Megapolis (with under 100 million people), investigators have identified the three relics from their perfectly preserved nameplates and ancient symbols as a FIAT® 500, a Rolls-Royce Phantom, and a Lamborghini Aventador, all manufactured by largely obsolete processes and technologies in 2013 (Figure 12–1)! A systematic process of reverse engineering, which included careful, complete, and meticulous visual examination along with assessment by a variety of nondestructive imaging, characterization, and analysis techniques, and even a series of virtual operational testing and experimentation, has come to the following conclusion, with supporting evidence stated parenthetically in square brackets [ ]:

- All three relics were, as originally suspected, conformed to be "automobiles," precisely dated by a variety of techniques to the year 2013, the common *role* of which was personal ground transportation over relatively short distances (estimated at less than 300 kilometers per day, due to their extremely limited speed capability, estimated at well below 300 kilometers per hour for two of the three). [*Supporting evidence:* All three relics employed four wheels driven by an unnecessarily complicated system of mechanical gears and/or fluid-driven discs, shafts, and so forth, by a rather crude heat engine that developed power from a very inefficient process of combusting a hydrocarbon fuel in pressure cylinders to force four or more pistons to drive a crude crankshaft to convert reciprocating motion into rotary motion for eventual propulsion of the vehicle through either two or all four driven wheels.]

- The *purpose* of each of the three relics, while similar, seemed to represent three common concerns in the twenty-first century: (1) economical transport of one or two adults (of larger than the average size of twenty-sixth-century adult humans at 1.6 meters) over short distances (estimated at typically under 50 kilometers) during what was known as "commutation" to places of common physical, more than mental, effort, typically in densely populated (for the time period) urban areas, where operational speeds were extremely low (e.g., averaging an estimated 50 kph or less) for the FIAT 500; (2) luxurious, if not ostentatious, transport of seemingly affluent adults, typically up to five, with nearly soundless and vibration-free movement at speeds of probably 100 to 160 kph, with the specially prepared skins of domesticated animals and extinct hardwood trees of the ancient African and South American rain forests, before deforestation for the Rolls-Royce Phantom; and (3) an apparent fascination of what the ancients seemingly perceived as "high speeds" and "sporty transport," by which they meant the violent onset of acceleration (by current standards) and harshness in the feedback of roughness of roadways and perception of lateral g-forces, which they seemed to

---

[3] An *automobile* is (or, in the preceding fantasy, was) a self-propelled passenger vehicle that usually has (or had) four wheels and an internal combustion engine that burns (or burned) hydrocarbon fuels (from long-extinct natural resources in the ground) for land transportation.

(a)

(b)

(c)

Figure 12-1 Photographs of a 2013 FIAT 500 (a), a 2013 Rolls-Royce Phantom (b), and a 2013 Lamborghini Aventador 700-4 Roadster (c). (*Sources:* FIAT is a registered trademark of Fiat Group Marketing & Corporate Communication S.p.A., used under license by Chrysler Group LLC; used with permission [a]; Rolls-Royce Motor Car Company [b], and Automobili Lamborghini S.p.A. [c], used with the permission of each, respectively.)

perceive as "sporty," and where technology, as crude as it was, seemed focused almost entirely on the propulsion system in the intriguing Lamborghini Avetador. [*Supporting evidence:* Small size, light weight, limited passenger capacity, small engine displacement and power, limited speed and acceleration capability, and nearly nonexistent cargo capacity for the FIAT 500 suggest utility and economy over comfort or performance. Greater interior roominess, soft, aesthetically pleasing, and noise-abating interior, wide variety of exotic natural materials as well as what were probably state-of-the-art synthetic materials, large engine, stability-enhancing heavy weight, and more than adequate cargo space (rivaling the size of some twenty-sixth-century residential rooms) attest to a concern for comfort with a modicum of performance for the Rolls-Royce Phantom. Limited room for two adults (of any size), difficult entrance and egress, a reclined driver position, a powerful (for the time) engine, elaborate transmission, stabilizing suspension, and brutal acceleration forces (without access to modern technologies to provide active abatement of jerk) indicate a focus, if not a fixation, on high performance (in the context of the period, of course) in the Lamborghini.]

- The *functionality* of the internal combustion engines, fuel control (all multipoint injection) systems, electric-spark ignition timing systems (all electronically controlled systems to vary the timing of combustion to optimize performance), the "manual" gear-operated and "automatic" fluid-driven transmissions, suspension systems, and so on, were all assessed by observations made during mechanical dissection, aided by nondestructive assessment techniques, careful dimensional measurements, and experimental operational measurements allowed characterization and differentiation of the three relics, with results for key items summarized in Table 12–1.

As fanciful as the preceding may seem (with the author taking a lot of liberties, albeit perhaps not enough, to project 500 years into the future), the significant points are these. First,

TABLE 12-1 Specifications for Vehicles Examined

	FIAT 500	Rolls-Royce Phantom	Lamborghini Aventador
Length	139.6"	230.0"	188.19"
Width	64.1"	78.3"	89.17"
Height	59.8"	64.5"	67.72"
Wheelbase	90.6"	90.6"	106.30"
Ground clearance	4.1"	5.87"	4.1"
Curb weight (lb)	2363	5840	4085
Fuel tank capacity (g)	10.5 (regular)	26.4 (premium)	23.8 (premium)
Engine			
Displacement (L)	1.4	6.7	6.5
Cylinders	in-line 4	V12	V12
Horsepower hp @ rpm)	101 @ 6500	435 @ 5350	710 @ 8250
Torque (ft-lb @ rpm)	98 @ 4000	531 @ 3500	510 @ 5500
	Multipoint fuel injection; variable valve timing	DOHC; multipoint fuel injection; variable valve timing	DOHC; multipoint fuel injection; variable valve timing
Transmission	5-speed manual	8-speed auto.	7-speed manual
Passengers	4	5 (comfortably)	2
Cargo volume (ft^3) (with seats in place)	9.5	16.2	5.3
Interior	Cloth seats; plastic trim; nylon carpet	Premium leather; matched rare woods; plush carpeting	Fine leather; masculine trim; nylon carpeting
Performance			
0–60 mph (s)	9.7	5.9	2.9
Top speed (mph/kph)	100/160	160/240	217/349

while there are enough similarities between and among automobiles, most (if not all) are recognizable as automobiles, just as doggies all seem to be recognizable to toddlers, whether dachshunds or dalmatians or Great Danes. Second, while similarities indicate similar roles, purpose and functionality need to be assessed by careful observation, measurement, and experimentation. Finally, the process can work both ways, or, if one prefers, either way. One can start with an object (such as the *Antikythera mechanism* of Chapter 8) or structure (such as the *Great Pyramid of Khufu* of Chapter 11) and use reverse engineering to identify or deduce role, purpose, and functionality from evidence, clues, and/or cues *or* start with the intended role, actual purpose, and scope and magnitude of functionality to deduce what the object or structure would have to look like.

12-2 Form, Fit, and Function

The term *form, fit, and function,* sometimes referred to as *FFF* or *F3,* is used within design, as well as in the manufacturing and technology arenas, to describe a physical entity's identifying characteristics. These include physical, functional, and performance characteristics that uniquely identify a part, component, structural element, device, mechanism, or subassembly or subsystem of a product, structure, or system. Of particular importance is if the criteria or specifications for form, fit, and function of a particular entity are met, then that entity can generally be considered interchangeable with other entities with the same criteria or specifications for design requirements.

The most common use of an assessment of form, fit, and function is to determine whether a proposed change to a part considered by a designer to ease manufacturing, reduce cost, circumvent some earlier-identified problems, or, perhaps, alter performance, up or down,[4] as indicated appropriate for the market need, represents a "minor" change, with little or no impact on the form, fit, and function, or a "major" change, which has significant effect on these factors. Beyond utility in design, FFF also is used for several aspects of management, including materials resource planning (MRP), product data management (PDM), and/or product life-cycle management (PLM) systems.

For the purposes of this book and chapter, form, fit, and function are used as a significant component of or step in reverse engineering to assess the suitability of a design to its intended role and actual, specific purpose and functionality. The correlation can actually be assessed in either direction, depending on what knowledge exists at the outset. If the role, purpose, and functionality of a particular item, entity, or structure are known at the outset, then observations made during reverse-engineering dissection can be used to assess whether those observations (as hard evidence or subtle clues) suggest that form, fit, and function for the item, entity, or structure are appropriate. If, on the other hand, the role, purpose, and functionality are not known at all or with sufficient certainty, then observations made during reverse-engineering dissection on form, fit, and function (from hard evidence or indicative or suggestive clues) can be used to assess what role, purpose, and function should (or could) be.

For clarification, following are brief descriptions of what is meant by each of *form, fit,* and *function*:

---

[4] Performance is altered up when its current level is considered unacceptable for some reason. Performance may be altered down to reduce costs or increase safety or product, structure, or system life, provided such alteration still results in an acceptable level of performance.

- *Form* refers to shape, size, and dimensions (constituting geometry), mass or weight, and/or other visually observable parameters that uniquely characterize an item, entity, or structure. In short, form defines the "look" of the item, entity, or structure. Weight, balance (as on a rotating shaft as an indicator of a need to prevent wobble or vibrations), and location of the center of mass (e.g., relative to the center of lift in an aircraft as an indicator of maneuverability capability) are often clues as to the quality of required performance. Color is seldom considered part of form, unless it imparts a specific meaning or function (e.g., red indicates warning, while patterns of earth tones may indicate need for camouflage).
- *Fit* refers to the ability of a part, component, device, or structural element to physically interface with or interconnect to or become an integral part of another part, component, device, or structural element in a mechanism, product, assembly, structure, or system. Fit relates to the *associativity of* the detail in relation to the assembly or structure or to other details, and involves tolerances on dimensions, as well as, perhaps, surface finishes at interfaces.
- *Function* refers to the action or actions that an item or detail is designed to perform. Often the reason for an item or detail's existence is the function it provides or is intended to provide (e.g., a pump is intended to move a fluid; a valve is intended to control flow of fluids; a fastener is intended to lock two or more parts together).

To tie together all of what has been discussed in this section, it is worth returning to the fantasy situation presented in Section 12–1, involving the discovery of three different perfectly preserved 2013 automobiles during an excavation in the twenty-sixth century and the conclusions drawn by expert cultural and technological archaeologists using a process of reverse engineering. So, here goes . . .

The FIAT® 500

The *role* of the FIAT 500 (as well as the common role of the other two "relics") was *to serve as a mode of personal wheeled ground transportation for two, or perhaps four, passengers* as the relic had the *form* of an "automobile" (see Footnote 3 for a definition). The specific *purpose* of this vehicle was deduced to be *for economical transport over typically short distances (perhaps up to 50 kilometers one way) at typically low speeds (perhaps 40 to 60 kph) in densely populated areas, where vehicular traffic density tended to be high, with lots of stop and go due to congestion, probably predominantly for use in commutation.* These deductions were based on *form* and *function,* with key evidence and clues being: small overall size, limited passenger capacity, very limited cargo space, low weight, and small (low horsepower but highly fuel-efficient) engine. *Fit* for this vehicle suggested high-volume/low-cost mass (probably highly automated) production based on the use of many purchased (versus manufactured) standard parts with a high degree of interchangeability.

The Rolls Royce Phantom

The *role* of the Rolls-Royce Phantom was also deduced *to serve as a mode of personal wheeled ground transportation,* consistent with the role of automobiles, of which it was a clear example. The specific *purpose* of this large, roomy, and elegant vehicle was deduced to be *for luxurious motor travel, more than simple transport, over all distances (from short commutation to long pleasure trips) at comfortably high speeds (up to 160 kph) on open highways.* The vehicle gave a sense of classic exterior styling, with a lush but tasteful interior, all clearly intended to project an air of success. These deductions were based primarily on *form,* but *supported by* handmade *fit* and attention to detail in a custom-manufacturing environment. *Function* was also *supported by* precise *fit,* in the power but quietness and smoothness of the engine and stability for rider comfort provided by the suspension. Custom assembly was clear from the lack of identical details, but always precisely fit.

The Lamborghini Aventador

The *role* of the Lamborghini Aventador was also deduced *to serve as a mode of personal wheeled ground transportation,* also consistent with the role of automobiles, of which this, too, was a clear example, despite some of its radical styling. The specific *purpose* of this highly aerodynamically streamlined vehicle with an extremely high power-to-weight ratio, was deduced to be *for high-performance driving on even challenging winding roads at speeds up to 349 kph!* The vehicle gave the appearance of speed in motion even while at rest. With room for only two passengers, in a near-supine position, and room for only the most essential cargo, the high-powered engine, seven-speed manual transmission, and highly stable suspension (to allow fast cornering), along with very wide tires screamed speed, speed, speed! Extraordinary performance was largely supported by *fit* and *function* of drivetrain components, as well as suspension components. Precision of fit of all components, mechanical and décor, evidenced manual assembly of near-handmade parts. *Form* of the body exterior suggested design for speed. *Form* of the interior suggested a masculine elegance that was rich but highly utilitarian.

Even with this flight of imagination, it should be clear that form, fit, and function is what one must focus on when attempting to identify or deduce the role, purpose, and/or functionality of an item, object, system, or structure during reverse engineering.

12-3 Using Observable Evidence and Clues to Assess Form, Fit, and Function

It is challenging to describe to an engineer how to use observable evidence and clues to assess form, fit, and function on the way to identifying or deducing role, purpose, and functionality during reverse engineering. Every situation is different. So how does an engineer do what the title of this section states? The short answer is simple: Look, think, and link! The detailed answer is similar: Be vigilant and see all there is to be seen; remain open-minded to all possibilities, not jumping to conclusions based on preconceptions; and use logic shaped by the context of time, place, and need. And, finally, remember Occam's razor, which, paraphrased, says: The most obvious answer is often the answer.

The approach to be used will be to employ familiar illustrative examples as representative cases of a variety of the most important categories, if you will. The evidence or clues to be employed in these categories include:

- The size and, for those loading conditions where cross-sectional shape matters (i.e., compression, bending, and torsion, or combinations of one or more of these with tension),

cross-sectional shape of structural members as an indicator of loading, load-carrying capability, and robustness[5]
- Material selection and method-of-construction from materials as an indicator of required serviceability based on predominant property(ies)
- Precision in manufacture as an indicator of need for accuracy and consistency
- Electrical and/or thermal robustness as an indicator of suitability for sustained or continuous (versus transient) operation without overloading
- Design details, an all-encompassing description, to deduce functionality (or purpose) from one or more key design features or characteristics

While almost certainly not all-inclusive, this list is surely both representative and illustrative of the general process and procedure.

The examples used for the preceding are, respectively, pickup trucks, automobile or truck tires, bolt threads, welding power supplies, and bulldozers.

Structural Size and Robustness (Pickup Trucks)

While a segment of Americans have always had a use for pickup trucks based on need, the last couple of decades suggest a romance that goes beyond utilitarianism. People now buy pickup trucks not because their work as farmers, tradespeople, or construction workers, requires them for hauling loads to conduct business, but, seemingly, because they find them fun (i.e., drivable and comfortable, having automatic transmissions, power steering, air-conditioning, and CD players) and useful on those occasions when they might need to haul something. In response to this expanded market—or, perhaps, driving it (no pun intended!)—manufacturers of pickup trucks have expanded and broadened choices. Chrysler Ram, Chevrolet, GMC, Ford, Toyota, and Nissan (as prime examples) all offer a range of models of pickup trucks intended to match design to need (i.e., provide required functionality), with two extremes being (1) lightweight or, more correctly, light-duty pickups and (2) super-heavy-duty pickups—often with one or more models between these extremes.

If one dissected one of each of these models (Figure 12–2a and b), there would be considerable evidence and numerous clues relating to form, fit, and function that would indicated roles (which is common, i.e., to haul stuff), purpose (a little lightweight stuff on occasion or a lot of heavy stuff all or most of the time), and functionality (capacity in cubic feet or yards, weight as tonnage, and pulling or towing capacity in thousands of pounds). Overall size is greater for the super-heavy-duty model, cross-sectional size is greater and shape often more complex for understructure/chassis members, the engine delivers more horsepower and torque (and, for the greatest robustness, may be diesel versus gasoline), the transmission offers more gears and lower ratios, the vehicle may have four-wheel drive with a low range (for extra pulling power), brakes will be more substantial, and suspension (springs, shock absorbers, stabilizing bars, etc.) will be more substantial. The need for robustness may even show up in the size and construction of the cargo bed, with thicker-gauge steel and added stiffeners. More thorough dissection would, almost certainly, reveal many other differences in materials (e.g., low-alloy steels versus plain carbon steels for through-thickness strengthening) and processing methods (e.g., forged versus cast connecting rods).

Table 12–2 compares the light-duty and heavy-duty pickups shown in Figure 12–2a and b.

---

[5] *Robustness is* defined as (1) "powerfully built; sturdy" or (2) "requiring or suited to physical strength or endurance."

Assessing Design Suitability   **269**

(a)    (b)

Figure 12-2 Photographs of two models of 2013 Ram pickup trucks: one the lighter-weight, lighter-duty Ram 1500 (a) and the other the heavy-duty Ram 3500. Heavier hauling and towing capability is enabled in the Ram 3500 by a larger, more powerful engine, as well as by more robust structural members, engine parts and drive-train components, and suspension members and components. (*Source:* Ram is a registered trademark of Chrysler Group LLC. the photographs were taken by A. J. Mueller. Permission for use by Mr. Mueller and Chrysler Group LLC is gratefully acknowledged.)

TABLE 12-2 Comparisons between Light-Duty and Heavy-Duty Pickup Trucks As Shown in Figure 12-2 (units as specified by the manufacturer)

	2013 Ram 1500 (light duty)	2013 Ram 3500 (heavy duty)
Width	79.4 in	96.4 in
Length	209 in	231 in
Height	74.6 in	73.6 in
Front track	68 in	68.5 in
Rear track	67.5 in	75.8 in
Wheelbase	120.5 in	140.5 in
Ground clearance	8.7 in	7.6 in
Curb weight (empty)	4,708 lb	10,100 lb

Base engine:
  Torque
  Horsepower
	4.7 l V8
	330 ft-lb @ 3,950 rpm
	310 hp @ 5,650 rpm
		5.7 l V8 HEMI*
		400 ft-lb @ 4,000 rpm
		383 hp @ 5,600 rpm

Payload capacity	~1,600 lb	~3,100 lb
Towing capacity	~7,000 lb	~14,250 lb

*A super-heavy-duty version of the Ram 3500 is available with a 6.7 l Cummins turbo diesel engine which provides a payload capacity of ~6,600 pounds and a towing capacity of ~30,000 pounds.

270 Chapter Twelve

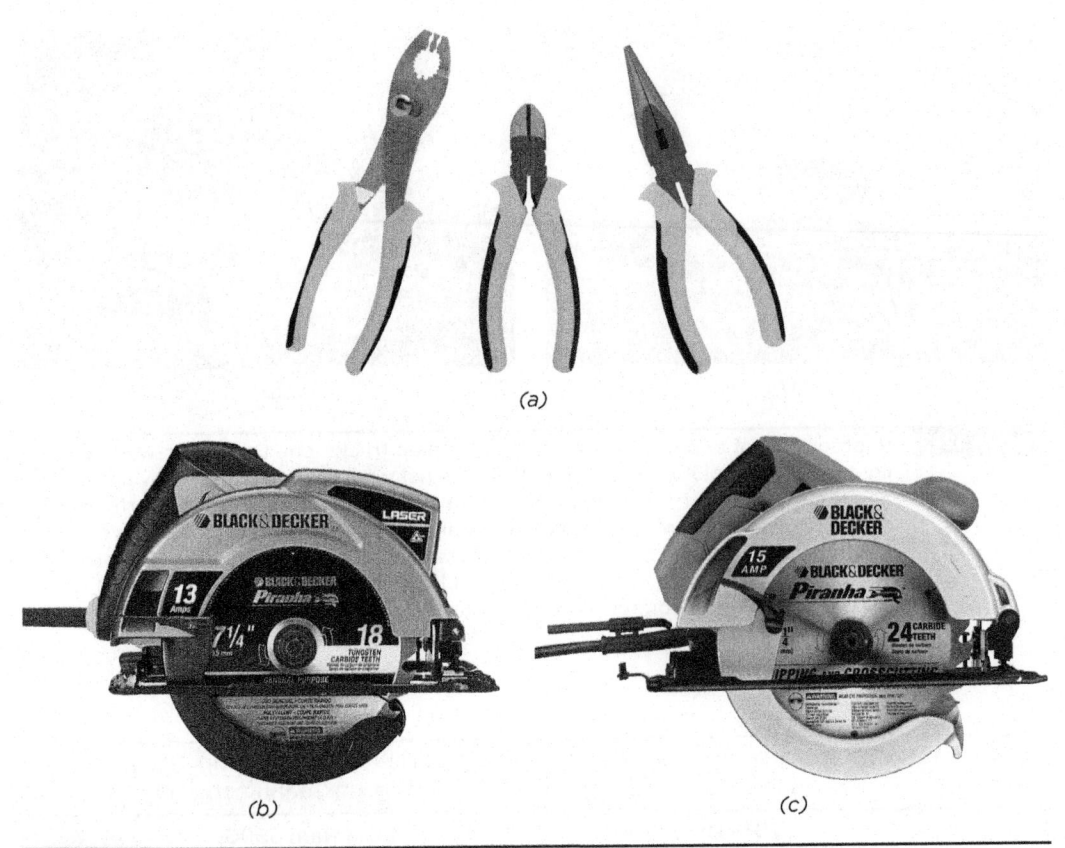

Figure 12–3 Photographs of heavy-duty hand tools, here a set of three different Stanley pliers forged from high-quality steel (*a*), and two different Black & Decker electric-powered circular saws, one a 13-amp model for general-purpose moderate-duty work (*b*) and one a 15-amp model for heavier-duty "ripping and cross-cutting" (*c*). Both types of tools exhibit form and fit that attest to heavy-duty capability but (*b*) even more than (*c*). (*Source:* Stanley Black & Decker, 1000 Stanley Drive, New Britain, CT 06053 (*b*), used with permission.)

In short, the pickup that needs to be stronger and more robust looks stronger and more robust. There are many other examples of light- and heavy-duty items, from hand and power tools for do-it-yourselfers versus professionals to home versus professional fitness center exercise equipment, sporting equipment, tractors and trucks, and many others.

Figure 12–3 shows examples of heavy-duty hand tools and a power tool.

Material Selection and Construction (Tires)

As discussed in Chapter 9, much can be learned about role, purpose, and functionality from the material(s) used in the manufacture or construction of parts, components, structural elements, devices, objects, products, structures, or systems. This also applies to form, fit, and function. A good example is automobile and truck tires.

Obviously, all tires have a common *role,* which is to serve as a covering for a wheel for land vehicles, including automobiles, motorcycles, trucks, vans, buses, tractors, and the like. Usually made of rubber (typically natural rubber, styrene butadiene rubber, butadiene rubber, isoprene rubber, or halogenated butyl rubber) reinforced with cords of fiberglass, nylon, polyester, or other materials, they are generally filled with compressed air (e.g., pneumatic tires or semipneumatic tires) for over-the-road vehicles, but can be solid or nonpressurized for some off-road vehicles. The construction of modern tires is actually quite complex, as they are highly engineered composites tailored in their design to provide a specific set of functional properties that depends on their intended *purpose.* Depending on whether their primary purpose is smoothness or softness and quietness of ride, high load-carrying capability (i.e., high strength), grip or traction from a combination of raised and/or recessed tread pattern, as well as better adhesion to a road surface, heat resistance (from sidewall flexing as the wheel turns and the tire shape flattens at the bottom), high speed (which must resist high heat), or durability, which could mean resistance to tread wear or resistance to losing air (i.e., going "flat") from a puncture, the formulation of the exact type of rubber, the degree of vulcanization,[6] and/or additives (principally, amorphous carbon black for heat resistance and strength) can vary considerably. In addition, service requirements also dictate the tire's construction in terms of the type and pattern of cord, the number of plies (i.e., layers of cord), and tread type and pattern.

Because tires are so complex in terms of material and structure or construction, much can be learned about their intended purpose (from form in FFF) and functionality (or function within FFF) by observation, including dissection, analysis (of rubber type), measurement, and test or experiment. Table 12–3 lists the major properties of and function afforded by various types of rubber used to manufacture tires.

Automobile tires (as an example, because similar situations exist for light and heavy truck and tractor tires) are described by an alphanumeric *tire code,* which is generally molded into the outer sidewall of the tire (sometimes with additional information on the inside sidewall). The code specifies the dimensions of the tire, as well as some of its key limitations, such as load-bearing ability and maximum speed. The code is quite elaborate and includes the following in the *ISO Metric tire code*:

- An optional letter (or letters) indicating the intended use or vehicle class for the tire: P for passenger car, LT for light truck, ST for special trailer, T for temporary (i.e., restricted for usage for "space-saver" spare wheels)
- Three-digit number giving the *nominal section width* of the tire at the widest point from sidewall to sidewall, in millimeters
- "/" character for character separation
- Two- or three-digit number giving the *aspect ratio* of the sidewall height as a percentage of the total width of the tire
- An optional letter indicating construction of the fabric carcass of the tire (B for bias belt, where the sidewalls are made of the same material as the tread, leading to a rigid ride; D for diagonal; R for radial; or no letter, indicating a cross ply tire)

---

[6] The process of *vulcanization,* invented by Charles Goodyear (1800–1860) in 1839, changes the strength and hardness of a particular rubber using additions (typically sulfur) that form cross-links between the long-chain molecules of the elastomeric polymer. The more cross-links, with more additive, the stronger and harder the rubber.

TABLE 12-3 Advantages and Disadvantages of the Major Types of Rubber Used in Tires

	Advantages	Disadvantages
Natural rubber (NR) (general use; TB [tread base])	Tear strength Wear resistance Impact resilience Low heat generation	Uniformity of quality Aging resistance Fatigue resistance Ozone resistance
Styrene-butadiene rubber (SBR) (PC tread)	Processability Uniform quality Aging by heat Frictional force	Impact resistance Heat generation
Butadiene rubber (BR) (sidewall)	Impact resistance Wear resistance Low temp. properties Fatigue resistance	Tear strength
Isoprene rubber (IR) (partially replace NR)	Tear resistance Wear resistance Impact resistance Low heat generation Consistent quality (versus NR)	Aging resistance Fatigue resistance Ozone resistance Cost is high versus NR
Halogenated butyl rubber (HBR) (inner liner)	High air impermeability Ozone resistance Fatigue resistance	Impact resilience Heat generation Adhesion

- Two-digit number giving the diameter of the wheel for which the tire is designed to fit, in inches (even for the ISO Metric tire code)
- Two- or three-digit number giving the load index
- One- or two-digit/letter combination giving the speed rating
- Additional marks for special conditions or capability

An example of a tire code using this format is: P215/65R15 95H M+S, the "M+S" indicating suitability for severe snow conditions, achieved from deep tread pattern.

Specific information on size, load range/ply rating, load index, and speed rating can be found on the Internet (e.g., Wikipedia.com under "Tire Code"). Similar codes exist for light truck, heavy truck, and tractor tires, albeit containing different letters and numbers and formatting.

The point to be taken away from this example is this: There is considerable information available on form (e.g., mostly from size and shape), fit (mostly from quality and consistency), and function (from properties and performance) from observation of materials-of-manufacture or -construction and processing or construction, as shown here for tires. A few familiar examples of other places materials provide key clues to form, fit, and function are drivetrain components for light- versus heavy-duty trucks and road cars versus race cars, hand tools, airplanes, sport equipment (e.g., tennis rackets and golf clubs), and many more.

Precision (Threads)

Another important design factor for some applications is precision. *Precision* includes both accuracy and consistency of internal details (as inputs, as it were) and required outputs. Precision most affects functionality (within role, purpose, and functionality) and function (within form, fit, and function), but appears in fit during reverse-engineering dissection.

Examples of where precision is important are timing devices, measurement devices, surgical instruments and medical equipment, devices requiring long-term dependability (e.g., deep-space probes), and many others. Because high precision costs money (for both the skill level and amount of labor required and for the capital equipment required), evidence or clues of its presence should be justified by what is known or believed to be required of the device or as a clue, in itself, for an unknown purpose.

A good example of how precision can vary for an item having the same general role but different specific purposes and required functionality is screw threads.

Screw threads are a helical detail consisting of peaks and troughs that cause translation via rotation. They are used on threaded fasteners such as bolts, machine screws, and self-tapping screws, and on power transmission devices (e.g., drive screws on moveable mechanisms, such as a vise). The latter, because they must transfer much high force, are visibly more robust in both size and shape. Figure 12–4 shows examples of both types, with the Square, Acme, Buttress, Knuckle, and Whitworth *thread forms* being power transmission types.

Threads on metal may be cut (i.e., machined), rolled, or forged or swaged, with the latter two methods producing smoother and generally stronger threads. Threads for fasteners are designated as "coarse," "fine," or "extra fine" based solely on their *pitch,* which is the number of threads per axial distance, which translates into amount of axial translation for one full rotation of the screw. Finer threads produce less axial translation per turn, so they allow greater control of axial motion. The designation in the Unified Thread Standard (UTS) of "Unified Coarse/UNC," "Unified Fine/UNF," and "Unified Extra Fine/UNEF" does not indicate lower quality or poorer tolerance. It strictly refers to pitch[7]! The tolerance of threads, which affects the tightness of fit between male and female threads, is designated with the *class* of the thread, with values of "1", "2", or "3" from most loose to most tight. An additional designation of "A" for external and "B" for internal threads is used with class.

By looking at the type and pitch of a threaded part (e.g., shaft) or fastener, it is possible to infer the need for tight (versus loose) fit and precision versus utility.

Electrical and Thermal Robustness (Welding Power Supplies)

The robustness of many electrical or electronic, as well as thermal, devices, products, or systems is limited by what is collectively known as *duty cycle*. A duty cycle is the percent of time that an entity spends in an active state relative to the total time under consideration. While the term is

---

[7] The other major standard to the Unified Thread Standard is the ISO Metric Standard, which, instead of using a "U" designation and stating pitch as threads per inch, uses an "M" designation and states pitch in millimeters per turn. The UNEF type thread is usually used for more than tight fit but, rather, to allow precise adjustment in measuring instruments.

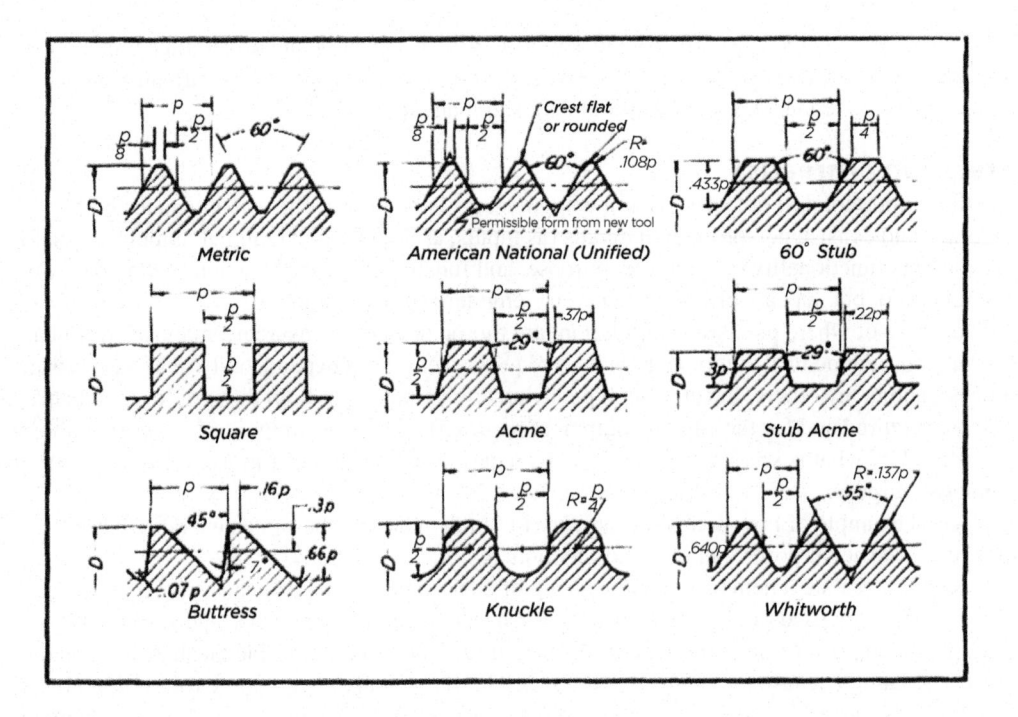

Figure 12-4 Schematic illustration of various types of screw threads or thread forms. The Metric and Unified types are found on threaded fasteners, such as bolts and machine screws, while the other forms are found on threaded shafts used for power transmission (e.g., between the jaws of a vise). The precision, in terms of the tightness of fit between male and female threads, is controlled by the *class* of the thread.

most often used in reference to electrical devices (motors, switching power supplies, etc.), it is also used in reference to other systems in which the buildup of heat causes an unacceptably high temperature (e.g., photovoltaic solar panels, some electric furnaces).

For an electrical device, a 60 percent duty cycle means the power is on 60 percent of the time and off 40 percent of the time, although the "on time" can range from a fraction of every second, over a period of minutes (as for welding power supplies, where duty cycle refers to the continuous "on time" over every 10-minute period), or days (e.g., for irrigation pumps). The device's *period* typically refers to how long it takes for the device to go through a complete on/off cycle. Therefore, the term *duty cycle* has no commonly accepted meaning for aperiodic devices.

An important aspect of an electrical or thermal device is its duty cycle, which would need to be determined by experiment/testing during reverse engineering. A good example of where duty cycle would be particularly important for serviceability is power supplies for electric arc welding, such as shielded metal arc (SMA), gas-tungsten arc (GTA or "TIG"), gas-metal arc (GMA or "MIG"). A do-it-yourselfer would generally only use a welding power supply on occasion and, even then, only for a relatively short period of time to get a particular job done. Therefore, welding power supplies intended for do-it-yourselfers operate on 120-volt residential service and only have to be designed to provide short duty cycles (e.g., 25 to 35 percent). They would also likely have to be relatively lightweight or allow easy portability to the work and return to storage when the work has been completed. A power supply intended for use in a steel fabrication shop, on the other hand, would be required to have a higher duty cycle and generally be more robust, as it is most likely to be used every day for hours a day. The most robust welding power supplies of all, in most cases, are those intended for use in the field, for pipeline installation or repair or other on-site welding where there is no electrical service. Such power supplies are either gasoline- or diesel-powered generators.

Obviously, one would expect to find many telltale differences in the size and durability of parts and components (both mechanical and electrical), materials-of-construction, and, perhaps, methods-of-manufacture. Taken individually and, particularly, collectively, these differences all provide clues on purpose and functionality, as well as form, fit, and function.

Examples of these three types are shown in Figure 12–5.

Design Details (Bulldozers)

This final example of form, fit, and function is somewhat of a catchall, as it uses details of a design that may be overt or subtle. There are also innumerable possibilities of "details." Here, as in so many other instances of reverse engineering, experience with the item being assessed is invaluable!

An example of how design details can be used to home in on purpose (and, perhaps, functionality) for different examples of the same basic thing, intended to fulfill the same basic role, is bulldozers.

Bulldozers, as almost everyone knows, move soil, rock, ore, or brush debris or refuse by pushing it with a blade using brute force (like a bull!). That's their common role.[8] Their specific purpose can vary from excavating a ditch or shallow hole, level or grading rough ground, removing trees and brush in preparation for excavation or grading, moving material to a preferred location to be picked up (as by a power shovel into a dump truck), or working a landfill to move solid trash for leveling, aeration, and/or compacting. The surface being worked can range from extremely hard (like solid rock in a quarry), to virgin compacted earth, to soft sand, to semistable garbage, to swampy or boggy ground. The required work (or duty) might be light (as on a small farm or homeowner's lot or driveway) to extremely heavy (as in a quarry or open-pit mine).

The clue to duty is sheer size and bulk (i.e., weight) and engine power. The clue to hardness or softness (or instability) of surface is to be found in details of the track for tracked bulldozers: wider tracks for softer, less stable surfaces, to spread the weight of the bulldozer over a greater area and reduce ground-pressure; heavier cleats for better grip or traction; special wear-resistant or higher-traction rubber grouser bars on track cleats or even huge rubber tires for rock (or wheeled bulldozers); and so on. Other design details can include protective cages for the operator for clearing trees or working a mine or quarry where there can be falling rocks, extensions on the tops of pusher blades for preventing entanglement of trash with the engine and studded steel wheels for compacting and aerating in landfill operations, and even additional armor for bulldozers used by the military (e.g., U.S. Army Corps of Engineers or U.S. Navy Seabees).

Figure 12–6 shows several examples of bulldozers intended for different purposes, demanding different functions.

As always with reverse engineering, the key is to be observant, to catch all of the subtle, as well as not-so-subtle clues.

---

[8] In fact, bulldozers might have another, similar role: to pull or tear things out of the ground or rip, tear, or gouge the ground itself to loosen it for another earthmoving machine (e.g., pay-loader or power shovel)

276 Chapter Twelve

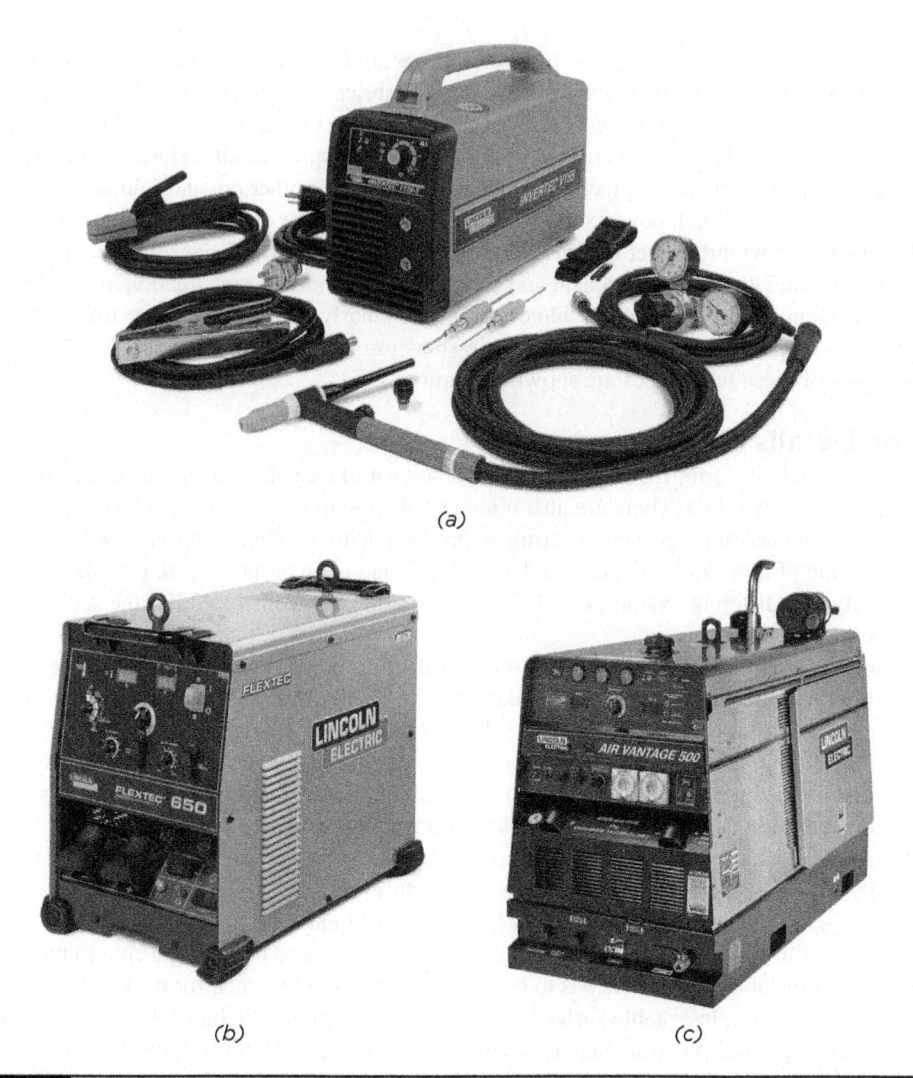

(a)

(b) *(c)*

Figure 12-5 Examples of three different types of power supplies for electric arc welding in different venues, requiring different levels of robustness, all produced by the world-renowned Lincoln Electric Company: (*a*) Invertec V155-S 120/230-volt input unit for SMAW (stick) in DC-straight or DC-reverse polarity and DC TIG (GTAW), weighing 15 pounds (6.8 kilograms) and having 100 percent duty cycle at 100 amps/24.0 volts or 30 percent at 145 amps/25.8 volts; (*b*) FlexTec™ 650 multiprocess welder for construction, fabrication, and other industrial welding with SMAW, DC TIG, MIG (GMAW), FCAW, SAW, and arc gouging requiring 380/460/575-volt 3-phase input and offering 100 percent duty cycle at 650 amps at a weight of 165 pounds (74.8 kilograms); and (*c*) Air Vantage Multiprocess (SMAW, DC TIG, MIG, FCAW, arc gouging) DC welder/AC generator/air compressor, which operates from a turbocharged four-cylinder Kubota diesel engine and offers 100 percent duty cycle for 500 amps, 20 kilowatts AC, or 60 cubic feet per minute (cfm)/600 pounds per square inch (psi) air at weight of 2200 pounds (1000 kilograms). (*Source:* Images provided courtesy of the Lincoln Electric Company, 22800 Saint Clair Avenue, Cleveland, OH 44117; provided courtesy of Bruce Chantry and used with permission.)

Assessing Design Suitability **277**

(a)

(b)

(c)

(d)

(e)

Figure 12-6 Examples of bulldozers which all have the same role but different purposes and functionality and, therefore, forms. The examples shown are: a Caterpillar D4H LGP, a relatively lightweight tracked model (a); a Caterpillar D9T, a heavy-duty tracked model (b); a Caterpillar 854K, a wheeled model (c) for grip on rock or, here, in a coal pit-mine; a specially equipped CAT for landfill operation (d); and a heavy-duty armored IDF CAT D9 able to work under heavy fire used by the U.S. Army Corps of Engineers (e). (*Sources:* Wikipedia Creative Commons, contributed by jha on 22 December 2008 [a]; courtesy of Caterpillar, Inc., Peoria, Illinois, used with their permission [b, c, and d]; and Wikipedia Creative Commons, contributed by Mathknight 29 April 2012 [e].)

12-4 Summary

The form (i.e., size and shape), fit (i.e., precision), and function (i.e., action or actions, including level of performance) of an item, entity, or structure helps determine (or deduce) its role, purpose, and functionality. During the process of reverse engineering, it is important to scrupulously examine the item, entity, or structure to find evidence or clues that help define form, fit, and function.

While literally every situation is different, key categories for careful observation (including measurement and experiment) include:

- Structural shape, size, and robustness (exemplified here by pickup trucks)
- Material choice and method-of-manufacture or -construction (exemplified here by tires)
- Precision of details (exemplified here by screw threads)
- Electrical or thermal robustness for service (exemplified here by the duty cycle of welding power supplies)

In short, the key is: Look. Think. Link.

12-5 Thought Questions and Problems

12-1 According to Wikipedia, a *time capsule* is "a historic cache of goods or information, usually intended as a method of communication with future people and to help future archaeologists, anthropologists, and historians." Time capsules are sometimes created and buried during celebrations such as a world's fair, a cornerstone laying for a public or government building, or at other events.

A special type of time capsule is a *space time capsule.* These are created to be "buried" in space, with the hope that they might be found by extraterrestrials so that they might be made aware of the existence of the Earth and "Earthlings." Currently, four time capsules are "buried" in space: two "Pioneer Plaques" and two "Voyager Golden Records" that were attached to spacecraft for the possible benefit of spacefarers in the distant future. A fifth major space time capsule is KEO, originally to have been launched in 2003, but delayed several times (2006, 2007/2008, 2010/2011, 2012) and presently scheduled for launch in 2014. KEO carries messages and information from the citizens of the Earth to Earthlings 52,000 years in the future, after KEO returns to reenter the Earth's atmosphere and land.

Besides messages from millions of ordinary people, KEO will also carry a diamond that encases (1) a drop of human blood (chosen at random); (2) samples of air, seawater, and earth; and (3) a microengraving of the DNA of the human genome on one of the diamond's facets. The satellite will also carry an astronomical clock (sound familiar?) that shows the current rotation rates of (4) several pulsars, (5) photographs of people of all cultures, and (6) "the contemporary Library of Alexandria," an encyclopedic compendium of current human knowledge.

In a thoughtful two- to three-page essay, explain from the standpoint of reverse engineering why items (1), (2), (3), and (4) were chosen, why the first three are encased in a diamond, *and,* more important, how future intelligent beings will "reverse engineer" them to extract knowledge and understanding of the Earth and its human inhabitants.

12-2 No one has a crystal ball, so no one can gaze into the future. There are, however, some people who are (and have been) rather remarkable for their ability to project from *what is* to *what is likely to be*. More than anywhere else, such people, known as *futurists,* are found among scientists, social scientists, and technologists or engineers. As defined in an article in Wikipedia, *futurists* or *futurologists* are "scientists and social scientists whose specialty is to attempt to systematically predict the future, whether that of human society, in particular, or of life on earth, in general." A surprisingly long "List of Futurologists" can be found on en.wikipedia.org. There are even two highly regarded think tanks in the United States that prognosticate the future—RAND and SRI International.

  a. In a sense, if not in point of fact, *futurologists* use reverse engineering, albeit backward! (This is interesting, since reverse engineering has been described in this book as a backward problem-solving technique; ref. Robert W. Messler, Jr., *Engineering Problem-Solving 101: Time-Tested and Timeless Techniques,* McGraw-Hill, 2013). By looking at the role, purpose, and function or functionality of something that exists now, futurologists attempt to systematically project forward to the likely new role, purpose, and function or functionality. Some might think this is just speculation or, at its best, simply *design,* but, as one "mentally dissects" *what is* to create *what might be,* it surely uses the principles of reverse engineering.

   Try your hand at prognosticating the future of technology in a thought experiment. Beginning with a given object or technology from the past, attempt to predict the present as the "future," from the perspective of the starting point, might have been. Try your best to project forward from what you would have known at the time (using references such as the Internet). Try not to taint your predictions by what you know by having been to this "future." The list includes:
   - Rocketry/space flight from the perspective of Robert H. Goddard (1882–1945)
   - Airplanes/air travel from the perspective of Orville (1871–1948) and Wilbur (1867–1912) Wright
   - Automobiles from the perspective of Nicolas-Joseph Cugnot (late eighteenth century) for the first steam-powered vehicle, or from that of Robert Anderson (ca. 1832–1839) for first electric-powered vehicle, or from that of Karl Benz (ca. 1885) for the first internal combustion engine-powered vehicle
   - ENIAC, the first electronic general-purpose computer (ca. 1946)
   - Atomic energy/nuclear weapons from the Manhattan Project (ca. 1939–1943)

  b. Choose *one* of the following and prepare a thoughtful essay of less than two pages describing the person or group, emphasizing the area of focus and process used for prognostication *and* the accuracy of some specific forecasts:
   - Arthur C. Clark (1917–2008)
   - Bill Joy (1954–)
   - Carl Sagan (1934–1996)
   - Michael Crichton (1942–2008)
   - RAND
   - SRI International

12-3 A key to successful reverse engineering involves interpretation of clues from the *form, fit, and function* of an item, object, device, product, structure, or system.

a. Briefly provide a definition of each of these terms in your own words.
b. This chapter of the book used the fantasy situation of the discovery of three different perfectly preserved 2013 "automobiles" (Section 12–1) to help understand form, fit, and function in Section 12–2. Try this yourself for *one* of the following fantasy situations involving the discovery of three or four related objects or entities, making your selection based on either your declared major or general/personal interest:
   (1) For civil, mechanical, or materials engineers:
   - a simple rope suspension bridge found in the wider Himalaya region or in South America
   - a wire-cable suspension bridge
   - a cable-stayed bridge
   (2) For aeronautical, mechanical, or materials engineers:
   - the Wright Flyer
   - the Ford Trimotor
   - the Heinke He178
   - the Boeing 787 Dreamliner
   (3) For electrical engineers:
   - the first television by John Logie Baird (October 2, 1925)
   - the first color TV (ca. 1953 in the United States)
   - the first LCD TV (ca. 1983 by Casio)
   (4) For biomedical or materials engineers:
   - the first moveable artificial limb (by Dr. Giuliani Vanghetti, 1898)
   - the first artificial arms with hook
   - a body-powered arm
   - Todd Khiken's prosthetic arm that "feels" (see YouTube video)

12-4 The point was made early in this book (Section 4–2) that good engineers, especially when engaged in reverse engineering or failure analysis (Chapter 6), must hone their *observation skills*. One needs to learn to use observable evidence and clues to assess form, fit, and function, in particular. These three aspects of a design are the best indicators of its suitability to purpose.

Key areas of or categories for making meticulous observations (in Section 12–3) are:
- size and shape (or geometry)
- materials-of-construction
- methods-of-manufacture
- precision in manufacture, quality, or workmanship
- electrical and/or thermal robustness
- design details

The first of these relates most closely to form. The second two relate most closely to function, as does the fifth. The fourth relates most closely to fit. The sixth and last relates to all three, i.e., form, fit, and function, depending on the detail(s).

Using the Internet to guide your search and selection, choose a *single* product, device, structure, or system and, within it, *two* specific examples (e.g., models) suitable to two different situations (demands, performance level, etc.) which exemplify each of the following, explaining your rationale for each pair chosen:
a. size *or* shape (e.g., overall external shape of cross-sectional shape of some key part)

  b. material-of-construction (which may necessitate a change in geometry)
  c. method-of-manufacture for the same material-of-construction (e.g., two ways to manufacture a part from an Al alloy)
  d. level of precision or fit *or* required quality
  e. electrical and/or thermal robustness

12-5 "Design suitability to purpose" may seem straightforward, *but* the original designer, as well as any engineer assessing a prior design by another engineer, must be sure he or she knows what the primary purpose is or was. Obviously, primary or principal purpose can and will differ from situation to situation, goal to goal, etc. Four common example purposes are:

- to provide *greater robustness* (i.e., durability), typically for professionals versus regular users, military versus civil, industrial versus consumer, etc.
- to allow *reduced cost,* typically for low-end users, light-duty, commodity products (i.e., made by the millions and considered expendable), etc.
- to provide *increased performance,* typically for special users, advanced users, professionals versus amateurs, etc.
- to allow or not allow *self-maintenance,* typically for protection of manufacturers warrantying their products

  Use the Internet to help find examples of any *two* of these purposes in a product. The product needs to be essentially the same within a single purpose but can differ from purpose to purpose. Within each chosen purpose, find a specific example of the product before and after or without and with the needed design modification (e.g., an electric circular saw for a do-it-yourselfer and for a professional carpenter).

  Prepare a report of less than one page on each chosen situation, being certain to identify the necessary changes to size and shape (geometry), materials-of-construction, method-of-manufacture, precision or quality of manufacture, and electrical and/or thermal robustness.

CHAPTER 13

Bringing It All Together with Illustrative Examples

13-1 Proverbs Make the Point; Pictures Fix the Lesson

An article by Frederick "Fred" R. Barnard appeared in the December 8, 1921, issue of *Printer's Ink,* an advertising trade journal, promoting the use of images as advertisements that appeared on the sides of streetcars. Along with the article was an ad entitled "One Look Is Worth a Thousand Words."[1] The consummate ad man, Barnard ran another ad in the March 10, 1927, issue with the altered wording "One Picture Worth Ten Thousand Words," and, to have people take the saying more seriously, he labeled it a Chinese proverb, replete with the saying in Chinese characters (Figure 13–1). The public soon (erroneously) attributed the proverb to Confucius (551–479 BC), a symbol of wisdom, if there ever was one.

One could debate whether the saying originated with or was simply copied from others by Fred R. Barnard, whether it said "a thousand" or "ten thousand," or whether Confucius might actually have uttered it himself at one time, but the simple truth remains. Nothing compares to seeing; in other words, "Seeing is believing."[2]

Another seemingly apropos proverb modified from a fourteenth-century saying (in old English), "Jt is ywrite that euery thing Hymself sheweth in the tastyng," says, "The proof of the

---

[1] The Russian writer Ivan Turgenev (1818–1883) wrote (in *Fathers and Sons* in 1862), "A picture shows me at a glance what it takes dozens of pages of a book to expound." (Wise advice for authors, present company excluded, of course.) The expression "Use a picture. It's worth a thousand words" appeared in a 1911 newspaper article that quoted editor Arthur Brisbane as he discussed journalism and publicity. A similar phrase, "One Look Is Worth a Thousand Words," appeared in a 1913 newspaper advertisement for the Piqua Auto Supply House of Piqua, Ohio.

[2] An idiom first recorded in 1639 but widely associated with Thomas the Apostle (later St. Thomas) when he said he would believe Jesus Christ had risen from the dead and appeared before the other apostles in Thomas's absence if he could "touch the nail marks in His hands and put his hand into His side." Its meaning seems obvious but could be debated: Physical or concrete evidence is convincing. There is always the issue of interpretation!

Figure 13-1 Ad by Fred R. Barnard in *Printer's Ink*, March 10, 1927, issue, using falsely claimed Chinese proverb: "One picture is worth ten thousand words." (*Source: Printer's Ink*, which after a change to *Marketing/Communications* in 1967, ceased to exist after 1972, and obviously could not be contacted for permission, despite due diligence by the author. Therefore, a portion of the original ad is reproduced here with no intent to circumvent any copyright.)

pudding is in the eating," although this author prefers a Spanish proverb by Miguel de Cervantes Saavedra (1547–1616) in *Don Quixote* (1615). The English translation (by Peter Anthony Motteux in 1701) says, "You will see it when you fry the eggs."

The point of this: The best way to fix all that has been written about reverse engineering in a reader's mind is to present some image-rich illustrative examples.

What follows in this chapter is intended to elucidate the following:

- *Observation (Section 4–2):* The examples presented rely on, and are intended to instill, the essential value of keen visual observation, without any aids of magnification (although low-power magnification can help in some instances). Readers will be coached on how to use hard evidence or subtle clues to find other possible evidence or clues.
- *Deduction of role, purpose, and functionality (Chapter 7):* For each example, the role of the item or object, and the purpose and functionality of each component or subassembly, will be deduced from evidence or clues. Readers will be coached on how to make the linkages.
- *Identification of materials (Chapter 9):* For each example, an effort will be made to identify at least the general type of material used in each component, and, to the extent possible by

observation alone, narrow down the material to a specific metal or alloy, polymer or composite, or ceramic or glass. Readers will be coached on how to use observable properties to guide identification.
- *Deduction of method-of-manufacture (Chapter 10):* For each example, an effort will be made to positively identify or logically deduce the most likely method used to create each component, as well as subassemblies and the overall assembly. Readers will be coached on how to use observable evidence or clues to guide identification.
- *Assessment of suitability of the design for purpose (Chapter 12):* For each example, an effort will be made to assess the suitability of each detail in the creation of the overall object or device to fulfill its needed role, intended purpose, and needed functionality. Readers will be coached on how to use observable information on form, fit, and function to deduce role, purpose, and functionality.

As this will be a "guided tour," there will, of necessity, be words to explain pictures, as opposed to pictures to expand upon words. In the end, as, hopefully, the reader looks again at the pictures and reflects upon the examples, both proverbs will ring true: A picture will be worth a thousand words, and the proof will be in the tasting.

Four different, but familiar, examples are presented, in order:

1. An inexpensive handheld electric-powered hair blow-dryer
2. An inexpensive older automatic electric coffeemaker
3. A modest-quality electric-powered leaf blower
4. A high-quality handheld electric-powered circular saw

These four items were chosen, as they represent a variety of interesting materials and methods-of-manufacture and involve some different detailed technical principles for their success.

Let's begin our guided tour.

13-2 Conair Electric Hair Blow-Dryer

Figure 13–2 shows a Conair 1875 (1875-watt) handheld electric hair blow-dryer; these devices, as a group, have a manufacturer's suggested retail price of from $14.99 for low-end consumer models to $69.99 or more for models intended for professional hairstylists, and typically can be found for a low of $10.99 to around $49.99.[3] The model shown was originally purchased (by one of the author's daughters) for $12.99 on sale. Production volumes for the low-end models surely exceed millions of units per year, so efficient manufacturing is essential. The particular unit lasted almost two years, receiving daily (if not more frequent) use.

This hair blow-dryer, like all hair blow-dryers, is an electromechanical device designed to blow cool, warm, or hot air over wet or damp hair in order to accelerate the evaporation of water particles and dry the hair during "setting" and "styling." Its *role* is to dry hair using blown, usually heated, air. The *form* of the unit suggests low weight (i.e., it is largely made of polymer), easy hand manipulation (i.e., it has a long, easily gripped handle and long power cord), and aerodynamic shape of the forward nozzle (i.e., to direct air in a concentrated stream). Some manufacturers are

---

[3] Conair is a U.S.-based multinational corporation that sells small electrical appliances, personal care products, and health and beauty products for consumers and professionals. The company acquired Cuisinart in 1989, Waring Products in 1998, and Pollenex after Jarden acquired Holmes in 2005.

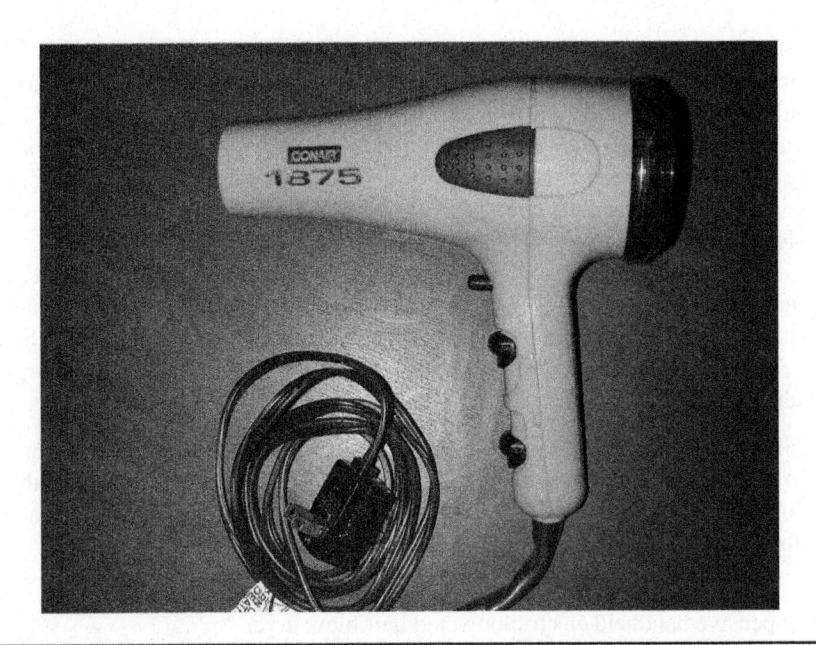

Figure 13-2 Photograph of a Conair 1875 (-watt) electric hair blow-dryer. (*Source:* Photograph taken by Donald Van Steele for the author, Robert W. Messler, Jr.; property of Robert W. Messler, Jr.) **Don't miss the color version of this figure, available at www.mhprofessional.com/ReverseEngineering.**

concerned about aesthetics, especially for higher-priced models. Professional models tend to be more utilitarian (without abandoning aesthetics) and much more robust in design and construction. The simplicity of the subject unit's exterior design (i.e., free of any "frills") suggests low cost.

Figure 13–3 shows the unit after disassembly, in an exploded view.[4] The major components, moving across the figure from left to right through the upper portion of the image are: (1) forward nozzle body unit with forward half of the integral handle, (2) electrical heater unit (top) and (3) heat shroud (bottom), (4) electric motor assembly with wiring to heater unit (to the left) and to the power cord (to the left and upward), (5) forward filter screen (top) and spring-wire holder (bottom), (6) rear body unit with rear half of integral handle, (7) rear filter screen, and (8) rear closure cap. To the left of the half handle for the forward nozzle body unit and right of the rear body unit and half handle are small parts associated with the on/off and air speed switches.

The device functions by drawing room-temperature (or ambient) air into the unit through openings in the rear closure cap using a fan (actually an impeller) at the aft end of a small two-speed (low and high) electric motor. Prior to reaching the motor/fan assembly, the air passes through a filter screen intended to prevent lint or other debris from reaching the motor and fan. The air is directed through the cylindrical casing surrounding and supporting the motor, cooling the motor

---

[4] As used here, and elsewhere in this chapter, "exploded view" implies an arrangement of component parts to show their relationship within an assembly, and in a manner that tends to reflect the general order of assembly. In this book, only major components were arranged to show order within the assembly, with minor parts (e.g., fasteners) not necessarily arranged to reflect their actual locations in the assembly.

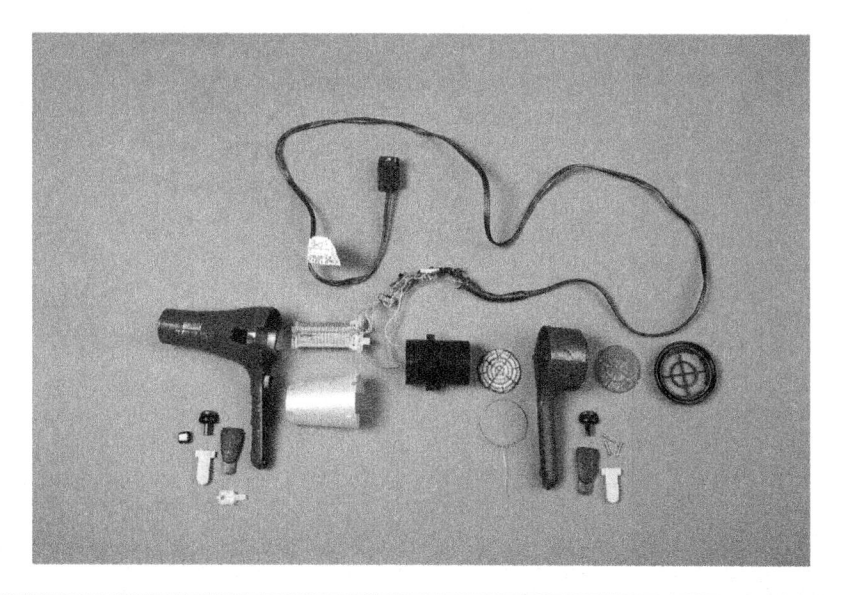

Figure 13-3 Photograph showing "exploded view" of the disassembled Conair 1875 electric hair blow-dryer. (*Source:* Photograph taken by Kris Qua of Kris Qua Photography, West Sand Lake, New York, for the author, Robert W. Messler, Jr.; property of Robert W. Messler, Jr.) **Don't miss the color version of this figure, available at www.mhprofessional.com/ReverseEngineering.**

on its way to the electric heater unit, where it is heated to the desired level (using a combination of low or high power levels and low or high air-flow rates, before being directed through the forward nozzle body to be concentrated and accelerated by the nozzle and directed onto the user's hair. Forward of the motor, just before moving air reaches the heater unit, is another filter screen.

Proceeding through the disassembled parts in what seems to be a logical, albeit not purely geometric, order, the following is observed:

In Figure 13–4*a,* the interior surfaces of the forward handle half (at the right) and rear handle half (at the left) are shown. These two parts are clearly made from a polymer, probably one that can tolerate modest impact (in the likely event the unit is accidentally dropped) and is thermoplastic to allow easy thermal molding. The use of molding is evidenced by numerous and varied geometric details, including (1) a stepped flange around the entire perimeter of the rear component (to help with *fit*); (2) a number of round, cylindrical posts to accept fastening screws to hold the two halves together or, on the rear component, to allow the motor assembly to be mounted; (3) a number of different details in the rear component to facilitate insertion, holding, and support of switch parts; and (4) additional details in both the forward and rear components to provide some structural stiffness to the handle (against bending). Also visible, particularly in the close-up of the rear handle half in Figure 13–4*b,* are marks from ejector pins used to help with removal of the newly molded but still warm and soft parts from dies. In short, these parts were probably injection-molded, as this process would allow intricate details to be created at high production rates and low cost (as the high cost of an intricate die is amortized over hundreds of thousands or millions of units).

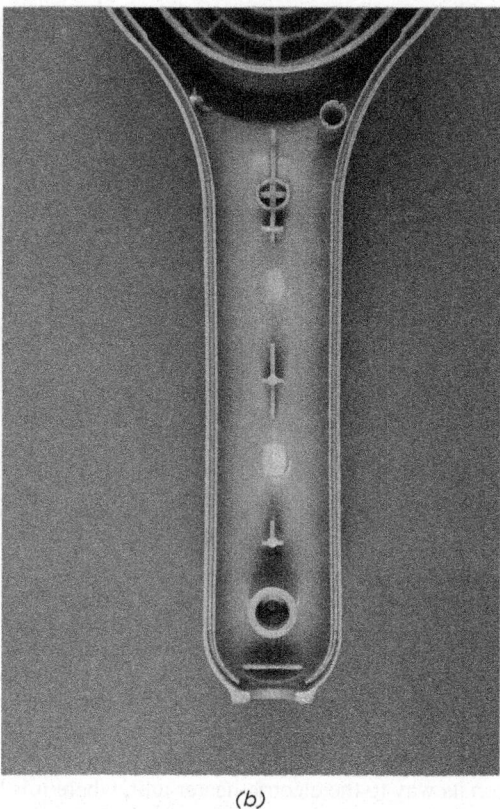

Figure 13-4 Photographs showing the interior surfaces of the forward (*a*) and rear (*b*) handle halves of the Conair 1875 electric hair blow-dryer. Ejector pin marks can be seen on both surfaces, but are particularly obvious in *b*. (*Source:* Photographs taken by Kris Qua of Kris Qua Photography, West Sand Lake, New York, for the author, Robert W. Messler, Jr.; property of Robert W. Messler, Jr.) **Don't miss the color version of this figure, available at www.mhprofessional.com/ReverseEngineering.**

Figure 13–5 shows the rear closure cap and two filter screen elements. The rear cap is obviously a polymer, again showing clear evidence (not shown in detail) that it (like the forward nozzle body halves) was injection-molded. Its design is obviously intended to keep the user's fingers from intentionally or accidentally entering the device and ever reaching the rear-mounted fan on the motor (although there is an intervening filter screen, too). The rightmost filter screen reveals a pattern of entrapped material that clearly reflects the geometric details of the rear cap, supporting that it was mounted just inside the cap. The leftmost filter screen contains much more entrapped fine lint residue and, upon careful observation, shows a pattern that reflects the details of the forward end of the motor (opposite the fan). This means the author arranged this screen incorrectly in the exploded view of Figure 13–2, as this filter screen was located just before the heater unit

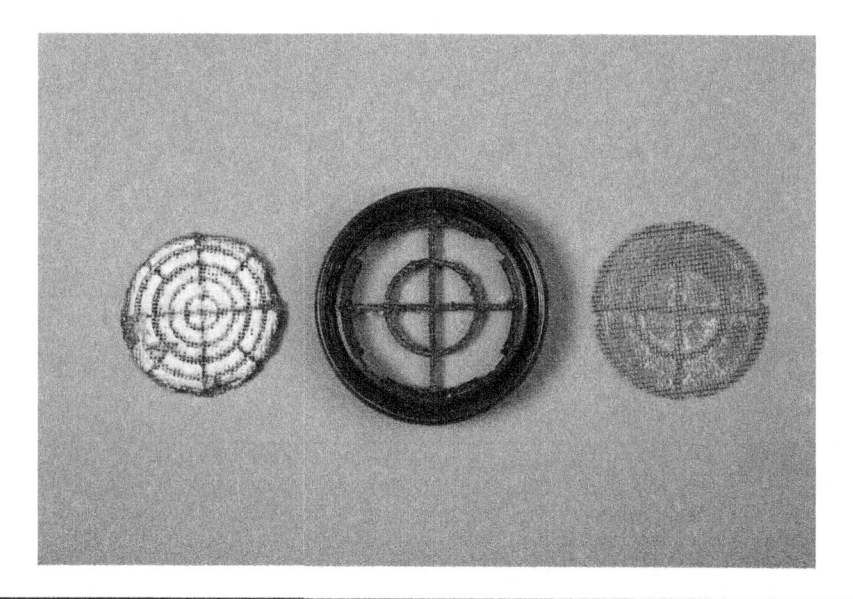

Figure 13-5 Photograph showing the rear closure cap and two filter screen elements. The filter element at the right is located just inside the cap, before intake air reaches the electric motor, while the element at the left is located downstream from the motor, just before air reaches the heater unit. (*Source:* Photograph taken by Kris Qua of Kris Qua Photography, West Sand Lake, New York, for the author, Robert W. Messler, Jr.; property of Robert W. Messler, Jr.) **Don't miss the color version of this figure, available at www.mhprofessional.com/ReverseEngineering.**

and not just before entering air reached the fan or impellor.[5] (This fortuitous, unintentional error by the author should prove that evidence never lies!) The reason for so much lint reaching this filter screen is not clear, as it is at odds with the general air flow direction. But it is what it is!

Figure 13–6*a* and *b* show the forward and aft ends of the motor assembly, respectively. While not disassembled, the casing and fan are clearly polymeric, although there are subtle clues (to a materials engineer) that the two are different types. The fan appears to be a molded thermoplastic (probably a nylon), while the hardness and stiffness of the casing, along with the fact that it must tolerate heat losses from the motor during operation, hint that this polymer may be a molded thermosetting polymer.[6] In all likelihood, the small electric motor is a standard-design purchased item, as it makes little economic sense for a company like Conair to get involved with this specialized area of design and manufacturing. The motor appears to be press-fit into the casing, held in place by the molded-in radial supports. The fan appears to have been press-fit onto the aft-pointing shaft for the motor.

---

[5] The wires from the power cord to the motor can be seen in Figure 13–6*a* to feed through a notch in the casing, around the filter screen element located at the forward end of the motor.

[6] Only more elaborate testing, as simple as a destructive test to see if the casing softens with heating (indicating it is made from a thermoplastic polymer) or chars (indicating it is made from a thermosetting polymer). To check whether a polymer is a thermoplastic, heat a straight pin until it radiates color and see if it causes a melted scar where it was caused to touch the part. If so, the material is a thermoplastic.

290 Chapter Thirteen

(a)

(b)

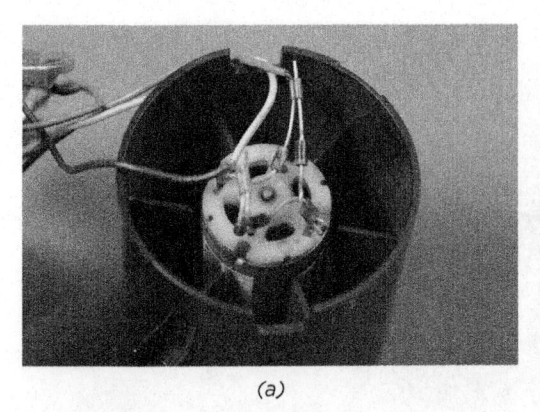

Figure 13-6 Photographs showing the forward (a) and aft (b) ends of the electric motor. A fan attached to the aft end cools the motor using intake air before it reaches the heater unit. (*Source:* Photographs taken by Kris Qua of Kris Qua Photography, West Sand Lake, New York, for the author, Robert W. Messler, Jr.; property of Robert W. Messler, Jr.) **Don't miss the color version of this figure, available at www.mhprofessional.com/ReverseEngineering.**

Figure 13–7 shows two views of the interesting heater unit, from the side in (*a*) and from the end in (*b*). Heat in the unit is generated by the resistance to current flow through long, thin high-temperature metal alloy wires, that is, from joule I^2R heating. The achievement of long length (to increase resistance) can be seen in the end view, where the wire is folded (bent) in and out to create a 16-pointed star pattern. The wires are wrapped around and supported by a material that must be resistant to high temperature, must be an electrical insulator (to prevent shorting), must be reasonably strong at high temperature (to hold its shape), and must be reflective to radiant heat. This latter conclusion comes from knowing that the wires literally glow from incandescence and that the heat must be directed radially outward to the air that is steaming by, so the radiant energy must be reflected by the central support structure. This evidence, combined with evidence found in the heat shroud shown in Figure 13–8, supports the fact that both the heater-wire support and shroud are made from the same material, and that material is a ceramic.[7]

Closer examination of the shroud reveals that it is very lightweight; is brittle; sloughs off fine, light-reflecting flakes; and appears to have a layered structure seemingly created by compacting together fine flakes to make the sheet material. A little digging on the Internet (unless one is an experienced materials engineer) strongly suggests the shroud is made from compacted mica.[8]

As far as suitability of design for purpose, everything about this item supports low-cost, high production volume, and modest operational life, at reasonable quality. Materials are simple

---

[7] Recall from Section 9–4 that ceramics are characterized as high-melting/heat-tolerant and electrical insulators.
[8] Mica actually refers to a group of naturally occurring sheet silicate (phyllosilicate) minerals that are brittle, layered (to allow easy splitting into thin sheets and/or crushing into small, thin flakes), high-melting/high-heat-resistance, and strong dielectric materials (or electrical insulators). The general chemical formula is $X_2Y_{4-6}Z_8O_{20}(OH,F)_4$, where X is usually K, Na, or Ca, Y is generally Al, Mg, or Fe, and Z is generally Si.

(a)   (b)

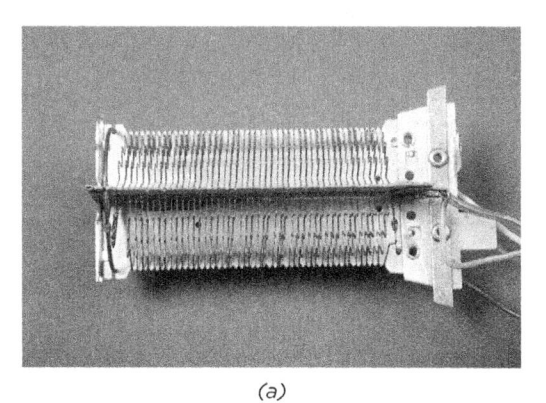

Figure 13-7 Photographs showing the heat unit from the side (a) and end (b). Wire wound in a 16-point star pattern is supported by and insulated by a heat-resistant, dielectric, reflective mica cruciform element. (*Source:* Photographs taken by Kris Qua of Kris Qua Photography, West Sand Lake, New York, for the author, Robert W. Messler, Jr.; property of Robert W. Messler, Jr.) **Don't miss the color version of this figure, available at www.mhprofessional.com/ReverseEngineering.**

and, as demonstrated by the heat shroud and heater wire supports, clever. Manufacturing is straightforward, using molding to amortize the cost of dies that impart complex geometry and intricate details over a large number of units. The required small electric motor was of a standard design and almost certainly a purchased (versus built) item.

While it won't be done for every illustrative example presented, a diagram that captures the force/energy flow and functional model for the hair blow-dryer is presented in Figure 13–9 (see Sections 5–5 and 5–6).

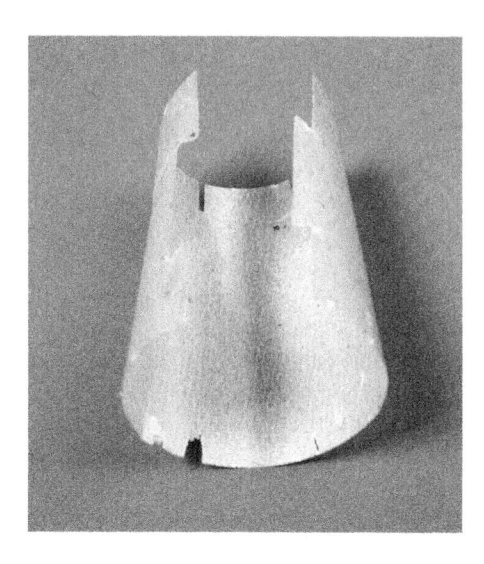

Figure 13-8 Photograph showing the mica shroud that surrounds the heater unit to reflect radiant heat from the unit and keeps it from reaching the thermoplastic body of the hair blow-dryer. (*Source:* Photograph taken by Kris Qua of Kris Qua Photography, West Sand Lake, New York, for the author, Robert W. Messler, Jr.; property of Robert W. Messler, Jr.) **Don't miss the color version of this figure, available at www.mhprofessional.com/ReverseEngineering.**

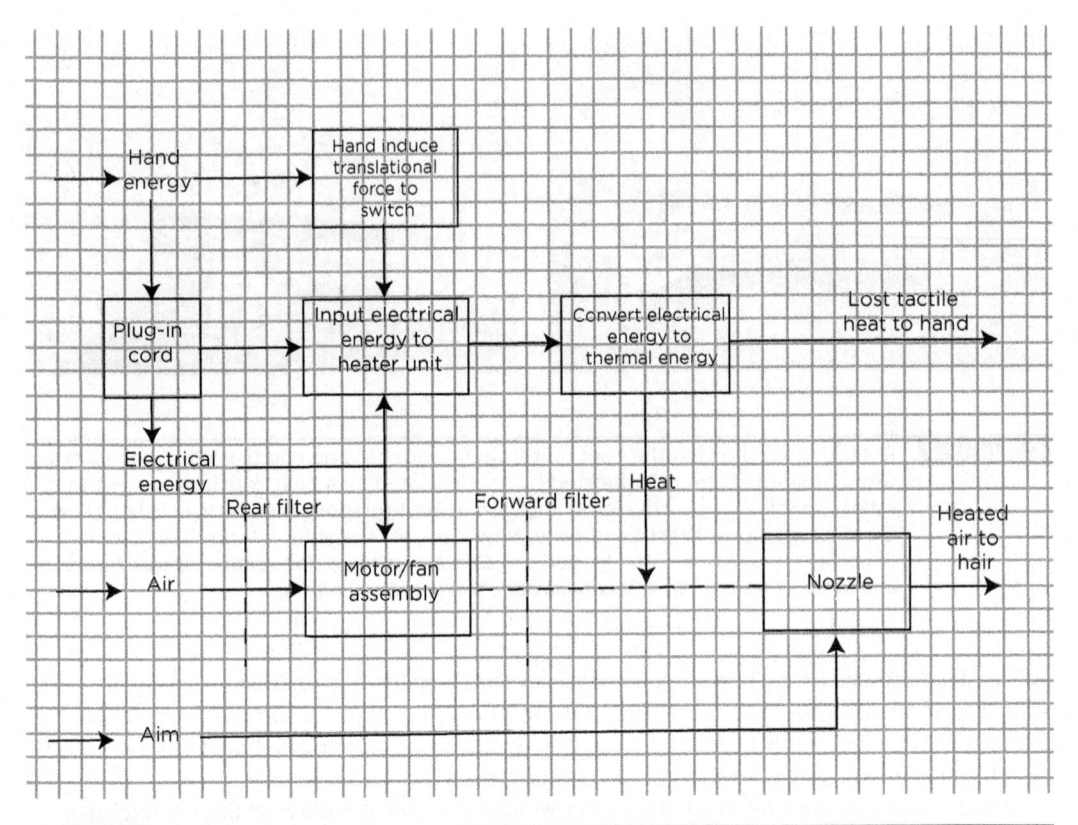

Figure 13-9 A combination force/energy flow diagram and functional model for the Conair 1875 handheld electric hair blow-dryer.

13-3 An Automatic Electric Coffeemaker

Figure 13–10 shows an early model of an unidentified manufacturer's automatic electric coffeemaker. These became a commodity countertop appliance almost as soon as they appeared on the market, selling in low-end, as well as midrange, department stores, which is not to say there weren't high-end manufacturers and models. Prices were probably in the $40 to $60 range, with units being sold at lower prices during special sales. Production volumes quickly became high (hundreds of thousands and, soon, millions per year), as everyone had to have this "modern" convenience that many saw as a necessity as life in America moved at a faster and faster pace, and a quick cup of coffee in the morning was the way many people coped. Prices became so competitive that traditional manufacturers of such appliances, such as Hamilton-Beach, Presto, Proctor-Silex, Norelco (part of North American Philips), and Sunbeam were quickly joined by a host of others like Black & Decker, Cuisinart, and, of course, Mr. Coffee, and manufacturing either moved offshore (to Mexico in the case of the unit shown here) or incorporated newly emerging design-for-assembly methods (principally relying on plastic snap-fits) better suited to robot assembly (ref. Messler).

Figure 13-10 Photograph showing an early model of an unidentified manufacturer's automatic electric coffeemaker. (*Source:* Photograph taken by Donald Van Steele for the author, Robert W. Messler, Jr.; property of Robert W. Messler, Jr.) ***Don't miss the color version of this figure, available at www.mhprofessional.com/ReverseEngineering.***

Automatic coffeemakers are a cooking appliance used to brew coffee from roasted and ground coffee beans placed in a paper or wire-mesh filter inside a funnel, which is placed over a glass, ceramic, or metal coffeepot or carafe. Cold water is poured into a separate chamber (reservoir), quickly heated to the boiling point, and directed to the funnel, where hot coffee is produced by the drip-brew process. The *role* of all automatic electric coffeemakers is simple: brew fresh coffee quickly and as needed. The general *form* (i.e., layout) of the device is fairly straightforward and similar. In the most modern devices, of which the one presented here is not an example, the size of the unit varies considerably from so-called one-cup units to units capable of making much larger quantities of coffee. The major difference in these units is the physical size of the system.

Figure 13–11 shows the coffeemaker after disassembly, albeit not in a perfect "exploded view." The major components are: (1) the body top component (with water reservoir) and (2) base component (top center, at the left, inverted, and at the right, respectively, in white, the latter with power cord attached); (3) the body top component cover and (4) base component bottom plate (left center, left, in white, and right, in metal, respectively); (5) metal up-feed tube, red rubber down-feed hose, and plastic fitting (left top, long thin parts); (6) filter holder/funnel (right top corner, in black); (7) carafe and cover (center, near bottom, to the left); (8) U-shaped heater unit and round metal support plate (center, near bottom, to the right); and (10) a sealing and insulating gasket (black ring at the lower right) and additional locking cap for coffeepot (upper left, round, near feeder pieces).

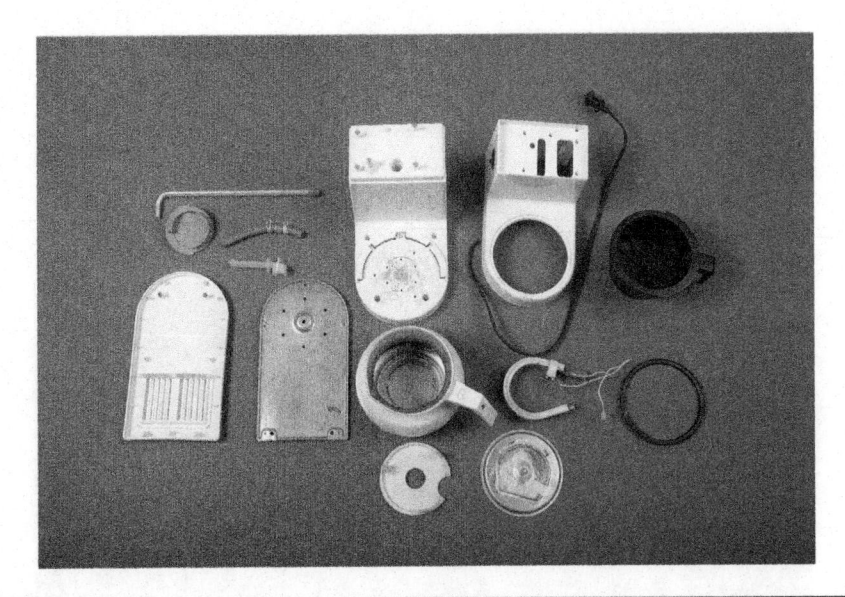

Figure 13-11 Photograph showing an "exploded view" of the disassembled automatic electric coffeemaker. (*Source:* Photograph taken by Kris Qua of Kris Qua Photography, West Sand Lake, New York, for the author, Robert W. Messler, Jr.; property of Robert W. Messler, Jr.) ***Don't miss the color version of this figure, available at www.mhprofessional.com/ReverseEngineering.***

The device functions by first placing the metal carafe (with its cover in place) on the round metal support plate, which is fitted into the large hole in the body base component, over the insulating and sealing gasket. Next, ground roasted coffee is placed in a paper filter (not shown) fitted into the black filter holder/funnel and the holder/funnel is set into the large hole in the body top component. Cold water is then poured into the reservoir of the body top component, where, when power is turned on, it is drawn down through the plastic fitting into the rubber down-feed hose to the U-shaped heater unit by convection forces, aided (once convection begins) by gravity. The water is heated as it passes through the channel portion of the U-shaped heater unit and passes back up through the metal up-feed tube to flow into the top of the coffee grind–containing filter/funnel assembly. The hot coffee then runs into the metal carafe, where it is kept hot by the heater unit lying beneath the metal support plate.

All of the plastic components shown in Figure 13–12 (body top and base components, removable top body component reservoir cover, and filter holder/funnel) were injection-molded from a thermoplastic that offers moderate impact resistance. Injection molding is indicated by the intricacy of the geometric details (e.g., peripheral stepped flanges, stiffening details, support and fastening posts for screws, and other details) and by marks from ejector pins that can be seen in the close-up image in Figure 13–12*b* and at numerous locations on the filter holder/funnel in Figure 13–12*d*.

The metal body base component cover plate (which closes the bottom of the body base component, and allows access to the heater element installation and possible replacement) is

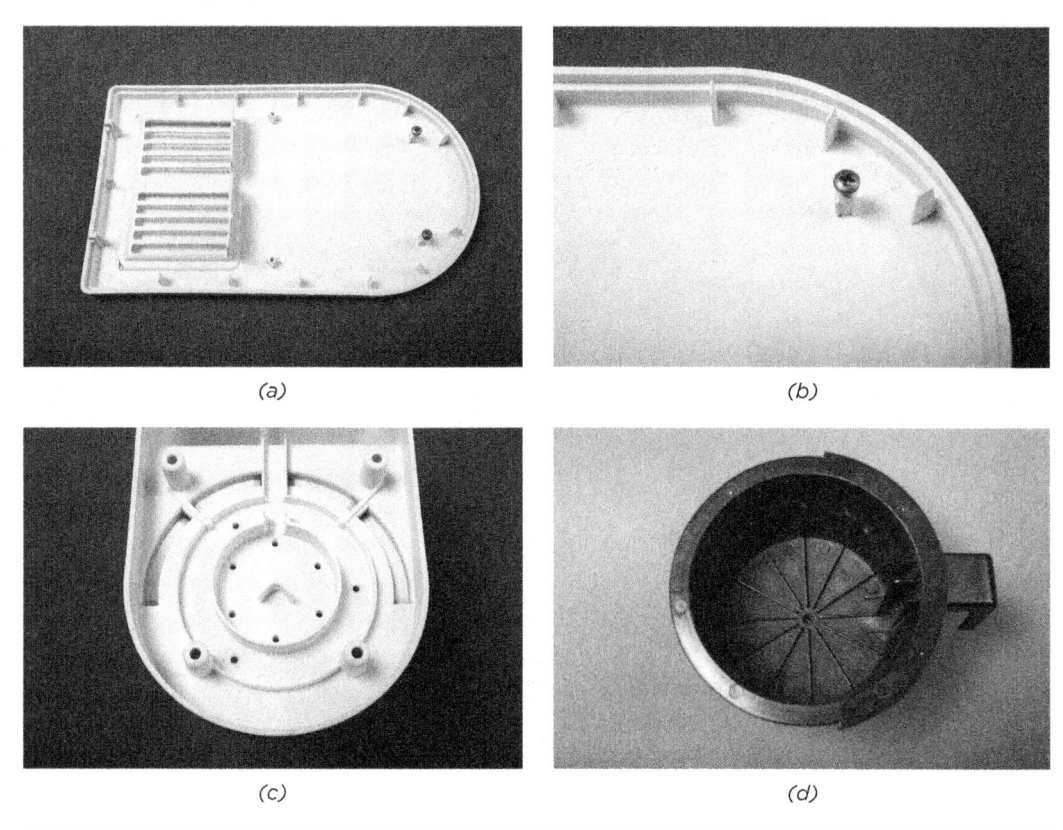

Figure 13-12 Photographs showing all of the plastic (polymer) components of the coffeemaker, including: body top (a) and base (b) components, removable top body component reservoir cover (c), and filter holder/funnel (d). Ejector pin marks found on all of these components are visible at several locations in (a), (b), and (d). (*Source:* Photographs taken by Kris Qua of Kris Qua Photography, West Sand Lake, New York, for the author, Robert W. Messler, Jr.; property of Robert W. Messler, Jr.) **Don't miss the color version of this figure, available at www.mhprofessional.com/ReverseEngineering.**

made by blanking (to cut the peripheral shape) and stamping (to cold form details into) sheet-gauge steel, which was galvanized with zinc to protect it from corrosion by water and/or coffee overflow or spillage (Figure 13–13). Small round-headed rivets inset from the bottom side of the plate served as "legs" to raise the coffeemaker slightly off a countertop (Figure 13–13*b*). These rivets were plastically deformed (i.e., were "upset" at their tail ends) to lock them in place (Figure 13–13*c*). The cold-forming operation on the blanked sheet steel was probably done simultaneously with blanking, as part of the stamping portion of the operation, and created a peripheral stiffening bead, as well as a recess for a portion of the heater unit and carafe support plate.

The two most interesting components of this item were the carafe and the U-shaped heater unit.

296 Chapter Thirteen

(a)

(b)

(c)

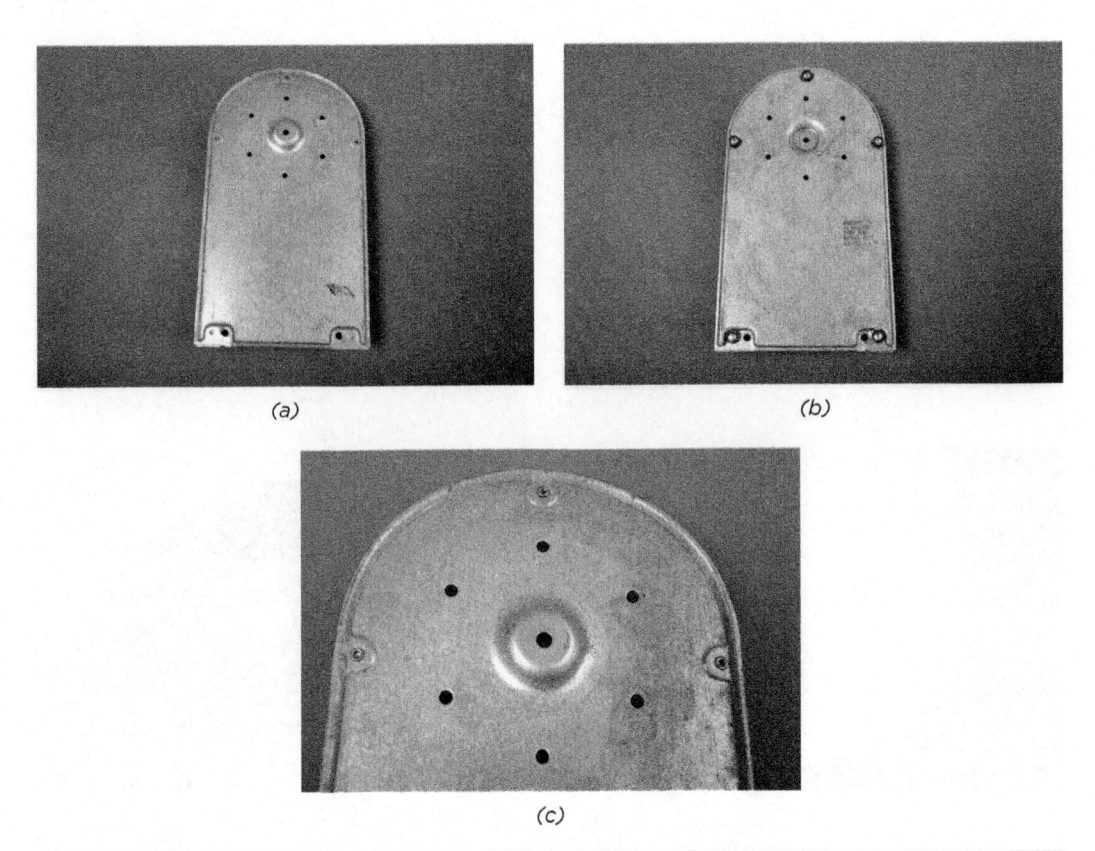

Figure 13-13 Photographs showing the metal body base component bottom cover or closure plate in top-side (*a*) and bottom-side (*b*) views, as well as in close-up of the top side (*c*). Round-headed rivets set in rubber grommets were installed (by upsetting) to serve as "legs" that lift the unit off a counter or table surface slightly. (*Source:* Photographs taken by Kris Qua of Kris Qua Photography, West Sand Lake, New York, for the author, Robert W. Messler, Jr.; property of Robert W. Messler, Jr.) **Don't miss the color version of this figure, available at www.mhprofessional.com/ ReverseEngineering.**

The carafe of this early model automatic electric coffeemaker was made of metal, so that it would be thermally conductive (Figure 13–14). In fact, the low density, whiteness, softness (to scratching with a screwdriver tip), nonmagnetic behavior, and sense of coolness (due to the rapid rate at which it extracted heat from one's hand) indicated it to be an aluminum alloy, but, most likely, pure Al, as AA1100.[9] Next to copper (which has the highest thermal and electrical conductivity of any engineering metal except silver, which may not, because of its cost, be considered an "engineering metal" in most cases), aluminum has the highest thermal conductivity

---

[9] The highest thermal, as well as electrical, conductivity for a particular metal occurs with its pure state, being lowered by any alloying addition.

(a)   (b)

Figure 13-14 Photographs showing the spin-formed aluminum alloy (probably commercially pure aluminum AA1100) carafe. Evidence of spin forming can be most easily seen on the inner surface of the lip of the carafe as closely spaced circumferential ridges (*b*). (*Source:* Photographs taken by Kris Qua of Kris Qua Photography, West Sand Lake, New York, for the author, Robert W. Messler, Jr.; property of Robert W. Messler, Jr.) **Don't miss the color version of this figure, available at www.mhprofessional.com/ReverseEngineering.**

of all engineering metals. Its choice for the design of this unit made sense, as the carafe was heated by the unit located under the metal support plate on which it sat. Normally, coffeepots or carafes are made from thermally insulating ceramic or glass, because hot coffee is put into them to stay hot without additional heating. In this early design, the intent seemed to be to keep the brewed coffee hot by active heating. Distinct circumferential ridges seen on the lip of the carafe (Figure 13–14*b*), as well as all along the inside surface of the vessel, give evidence that the carafe was made by cold metal spin forming.[10]

The U-shaped heater unit is especially fascinating. It consists of a star-shaped channel in a round body piece integral to a rectangular body piece with a central round hole (Figure 13–15*a*). Cold water flows into the star-shaped channel through a rubber hose, is heated by a resistance-heated core wire embedded in electrical insulating ceramic contained within a soft metal sleeve, with the entire assembly being inserted snugly into the hole of the rectangular body piece. Joule (I^2R) from what is likely a high-resistance, heat-resistant core wire element passes into the rectangular body piece and then into the integral round body piece. Heat in the round body piece is then conducted to the water, with greater efficiency due to the large surface area created by the star-shaped channel (Figure 13–15*b*). It is unlikely the water is heated to the necessary temperature to drip-brew the coffee in a single passage, but it might be. If not, it simply recirculates until it is hot enough. The complex shape of what is likely a pure aluminum (i.e., AA1100) heater unit

---

[10] *Metal spin forming* is a metalworking process by which a disk or tube of metal is rotated at high speed, usually on a power lathe, and formed into an axially symmetric part using the tip of special hand tools levered on a mounting piece or post. The process is especially amenable to soft, ductile (e.g., face-centered cubic) metals and alloys, of which aluminum is a particularly well-suited example. The process was used in the aircraft industry and was and is used to make some aluminum pots and pans (if they are not deep-drawn).

298 Chapter Thirteen

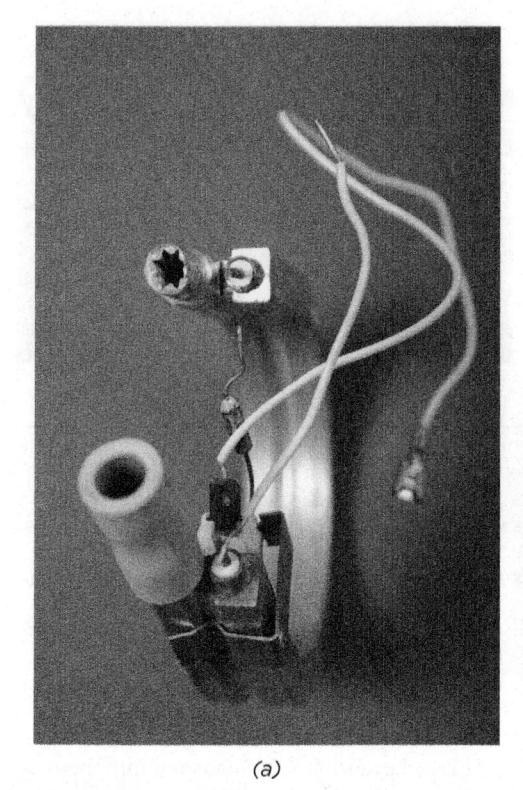

Figure 13-15 Photographs showing the interesting and clever U-shaped pure aluminum heater unit (a). The close-up (b) shows details of the extrusion used to produce the rectangular portion containing a round hole that contains a sleeved heater core wire and a round portion that contains a star-shaped channel to more effectively transfer heat to water as it passes through the channel. (*Source:* Photographs taken by Kris Qua of Kris Qua Photography, West Sand Lake, New York, for the author, Robert W. Messler, Jr.; property of Robert W. Messler, Jr.) **Don't miss the color version of this figure, available at www.mhprofessional.com/ReverseEngineering.**

suggests it was made by hot extrusion, by which both the external shape and internal shaped holes could be produced in a single operation.[11]

As far as suitability to purpose, the design seems adequate. The injection molding of polymer parts and blanking and stamping of the base cover plate keep production costs low and are well-suited to high-rate, high-volume production. The use of an aluminum extrusion for the heating unit was clever and efficient, as pieces of the required length could be cut from very long extruded stock and cold-formed into a U-shape once the sleeved heater core wire was inserted. The sleeved

---

[11] *Extrusion* is a process used to create objects of fixed cross-sectional profile. A material is pushed or drawn through a die of the desired cross section. Very complex shapes can be made, with very high quality surface finishes, and the compressive stresses used allow even relatively brittle materials to be processed. Aluminum and Al alloys are widely extruded, with extrusion being done hot, so the metal or alloy is soft and plastic but fully solid.

core wire was likely purchased from a manufacturer of such items. The carafe processing was clever, although it is clear (from Figure 13–14) that it was not easy to keep the inside of the carafe clean, as it stained badly. The overall design using simply convection forces to move the water is questionable, as it might be better to use a small pump unit, although no one could argue that natural convection (a) works and (b) is cheap. While the design seemed effective, changes for later models suggested it was less than optimum.

13-4   Toro Electric Leaf Blower

Figure 13–16 shows a Toro Model 850 electric leaf blower (or sweeper). The particular model subjected to reverse-engineering dissection and analysis had a low and high speed setting as an integral part of the on/off switch and operated on 120-volt residential service, typically using a 100-foot-long outdoor-rated, three-prong, grounded power extension cord (not part of the unit).

Toro, like many other manufacturers of lawn and gardening equipment such as leaf blowers, edger-trimmers, and weed whackers, offers a wide range of models that are either electric or gasoline-engine powered, have either simple blowing action for sweeping leaves from lawns or gardens or dust and light dirt from walkways and driveways, or both blowing and vacuum capability, the latter for sucking leaves up into a removable collection bag. Most have two speeds (like the dissected unit) but are available with different "wind" velocities from around 158 to over 240 mph; the higher-velocity units cost more, as do the combination sweeper/vacuum units.

Figure 13-16   Photograph showing the Toro Model 850 electric-powered leaf blower/sweeper. (Source: Photograph taken by Donald Van Steele for the author, Robert W. Messler, Jr.; property of Robert W. Messler, Jr.) **Don't miss the color version of this figure, available at www.mhprofessional.com/ReverseEngineering.**

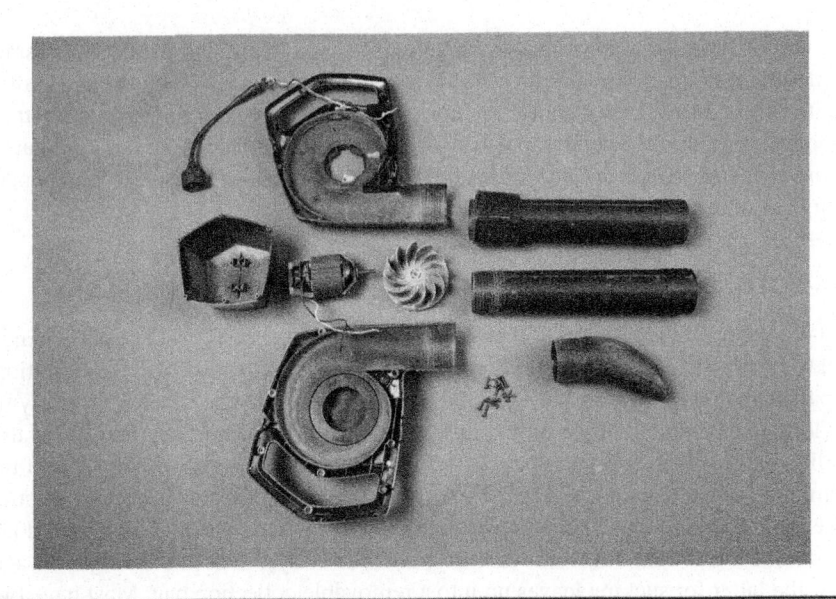

Figure 13-17 Photograph showing the disassembled Toro leaf blower. (*Source:* Photograph taken by Kris Qua of Kris Qua Photography, West Sand Lake, New York, for the author, Robert W. Messler, Jr.; property of Robert W. Messler, Jr.)

Gasoline-powered units are all more expensive than electric-powered units but offer greater blowing/drawing power and freer movement without an extension cord, albeit at greater weight. Manufacturer suggested retail prices for electric sweepers run from about $45 to $90, with prices running from $36 to $70 during sales. Annual production probably amounts to hundreds of thousands of units of each of around three models—low, medium, and high end. The subject unit cost about $55 and operated well for seven or eight seasons, before the switch began to operate intermittently.

This leaf blower, like all leaf blowers or sweepers, is a powered device (i.e., using an electric cord or gasoline) that produces a high-velocity concentrated stream of air that can be shaped in some units by different nozzle tips to blow leaves or other light debris from lawn, garden, or paved surfaces using a sweeping action. Its *role* is to blow leaves or other light debris from surfaces. The *form* of the device is a good indicator of its blowing (or, in some models, vacuuming) power and weight—higher-velocity air streams (or vacuum) resulting in greater size and weight, with gasoline-powered devices generally being considerably heavier and, therefore, being back-mounted as opposed to handheld. The subject unit weighed about 5 to 6 pounds and produced an air-stream velocity of about 160 mph. Most manufacturers are concerned about aesthetics, at least to give their products easier brand recognition.

Figure 13–17 shows the unit after disassembly, in an "exploded view." The major components, starting with the main body housing halves (top and bottom, left of center) and then working across from left to right, are: (1) the main body halves with integral handle, the left-hand side of which (at the top) has the power cord attached; (2) the rear body-closure cap; (3) the electric motor; (4) the two-sided impellor/fan; (5) the extension tubes (of which there are two); (6) the nozzle end piece; and assorted screws.

(a)

(b)

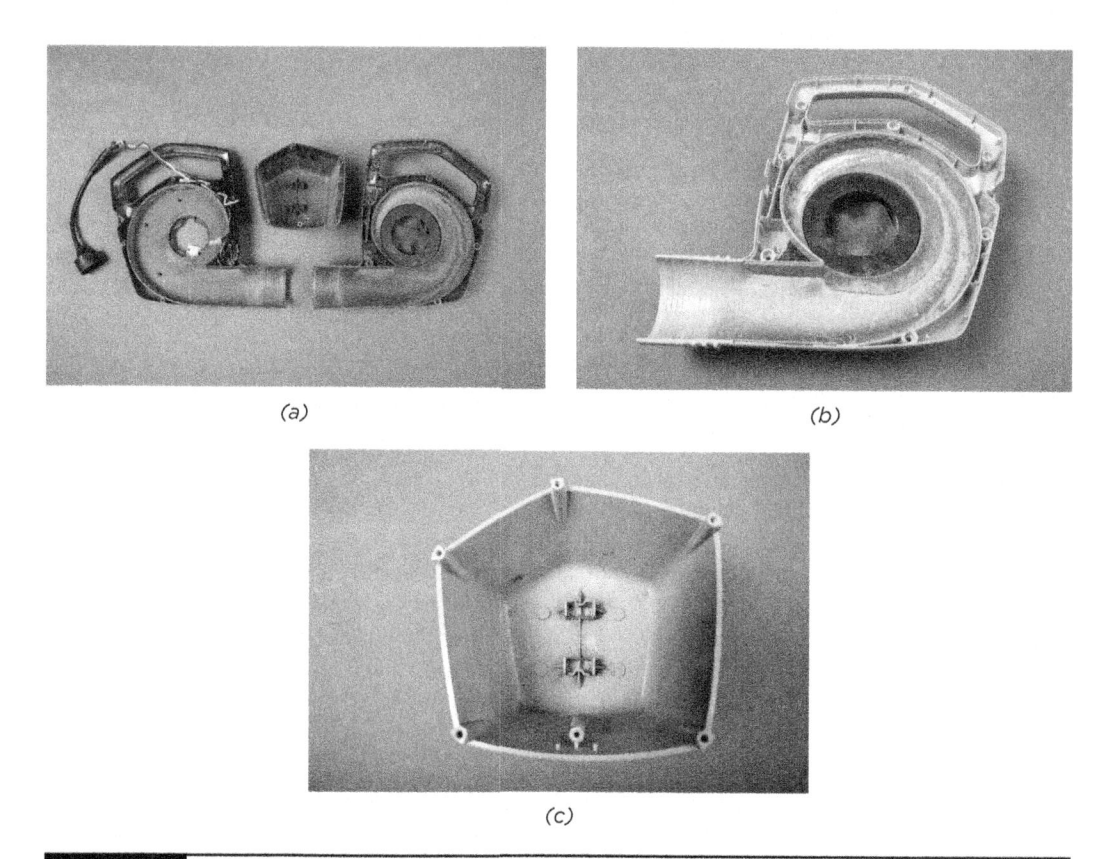

(c)

Figure 13-18 Photographs showing the main body halves (*a* and *b*) and rear closure cap (*c*), all of which were made of an injection-molded thermoplastic polymer. Telltale ejector pin marks can be seen on the interior surfaces of the parts in (*b*) and (*c*), and, while not visible in (*a*), were also present in the component. (*Source:* Photographs taken by Kris Qua of Kris Qua Photography, West Sand Lake, New York, for the author, Robert W. Messler, Jr.; property of Robert W. Messler, Jr.) **Don't miss the color version of this figure, available at www.mhprofessional.com/ReverseEngineering.**

Figure 13–18*a* shows the main body halves and rear closure cap, all of which were made of an injection-molded thermoplastic polymer that offered impact resistance against accidental dropping and inevitable banging. The right-hand main body half shown in Figure 13–18B reveals that it contains a foam rubber dust and dirt filter element and black soft rubber gasket for sealing. The swirling pattern of the air in the plenum chamber as it is propelled out of the impeller/fan to the blower's tube-and-nozzle assembly can be seen from dirt residue. That these main body halves and end closure cap were injection-molded is suggested by the geometric complexity of the right body half, consisting of various stiffeners and mounting posts to accept screws that join the two body halves. Positive evidence that injection molding was used can be found in marks from ejector pins, particularly evident on the inner surface of the end closure cap (Figure 13–18*c*).

A standard-design, small but powerful high-speed electric motor, almost certainly a purchased item not made by Toro, is shown in Figure 13–19. Rust on the outer surface of the iron stator/

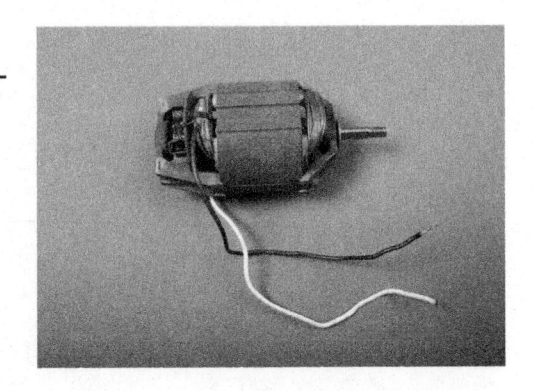

302 Chapter Thirteen

Figure 13-19 Photograph showing the standard-design, small but powerful electric motor for the leaf blower. (*Source:* Photographs taken by Kris Qua of Kris Qua Photography, West Sand Lake, New York, for the author, Robert W. Messler, Jr.; property of Robert W. Messler, Jr.) **Don't miss the color version of this figure, available at www.mhprofessional.com/ReverseEngineering.**

case of the motor attests to the leaf blower's age of even to eight years, in which it was, inevitably, exposed to rain during fall leaf cleanup in upstate New York. Most small appliance and lawn and garden power equipment manufacturers buy, rather than make, electric motors from motor manufacturers, as the design and fabrication of such items is specialized and best left to experts.

Figure 13–20 shows the impellor side (*a*) and fan side (*b*) of the impellor/fan component. The impellor draws air through openings in the right half of the main body component, through the foam rubber filter, and into the plenum of the joined main body halves, to be expelled down the extension tube and nozzle end-tip assembly. Erosive wear at the ends of all of the vertical-standing

(a)   (b)

Figure 13-20 Photographs showing the intake impellor (*a*) and cooling fan (*b*) of the integral impellor/fan component. Subtle evidence that the part was injection-molded, probably from a thermoplastic nylon, was found but is not visible in the photographs. Wear at the ends of the vanes of the impellor (*a*), as well as dirt residue on vane faces, support that this was an impellor which drew air into a plenum, where that air was then accelerated to create a fast-moving stream that was further accelerated as it passed through the extension tubes and nozzle. No wear and only light black dust was found on faces of the fan blades, supporting that this was the fan side of the component. (*Source:* Photographs taken by Kris Qua of Kris Qua Photography, West Sand Lake, New York, for the author, Robert W. Messler, Jr.; property of Robert W. Messler, Jr.) **Don't miss the color version of this figure, available at www.mhprofessional.com/ReverseEngineering.**

(a)   (b)

Figure 13-21 Photographs showing the interlocking extension tubes (a) and nozzle endpiece (b). The extension tube sections were made from a molded thermoplastic, evidenced by fine parting lines that run the length of each tube on diametrically opposite locations. These parting lines can be seen near the tops of the tubes in (a). Extensive scratching and abrasive wear at the end of the nozzle (b) are evidence that this portion of the extension tube rubs on the ground. (*Source:* Photographs taken by Kris Qua of Kris Qua Photography, West Sand Lake, New York, for the author, Robert W. Messler, Jr.; property of Robert W. Messler, Jr.) **Don't miss the color version of this figure, available at www.mhprofessional.com/ReverseEngineering.**

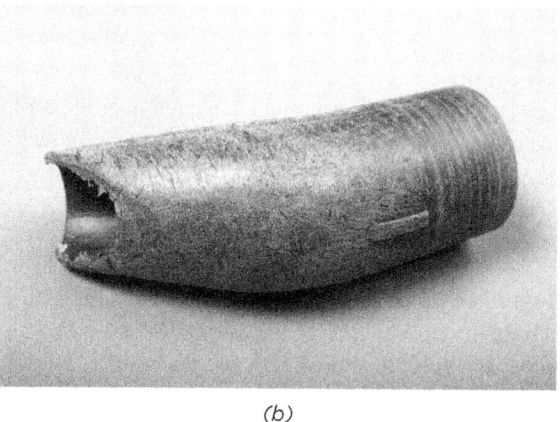

vanes on the impellor give evidence that it is an impellor (to take air in) and that the filter is not perfect (or was not properly cleaned periodically). There is also considerable dust residue on the concave faces of all vanes. The fan side of the impellor/fan component is free of silt residue and erosive wear but does show some black soot deposit, probably from the motor. The fan, being integral with the impellor, spins at the same speed and provides moving air to cool the electric motor.

Figure 13–21a shows the two sections of the tube extension, which couple using raised interlocking rigid design features (not visible, but located at the left of Figure 13–21a) using a push and clockwise half-twist (ref. Messler). The coupled tube sections attach to the blower body using the large knurled, internally threaded coupling (at the right of Figure 13–21a). There is evidence that these were either pressure or blow molded from a thermoplastic polymer, in the form of a parting line that runs along the length of the outer surface of each section at the midplane of

304 Chapter Thirteen

the tube sections in Figure 13–21a. The nozzle end tip shown in Figure 13–21b seemingly was made from a tougher, more wear-resistant polymer, so that it could tolerate frequent contact with and rubbing on the ground or pavement. Evidence of severe scratching and abrasive wear can be seen on the tip. While this part may have been injection-molded using a removable core to form the interior of the piece, the polymer may be a thermosetting type, as these are, on average, stronger and tougher and harder (and more wear-resistant) than thermoplastic polymers. For thermosetting polymers, the preferred molding process is sometimes referred to as "reaction transfer molding."

As far as suitability of design for purpose, this product gave evidence of sensitivity to cost, but not if it compromised fit and function. The leaf blower performed flawlessly for seven or eight seasons, spring (for removing twigs from the lawn), summer (for blowing lawn clippings and dirt off the driveway), and fall (for blowing leaves). Many details of the design showed concern for user safety, including a protective cage over the air intake on the right-hand main body half and weather-sealed power cord connection to the blower and to the male plug and an electrical-ground post on the three-prong power cord plug.

13–5 Skil™ Handheld Electric Circular Saw

Figure 13–22 shows a Skil handheld 120-volt electric 7¼-inch, 2⅛-horsepower circular saw. These most popular electric power saws are made by several manufacturers, of which Skil is probably best known—the name "Skilsaw" is often used generically to refer to handheld electric circular saws.[12] Various models, from low-end models for do-it-yourselfers to high-end models for professional carpenters, range from $49.99 to over $149.95 retail, with units available on sale for as low as $34 to $36. The subject model was purchased more than 20 years ago for around $45 and is still capable of operating. With nearly every homeowner eventually buying a handheld electric circular saw, production volumes for such saws are in the millions per year, overall, with Skil being a leader.

An electric circular saw is a machine that uses a toothed circular metal cutting disc or blade. The blade is a tool for cutting wood or other materials, albeit with metal or hard ceramics requiring special fine-toothed blades with hard carbide tips or inserts. Machines can be handheld or table-mounted, with the handheld types obviously being lower powered and lighter for portability. The cuts produced are straight, with a narrow kerf and good cut surface. The *role* of the device is to produce straight cuts or controlled-depth grooves in wood or other materials. The *form* of a particular circular saw indicates a saw's cutting capability (with blade size) from size and durability (from the robustness of components), with professional models, which receive much more use, being heftier.

Figure 13–23 shows the subject unit after disassembly, in an "exploded view." The major components, moving across the figure from left to right, are: (1) side-mounted motor closure cap; (2) main body component with integral motor housing and attached power cord (top); (3) 2⅛-horsepower electric motor with attached metal fan (center); (4) half-handle component (bottom),

---

[12] In fact, the electric circular saw was invented in 1923 by Edmond Michel, a French immigrant to New Orleans, and was shortly thereafter produced and marketed by him and partner Joseph Sullivan in their Chicago-based Michel Electric Handsaw Co., in 1924. The company name was changed to Skil in 1926, and Sullivan developed the name "Skilsaw" after Michel left the business around that time.

Figure 13-22 Photograph showing the Skil 7¼-inch (diameter of the blade), 2⅛-horsepower handheld electric circular saw. (*Source:* Photograph taken by Donald Van Steele for the author, Robert W. Messler, Jr.; property of Robert W. Messler, Jr.) ***Don't miss the color version of this figure, available at www.mhprofessional.com/ReverseEngineering.***

which joins to the main body component; (5) drive gear for the cutting blade (direct center, right of motor); (6) upper metal fixed blade guard (top, right of center); (7) lower metal retractable blade safety guard (bottom, right of center); (8) blade (center, right of center); (9) blade mounting washers (right of center); and (10) adjustable guide footplate (extreme right).

The device functions by aligning the blade to the intended cut-line, forcing the spring-loaded lower retractable safety guard backward to expose the blade, and placing the guide footplate onto the surface of the piece to be cut. When one is ready to make the cut, the trigger-switch within the trigger guard of the handle is depressed, causing the blade to rotate at full speed (often with a sensible jerk). The entire unit is then slid forward, pressing downward on the handle to keep the guide footplate tight against the work, and sighting into a small rectangular notch in the leading edge of the guide footplate to keep a penciled or scribed cut-line centered in the notch while pushing the saw forward at a slow but steady rate. (For safety's sake, it is best to keep both hands visible—one on the handle, keeping the trigger depressed, and the other on top of the fixed upper blade guard.) When the cut is completed, and the saw blade exits the work or reaches a predetermined line, the trigger is released and the saw lifted to pull the blade from the cut while rotation is stopping.

Proceeding through the disassembled parts in what seems to be a logical order, the following is observed: In Figure 13–24, the inside surface of the half handle component (*a*) and the outside surface of the main body component with integrated half-handle, looking down into the integral

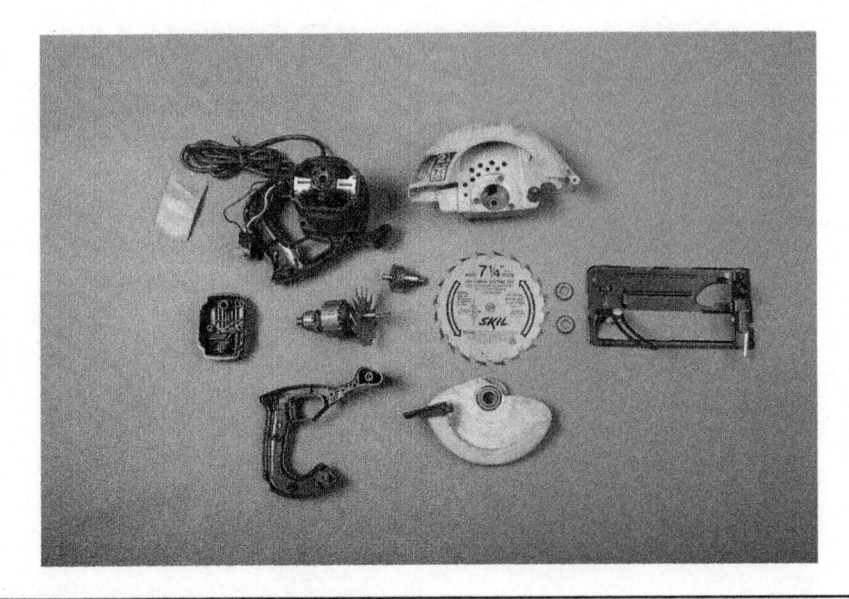

Figure 13-23 Photograph showing an "exploded view" of the disassembled Skil handheld electric circular saw. (Source: Photograph taken by Kris Qua of Kris Qua Photography, West Sand Lake, New York, for the author, Robert W. Messler, Jr.; property of Robert W. Messler, Jr.) ***Don't miss the color version of this figure, available at www.mhprofessional.com/ReverseEngineering.***

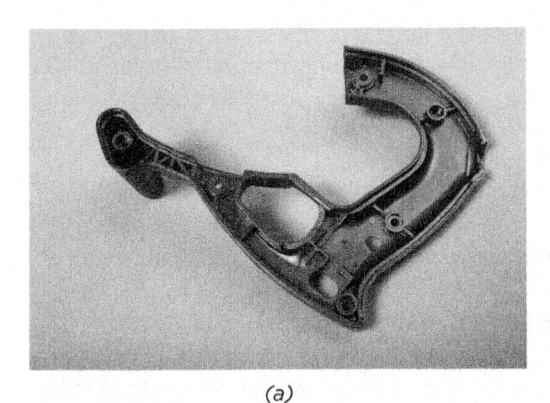

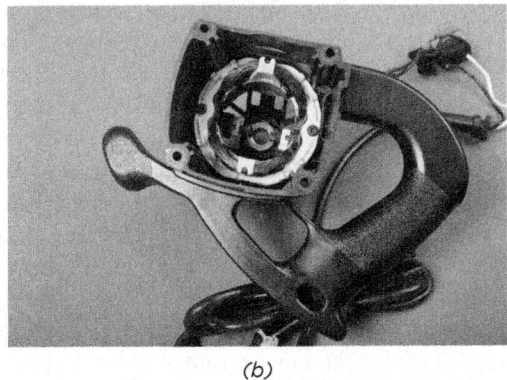

(a)

(b)

Figure 13-24 Photographs showing the injection-molded thermoplastic handle half components of the saw. The half shown in (b) includes an integral housing for the saw's $2\frac{1}{8}$-horsepower electric motor, the stator for which can be seen down inside the housing. (*Source:* Photographs taken by Kris Qua of Kris Qua Photography, West Sand Lake, New York, for the author, Robert W. Messler, Jr.; property of Robert W. Messler, Jr.) ***Don't miss the color version of this figure, available at www.mhprofessional.com/ReverseEngineering.***

motor housing to see the copper-wire stator windings (*b*), are shown. Both were obviously made (rather typically for such thin-shell housings and intricate parts) by injection-molding a high-impact-resistant thermoplastic polymer. Evidence that injection molding was used appears on the inside of the half handle and includes a stepped peripheral flange, numerous and varied stiffening elements, cylindrical posts for accepting screws that join the two handle halves together, as well as telltale marks from ejector pins, as smooth circular flat spots here and there. Comparable geometric details are found in the larger main body half component, along with telltale ejector pin marks on the inside surface, but also intricate texturing of the outer grip surface of this and the other handle half. All of these details are faithfully replicated in molding using one-time-expense permanent metal dies that allow design and tooling fabrication costs to be amortized over many thousands, tens of thousands, or hundreds of thousands of units.

Figure 13–25 shows the various metal components constituting the fixed upper blade guard from the blade side (*a*) and outer side (*b*), and the lower retractable blade guard from the outside (*c*). The relative light weight (for their size), from lower density (around 2.9 g/cm^3; one-third that of steel at about 7.9 g/cm^3), whitish-silver color, relative softness (to scratching), nonmagnetic behavior, and relative coolness to touch (from three times higher thermal conductivity of Al alloys versus magnetic steel) all indicate the material-of-construction is an Al alloy (for strength and lightweight portability). Furthermore, all three of these components were made by casting, probably using permanent dies, to amortize high die costs over many units. This is evidenced by four observations: (1) complex geometry (radial stiffeners on the retractable guard, stiffeners, mounting guides on the outside of the upper unit, etc.) with intricate details (like the logo "Skilsaw" on the upper fixed guard unit); (2) numerous raised short circular bosses for ejector pins to push against in order to help extract solidified but still hot and soft parts from dies (to speed production); (3) a few recessed marks from ejector pins of solidified but still soft metal (visible at 7 and 5 o'clock locations below the copper bearing seen in the center of the outside view of the upper unit); and (4) some evidence of untrimmed flash along inside edges of the upper guard unit and a subtle (but visible) parting line near the midlength of the barrel-shaped integral hinge on the fixed upper guard unit in (*d*).

Figure 13–26 shows the front face of the blade for the 7¼-inch Skil circular saw (*a*), along with the inside and outside mounting washers (the use of which is shown in a printed image on the blade), and an edge view of the blade to try to show how alternating teeth in the blade are bent slightly inward or slightly outward from the plane of the blade so as to create a cut (or kerf) slightly wider than the blade body to prevent binding and difficult retraction. A small magnet revealed the blade to be a ferromagnetic steel, probably a carbon steel with about 0.30 to 0.40 wt.% C to keep the body of the blade strong but tough (i.e., to resist impact or shock loading in normal use, if the blade catches or binds). The fact that the teeth appear much darker, with no evidence of any weld attaching the teeth tips to the blade body, indicated the teeth were rapidly heated locally by rotating the blade between high-frequency induction coils to bring them into the austenite region for the steel, and then rapidly cooling (i.e., quenched) in water to produce a hard, wear-resistant martensitic microstructure just in the tips. (As soon as the blade is used, the teeth are heated by friction and are slightly tempered to reduce brittleness after quenching and impart toughness to tolerate shock loading.)

The close-up of the face-mounting washers for the blade (in Figure 13–26*c*) suggests they were forged, and, because of their small size and the intricacy of the raised lettering around the washer's slanted face, the author suspects (based on experience) that the method-of-manufacture was

308 Chapter Thirteen

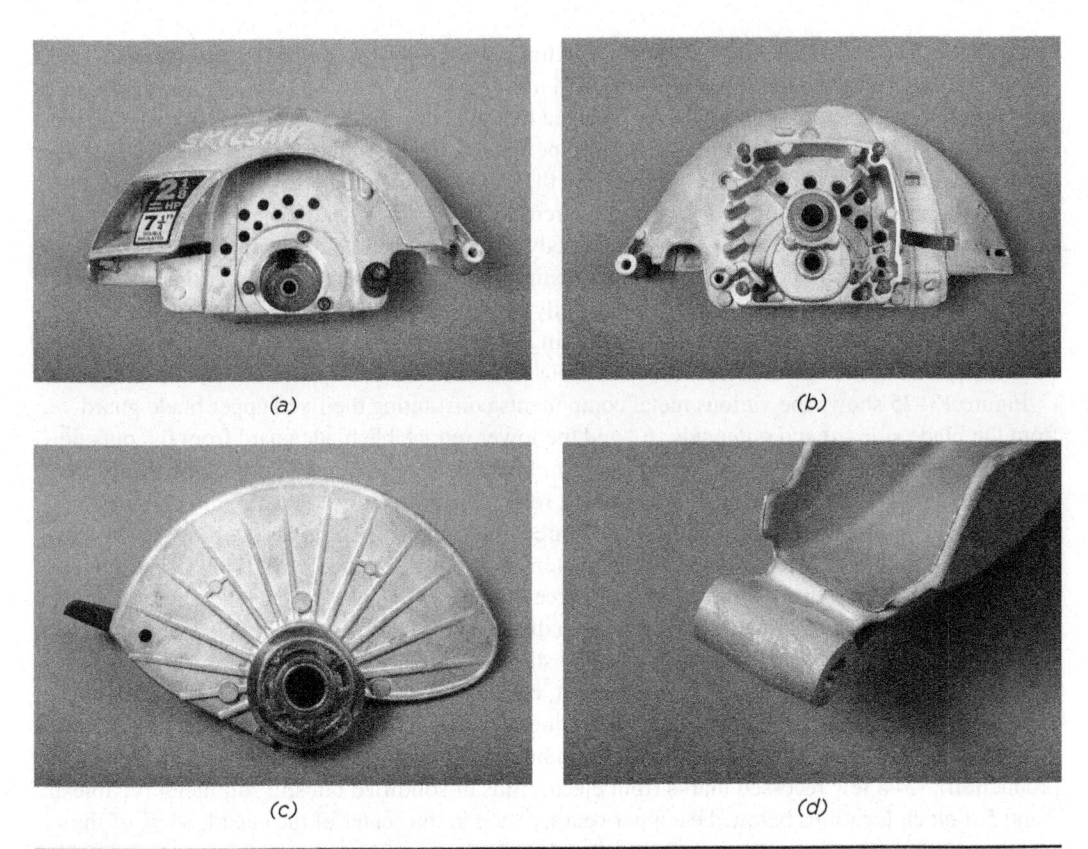

Figure 13-25 Photographs showing the cast aluminum alloy fixed upper blade guard component, from the blade side (a) and opposite side (b) and retractable lower blade guard component (c), with detail showing boss (d). Short round bosses, which can be seen in (a) and (d), are used to support the ejector pins used to help extract the solid but still very hot parts from their dies. Evidence of ejector pin marks can also be seen on the surface of the round feature at the bottom of (b). Untrimmed burrs on the edges of the fixed upper blade guard casting can be seen in (c), as can evidence of the parting line of the die on the circumference of the body of the integral hinge detail. (*Source:* Photographs taken by Kris Qua of Kris Qua Photography, West Sand Lake, New York, for the author, Robert W. Messler, Jr.; property of Robert W. Messler, Jr.) **Don't miss the color version of this figure, available at www.mhprofessional.com/ReverseEngineering.**

powder processing using cold compacting and sintering. (The only way to differentiate between forging and powder compacting/sintering would be to metallurgically section, polish, etch, and examine the microstructure of the washer. A forged part would show subtle flow lines and no porosity, while a powder-processed part would be free of flow lines but contain some porosity (typically, 2 to 5 vol.%).

Figure 13–27a and b show the top and bottom faces of the guide footplate. Again, as this piece exhibited strong attraction to a magnet and had a density (of about 7.9 g/cm3) that suggested iron,

Bringing It All Together with Illustrative Examples **309**

(a)

(b)

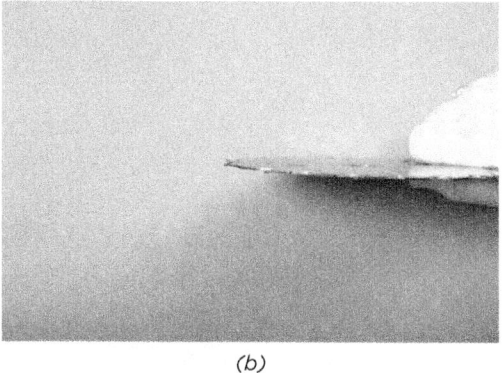

(c)

Figure 13-26 Photographs showing the circular saw blade from its face (*a*) and edge (*b*). The latter view shows how alternating teeth were formed slightly inward or outward from the plane of the blade to produce a kerf that is wider than the blade to prevent binding. A close-up of one of the washers used to mount the blade (see in [*a*]) is shown in (*c*). Subtle details (e.g., the intricacy and faithfulness of details like the lettering) suggest this part was made by powder processing. (*Source:* Photographs taken by Kris Qua of Kris Qua Photography, West Sand Lake, New York, for the author, Robert W. Messler, Jr.; property of Robert W. Messler, Jr.)

it was concluded to be a common, low-cost plain carbon steel, probably with around 0.10 to 0.15 wt.% C, for easy-formability and only lower loading (and stress).[13] Several features gave evidence that the part was fabricated by simultaneously cold blanking, stamping, piercing, and forming the starting metal blank. The blanking cut the outer rectangular shape, stamping and piercing formed the internal cutouts (with some material trimmed and removed, as for the large, long rectangular cutout, and some material left attached along one or two edges to allow forming), and forming

---

[13] The carbon content of a steel not in the martensitic condition (e.g., from water quench hardening) can be estimated from the relative proportions of white-etching ferrite and dark-etching pearlite using the "lever rule" (ref. Messler2).

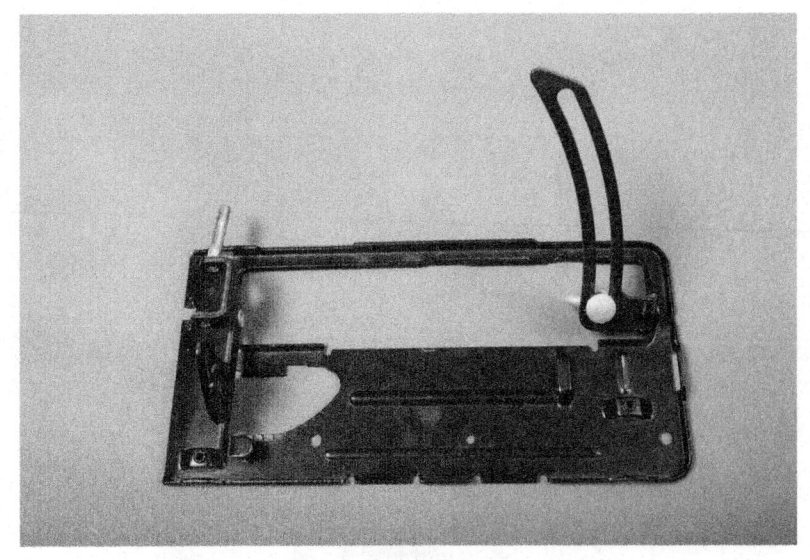

(a)

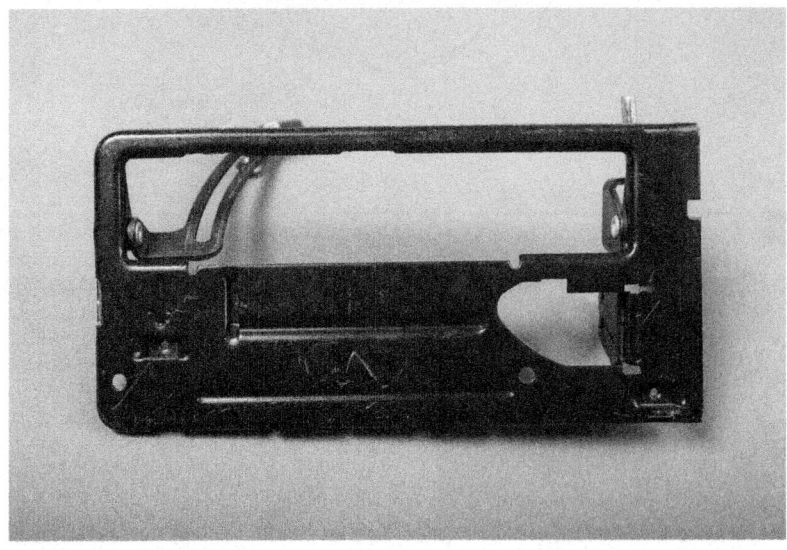

(b)

Figure 13-27 Photographs showing the sheet-gauge plain carbon steel slide plate, from the top (*a*) and bottom (*b*). Details of this component suggest it was blanked or punched, to create the outer perimeter, and simultaneously stamped and formed to create interior details (e.g., flanges, beads, and out-of-plane features). (*Source:* Photographs taken by Kris Qua of Kris Qua Photography, West Sand Lake, New York, for the author, Robert W. Messler, Jr.; property of Robert W. Messler, Jr.) ***Don't miss the color version of this figure, available at www.mhprofessional.com/ReverseEngineering.***

Figure 13-28 Photograph showing the small but powerful (2⅛-horsepower) electric motor for the circular saw. This motor was almost certainly purchased as a standard model from a manufacturer of small electric motors, as part of what is known as a "make or buy" decision by a manufacturer like Toro. (*Source:* Photograph taken by Kris Qua of Kris Qua Photography, West Sand Lake, New York, for the author, Robert W. Messler, Jr.; property of Robert W. Messler, Jr.) **Don't miss the color version of this figure, available at www.mhprofessional.com/ReverseEngineering.**

created the folded peripheral stiffening flange and lengthwise stiffening beads. Of particular interest, as it is clever, is how a few details (e.g., the upstanding tab at the left of Figure 13–27*a* and the short, raised rectangular feature just inside the right-hand edge of Figure 13–27*a*) were stamped and pierced (i.e., cut-through) along some edges but left uncut along one end or two parallel edges, to allow forming of the release feature. The purpose of this entire component is to allow tilt adjustment up and down in the direction of cutting or side-to-side to the direction of cutting to create a bevel cut.

Finally, Figure 13–28 shows the small but powerful (i.e., 2⅛-horsepower) electric motor, with an attached aluminum alloy (whitish-silver, soft, nonmagnetic) fan for keeping the motor cool. As is usually the case for small appliance and power tool manufacturers, among other industries, this motor is almost certainly a purchased (not made) part, as specialty motor manufacturers offer low-cost standardized designs of great reliability.

As for suitability of the design for purpose, this saw was a wonderful example of how a quality product can be made economically for homeowners/do-it-yourselfers. Materials were entirely appropriate for the intended use. Fit was very good, especially for a low-end model. Finish was somewhat lacking, as evidenced by the burrs on the inside edges of the fixed upper blade guard. In a separate exercise conducted in a class of design at the engineering school where the author was a faculty member for three decades, a professional model Skil circular saw exhibited more substantial construction, with heavier-gauge steel, finer finish castings, and heavier-duty molded polymers.

13-6 Lessons Learned

Hopefully, after reading Chapters 7, 9, 10, and 12, and reflecting on what is presented in the preceding four illustrative examples, you will have learned some lessons. To help you, here are some key lessons that *should* be learned, if not from a first reading, then from a second or third reading:

- Look at the item, entity, or structure as it is at the outset (i.e., either intact or in ruin), to get a sense of its overall form as an indicator of role and purpose and, at least to some degree, its function(ality).
- Look at and consider the workmanship, to get a sense of concern for quality, whether by need (for fit and function) or simply pride of ownership by the creator.
- Disassemble or dissect the item, entity, or structure systematically, to see what goes where and to begin to assess how parts, components, or structural elements interact.
- Arrange the disassembled item, entity, or structure (or rearrange an already broken, disassembled, or ruined item, entity, or structure) as it would appear in an exploded view, to see how everything fits together by location in the whole, orientation relative to other details, and fit with abutting details. (This and the previous step help create a force/energy diagram and functional model, if these are deemed necessary.)
- Carefully examine each and every part in some logical order based on the exploded view (as exploded views should reflect how things were assembled), to get a sense of the form, fit, and function (and purpose and functionality) of each.
- Observe everything there is to see or be sensed by touch or sound for density, thermal conductivity, and modulus of elasticity (through the transmission of sound), to identify or deduce the material-of-manufacture or -construction.
- Use knowledge of the material-of-manufacture or -construction, and the inextricable interrelationship among structure-properties-processing-performance, as clues to the function (or functionality), as materials are selected for properties to fulfill function and properties influence the choice of method-of-manufacture or -construction.
- Conduct simple tests of heft (for density relative to a familiar material such as steel), luster and color (for metals), sense of coolness (for thermal conductivity relative to a familiar material like steel), sound with tapping (for modulus of elasticity relative to a familiar material like pine or polymer or steel), flexibility in bending (for modulus of elasticity relative to a familiar material like steel, accounting for any shape factor), and hardness or softness by scratching with a car key or knife blade (for hardness relative to a nearby familiar material like annealed low-carbon steel, or location to location on the same part, to test for surface hardening for wear resistance).
- Look for clear evidence or subtle clues associated with certain methods-of-manufacture or -construction—specifically, as examples:
    - ✓ Geometric complexity (to indicate use of a process amenable to such complexity, like machining or casting of metals or molding of polymers)
    - ✓ Intricacy of details, especially on surfaces (to indicate use of a process amenable to such details, like casting of metals, molding of polymers, or powder processing of metals or ceramics)
    - ✓ Machining marks (to indicate machining or grinding or polishing)

- ✓ Ejector pin marks (to indicate casting of metals or molding of polymers)
- ✓ Parting lines (to indicate casting of metals or molding of some polymers)
- ✓ Flashing or flash (to indicate forging of metals or, sometimes, molding of polymers)
- ✓ Flow lines (to indicate forging of metals or molding of polymers, if not done properly)
- ✓ Surface texture (to indicate as-cast condition of metals, as well as specific process, e.g., rough, low-cost sand casting versus smooth precision die casting)

- Look at form and fit, quality of workmanship, and cost of the materials and methods-of-manufacture or -construction, as guides to the assessment of suitability of design for purpose.
- Draw on past experience and do whatever you can to gain experience.
- Look. Think. Link.
- *Think!*

13-7 Summary

"A look [or picture] is worth ten thousand words." "The proof of the pudding is in the tasting." So go two proverbs, and so it is for reverse engineering. This chapter was intended to pull everything together that was presented in Chapters 7, 9, 10, and 12. The approach was to present four illustrative examples of familiar products that were dissected and analyzed, including: (1) a handheld electric hair blow-dryer, (2) an automatic electric coffeemaker, (3) a handheld electric-powered leaf blower or sweeper, and (4) a handheld electric-powered circular saw. Each was systematically dissected and every major component's purpose and function were identified, materials-of-construction were identified (even if only by generic type, e.g., thermoplastic polymer, or alloy such as cast aluminum alloy), method-of-construction (including assembly), and suitability of design to purpose. Key lessons learned were listed in a final section.

13-8 Cited References

Messler, Robert W., Jr., *Integral Mechanical Attachment: A Resurgence of the Oldest Method of Joining,* Butterworth-Heinemann/Elsevier, Burlington, MA, 2006.

Messler2, Robert W., Jr., *The Essence of Materials for Engineers,* Jones & Bartlett Learning, Burlington, MA, 2011.

13-9 Thought Questions and Problems

13-1 A full and complete (i.e., proper) effort in the reverse engineering of a mechanism, structure, system, or material should always include the following five aspects:
- *observation* (involving all senses, with the logical caveats concerning smell and, especially, taste) (Section 4–2)
- *deduction of role, purpose, and function or functionality* (Chapter 7)
- *identification of materials-of-construction* (Chapter 9)
- *deduction of methods-of-manufacture* (Chapter 10)
- *assessment of suitability of design to purpose* (Chapter 12)

In a thoughtful essay of two pages or less, explain the importance of each of these five aspects of a proper effort in reverse engineering. Specifically, relate your responses to the interrelationship among structure ⇔ properties ⇔ processing ⇔ performance (Section 9–2) that underlies materials engineering as well as all engineering disciplines that use engineering materials to create a product, device, structure, or system.

13-2 Referring to the five aspects reiterated in Question 13–1 (from Section 13–1), suggest the most likely implication of each of the following findings during a reverse engineering effort, defending your answer in each case.

a. A structural member or element extracted during the reverse engineering of a civil structure resulted in the following *observation*: An element with a rather high (aspect) ratio of length–to–cross-sectional area reveals an I-shape with a long vertical web compared to horizontal caps.

b. A mechanical device or mechanism contains a pair of precision (versus heavy-duty or robust) gears: a drive gear with a large diameter and large number of teeth and a coupled gear with a much smaller diameter and fewer teeth. (Deduce the purpose and function.)

c. The basically trapezoidal tubular frame of a bicycle has been changed from an Al alloy to a graphite fiber-reinforced epoxy composite material. (Compare pertinent properties.)

d. The electric shielded-metal arc fusion welded steel framework or support structure for a high-rise office building in San Francisco, California, is changed to high-strength bolted shear-type friction joints in a replacement building of otherwise similar geometry.

e. Two seemingly similar, if not identical, V-8 internal combustion gasoline automobile engines are completely "dissected"—one from a state police cruiser and one from a normal consumer's vehicle. The connecting rods of pistons in the police car are found to be: (1) forged versus cast, (2) low-alloy versus plain carbon steel, and (3) slightly larger in cross-sectional area.

13-3 One of the most interesting details of the Conair electric hair blow-dryer (Section 13–2) is the heat-shielding shroud that surrounds the resistance-heated heater unit. The material-of-construction is very unusual, as is the method-of-manufacture.

Prepare a brief but thoughtful essay of about one page that describes and discusses: (1) mica, a material to deal with the thermal and electrical insulation in general, and radiant energy shielding in particular, citing specific relevant properties, and (2) a logical method-of-manufacture, given that mica is an inherently brittle and soft (check its MOH hardness) layered mineral with poor through-thickness strength.

13-4 One of the most interesting details of the automatic electric coffeemaker (Section 13–3) is the U-shaped, complex cross-sectioned metal heat exchange element. The geometry, material-of-construction, and method-of-manufacture, as well as a subtlety in the design, all attract one's attention.

Prepare a brief but thoughtful essay of about one page that describes and discusses: (1) the geometry (i.e., U-shape and external and internal cross sections); (2) the choice of material, supported by relevant property(ies); (3) method-of-manufacture; and (4) subtle detail of the cross section. Support all of your responses.

13-5 Two interesting findings in the Skil handheld electric circular saw (Section 13–5) are: (1) details pertaining to the teeth of the cutting blade and (2) details pertaining to the mounting nuts for the cutting blade.

a. The teeth of the cutting blade are (1) angled out of plane in opposite directions for alternating teeth and (2) demarcated by a pronounced region of "heat tint," but no apparent weld.

Prepare an essay of less than one page on (1) why the teeth are so fabricated and how that might have been accomplished in production and (2) the reason for the "heat tint." Specifically describe alternative methods by which hard-wear-resistant steel teeth could be created on a blade, citing a likely steel composition for the body of the blade and the teeth. Explain why you think the method used was chosen.

b. The mounting nuts for the cutting blade give a seasoned materials engineer (like the author) pause. A less-experienced engineer might think they were cast, which they almost surely were not! An experienced engineer would suspect they were forged, which they certainly could have been. But the intricacy of surface details and the absence of any clue relating to forging raise the possibility of fabrication by powder processing.

Prepare an essay of less than one page on deducing the method-of-manufacture for this interesting detail of the saw. Include the visual clues one would look for to hint at each of the possibilities of casting, forging, and powder processing. State what one would have to do with the nuts to provide incontrovertible evidence of one method over another, describing how each method would be identifiable. (You will need to use the Internet or a good reference on manufacturing methods; see Cited References for Chapter 10.)

CHAPTER 14

Value and Production Engineering

14-1 Manufacturability

It was stated in Section 9–1, and has been stated again and again, that *functionality* is or should be the principal goal of all engineering design, because if something doesn't do what it is intended to do, it doesn't matter that it was easy to manufacture (or construct), looked good (i.e., was aesthetically pleasing), or was economical to create and inexpensive to operate (i.e., its cost was reasonable).[1] Now, although there may be situations where *aesthetics* is especially important, such as the overall lines of an expensive oceangoing yacht as well as its interior décor, such a goal would never be more important than functionality (e.g., quietness, power, speed, seaworthiness, and reliability for a yacht). Likewise, there are absolutely situations where *cost* is a major consideration, based on what the market will bear. But, again, cost is never more important than the level of functionality expected, even if there need to be some compromises to keep costs down.

Section 9–1 also made the point that once the type and level of functionality and performance have been decided upon by the design engineer, the next most important factor is *manufacturability*.[2] To repeat what has been said before: The most creative or innovative design is of no value if the item or entity cannot be built.

---

[1] It is worth repeating: The cost associated with an engineering creation includes the cost to create (i.e., design and build) *and* the cost to own, operate, and, ultimately, dispose of the entity. The latter constitutes *life-cycle cost*. The goal with cost is affordability, by which is meant reasonable and appropriate cost, based on the perceived or target market. "Cheap" has a negative connotation, and is seldom what consumers really want or will confess they want.

[2] As used here and elsewhere in this book, *manufacturability* includes: (1) ease, quality, and consistency of the fabrication and processing of each part, component, detail, or structural element and (2) ease, quality, and consistency of the assembly of parts, components, details, or structural elements into devices, products, mechanisms, subassemblies, structures, or systems.

Several factors need to be considered relative to an item, object, structure, or system's manufacturability, including:

- Ease of manufacture (as fabricability and assemblability)
- Cost of manufacture (not including costs to operate, repair, or dispose of)
- Methods-of-manufacture (as process operations and steps or operations and sequence)
- Producibility (to facilitate faster production rates and better yield, with fewer rejections)

While each of these factors (as a subject) could merit a book, the objective of addressing them in this chapter is to indicate the role that reverse-engineering dissection could play for each. Each of these factors is therefore addressed by a separate section.

So let's begin!

14-2 Design for Manufacturability

The design stage for an object, product, structure, or system is extremely important. The majority of an entity's life-cycle costs are established or committed at the design stage. In fact, decisions made during the design stage are generally considered to contribute 70 percent of a product's final cost, with materials, labor, and overhead accounting for the other 30 percent. A careless job during design or, worse, an error, carries forward to affect all future costs for manufacturing/construction, operation, maintenance service or repair, and even ultimate disposal via recycling.

A product's design is not based solely on good design for function or purpose, but it should facilitate manufacturing as well.[3] It is not unknown (although it should be rare) that an otherwise good design in terms of function is difficult, if not impossible, to produce economically, if at all. To avoid this pitfall, a design engineer will often create a preliminary design and/or physical model (i.e., a prototype) that can be provided by engineering to the manufacturing organization for review and invited feedback. This process is called a *design review*. If not done diligently, a product could fail during manufacturing.

Design for manufacturability—sometimes referred to as *design for manufacture* (DFM)—is the general engineering practice (some might say "art") of designing products in such a way that they are easy to manufacture or construct. This practice encompasses the ready fabrication or creation of detail parts, components, or structural members, as well as the facilitated assembly of these into devices, products, mechanisms, structures, or systems; the latter being called *design for assembly*. Failure to follow DFM guidelines can result in a need for redesign, loss of manufacturing time, and longer time to market.[4] There are several good references on the topic (ref. Bralla; Poll).

A nice example is shown in Figure 14–1 for the fork end of a pneumatic piston. A variety of design variations are shown using different manufacturing approaches. The impact on recurring piece-part cost and nonrecurring tooling cost is shown, with some dramatic savings for parts made from sheet metal and molded polymer.

---

[3] *Product* is being used in this chapter to refer to any designed and built physical entity.
[4] DFM guidelines are generally, and best, developed by and provided to design engineers by the company in which they work, although there are some generic guidelines such as: (1) minimize the number of parts; (2) keep the geometry of parts as simple as possible to meet functional requirements; (3) use standardized parts and, especially, fasteners, wherever and whenever possible; etc.

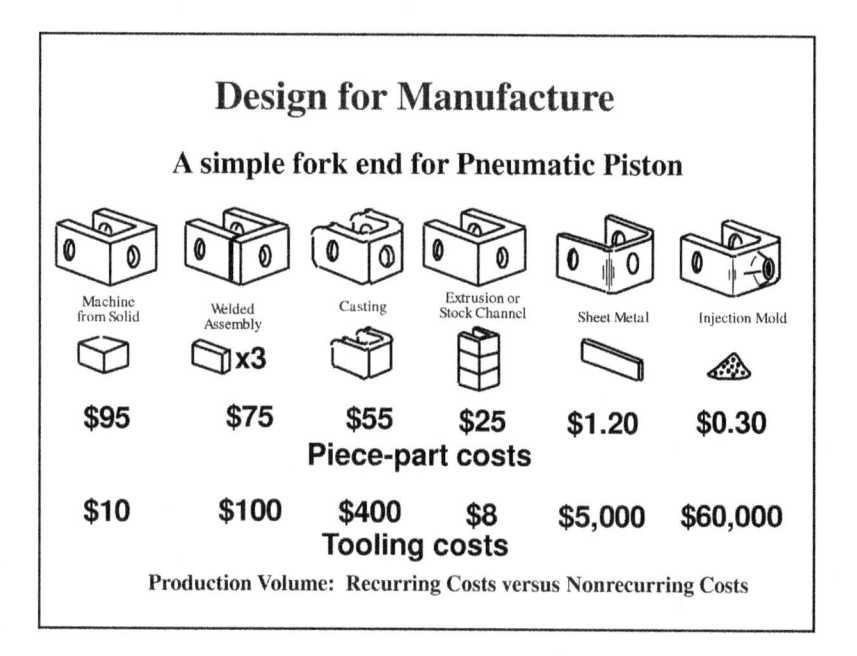

Design for Manufacture

A simple fork end for Pneumatic Piston

Machine from Solid	Welded Assembly	Casting	Extrusion or Stock Channel	Sheet Metal	Injection Mold
$95	$75	$55	$25	$1.20	$0.30

Piece-part costs

| $10 | $100 | $400 | $8 | $5,000 | $60,000 |

Tooling costs

Production Volume: Recurring Costs versus Nonrecurring Costs

Figure 14-1 Design for manufacturing options for a simple fork end of a pneumatic piston, showing recurring versus nonrecurring costs for each. (*Source:* From ME170 Design for Manufacture; used with permission of Dr. Mike Philpott, University of Illinois at Urbana-Champaign.)

Design for manufacture involves both design for process (during the fabrication stage for details) and design for assembly (during the assembly or, for construction, erection stage). Let's look briefly at each.

Design for Process

The objective of this stage of DFM is for the designer to decide upon the preferred method(s) by which parts, components, or structural elements will be fabricated, and to take the method(s) into account when deciding on certain details or features. A few examples should suffice to make the point. Readers desiring more information on this topic are encouraged to seek specific references on specific processes (ref. Campbell on casting; ref. Sheridan and Unterweiser on forging; ref. Malloy on molding). For the time being, here are some examples:

- Parts to be cast should avoid having sharp radii (to avoid stress risers), avoid drastic section changes (to preclude early molten metal freeze-off in the mold or die), avoid reentrant angles (to allow extraction of the newly cast part from the mold or die), try to avoid a requirement for close dimensional tolerances in more than two orthogonal directions (to avoid high mold or die costs), and provide some logical parting plane (to allow easy part extraction from split molds or dies).
- Parts to be forged should avoid excessive geometric complexity (to avoid high die costs), consider metal plastic flow (to avoid drastic section changes), consider "blocking" stages (to

allow progressive stages for metal movement toward the final form), provide some logical parting plane (to allow easy part extraction from split dies), and try to take advantage of metal plastic flow lines (for added strength).
- Parts to be welded should be made from weldable metals or alloys. Facilitate nondestructive inspection of internal defects (e.g., via line-of-sight paths for x-rays), and not require disassembly of the weldment.

There are many other design guidelines for other processing methods such as molding, powder processing, composite lay-up, machining, brazing, and ease of inspection via nondestructive examination.

Design for Assembly

Not surprisingly, *design for assembly* (DFA) is a process by which products are designed with ease of assembly of detail parts in mind. Design for assembly philosophy along with various rules and recommendations were first formally proposed in the 1960s and 1970s in order to help design engineers consider and avoid assembly problems largely during robotic or other automated assembly during the design stage. The methodology was greatly formalized in 1977 by Geoffrey Boothroyd at the University of Massachusetts at Amherst. Boothroyd's methodology allowed estimation of the time for automated assembly versus manual assembly, breaking assembly down into 10 steps, as shown in Table 14–1. Since every step added time and cost, as well as posed additional risk or other problems to assembly, eliminating steps became a major goal to make assembly easier.

Boothroyd went on to introduce three simple criteria that could be used to determine, theoretically, whether any parts in the assembly could be eliminated or combined (ref. Boothroyd). These three criteria are:

TABLE 14-1 A Simple Model of the Typical Assembly Process (after Geoffrey Boothroyd)

- Parts are purchased and placed in an inventory system (procurement system).
- Parts are manufactured and placed in an inventory system (manufacturing system).
- Parts are inspected for quality (quality assurance).
- Parts are presented (positioned and oriented) to the operator or automated mechanism (material handling).
- A partially completed assembly arrives at the workstation (AGV or conveyor system).
- The partially completed assembly (base) is presented to the operator or automated mechanism (fixturing).
- The part is picked up ⇒ oriented ⇒ guided into place ⇒ fit.
- A fastener is picked up ⇒ oriented ⇒ guided into place ⇒ fastened.
- The operator picks up a tool ⇒ the tool is oriented ⇒ the tool is used ⇒ the tool is set aside.
- The installed part is inspected for proper alignment, orientation, and quality of fit.

*Every step adds time and cost,
and may introduce risk or additional problems.*

TABLE 14-2 Three Keys for Easier Design for Assembly

Simplify design:
- Reduce inventory.
- Reduce number of vendor/suppliers.
- Lower material costs.
- Seek standardized parts.
- Reduce assembly time.
- Simplify assembly.

Simplify assembly operations:
- Provide a stable base (to which other parts are added).
- Strive for z-axis insertion (i.e., vertical push).
- Consider ease of handling and ergometrics.
- Seek self-locating features.
- Consider part symmetry (so there is no upside down).

Exploit material properties:
- Use moldable polymers (to incorporate snap-fit features).
- Use sheet metal (which can be bent, cut, formed to create attachment).

1. Does a particular part *move relative to all other parts* in the assembly? Only large motions need to be considered, not small movements, deflections, or hinging motions.
2. Must a particular part be *made of a different material than other parts* in the assembly, or *must it be isolated from other parts*? Only fundamental reasons concerned with required material properties are acceptable for using a different material.
3. Must a particular part be *separate from other parts* of the assembly in order to make assembly, disassembly, or maintenance possible?

If the answer to any one of these three questions is *no,* the particular part is a good candidate for elimination, with the reduction of part count being a major goal of DFA.

Together with Dewhurst and Knight (ref. Boothroyd, Dewhurst, and Knight), Geoff Boothroyd proposed three key guidelines to facilitate design for assembly, namely (1) simplify design, (2) simplify assembly operations, and (3) exploit material properties. Table 14–2 summarizes these three guidelines and provides some details.

There are many notable examples of good design for assembly, with several available on the Internet. Two particularly notable examples that might be looked at are Sony's Walkman and Swatch's watches.

At Rensselaer Polytechnic Institute (RPI), Troy, New York, the author, while technical director of the Center for Manufacturing Productivity, led a consortium of six to eight companies in the study of integral mechanical attachment using elastic snap-fit design features molded into thermoplastic polymer parts. The goal was to drastically simplify assembly particularly using robots, for which assembly motions need to be very simple. Preferred motions, in descending order of preference, were (and are) vertical (z-axis) push, horizontal (x- or y-axis) or vertical (z-axis) slide (i.e., like a hinge), horizontal or vertical slide, or combined vertical push and immediate subsequent rotation (as with the cap of a plastic pill bottle). A variety of snap-fit features that relied on elastic deflection and recovery or, alternatively, which operated by remaining rigid, were studied and mathematically analyzed for required assembly and subsequent retention forces.

TABLE 14-3 Summary of the Major Types of Integral Mechanical Attachments (by Interlocking Mechanism)

Rigid Interlocks (using rigid designed-in features):
- Tongues and grooves
- Dovetails and grooves
- Mortise and tenons
- T-slots and Ts
- Shaped rails and ways
- Wedges and Morse tapers
- Shoulders and flanges
- Bosses, lands, and posts
- Tabs and ears
- Integral keys and splines
- Integral threads
- Knurled surfaces

Elastic interlocks (using elastic deflection/recovery of features):
- Integral snap-fit features (e.g., cantilever hooks)
- Integral spring tabs
- Snap slides and clips
- Clamp features and clamps
- Interference press and thermal-shrink fits

Plastic interlocks (using plastically formed features):
- Setting
- Staking (punch and thermal types)
- Crimping
- Hemming
- Formed tabs and slots
- Beaded-assembly parts
- Metal clinching

A significant contribution to the body of knowledge of integral mechanical attachments was made by developing a unique classification scheme (ref. Messler). Table 14–3 summarizes major types of integral mechanical attachments classified as rigid, elastic, or plastic.

The advantages of elastic snap-fits, in particular, include ease of assembly via simple assembly motions and low engagement forces, ten- to twentyfold retention forces, and dramatically reduced part count by eliminating fasteners.

14-3 Value Engineering

The saying goes: "Necessity is the mother of invention."[5] So it was during World War II. America was faced with (and had to cope with) shortages of raw materials, component parts, and skilled

---

[5] While the author of this proverb is not known, it is sometimes ascribed to Greek philosopher Plato (424/423–348/347 BC). This phrase was familiar in England, but in Latin, not in English. In 1519, the headmaster of Winchester and Eton, William Horma, used the Latin phrase *Mater artium necessitas* in his book *Vulgaria*. In 1545, Roger Ascham used a close English version of "Necessitie, the inventour of all goodnesse" in his book *Toxophilus*. In 1608, George Chapman also, in

labor. In response, Lawrence Miles, Jerry Leftow, and Harry Enlicher at The General Electric Company (GE), looked for acceptable substitutes. In the process, they noticed that many of these substitutes improved the product or reduced cost or both. What began of necessity was turned into a systematic process they called "value analysis," often performed in *value engineering.*

Most often performed by industrial engineers as a project management task, *value engineering* (sometimes abbreviated VE) is a systematic method to improve the "value" of goods or products (or, more broadly, services) in terms of their function. In this context, *value* is defined (and assessed) as the ratio of function to cost. Value can be increased by either improving the function or by reducing the cost, with a basic tenet of value engineering being to preserve and not reduce function as a consequence of pursuing improvements in value.

Value engineering follows a structured thought process based entirely on "function," what something actually does, not simply what it is. The method uses rational logic, typically employing a "how" and "why" questioning technique, together with the objective analysis of function to identify cause-and-effect relationships that increase value. A key is to describe some detail of an object with such clarity that it can be described with one active verb (to describe action) and some measurable noun (i.e., a thing). An example should help.

If one considered the function of a pencil, the most terse and accurate description of its function would be "to make marks." This description allows comparison of a pencil to other things that can make marks—for example, chalk, crayon, lipstick, spray can (of paint), diamond scribe on glass, or stick in soft sand. At this point, the engineer can clearly identify and choose between alternatives that also make marks to determine which (if any) may be more (or, ideally, most) appropriate. In most cases, the goal of value engineering is to identify and eliminate all unnecessary expenses and, in so doing, increase the value of the entity for the manufacturer and/or its customers. The means of doing this is either product tear-down (Chapter 4) or reverse engineering, as the latter, involving mechanical dissection to identify role, purpose, and functionality is, in practice, the logical technique to be used.[6]

Like reverse engineering, value engineering, which has a much narrower goal (of reducing cost) than reverse engineering (of gaining knowledge and understanding), should be performed by a systematic procedure following a multistep plan originally suggested by Lawrence Miles in six steps, added to later by others to create eight steps, as follows:

1. Preparation (getting ready, formulating the goal)
2. Information (gathering background data or needs)
3. Analysis (studying the dissected entity)
4. Creation (coming up with the lower-cost alternatives)
5. Evaluation (assessing the impact of changes)
6. Development (making necessary refinements)
7. Presentation (to management)
8. Follow-up (to make appropriate changes).

---

his two-part play *The Conspiracy and Tragedy of Charles, Duke of Byron,* used a very similar phrase—"The great Mother, of all productions (graue Necessity)." But the earliest actual usage of the proverb "Necessity is the mother of invention" is sometimes ascribed to Richard Franck, who used it in his book *Northern Memoirs, Calculated for the Meridian of Scotland.*

[6] Recall the "Father of Value Analysis," Yoshihiko Sato (Chapter 5, Footnote 4).

TABLE 14-4 The Structured Thought Process Used in Value Engineering

Gather information:
1. What is being done now?
   - ✓ Who is doing it?
   - ✓ What could it do?
   - ✓ What must it do?

Measure:
2. How will the alternatives be measured?
   - ✓ What are the alternate ways of meeting requirements?
   - ✓ What else can perform the defined function?

Analyze:
3. What must be done?
   - ✓ What does it cost?

Generate:
4. What else will do the job?

Evaluate:
5. Which ideas are the best?
6. Develop and expand ideas
   - ✓ What are the impacts?
   - ✓ What is the cost?
   - ✓ What are the performance?
7. Present ideas

For purposes of this section of this chapter, the preceding list can be reduced to four steps:

- *Information gathering.* Asking what are the essential requirements of the object. The specific method for this step is to create a functional model (Section 5–6) using functional analysis in an effort to determine which functions and/or performance characteristics are most important. Table 14–4 gives the questions that need to be addressed and answered.
- *Generating alternatives (i.e., creation).* Here, one needs to ask what alternative ideas, principles, techniques, geometry, materials, methods-of-manufacture, and so on, will perform the desired functions better at the same cost or as well at a lower cost.
- *Evaluating alternatives.* Here, all alternatives generated in the previous step are objectively assessed for how well they will meet the required functions and how great the cost savings will be, or how much the *value* will be increased as a ratio of functional performance to cost.[7]
- *Presentation.* Here, the best one or two alternatives must be chosen, refined as necessary, and presented to stakeholders (management, marketing, clients, etc.).[8]

To reiterate, Table 14–4 nicely captures and summarizes the questions that should constitute a complete structured thought process during a value engineering exercise.

---

[7] Assessing functional performance requires experimentation and measurement, with the objective of quantifying it.
[8] Obviously, for complex entities, composed of many parts, components, structural elements, etc., this step can be challenging, as it really requires choosing one, two, or perhaps three alternative integrated configurations.

As a final comment, the following anecdote from the author's time working in the military aerospace industry should be of interest.

When any new design for an airplane, particularly (but not only) an advanced military fighter aircraft, has reached the point that a flying prototype has been built and test flown, an ever-present primary goal (and anxiety) for achieving performance (i.e., function) begins to shift toward a new goal to reduce costs—at least, if the aircraft performs as required. Knowing the role, purpose, and functionality of a fighter aircraft is critical, and performance at minimum weight (for high thrust-to-weight ratio) is paramount, but cost is usually secondary. The airplane must perform, so every pound of unnecessary weight becomes very valuable![9] Provided the engine manufacturer has met the thrust (or power) goals and the airframe manufacturer has kept weight down as much as practical (if not feasible), the target thrust-to-weight ratio (T/W) will be met and the aircraft will perform better than expected. If this is the situation, a serious effort of value engineering will begin.

For the Grumman F14 Tomcat (Figure 14–2a), the flight-test aircraft exceeded expectations, but so, too, had costs![10] Where extremely expensive precision castings of exotic (at the time) lightweight titanium alloys (density of 4.3 g/cm^3) seemed the only viable choice for some components (e.g., pumps, valves, fittings) based on their specific strength or strength-to-weight ratio, with weight no longer a nagging concern, structural design engineers gave a second look at similar castings produced from three to five times less expensive precipitation-hardening stainless steel alloys (with an average density of around 8 g/cm^3). As a result, dozens of pump housings, valve bodies, fittings, and so forth, were redesigned in stainless steel. The thousands of parts that make up such a complex aircraft (Figure 14–2b) provided ample opportunity, as well as a substantial challenge, for balancing function and performance against cost.

These and literally hundreds of other details were rethought and made less expensive by choosing and using a less expensive material, a less expensive process, or a less stringent (in terms of precision) design. Through this effort, as well as later efforts relating to producibility (see Section 14–4), the cost of the F14 was reduced to a more favorable level.

14-4 Production Engineering

The Business Dictionary (at http://www.businessdictionary.com) defines *production engineering* as:

> design and application of manufacturing techniques to produce a specific product. It includes activities such as (1) planning, specification, and coordination of resources [generally the responsibility of industrial engineers]; (2) analysis of producibility, production processes, and systems and (3) application of methods, equipment, and tooling [all of which are generally the responsibility of manufacturing engineers]; (4) controlled introduction of engineering changes [generally the coordinated responsibility of design engineering and manufacturing engineers via production management]; and (5) application of cost control techniques [generally the coordinated responsibility of value engineering and project and program management].

---

[9] This is why readers of news articles are shocked to hear that a certain component on a spacecraft, such as a toilet seat, cost hundreds of dollars. Low weight is worth a great deal!

[10] While readers will recall that tests A/C #1 and A/C #2 eventually crashed, the former from a design fault in the hydraulic control system, the latter from pilot error, all essential data on flight characteristics exceeded all expectations.

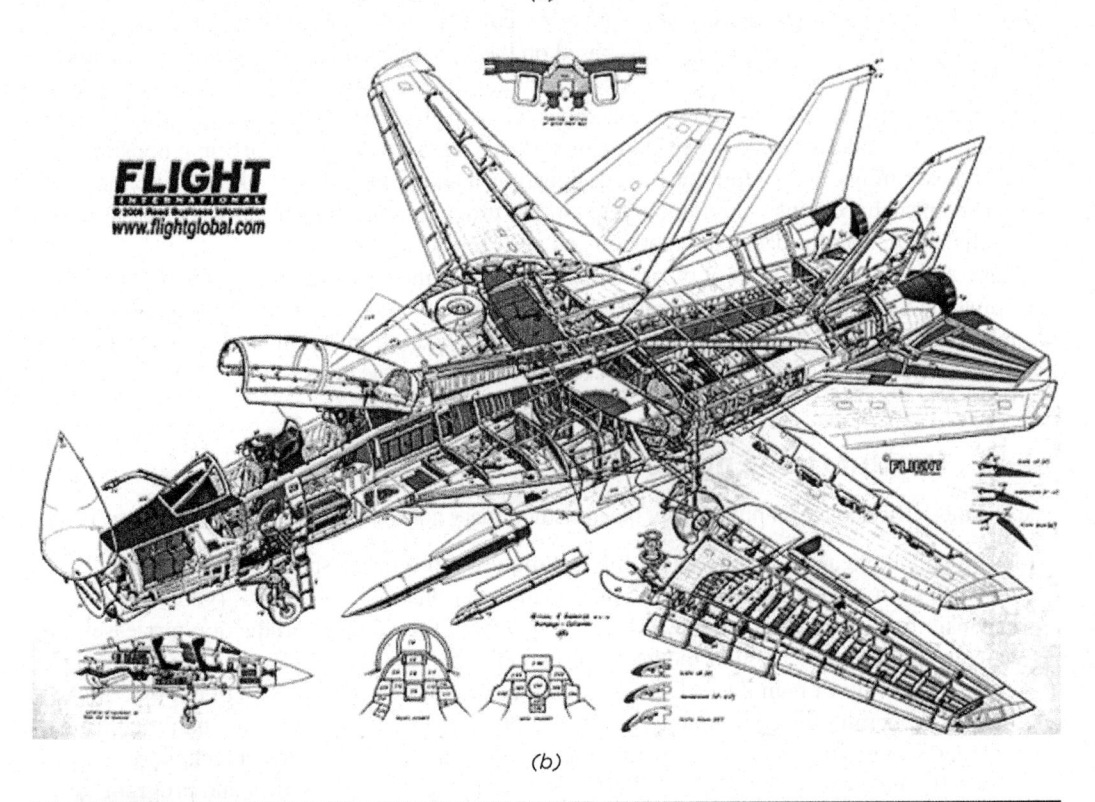

Figure 14-2 Photograph of the Grumman F14 Tomcat in flight, with wings extended during in-flight refueling (a). ***Don't miss the color version of this figure, available at www.mhprofessional.com/ReverseEngineering.*** Cutaway view schematic drawing of the F14 showing the thousands of parts that need to be designed for both function and cost (b). (*Sources:* Wikipedia Creative Commons, contributed by Dual Freq on 1 January 2008 [a]; and from Flight Global at www.flightglobal.com, used with permission [b].)

For the purposes of this book, that is, reverse engineering, two of these five activities commonly guided by the knowledge gained from reverse engineering are (1) *methods engineering* or *methodizing* and (2) *producibility*. Let's look at each, briefly.

Methodizing

While *methods engineering* covers a broad spectrum of activities directed at making manufacturing easier and more cost effective, *methodizing,* as used herein by the author, is focused on identifying the chronological serial (and/or parallel) steps needed to create a manufactured product. By studying the product's design, an experienced engineer, often with the assistance of skilled and experienced production workers familiar with the product (e.g., a civil or military airplane, an automobile, or a computer), divides the required manufacturing or production process into individual major or generic tasks, with each task adding value as the product to be produced progresses toward completion. This effort often begins with a functional model (Section 5-6).

Without prior experience with manufacture of the product, methodizing must be done by building a prototype, often by hand, and then reverse engineering the product to methodize operations.

As a very simple (and familiar) example, consider making a sausage pizza. The sequence of operations and steps, some of which could be performed in parallel by different workers, are as follows:

1. Make dough
   a. Add measured amount of flour to a bowl
   b. Add measured amount of yeast to the bowl
   c. Progressively add tepid water to the bowl
   d. Mix to create ball of dough with proper consistency
   e. Allow dough to rise (proof)

1. Make sauce
   a. Add tomatoes or tomato sauce to a pot
   b. Add measured amount of tomato paste to the pot
   c. Add seasoning (e.g., salt, pepper, garlic, oregano) to the pot
   d. Simmer for 3 to 4 hours over low heat, stirring frequently
   e. Allow sauce to cool

2. Prepare raw pizza
   a. Flour tabletop and brush olive oil onto pizza pan
   b. Roll appropriate amount of dough flat to create round or rectangular shape of desired size and thickness
   c. Place rolled-out dough on pan or stone
   d. Apply desired amount of sauce with ladle or brush; spread over dough
   e. Add desired amount of mozzarella cheese
   f. Add sausage and spices
3. Bake pizza
   a. Place prepared raw pizza in preheated oven at 500°F/300°C
   b. Bake until cheese melts and browns and crust begins to brown at edges and bottom
   c. Remove baked pizza from oven
   d. Cut into desired number of wedges or rectangular pieces
4. Eat

It should be obvious from this simple example that methodizing creates a "recipe" for producing the desired product. With prior experience with the product or a similar product, methodizing can be done from an engineering assembly drawing prior to a prototype, without requiring reverse engineering. Without such experience, a prototype must usually be built and then reverse engineered to dissect it into logical progressive operations and steps.

While the author was working as a group head in Advanced Materials and Processes Development at Grumman Aerospace Corporation in the 1970s, he served as the materials expert on a team of a half dozen industrial engineers, manufacturing engineers, and electrical and mechanical engineers to methodize a prototype segment of a linear superconducting magnet designed and hand built by a team of doctoral-level theoretical and nuclear physicists at the Brookhaven National Laboratory in Upton, Long Island, New York. Not surprisingly, these scientists (as opposed to engineers) had only a general design, which they revised as they built the magnet, making adjustments on geometry and dimensions as they went, to achieve fit. By the time they were done, they had a wonderful functional superconducting linear magnet for a proton storage ring system for use in constructing a high-energy particle accelerator for a new ISABELLE hadron collider (Figure 14–3).[11] Now all they needed were another 1599 identical 5-meter segments to complete the large circular countercirculating storage rings.

Of course, these same physicists were not going to build any more—no less, 1599 more—magnets. That job would be left to a contractor. The problems to be overcome, however, were multifold before fabrication could begin. These problems included: (1) no accurate plans, not even an accurate undimensioned rendering; (2) no meaningful dimensions, with reasonable tolerances, as (a) each part had been custom fit and, as physicists are inclined, they believed they needed the dimensional accuracy they had achieved by hand working parts to three to four decimal places in meters; (3) no bill of materials or parts; and (4) no record of the step-by-step procedure by which the prototype magnet was built. So along came Grumman Aerospace engineers, including the author.

What was required first was an accurate engineering drawing of the magnet, piece by piece. Second, a logical methodology was needed for building the magnet step by step. Third, the team of engineers had to come to some consensus, and some compromise with the physicists, as to the degree of accuracy that was actually required for the dimensions of various parts to allow function and to achieve performance while allowing realistic production of the required number of magnets at an achievable cost. The Grumman team would methodize magnet construction.

To begin, the just-completed magnet was completely and systematically disassembled, down to the smallest details and fasteners. Drawings were created for each part and for each subassembly, as well as for all major assemblies and for the overall magnetic, as the magnet was torn down (see Chapter 5). Each part was carefully measured to establish key dimensions required for form and function, and, for each such dimension, the degree of precision required for fit was also determined between the Grumman and the Brookhaven teams. Elaborate force/energy flow diagrams and intricate and complex functional models were made for major subassemblies and for the overall integrated assembly.

---

[11] ISABELLE stands for "Interacting Storage Accelerator" + "belle," for "beautiful." It was to be a 200 + 200 GeV proton-proton colliding beam accelerator partially built by the U.S. government at Brookhaven National Laboratories in Upton, Long Island, New York, and completed in France, before the project was canceled in July 1983. ISABELLE was part of an international program to develop a relativisitic heavy ion collider (RHIC).

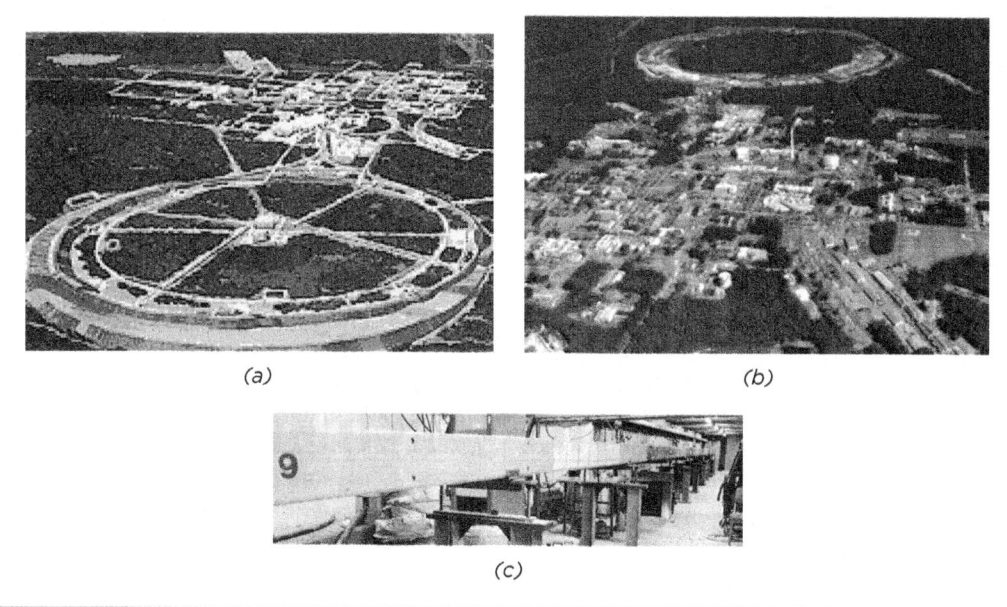

Figure 14-3 Artist's concept for ISABELLE proton-proton collider (a), early construction site (b), and joined linear segments of the superconducting magnet for the proton storage rings (c). (*Source:* www.bnl.gov, as a government national laboratory, use is free.)

Once form, fit, and function were determined by measurements and operational tests (or experiments) by the engineers, the materials used to create each part were reconsidered in an effort to (1) select materials most appropriate for each part to provide essential properties and (2) to (a) replace unnecessarily exotic and/or expensive materials, (b) replace deficient materials in terms of robustness for the planned service life of the accelerator system, and (c) substitute for any materials that unnecessarily complicated manufacture.

The next major task was for the team of engineers, led by manufacturing specialists, to decide exactly how best to make each part and each subassembly or subsystem on the way to constructing the entire magnet. Each and every operation was carefully identified and the proper sequence of operations required to create each subsystem was determined. Decisions were made as to which subsystems could be built in parallel (as separate entities), along with a system for ensuring such independently built subsystems would properly integrate to create the magnet segment.

The entire effort to methodize one segment of the large accelerator magnet took a couple of months. Reverse engineering had put the Brookhaven scientists into a position to have magnet segments built "to print," with assurance that the completed accelerator would (1) work as intended, (2) be as simple to assemble as possible, (3) be as economical as possible to build, and (4) be as robust in service as feasible. But was there a way to further simplify the planned project? Was there a way to "productionize" the manufacture of 1600 magnet segments? It was time to consider "producibility" beyond feasibility.

Producibility

With a method for fabricating each and every detail, for assembling details into devices, substructures, subassemblies, and subsystems, and for assembling these into the overall magnet established and thoroughly documented, the next task in manufacturing is to decide how to build multiple copies as cost effectively and efficiently as conceivable.[12] The task at this stage is to establish producibility.

Producibility is defined as follows by the Business Dictionary (at www.businessdictionary.com): "ease of manufacturing an item (or group of items) in large enough quantities [that economics and efficiency matter]." The Business Dictionary goes on to say: "It [producibility] depends on the design features and characteristics of that (those) item(s) that enable economical fabrication and processing (e.g., heat treatment), assembly, and inspection or testing by using available or existing processes, techniques, equipment, etc."

Whenever a new item, product, structure, or system is initially manufactured (or, for a structure, constructed), a manufacturing plan must be developed; that is, the entity's manufacture (or construction) must be methodized. But once the initial "bugs" are sorted out using early versions of the first few items of a product for which there will be many (possibly hundreds or thousands of) units, it is almost a certainty that ways will be sought to make manufacturing easier, more efficient (i.e., requiring less time with fewer rejects), and more economical. This task involves improving the item's producibility.

The starting point is a disassembled early model and the use of reverse engineering to address the "suitability" of the design (see Section 12–2). Without attempting completeness in the context of this book, here are a few areas to be assessed:

- Opportunity for reducing the number of parts (i.e., reducing part count) by their elimination or combination
- Opportunity to use standardized parts (e.g., small electric motors, gauges, valves, and, especially, fasteners)
- Opportunity to reduce or eliminate fasteners in favor of elastic snap-fit or other integral mechanical attachment features
- Opportunity to eliminate or combine operations or processes
- Opportunity to manufacture some parts, devices, mechanisms, subassemblies, or subsystems in parallel (versus serially, i.e., one after another)
- Opportunity to use automation or mechanization
- Opportunity to change materials or process(es) to increase production rate, reduce costs, and/or increase material utilization to reduce scrap losses
- Opportunity to prevent process-induced defects (to increase yield of acceptable units)
- Opportunity to ease up on unnecessarily tight tolerances

Returning to the superconducting linear magnet segments for ISABELLE, the greatest improvements in productivity came from (1) softening unnecessarily tight tolerances, (2) using standardized, off-the-shelf parts, (3) combining parts, and (4) finding opportunities for interchangeability among segments.

---

[12] A capable manufacturer can build one of anything! But the cost would be high! The trick for making a profit as a manufacturer is to build more than one with consistently high quality and at a profit. This requires productionizing.

14-5 Summary

The best design, offering the best functionality for the design's purpose, means nothing if the object, device, product, structure, or system cannot be built. The first need, and first of what will likely be but one of a few iterations to optimize the design, is to consider *manufacturability*. A key for success with the first (or early) units produced is that the engineer take into account design for manufacturability as well as design for assembly.

Once any "bugs" associated with the initial design have been worked out, often using a full-scale prototype (ref. Messler2), the steps or operations to transform raw materials, input energy, and input information (or data) to a create a suitable entity must be decided upon, including the proper sequence of operations in the overall serial flow, as well as possible parallel paths, each with its own sequential steps. This is accomplished with a process known as *methodizing*.

Reverse engineering of a competitor's product or of one's own product is the logical way to improve all three of (1) design for manufacture, including design for assembly, (2) methodizing, and (3) producibility, if one doesn't have considerable prior experience with that or a similar product.

14-6 Cited References

Boothroyd, Geoffrey, *Assembly Automation and Product Design (Manufacturing Engineering & Materials Processing)*, 2nd edition, Taylor and Francis, Oxford, UK, 2005.

Boothroyd, Geoffrey, Peter Dewhurst, and Winston A. Knight, *Product Design for Manufacturing and Assembly*, 3rd edition, CRC Press, Boca Raton, FL, 2010.

Bralla, James, *Design for Manufacturability Handbook*, 2nd edition, McGraw-Hill, New York, 1998.

Campbell, John, *Complete Casting Handbook: Metal Casting Processes, Techniques, and Design*, Butterworth-Heinemann, Burlington, MA, 2011.

Malloy, Robert A., *Plastic Part Design for Injection Molding: An Introduction*, 2nd edition, Hanser Publications, Cincinnati, OH, 2010.

Messler, Robert W., Jr., *Integral Mechanical Attachments: A Resurgence of the Oldest Method of Joining*, Elsevier/Butterworth-Heinemann, Burlington, MA, 2006.

Messler2, Robert W., Jr., *Engineering Problem-Solving 101: Time-Tested and Timeless Techniques*, McGraw-Hill, New York, 2013.

Poll, Corrado, *Design for Manufacturing: A Structured Approach*, Butterworth-Heinemann, Burlington, MA, 2001.

Sheridan, S. A., and Paul M. Unterweiser, *Forging Design Handbook*, American Society of Metals, Materials Park, OH, 1972.

14-7 Recommended Readings

Kaufman, Jerry, *Value Analysis Tear-Down*, Industrial Press Inc., New York, 2004.

Park, Richard, *Value Engineering: A Plan for Invention*, CRC Press, Boca Raton, FL, 1990.

Priest, John W., *Engineering Design for Producibility and Reliability,* Marcel Dekker, New York, 1988.

Tres, Paul, *Designing Plastic Parts for Assembly,* 6th edition, Hanser Publications, Cincinnati, OH, 2006.

14-8 Thought Questions and Problems

14-1 A good definition of *manufacturabilty* is "the extent to which a good can be manufactured with relative ease at minimum cost and maximum reliability." This may appear easier to achieve than it actually is. Succeeding requires that the definition be "dissected" to understand what is really involved.

a. *Relative ease* is an ambiguous term. (Notice that the author refrained from writing "a relative term." Well, he nearly refrained!) Relative to what?, is the question. Compared to how others do it? Compared to how it used to be done in the past? Compared to how your organization did it the last time they did it?

In a brief but thoughtful essay of about one page, address the following in a cogent narrative (versus rambling discourse or a bulleted list):
- Why might it be generally easier to manufacture a particular (i.e., nearly the same) article now than it was in the past? (*Hint:* Think about the impact of technology.)
- Why might it be easier for the same manufacturer than it was before? (*Hint:* Think about more than just technology!)
- Why might it be easier for one manufacturer than for another? (*Hint:* Don't get bogged down solely with technology.)

b. *Minimum cost* seems to be absolute. It almost certainly isn't!

In a brief but thoughtful essay of about one page, address the following in a cogent narrative:
- How would one quantify "minimum"?
- Is minimum cost, as an absolute, what a manufacturer—or a consumer, customer, or patient—really wants?
- What should a particular manufacturer mean by "minimum cost"?

c. *Maximum reliability* seems to be another absolute. Again, it almost certainly isn't!

In a brief but thoughtful essay of about one page, address the following in a cogent narrative:
- What factors associated with manufacturing contribute to reliability in a positive way?
- For each factor you identify, how does a manufacturer realistically make progress with that factor?

14-2 *Design for manufacturability* or, more simply, *design for manufacture,* involves two aspects of manufacturing: (1) design for process and (2) design for assembly.

a. Once an engineer knows how something he or she designed will be manufactured (or constructed), manufacturability can only be improved (no less, optimized) by readdressing details of the design in light of the actual planned process to be used.

Identify, describe, and explain what the designer needs to consider—and possibly modify—for each of the following general processes:

Value and Production Engineering   **333**

   (1) casting, to achieve net shape in a metal part
   (2) machining, to achieve net shape and close tolerances in a metal part
   (3) molding, to achieve net shape and intricate details in a polymer part
   (4) welding, to assembly metal parts
   b. Most people, including most engineers, would be surprised by how much (as a percentage of the total) assembly of details contributes to the cost of a final product, structure, or system. *Design for assembly* is important for keeping assembly costs under control and is essential when assembly is to be fully automated.
   In a brief but thoughtful essay of about one page, discuss the following in a cogent narrative:
   (1) What are the key factors associated with assembly being costly?
   (2) Why is it *essential* to make assembly easy when full automation is employed?

14-3 *Value engineering* is "a systematic method to improve the 'value' of goods or products or services by using an examination of the cost," while, as a key tenet, preserving and not reducing basic functions as a consequence of pursuing value improvements.
   With this definition, and key tenet, in mind, address the following in brief but thoughtful narrative answers:
   a. The role of attempting to reduce part count in assemblies
   b. The role of replacing mechanical fasteners with some other method of joining
   c. The value of "foolproof" assembly in which part orientation either doesn't matter or is clearly fixed
   d. The role of using "standardized" fasteners, parts, and/or materials (i.e., generic, as opposed to proprietary materials)
   e. The role of "make-versus-buy" decisions for parts, subassemblies, etc., that could be purchased for less than they could be made

14-4 *Methodizing* involves breaking a complicated product, structure, or system that needs to be manufactured or constructed into logical operations or steps. In modern manufacturing, often involving some level of automation or an assembly line involving small teams of workers, each step must be able to be accomplished with minimum non-value-adding movements and in a fixed (and usually short) period of time.
   Methodize each of the following as best as you can by looking on the Internet for any assistance with the object:
   a. Methodize the manufacture of your cell phone or smartphone, with a line to make each major subassembly or sub-system (e.g., the populated circuit board) and a line to assemble the overall device.
   b. What are the key factors associated with assembly that contribute to cost?
   c. Methodize the manufacture of your computer desk or another "easy-to-assemble" piece of furniture.
   d. What are the key factors associated with assembly of the item you chose to methodize that contribute most to cost?
   e. How were assembly costs of the items in parts (a) and (c) minimized?

14-5 Once a product, etc., has been in production for some time (usually a short time), to "iron out wrinkles" or "work out bugs," a manufacturer commonly seeks to improve *producibility*.
   Choose a manufactured device, product, object, or item in your possession or residence and mentally (if not physically) "dissect" it to allow examination of component parts

and other details, to suggest, in an essay of about one page, how it could be made more producible. (*Hint:* It might be easier to do this for a product for which the production volume is relatively low now but would need to be dramatically increased. Another possibility is to consider a product that is probably assembled manually now but which it would be preferable to assemble using automation.)

CHAPTER 15

Reverse Engineering Materials and Substances

15-1 Flattery or Forgery

Japanese fashion designer Yohji Yamamoto (1943–) once said: "Start copying what you love. Copy, copy, copy. At the end of the copy, you will find yourself." Really? Not for academicians, who'd find themselves out of work, having had their tenure revoked! No transgression is more serious for an academician than plagiarism. Long before Yamamoto, Charles Caleb Colton (1780–1832), an English cleric, writer, and collector, well known for his eccentricities, made one of his more famous quotes when he said: "Imitation is the sincerest form of flattery."[1] Perhaps. But that's not saying much for originality!

Whenever reverse engineering is abused to create artless copies, usually without proper attribution of origin, the ethical engineer should be angered. After all, the artless copy could have been of his or her original creation. And, while inappropriate (and, often, unethical) use of reverse engineering occurs in too many instances (see Section 2–3), nowhere does it seem to be more rampant than in the area of materials and substances.

The dictionary defines a *material* as "a substance or substances out of which a thing is or can be

---

[1] The actual quote is: "Imitation is the sincerest [form of] flattery," with "form of" being inserted, erroneously, but fittingly, by others more anonymous even than Colton. Colton's books (e.g., *Lacon, or Many Things in Few Words,* 1820, and *Lacon, Vol. II,* 1822) included collections of "epigrammatic aphorisms" (i.e., original thoughts, spoken or written in a concise, clever, amusing, and memorable form) and short essays on conduct. Though now mostly forgotten, these were phenomenally popular in their day. In fact, Colton may have been following his own advice when he made the quote. The first of alternate versions of this well-known proverb is found in a biography of Marcus Aurelius (AD 121–180), *Emperor Marcus Antoninus: His Conversations with Himself,* 1708, by Jeremy Collier and Andre Dacier: "You should consider that imitation is the most acceptable part of Worship, and that the Gods had much rather Mankind should Resemble than Flatter them." A closer version appeared in the English newspaper *The Spectator*, in 1776, when Joseph Addison wrote: "Imitation is a kind of artless flattery."

made," in essence, "a component or constituent of matter," as a *raw material*. The same dictionary[2] defines *substance* as "a material of a particular kind or constitution," or, alternatively, "a specific type of matter; esp. a homogeneous material with definite composition." These definitions are difficult to differentiate. So, too, are the terms *material* and *substance* difficult to differentiate. But let's forgo the dictionary for some definitions by a materials engineer—at least, by one materials engineer: the author!

In the context being considered in this chapter of this book, *materials* and *substances* are differentiated thus:

A *material* is a solid with a structure that determines its various properties, mechanical, electrical, thermal, optical, magnetic, chemical, and so on (see Section 9–3). As a solid, the material has a particular volume for a given quantity of mass, the volume has a shape of its own (i.e., not defined by any container), and the shape can resist or support an applied load or force without any change, up to some limit. The structure begins at the level of atoms and is characterized by the way in which those atoms are bonded to one another (e.g., by ionic, covalent, or metallic strong primary bonds, or by much weaker secondary van der Waals bonding).[3] At the atomic level, the structure may exhibit a particular repeating arrangement over many tens, hundreds, thousands, or much greater atom distances; in other words, it may exhibit long-range order, or not.[4] The former materials are known as *crystalline materials* and include metals, ceramics, many semiconductors, and some polymers. The latter materials are known as *amorphous materials* and include glasses, some semiconductors, and many polymers.[5] Materials can consist of only one atomic species (i.e., they can be elemental) or they can consist of either random mixed, reacted, or ordered mixed combinations of two or more atomic species (e.g., alloys, ceramic or intermetallic compounds, or ordered alloys, respectively) or simply randomly mixed atomic species (e.g., normal or nonordered alloys and glasses[6]). Solid materials of particular interest to engineers are known collectively as *engineering materials,* as they are capable of being reproduced to give the same set of functionally specific properties. When two or more different materials (two different metals or alloys, a metal or alloy and a ceramic or a polymer, a metal or alloy and a ceramic and a polymer, etc.) are combined to create a specific set of properties in an engineered mechanical mixture, these are known as *composite materials.*

Besides having a structure at the atomic level from which properties arise, engineering materials have a nanoscale (i.e., 10^{-9} to 10^{-7} meters) and/or a microscale (i.e., 10^{-6} to 10^{-3} meters) structure from which other properties arise. These nanostructures and/or microstructures can be manipulated to produce a specific set of properties via processing (e.g., at the material producer's site or during manufacturing).

A *substance,* in the context of this book, is often a liquid, occasionally a semisolid (e.g.,

---

[2] As stated before in this book, definitions presented all come from *The Free Dictionary* at www.thefreedictionary.com.
[3] Some solid materials, known as *polymers,* are composed of large, long, chainlike molecules—or macromolecules—that are, in turn, composed of atoms. The solid material, or polymer, exhibits properties that arise from the way these macromolecules are bonded to one another and entangled with one another, as well.
[4] When there is no long-range order, there may be still be some short-range order (e.g., up to several atom diameters), or there may not be even much short-range order.
[5] It is possible for some polymers to exhibit a mixed structure with interspersed crystalline and amorphous regions. These are known as *semicrystalline* polymers.
[6] In fact, the different atomic species that make up a glass are not entirely random, as there is a basic imperfect network, often based on covalent bonded Si atoms, and various other atomic species interspersed within or among the network. Over some range, a particular composition would be found to exist from one location to another within the glass.

viscoelastic material such as putty or Jell-O), or, sometimes, a solid with a particular chemical formulation[7] but with no discernable, significant, or important structure. Useful properties or, more popularly, characteristics, arise from the chemical formulation much more than from the structure, in most cases. Examples of *substances,* in this context, include flavors or flavorings (e.g., vanilla, strawberry, cola), fragrances (e.g., vanilla, lavender, banana, jasmine) and perfumes (e.g., Chanel No. 5, Dolce & Gabbana Pour Homme), soft drinks (e.g., Coca-Cola, Mountain Dew), sweeteners (e.g., saccharin, Splenda), fermented beers and ales (e.g., Coors Lite and Dinkel Acker), distilled whiskeys (e.g., Wild Turkey and Glenfiddich 18), honey, Greek yogurt, lemon Jell-O, Silly Putty, over-the-counter drugs (e.g., buffered aspirin, Vick's VapoRub), legal pharmaceuticals (e.g., Lipitor, Viagra), and many more.

From this list, it should be apparent that most substances are developed as commercial products for the general consumer market, with formulations being highly proprietary to maintain competitive advantages. This is distinctly different than is the case for most materials, particularly for engineering materials. These are developed by specialty producers of materials (e.g., metals companies like Alcoa, Nucor, and TiMet; ceramics producers like Coors and Kyocera; and polymer producers like DuPont, Dow, and Goodyear) to specifications of chemistry or, alternatively, properties.

As will be shown in the next section, both engineering materials and substances of all kinds have been and continue to be the subject of reverse engineering, usually for profit and not for flattery, and, surely, not to allow copycats to find their own designs.

15-2 Motivations for Reverse Engineering Materials and Substances

The motivations for reverse engineering are many and varied (see Section 2–3), with most being both legal and ethical (see Chapter 17). It should come as little surprise that materials, as engineering materials, and substances, largely as consumer products, have been the object of reverse engineering for many of these motivations, legal and ethical, legal but of questionable ethics, outright unethical, and outright illegal.

Materials, particularly those deemed especially important for engineering (e.g., for a needed set of functionally specific properties and for optimum performance), have been the subject of reverse engineering as part of military espionage, which has been routine since the first organized battles between sovereign armies. Three modern examples include new high-temperature alloys for use in the jet engines of fighter and supersonic bomber aircraft,[8] advanced ceramic and composite ballistic armor for battle tanks, and radar-absorbing materials (RAM) for coatings on skins of stealth aircraft. Reverse engineering has also been used as part of commercial or industrial espionage, with modern examples being monitoring of new environmentally friendly

---

[7] The use of *formulation* in this chapter and book is intended to indicate a more sophisticated situation than simple chemical composition. For example, formulations depend greatly on the specific form (blended elemental metals or a master alloy, pure reagent-grade compounds or natural compounds or minerals, etc.) and source of the raw materials or ingredients.

[8] Such alloys are commonly referred to as *superalloys,* for their "super" mechanical strength and resistance to oxidation at elevated temperatures. In order of increasing temperature capability, these have included Fe-, Ni-, and Co-based types. Next-generation alloys are being developed based on near-refractory Nb (melting point of 2469°C/4476°F versus 1538°C/2795°F for Fe, 1455°C/2651°F for Ni, and 1495°C/2763°F for Co, as well as thermal-stable oxide dispersion strengthened Fe and Ni ODS superalloys.

(i.e., lead-free) solders to appear in commercial electronics, high-strength/low-weight composites for Olympic bobsleds, and high-stiffness/low-weight masts for America's Cup sailing yachts. Use of reverse engineering for military espionage is probably never legal, as objects to be dissected are always obtained surreptitiously, but it is seldom ever considered unethical by the engineers involved. After all, what would one not do for the security of one's country? Use for commercial espionage, on the other hand, may not be legal, depending on how a competitor's product is obtained, and may or may not be considered unethical by an engineer, depending on the circumstances.

In a similar vein, materials have frequently been the subjects of more open competitive technical intelligence gathering as one company dissects a competitor's product to determine what new materials might be being used. Common examples abound in medical products and devices, including when artificial replacement joints shifted from 304 or 316 austenitic stainless steel to new titanium alloys and when Teflon and then ceramics appeared in the acetabular (ball-and-socket) components of replacement hip joints. In most of these cases, reverse engineering is legal, provided objects subjected to dissection were obtained by legitimate means (see Section 17–3), with the ethicality being dependent on the motivation and ultimate use of knowledge gained.

There have also been examples of illicit use of reverse engineering for creating unlicensed or unauthorized copies of competitor's products. A list of various uses, some of which are legal and ethical and some of which may be or are not ethical or, perhaps, legal, includes:

- Substitutes
- Replacements
- Imitations
- Copies
- Look-alikes
- Knockoffs

While sometimes used inconsistently, and with some overlap, here are reasonably well-accepted definitions of each of these terms as goals, as well as some familiar examples:[9]

- *Substitutes:* The plural form of "one that can take the place of another," perhaps with some discernable difference(s) but usually with no unacceptable sacrifice or compromise of capability or performance. Two good examples are sugar substitutes and margarine (or oleomargarine), the former of which the sugar industry prefers be called "artificial sweeteners" and the latter of which the American Dairy Association prefers be called "butter substitute."[10] A key for substitutes is that they are usually intended to offer an alternative with some perceived benefit. For sugar substitutes it's little or no calories and freedom from carbohydrates, while for the margarine it's healthful vegetable (versus animal) fats and, in some cases, less cholesterol.
- *Replacements:* The plural form of "one that replaces another," with discernable differences

---

[9] All definitions, in quotation marks, are from www.thefreedictionary.com.
[10] Interestingly, when oleomargarine first appeared on the American market, it came in a soft plastic sleeve with two partitioned compartments. One contained the stark white oleomargarine, the other a school-bus-yellow dye, which had to be kneaded into the white margarine to give it a more realistic "butter yellow" appearance. The American Dairy Association fought to have the dye changed to an intense Kelly green, so that the resulting mixture would, in their words, "be distinguishable from real butter." Of course, the real motive was for the consumer to be faced with an unappetizing green slime!

from the original and often some sacrifice or compromise of capability or performance that may be unacceptable or, at least, nonoptimal. Two good examples are replacement hip joints and replacements for natural asbestos, a known carcinogen. There is no question that the best artificial hip (or other) joint is never as good as a healthy original, and, in fact, no replacement has been found for natural asbestos with anywhere near the insulating quality of natural asbestos.[11] Replacement materials are often sought to overcome depletion (such as a replacement for oil), extinction (such as a replacement for natural elephant tusk ivory for piano keys), or scarcity (such as a replacement of cobalt-based alloys in the United States, which has no natural resource in cobalt, which, in superalloys, would be a strategic material). The key difference between *substitutes* and *replacements* is found in the words "can take the place of" in the definition of the former. A substitute can take the place of the original with the user usually not being troubled, as the substitution is often by choice. Not so for replacements!

- *Imitations:* The plural form of "something derived or copied from an original," rarely with any attempt to deceive the user, although perhaps intended to deceive others. Good examples are imitation mink coats, imitation diamonds, and imitation or artificial vanilla. The minks may be happy with a faux mink coat but would never mistake one for a former relative. Likewise, the fashion-conscious, but frugal, socialite might be happy with her 10-carat cubic zirconia faux diamond ring, and her casual acquaintances might not recognize it for what it isn't, but it falls far short of a real diamond for many properties, not only sparkle. Budget-conscious cake bakers are more than satisfied with imitation vanilla, with a single real vanilla bean costing more than a good porterhouse steak. Imitations are often used as a much-lower-cost alternative.

- *Copies:* The plural form of "an imitation or reproduction of an original; a duplicate," which, while very similar, almost always lacks some quality or qualities (albeit, perhaps of secondary importance) possessed by the original. This term is the most difficult to define unambiguously, as it is used to denote what several other terms denote better (e.g., imitations, replicas, reproductions). The best example of a copy is a xerographic copy of a document, which possesses the essential features of the original but lacks details requiring higher resolution. There were 26 paper copies made of the Declaration of Independence, the original of which was hand written in ink on parchment. Known as the "Dunlap broadcasts," they were printed the night of July 4, 1776, with the intention of being read to the pubic in various locations around the colonies, making them contemporaries of the original, but copies nonetheless. None were ever intended to defraud, and, today, 21 copies are held by American institutions, 2 copies are held by British institutions, and 3 copies are held by private collectors.

An important type of copy that began to flourish in the 1970s is a clone. *Clones* are defined in the dictionary, albeit in the singular form, as "one that copies or resembles another, as in appearance and/or function." The most familiar example (discussed in Section 17–4) involves clones of IBM's first personal computer (PC), the IBM 5150. Known, quite specifically, as "IBM clones," these were copies created by emerging competitors anxious to cash in on an emerging market boom. The way they were created was for the imitator to dissect an IBM 5150 and reverse engineer the design. The process had been facilitated, unknowingly, by IBM when it used off-the-shelf components, most significantly the Intel 8088 CPU (computer processing unit). IBM had

---

[11] Not surprisingly, other fibrous minerals resembling asbestos in appearance and insulating character are also carcinogens. After all, if it looks like a duck, walks like a duck, and quacks like a duck, it's probably a duck. Or it's enough like a duck that it has other characteristics of a duck, some of which might not be endearing—such as that it poops like a duck.

filed a patent for their design and subsequently set up a complex net of legal restrictions, to little avail. It turned out (after numerous attempted lawsuits by IBM) that nothing legally prevented would-be competitors from obtaining a 5150 on the open market, dissecting it, and copying it, perhaps with some original features or characteristics—so long as the copycat didn't attempt to pass off their PC as IBM's, which they did not do, anxious to have their own name become familiar to a rapidly growing body of consumers eager to have a PC, any PC!

- Another important and valuable example of copies is so-called generic drugs. A generic drug replaces or competes against a drug for which there was a patent (albeit expired once the generic appeared). Every drugstore chain (CVS, Walgreen, Rite Aid, etc.) offers its own generic versions of over-the-counter, nonprescriptive drugs such as buffered aspirin, ibuprofen, multivitamins, and antihistamines. Large pharmaceutical firms offer generic versions of prescriptive medications made famous by the original, such as generic statins for Lipitor and generic versions of Viagra. More is discussed about *generics* in Section 15–4. An impressive cross-listing of generics and brand-name pharmaceuticals is given at www.needymeds.org/generic_list.taf.

- *Look-alikes:* These are no less than their name implies. They are intended to look like a famous or popular brand-name product; however, there is no attempt to mislead or deceive the consumer. Christian Louboutin (1963–), the French designer of women's footwear, set the world of high-end women's shoes on fire with the introduction of shoes with shiny, red-lacquered soles that became his signature, as well as a symbol of haute fashion tastes (search the Internet for innumerable examples under "Christian Louboutin shoes"). With retail prices starting at around $700 and easily reaching $2000, low-cost look-alikes soon appeared. Fortunately for the average woman desiring to appear fashionable but without the resources to afford the best, there are many low-cost imitators and their look-alikes that can be found with a search of the Internet. Some market their wares as "budget-conscious," while others more blatantly proclaim their offerings as "chic for cheap."

Such copycat behavior may be artless, as it reflects both Yohji Yamamoto's and Charles Colton's quotes at the beginning of this chapter, but it is (1) very profitable and (2) perfectly legal—so long as there is no attempt to deceive the buyers into believing they are purchasing an original. The goal is to fool the casual observer of a young woman strutting down Fifth Avenue in New York City in her red-soled pumps, not the buyer. You need to decide whether what was done was ethical, as it can differ from situation to situation.

The world is full of people who knowingly buy look-alike products, from brightly patterned Vera Bradley look-alike handbags online to look-alike Coach bags with their pattern of giant "C"s at outlet store complexes, to look-alike Gucci faux leather handbags from natives of Africa along the dock at St. Mark's Square in Venice. Look closely, though, and you'll see the real manufacturer of a red-soled Louboutin look-alike was *not* in Paris, France!

No substance has been the victim of more reverse-engineering attempts than Coca-Cola. Originally created as a patent medicine by John Pemberton in 1886, Coca-Cola was brought out as a brand-name product in 1888 by Asa Griggs Candler. His marketing tactics with Pemberton's cocaine-laced, flavorful sparkling elixir led to Coke's dominance of the soft drink industry worldwide. Dozens, if not hundreds, have tried to buy, steal, or reverse engineer Coca-Cola with limited success. After all, maybe no other cola soft drink tastes like Coke, but enough come sufficiently close to have garnered a healthy profit, even if only a small market share. The moral of this story is *not* "Copy, copy, copy" à la Yamamoto, but "A small percentage of a very large number yields a large number." Coke imitators, while trying to taste alike as opposed to look alike, have made billions off Pemberton's secret elixir.

So long as the buyer knows that what he or she is buying is a look-alike, no one gets hurt. But this is not always the case . . .

- *Knockoffs:* The plural form of "an unauthorized copy or imitation," especially a cheap copy of a costly original. The three key differences between a knockoff and a look-alike are deception, deception, and deception. Rather than just trying to look like the original, but without misrepresenting the copy as the original, a knockoff attempts to faithfully reproduce superficial, but highly visible, appearances and mislabel the fake as the original. A vivid example seen by the author was the subject of a featured story on *60 Minutes* by the late Mike Wallace.[12] Wallace showed a full set of three woods, seven irons, and a putter made by the highly regarded, and expensive, Callaway Golf Company, in a fashionable Callaway bag, with signature Callaway head covers for all of the clubs. The problem was, they were *not* made by Callaway but, rather, by an unknown manufacturer in the People's Republic of China. Wallace had purchased them—and a duplicate set—for "under $300 each." Knowledgeable—and envious—golfers know that a Callaway driver alone can exceed that price![13] What's the story?

Wallace had two sets of Callaway knockoffs, one of which he brought to Callaway's headquarters, where he met with executives, marketing managers, and design and manufacturing engineers without telling them the purpose of his visit. All of them confidently and proudly acknowledged the clubs to be theirs. But Wallace knew they were not, and he told them so, at which point he took a #1 driver, a #3 wood, and a #7 iron to Callaway's test facility. After having a robot drive 50 balls with each of Wallace's clubs and the company's own clubs, they compared scatter patterns. To their amazement and dismay, the PRC knockoffs averaged slightly shorter range for each club but, more important to most golfers, a comparable pattern of scatter about a straight line down the center of the range.

Complete destructive dissection of clubs donated by Wallace revealed a remarkable job of reverse engineering by the Chinese. While several materials and manufacturing methods differed between the genuine Callaway clubs and the knockoffs, engineers were astounded! Of course, the hunt for the culprit of the scam had shifted from being Wallace's to being Callaway's task.

The moral: Caveat emptor! ("Let the buyer beware!"). In the modern world of computer-based design and manufacturing, including rapid prototyping (ref. Messler), little is safe from reverse engineering to create an artless copy by unscrupulous people. Some may call what they produce and offer "replicas," but these are often intended to fool the buyer and the viewer alike. Replica watches that blatantly copy the work of world-renowned Swiss manufacturers, regrettably, abound. Knockoffs can be found under brand names such as Rolex (Figure 15–1). But look closely at form and fit, and particularly materials and workmanship, and the sham is revealed.

---

[12] *60 Minutes* is a weekly magazine-style news program produced by CBS. Myron Leon "Mike" Wallace (1918–2012) was an American journalist known for his straightforward, no-nonsense, in-your-face probing interview style.
[13] A Callaway RAZR Fit Xtreme Driver listed for $399.99 on the Internet in 2013. So, what Mike Wallace paid for an entire set of knockoff Callaways was what one would pay for a single real Callaway RAZR Fit Xtreme Driver!

342 Chapter Fifteen

(a)   (b)

Figure 15-1 Photographs of a genuine Rolex watch, one of the most highly esteemed watches in the world, here with a blue dial face and gold and stainless steel bracelet retailing for between $10,000 and $13,000 (a) and an unscrupulous knockoff, which, more than imitating the look of the original, attempts to pose as the original by using the Rolex name and signature logo and the untrue imprint "Swiss made," here with an advertised price of about $150 (b). While the website mentions this is a "replica," nothing on the watch makes this apparent. (*Sources:* Wikipedia Creative Commons, contributed by Dosto on 10 September 2007 [a]; and from the website http://rolex.china-direct-buy.com, without permission, as no contact could be located, despite several attempts. There is no attempt to circumvent any copyright that might exist [b].)

Before leaving this section, two other terms merit defining: *improvements* and *proprietary materials*.

An *improvement*, whether of a material or substance or other product, is a refinement of an earlier model for some intended change for the better. The improvement may be in the form of expanded function, extended performance, greater robustness in service, more favorable (usually lower) cost, or better appearance. The key point is: Something should be discernably better. One of the primary motivations for reverse engineering, particularly when used by a designer/manufacturer on their own product, is to learn how that product can be improved, in some fashion.

A *proprietary material* (or, more broadly, a proprietary product) refers to a material that is "exclusively owned by a private individual or corporation under a trademark or patent." Many materials begin their life, as they enter the marketplace, as proprietary in order to give the creator/

developer a sales edge for some period of time, as a way of recuperating money invested in the material or product. Pharmaceutical firms are renowned for introducing proprietary drugs for which they hold a patent and, hopefully, market exclusivity, for some period of time. But material producers also create and market proprietary alloys, for example. Particularly good examples are alloys marketed under the trade name Inconel by the International Nickel Company (INCO). One specific example is INCONEL 600, a Ni-Cr-Fe solid-solution strengthened alloy offering good corrosion and heat resistance with excellent mechanical properties and good workability. Once the patent on this alloy expired, product with exactly the same composition became widely available as Alloy 600, which is not proprietary, as it is now produced by several metal producers under the international designation UNS N06600.

While Alloy 600, product form for product form, is the same for all producers, the author thinks there is an advantage to buying INCO's INCONEL 600: No one knows more about it, so no one is able to provide the breadth and depth of technical support. As seasoned engineers know, and young engineers learn, when it comes down to it, it's all about access to strong technical support from material producers!

Table 15–1 summarizes the preceding terms.

TABLE 15-1 Summary of the Various Motivations for Reverse Engineering Materials and Substances

Motivation Type	Key Characteristics	Examples
Substitutes	To be able to take the place of an original, with some perceived benefit to the consumer	Sugar substitutes Salt substitutes Wood veneers
Replacements	To replace an original for some purpose, with some recognizable sacrifice/compromise for the consumer or user	Replacement joints Asbestos replacement Co-free hard-facing alloy
Imitations	Derived or copied from an original, usually for a cost savings, but not to fool the user	Imitation flavors Imitation diamonds (e.g., cubic zirconia)
Copies	Reproductions of an original or a duplicate, which almost always lack some details; usually not meant to defraud the consumer	Xerographic copies Replica cars (e.g., Ford Model Ts)
Clones	Virtually identical copies at lower cost	PCs
Generics	Exact copies at lower prices	Drugs; packaged foods
Look-alikes	Close facsimiles to appear like the original at lower cost; rarely deceiving the consumer	Faux furs Faux perfumes Faux fashions
Knockoffs	Cheap copies, often, but not always, intended to defraud the consumer, by using brand name and logos illegally	Cheap replica fine Swiss watches Cheap brand-name golf clubs Cheap designer handbags

15-3 Finding Substitute and Replacement Substances and Materials

There are circumstances and situations that demand that a substitute or replacement substance or material be found. Recall that a *substitute* is generally fully able to replace an original, perhaps with some discernable differences, but without any unacceptable sacrifices or compromises for the user. In fact, a user often elects for a substitute even though an original may be available. A *substitute substance* or *material* might provide a cost savings without any loss of needed functionality or performance, or it might, in fact, offer some advantage not offered by the original version.

Examples of substitute substances (i.e., sugar substitutes and margarine versus butter) were provided in Section 15–2, but there are many others such as low-calorie soft drinks and sugar-free cookies (for dieters and diabetics) and light (lower-calorie) beers or nonalcoholic beers. As far as substitute materials are concerned, the same is true: the substitute is discernably different *but* offers some perceived advantage over the original. So-called plastic lumber, made from recycled polymer waste, feels and nails differently than wood but offers much better environmental durability (i.e., it doesn't rot). It is not a shortage of real wood, nor is it lower cost (because the cost of plastic wood is often higher than that of real wood) that leads to the choice of plastic wood over real wood *but* some property (e.g., rot resistance) or feature (e.g., social and environmental consciousness to recycle waste) perceived as valuable or useful by the consumer.

Replacement materials (as well as substances) are created based upon a different motivation. They offer discernable differences and, almost always, come with some sacrifice or compromise in functionality or performance that is often seen as unacceptable but inevitable. The driving force for creating a replacement material is often loss of access to the original material. This might be depletion of a natural resource (as when the world finally runs out of oil, as it will inevitably do at some point in the future, albeit much further in the future than once touted and shouted) or extinction or risk of extinction (as when mammoth, whale, and elephant ivory, tortoiseshell, rhinoceros horn, whale oil, etc., became illegal on the world market) or that a material is simply too scarce to be viable.

Two examples of how scarcity can or could force replacement materials are (1) replacements for cobalt superalloys in the United States and (2) potential shortages of rare-earth metals in the United States, with no viable replacement. In the former case, the United States has no natural domestic resource of cobalt, so it was forced (or, perhaps more logically, decided) to seek replacements for cobalt-based and cobalt-bearing superalloys and specialty (e.g., high-speed) tool steels. With a growing dependence on securing cobalt from unfriendly and unstable sources (e.g., warlord-ravaged African countries) and nervousness about a forced shortage of what was becoming a strategically essential material, the United States sought cobalt replacement materials. In the latter case, with only one U.S. mine with resources of mixed rare-earth metallic elements used in magnets essential for the operation of innumerable computer and other electronic devices, the United States has become potential hostage to the PRC, which has virtually cornered the market on these strategically essential metals. What will happen remains to be seen—and resolved!

A key point about finding substitute and replacement substances and materials is that the logical way to proceed often involves some reverse engineering. For example, based on what the role, purpose, and functionality of the original were, what might be substituted or replaced, and with what sacrifice, compromise, and/or cost?

15-4 Creating Generic Materials (Generics)

The dictionary defines the word *generic* as "relating to or descriptive of an entire group." The term can be applied anywhere or to anything for which this definition holds but is surely best known in the widely recognized term *generic brands*. These are found in virtually every supermarket in the United States. Often, generic brands are identified by rather plain containers, with rather plain labels, free of any major brand's logo (e.g., Campbell's familiar red-and-white label with gold medal logo and black signature). Besides Kellogg's, Post, and General Mills breakfast cereals, the supermarket chain will often have shelf space devoted to its own brand of familiar breakfast favorites like flakes made from corn flour, oat cereal with holes like tiny donuts, bran flakes with raisins, and artificial fruit- or chocolate-flavored and overly sweet corn puffs. They will offer similar store-brand canned vegetables; fruits; soups; cardboard containers of whole, 2 percent, 1 percent, and skimmed milk; salt; wash detergent; facial and toilet tissue; and paper towels. Most even have their own, generic, brands of buffered aspirin, antibiotic ointment, multivitamins, and bandages with nonsticking pads (i.e., generic versions of Band-Aids).

Generic "Band-Aids" are an excellent example of how an original product can become so successful and familiar that it comes to represent to consumers anything that closely resembles it. *Band-Aids* is actually the trade name still held by Johnson & Johnson, an old and highly regarded multinational manufacturer of medical products, principally (but not exclusively) consumables. Much to Johnson & Johnson's dismay, its patent on its most familiar product expired long ago. However, so familiar is J&J's product that anything that now resembles it is referred to as "Band-Aids." Similar things have happened with "Coke" in reference to a carbonated cola soft drink, "Kleenex" in reference to soft paper facial tissues, and, in Grandma's and Grandpa's heyday, "Frigidaire" in reference to a refrigerator[14] (Figure 15–2).

While certainly the most flattering form of imitation, having one's brand name become a household word doesn't ring up on the cash register or show up on a company's bottom line.

Perhaps the most familiar—and lucrative—use of generics occurs with licit drugs, including nonprescriptive over-the-counter and prescriptive types.

A generic drug ("generics," for short) is "a drug product that is comparable to brand/reference listed drug product in dosage form, strength, route of administration [oral, suppository, transdermal, or injectable], quality and performance characteristics, and intended use." It has also come to refer to any drug marketed under its chemical name without advertising. While they may

---

[14] The first self-contained refrigerator was invented by Nathaniel B. Wales and Alfred Mellowes in 1916 and was offered by their Guardian Frigerator Company located in Fort Wayne, Indiana, soon after. The name "Frigidaire" was adopted when William C. Durant, one of the founders of General Motors, personally invested in Wales and Mellowes's new company, or, really, their invention. GM bought the company in 1919 and produced "Frigidaire" appliances, starting with the novel refrigerator, until 1979. The Frigidaire division of GM's Delco-Light subsidiary, and, later, independent division operated out of Dayton, Ohio, was purchased by the White Sewing Machine Company (later to become White Consolidated Industries) in 1979, which was, in turn, acquired by Electrolux in 1988. The actual Frigidaire self-contained refrigerator was extremely popular, as was the "Frigidaire Girl," who was an early example of the use of sex in advertising (Figure 15–2).

Figure 15-2 A photograph of a 1927 advertisement for a Frigidaire, the first self-contained refrigerator marketed in the United States, and the famous "Frigidaire Girl." (*Source:* From the Kettering University Archives, with their kind permission through David C. White, Archivist.)

not be associated with a particular company, generic drugs are subject to the regulations of the relevant government agencies of countries where they are produced and dispensed. Generic drugs are labeled with the name of the manufacturer and the adopted, nonproprietary name of the drug. It is essential that a generic drug contain the same active ingredient(s), by chemistry and dosage.

From a legal standpoint, generic drugs can be produced without patent infringement when: (1) the patent has expired; (2) the generic company certifies that the brand company's patents either are invalid, are unenforceable, or will not be infringed (all of which may need to be proven in court); (3) for drugs which never held patents; or (4) in countries where the drug does not have current patent protection.

The obvious benefit of generics to consumers is lower cost, which is enabled by the drug having been copied, without the very real burden and substantial cost of research and development, including, in the United States, Food and Drug Administration (FDA) approval. Another advantage can be greater availability, as the original manufacturer either may not have production capacity or may choose to limit supplies as a way of controlling or justifying price.

How does a generic arise? The short answer is: via reverse engineering. But this is easier said than done. It's one thing to copy a cornmeal-based flake (which is a *substance* in the context of this chapter) or a "Band-Aid" (which is an object or product that consists of a flesh-toned adhesive-backed tape with a sterile, nonstick, gauze pad beneath a breathable covering), and quite another to attempt to copy Coca-Cola, One-A-Day multivitamins (with a "1" in shading behind the "A"), Viagra, or a coated arc welding electrode. How so?, you ask.

As consumers who read labels know, and those who have tried to copy Coke know only too well, a content label lists the major ingredients constituting a substance (product) but not *everything* and not the form of the ingredient(s) or the exact amounts, nor the detailed procedures used to create the product. So, in Coke, there are listed the following ingredients (by convention, from greatest to least content):[15] high-fructose corn syrup *or* sucrose derived from cane sugar; caramel color [?]; caffeine [?]; phosphoric acid [?]; coca extract [?]; lime extract; vanilla; and glycerin. Oh, and water and carbon dioxide gas, not listed. But is high-fructose corn syrup or sucrose "derived from" cane sugar (whatever that means!) used? One or the other or both? (For consistency, it's surely both!) How much of each? What are the source, form, and potency of "caramel color"? Of "caffeine"? Of "phosphoric acid"? And what, exactly, is "coca extract"? What it is not is kola nut extract![16] In lieu of any kola nut extract, "coca extract" (an invention of the inventor) is used to develop Coke's distinctive taste. It is a secret *formulation* of principally vanilla and cinnamon, with trace amounts of orange, lime, lemon, and spices, such as nutmeg. But how much of each? In what form and concentration? What spices beyond nutmeg?

As iconic as the Coca-Cola and Coke brands have become around the world, the company's product red-and-white logo cannot be shown in any form without payment of licensing fees. Hence, there is no picture of a Coke in this book, no matter how refreshing that might be.

Oh, and there's water in Coke. As aficionados of fine beer—as well as bottled spring waters, with or without "gas"—know, it's all about the water! And then there's the "gas," which is carbon dioxide. But those who prefer Pepsi-Cola over Coca-Cola, or vice versa, know, Pepsi seems to have more gas, as well as what many feel is a less sweet, and an overall different, taste than Coke.

You see, one really knows nowhere near enough to duplicate the distinctive taste that has endeared and endears hundreds of millions, if not billions, of people to Coca-Cola. Just as your father's mother may have not been completely forthright—no less forthcoming—to your mother, her daughter-in-law, with her recipe for Italian tomato sauce with which your father grew up and loved as much as life itself, so, too, does Coca-Cola have secrets no one will ever know. Every

---

[15] In fact, the soft drink is produced from syrup available for restaurants, soda shops, bars, and other vendors as "Coke syrup." It is based on a formula that is a trade secret for which there has never been a patent, so that details on ingredients are never divulged. It was originally invented as "French Wine Coca" by Colonel John S. Pemberton (1831–1888) to overcome an addiction to morphine he developed when wounded in the Civil War. As a pharmacist in Atlanta, Georgia, Dr. Pemberton created his formula in 1886 and modified it into a nonalcoholic beverage soon thereafter. It became the most recognizable soft drink in the world.

[16] The *kola nut* is the caffeine-containing fruit of the kola tree, a genus (Cola) of tree found in the tropical rainforests of Africa. It is chewed by native people to initially induce energy, but it is addictive.

recipe involves, in order, (1) the basic ingredients, (2) the form of the ingredients (e.g., sweet or salted butter, margarine, Crisco, or lard when the recipe calls for "shortening"), (3) the order in which ingredients are added, (4) the procedure (e.g., "sauté one large, finely chopped Vidalia onion in extra-virgin olive oil until translucent, adding a mixture of ground sweet and hot Italian sausage, and stirring until the sausage is cooked", etc.), (5) the "secret ingredients," (6) the technique-sensitive details (e.g. . "fold, don't mix," "don't overwork the pie crust dough," etc.), and, of course, (6) the "love" of the chef.

This may sound ridiculous, but it is not. Ingredients matter. Form of ingredients matters. Procedure matters. Technique matters. Caring—a form of workmanship (see Section 7-3)—matters. Reverse engineering to find chemical composition, by elemental analysis and then form (e.g., elemental or as a compound), is relatively straightforward. Not so for the source of ingredients that are occasionally a single compound but are more often mixtures of compounds (e.g., feldspar, a mineral commonly found as an ingredient in the coating of welding electrodes). This level of the *formulation* (versus composition) of a substance or product is usually indeterminate. As for how the substance was made, that is usually totally indecipherable without tremendous experiential knowledge of similar substances or products.

So reverse engineering is a good starting point, but not the total answer.

15-5 Synthesizing Natural Materials and Substances: Biomimicry

Scientists and engineers alike have—again—become fascinated by the achievements of Nature. Just as Daedalus and his son, Icarus, were so fascinated by and envious of the ability of birds to fly that, according to Greek myth, they stuck feathers to their arms using beeswax, leaped off a cliff along the shore, and flew over the sea, until Icarus, against his father's warnings not to do so, flew too close to the Sun and fell into the sea when the beeswax melted, so, too, have scientists and engineers across the millennia used things observed in the natural world as inspiration for their own creations and designs. Examples abound with things from swim fins for easier and more efficient propulsion in the water to morphing wings for advanced, experimental aircraft.[17]

The author has, for many years, advocated that students in his design classes look to Nature for inspiration, the reasons being: (1) what one sees in surviving species is Nature getting it right (or the species wouldn't have survived), (2) Nature never (in the long term, at least) allows any of her designs to damage the environment, and (3) if one believes in evolution (for which there is evidence), a surviving design has likely evolved over time to become increasingly better suited to prevailing conditions and needs. These are important goals for designers!

Human beings imitate Nature by reverse engineering what they observe. They consider role and purpose and functionality (Chapter 7); look at form, fit, and function (Section 12–2); consider materials-of-construction (Chapter 9); and even consider method-of-construction or creation (even

---

[17] *Morphing wings* change their geometric configuration in flight as desired for optimal flight characteristics in different regimes, just as birds do. Wings extended straight out from the body, and spread, maximize lift for soaring. Wings pulled back to a severe angle of sweep allow high-speed dives.

if not manufacture) (Chapter 10). If one accepts the author's premise that Nature always gets it right, there is no need to reflect too long on suitability of design for purpose (Chapter 12). The design is suited to the purpose by evolution!

An attempt to emulate Nature has most certainly found its way to materials and substances, too. Efforts to emulate and imitate materials are discussed in Section 15–6. This section considers efforts to synthesize natural substances.

The study and practice of imitating natural materials, substances, processes, mechanisms, structures, entities, and even systems falls under the descriptive term *biomimicry,* as well as the less obvious term *biomimetics*. In all cases, the actual approach involves systematically "dissecting" the natural item to learn how it works and then trying to come up with a reasonable copy—in other (and fewer) words, to reverse engineer it.

Three important and, one hopes, interesting examples are (1) synthetic diamonds (a natural material), (2) artificial spider silk (a natural material or substance, depending on the particular silk and viewpoint), and (3) synthetic bioadhesives (usually a natural substance). Let's look briefly at each.

Synthetic Diamonds

No sooner had Sir Humphry Davy (1778–1829) discovered that diamonds are, in fact, composed of pure carbon in a particular crystalline form, than other chemists, by the dozens, no doubt, frantically sought ways to synthesize these beautiful and highly sought-after valuable gemstones, which also were valued for their phenomenal hardness for use in industrial cutting tools. But, with a few false-claims in between, it wasn't until 1954 that four collaborating scientists at the General Electric Corporate R&D Center (Schenectady, New York) succeeded. On December 16, 1954, H. Tracy Hall (1919–2008) created the first reproducible synthetic diamonds using a modification of a Bridgman anvil press developed in 1951 at the GE laboratories. Hall's success was enabled by coworkers Francis P. Bundy (1910–2008), Herbert M. Strong (1908–2002), and Robert H. Wentorf, Jr. (1926–1997).

Synthetic diamond, also known as "laboratory-created diamond," "laboratory-grown diamond," and "cultured diamond" or "cultivated diamond," is crystalline carbon with a diamond-cubic structure, in which each carbon atom is covalent bonded to four other carbon atoms in a phenomenally strong tetrahedron, produced in an artificial process, as opposed to the natural diamonds, which are created by high-pressure, high-temperature geological processes over millennia. Synthetic diamond comes in two forms, designated HPHT diamond and CVD diamond, after the methods of their synthesis. HPHT stands for "high-pressure, high-temperature," while CVD stands for "chemical vapor deposition." While, for decades, the resulting diamonds were fine for industrial purposes (being hard), their perfection was not sufficient for gem-quality examples to be produced. More recently, gem-quality synthetic diamonds have been produced.[18]

---

[18] It is now possible to synthesize gem-quality rubies, sapphires, and emeralds. In fact, these are so perfect that the purchaser of colored precious gemstones needs to be very careful! There are more synthetic stones than genuine stones. These are not imitations; they are the real thing, only synthesized in the controlled environment of a laboratory as opposed to in Nature's cauldron. Caveat emptor!

Artificial Spider Silk

The motivation for creating synthetic diamonds is clear (no pun intended!), that is, an economical source of industrial-quality diamonds for use in abrasive cutting, grinding, and polishing. Not so clear is the motivation to create artificial spider silk. The mechanical, physical, and chemical properties of spider silk are actually quite remarkable. With a tensile strength comparable to that of steel (at 500 to 2000 MPa), but with a density (at 1.31 g/cm^3) that is one-sixth of steel (at about 7.9 g/cm^3), the strength-to-weight advantage of spider silk is very intriguing. Combine this with an ability to extend five times in length, tremendous toughness, retention of strength and toughness from −40°C/−40°F to 220°C/364°F, and an unusual ability to contract in the presence of water, an artificial method of producing the material is very interesting.

While ancient people used spider silk to cover wounds, and some fishermen in Asia and Southeast Asia used spider silk to catch small fish, the driving motivation at the moment is use in mammalian neuronal regeneration—that is, the regeneration of severed nerves. This alone is reason enough to continue to pursue this difficult prize.

While spider silk is a protein fiber created inside and spun from a set of special organs possessed by all spiders, replicating the complex conditions required to produce comparable fibers is daunting. The challenge is further complicated by the fact that all spiders seem to be able to produce, at will, seven different types or varieties of silk, including types optimized for capturing and immobilizing prey (with a combination of strength, toughness, and stickiness), producing nests and cocoons (which are not sticky, but are stringlike and waterproof), creating optimum properties for guide lines, drop lines, anchor lines, alarm lines, and pheromonal tracks for their webs (see Figure 15–3).

Chemists bent on solving the mystery of spider silk have a special reverse-engineering challenge.[19]

Synthetic Bioadhesives

There are innumerable species in the natural world that produce adhesives, used to bond one thing to another. In fact, a high-level classification scheme divides adhesives into natural and synthetic categories (ref. Messler2). Familiar *natural adhesives* are produced from fish oil or skin, animal protein (e.g., rendered bones, hooves, and fat), milk (e.g., casein), insects (e.g., lac beetles), and plants (e.g., pine-tree pitch). Other types are produced by bacteria, fungi, algae, and certain worms. While often not as strong as many synthetic adhesives, the animal-derived types tend to be biocompatible. In fact, biocompatibility is *the* major motivation for seeking to reverse engineer natural adhesives or bioadhesives.

It should come as no surprise, however, since there is "nothing new under the Sun," that there are some remarkably strong adhesives to be found in Nature.[20] Several examples are produced by mollusks and crustaceans, with those produced by mussels and barnacles being especially interesting.

Mussels and barnacles produce a protein, which is an organic monomer capable of forming an extremely strong polymer via polymerization in the presence of water, including salt water for many varieties. Anyone who has owned a boat that operates in salt water knows how tenaciously

---

[19] A Wikipedia article entitled "Spider silk" lists five different approaches being pursued to create an artificial silk.
[20] The proverb "There is nothing new under the Sun" is taken from Ecclesiastes 1:9 (New International Version of the *Holy Bible*): "What has been will be again, what has been done will be done again; there is nothing new under the sun."

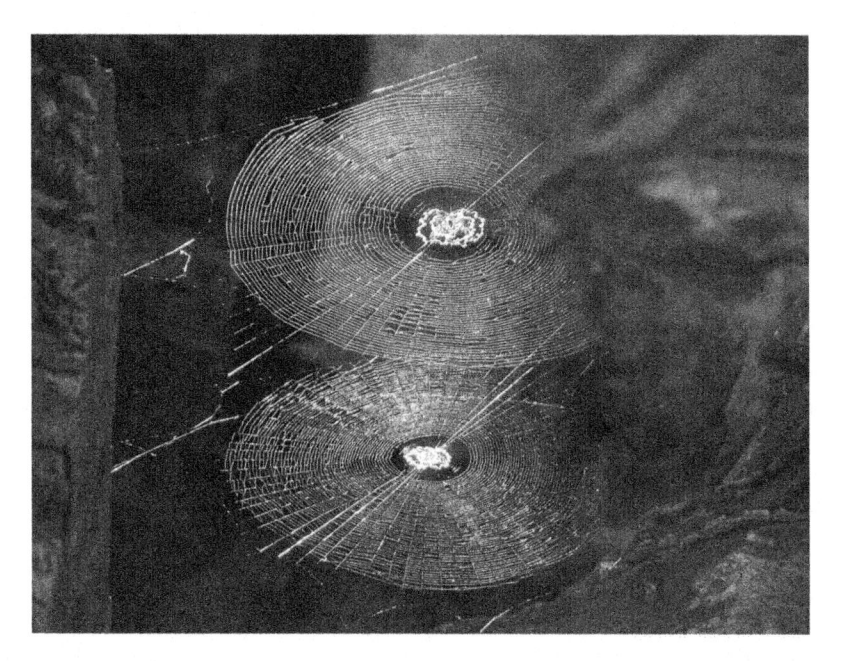

Figure 15-3 A photograph of a wonderful example of the webs made by spiders (here, an orb-weaver spider) using the protein-based silk they manufacture and spin with a special set of organs in their bodies. (*Source:* Wikipedia Creative Commons, contributed by Uspn on 79 July 2009.) **Don't miss the color version of this figure, available at www.mhprofessional.com/ReverseEngineering.**

barnacles adhere to the hull of those boats, whether wood, fiberglass, aluminum alloy, or steel. They even adhere to the skin of great whales. Similarly, if one is observant, one has seen mussels attached to wood pilings, concrete and steel supports, concrete walls, rocks, and ropes, on which they are cultivated in "farms." The attraction of these phenomenally strong adhesives, triggered or catalyzed by saline water (not unlike our own blood), is that they are biocompatible. As such, they are intriguing for use as surgical adhesives for both soft (e.g., skin, muscle, tendon, ligament, and nerve) tissue and hard (e.g., bone) tissue.

Figure 15–4 shows, rather beautifully, how barnacles and mussels adhere to rock, wood, metal, and the like.

Above and beyond this potentially valuable application, these natural adhesives work under water and stick to low-surface energy, nonpolar material surfaces, like Teflon and other polymers.

Once again, if there is to be further success, it will be enabled by reverse engineering.[21]

---

[21] While there are a number of interesting articles on recent developments pertaining to synthetic adhesives that mimic the mussel, chemical engineers at Holland's Delft University of Technology were the first to figure out that the active protein used by mussels is Mefp-1, which, besides water, needs a certain amount of oxygen and a low acidic environment to work. This is very close to what is found in the human body, making use of such an adhesive very attractive for surgery and tissue engineering.

352 Chapter Fifteen

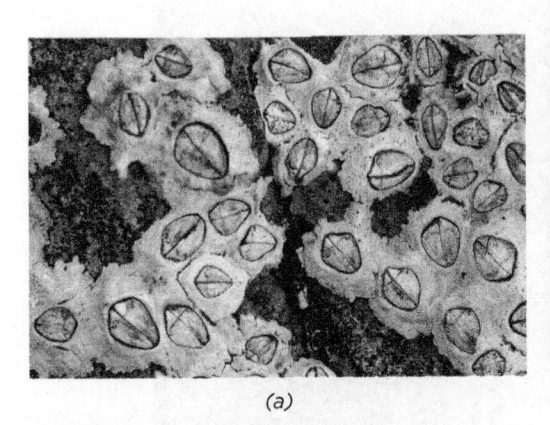

(a)   (b)

Figure 15-4 Beautiful photographs (which need to be seen in full color to be appreciated) of barnacles and mussels, as they adhere to virtually anything using a natural bioadhesive they produce, and which engineers and scientists are seeking to imitate. (*Sources:* Wikipedia Creative Commons, photograph by Michael Maggs taken on 2 August 2007 and posted by BetaCommandBot on 13 June 2008 [*a*], and photograph taken by Mark A. Wilson, Department of Geology at The College of Wooster and posted by Magnus Manske on 30 March 2008 [*b*], with both photographers being acknowledged for their superb work.) **Don't miss the color version of this figure, available at www.mhprofessional.com/ReverseEngineering.**

15-6 Imitating Natural Materials

There is a particular challenge when it comes to reverse engineering materials as defined earlier in this chapter (Section 15–2) as *engineering materials*. For these materials, chemical composition alone is not enough, and, as described in the previous section, getting a composition right in a formulation, such as found in substances, usually requires identifying and replicating the right form or source (e.g., a natural mineral as opposed to a "reagent grade" compound) of the ingredient, not just the chemistry. But it gets more challenging for engineering materials, as these all develop many of their functionally specific properties from their nano- and/or microstructure, beyond chemistry.

While characterizing the nano- or microstructure of an engineering material is not too difficult, reproducing it can be much more challenging. Characterization for a metal, alloy, or ceramic, for example, as all are crystalline materials, involves, in order, identifying and characterizing all of the following:

- Individual solid-state phases present (whether elemental, disordered, or ordered solid-solutions, compounds, second-phase precipitates or dispersoids, amorphous regions, etc.)
- Grain size of phases or phase regions, or, for eutectics and eutectoids, colonies
- Phase or grain shape or morphology
- Presence or absence of phase alignment

- Presence of grain-boundary 2nd phases (e.g., carbides) or segregates
- Orientation of crystallographic planes from grain to grain (i.e., an orientation texture)
- Presence, volume fraction, size, and distribution of pores (i.e., porosity) or nonmetallic impurities

For engineered composite materials, in addition to the preceding for both the continuous, surrounding matrix phase and reinforcing phase(s), one needs to further characterize the following on a more macroscopic scale (e.g., 10^{-4} to 10^{-1} meters):

- Type of matrix (e.g., polymeric, metallic, ceramic, or carbonaceous)
- Type of reinforcement (e.g., metal, alloy, ceramic, intermetallic, glass, or carbon graphite)
- Type of composite (e.g., particulate, random or aligned chopped fiber or whisker, aligned, unidirectional continuous fiber or cross-plied continuous fiber or woven continuous fiber, laminate)
- Volume fraction of reinforcement
- Size and distribution of reinforcement
- Interface strength between reinforcement and matrix phases

The problem with attempting to replicate a microstructure via reverse engineering is that the microstructure (as well as finer-scale nanostructure) is the result of the processing the material underwent. For metals and alloys, which exhibit inherent ductility, the microstructure that develops is a complex interaction among heat (i.e., temperature, time, and heating and, especially, cooling rates during heat treatment) and strain (i.e., strain rate, degree of strain, direction of strain during plastic working). It is possible to develop very similar appearing microstructures via different time-temperature-strain paths or histories, yet each may not result in precisely the same structure and properties. For ceramic materials, which tend to lack ductility, time-temperature history is very important, and pressure during any compacting process may also be important. For polymeric materials, the nearly limitless possibilities that exist among types of polymer units in a polymer chain, chain length and/or degree of polymerization (e.g., molecular weight), chain configuration and geometry (e.g., homopolymer or copolymer, the latter being random or blocky, side branching or grafting, chain cross-linking), degree of crystallization, and chain alignment via strain processing all matter.

In short, attempting to reverse engineer the microstructure of an engineering material is extremely complex and can be intractable. Remember, in engineering materials, structure determines properties, and processing affects both to create performance (see Section 9–2).

There are many examples of materials attempting to imitate natural materials that are superficial (e.g., simulated wood in plastic, simulated brick or stone in plastic), as the imitation is "only skin-deep," if you will—that is, it is purely aesthetic. Of much greater interest and significance are examples where materials are synthesized or, more accurately, *engineered,* to imitate natural materials. While there are too many examples to cover herein, three particularly important and rich examples that will be briefly discussed are (1) engineered wood, (2) synthetic stone, and (3) synthetic fibers.[22] Let's look at each.

---

[22] While one could consider synthetic diamonds under this section heading, these were considered in Section 15–5.

Engineered Wood

Engineered wood, also known as "engineered lumber," "composite wood," "man-made wood," and "manufactured board," includes a range of derivative wood products which are manufactured by bonding together particles, fibers, strands, chips, fragments, pieces, or veneers of real wood using adhesives to create a composite material.[23,24] The motivation for creating these engineered materials varies from situation to situation and has varied over time, but major motives include (1) conservation of a precious natural resource,[25] (2) more efficient utilization of a precious natural resource, (3) lower cost via the use of what would otherwise be sawmill scrap or waste, and (4), in the best but rarest cases, to create a material with superior properties to natural wood.

The types of engineered wood include the following:

- Plywood
- Particleboard or chipboard
- Oriented strand board (OSB)
- Glued laminated timber (Glulam)
- Laminated veneer lumber (LVL)
- Cross-laminated timber (CLT)
- Parallel strand lumber (PSL)
- Laminated strand lumber (LSL)

The motivations for creating engineered wood may have varied, but the original inspiration came from Nature and was enabled by reverse engineering what was observed.

Figure 15–5 shows an engineered wood or lumber product known as Glulam, which, pound for pound (or kilo for kilo) offers exceptional load-carrying capability compared to natural timbers.

Synthetic Stone

Synthetic stone, or "artificial stone," is the name of various kinds of man-made stone products used since the eighteenth century. It is used in building construction, in civil engineering works, and in industrial products, such as grindstones and support bases for machines and metrology equipment (e.g., coordinate measuring machines). Two important examples are (1) synthetic marble and (2) synthetic granite.

Synthetic marble is a solid material made by mixing, molding, and curing marble dust and bauxite with an acrylic or polyester resin. It is most commonly used in countertops. While it is essentially as hard as marble, it is more durable in that it better resists chipping and breaking and better resists staining by precluding absorption of colored liquids into normally porous natural marble (via the resin-matrix binder phase). Synthetic granite is made from a mixture of powdered

---

[23] *Wood* is itself actually a composite, albeit a natural composite, material. Both softwoods and hardwoods contain a mixture of a softer, weaker, tougher phase known as *cellulose* and a harder, stringier, but less tough phase known as *lignin*. As in all composite materials, these two phases, with different properties, act synergistically to create a superior combination of properties in the composite.

[24] The adhesives used in creating engineered wood products include urea-formaldehyde (UF), phenol-formaldehyde (PF), melanine-formaldehyde (MF), methylene diphenyl diisocyanate (MDI), and polyurethane (PU).

[25] One should not lose sight of the fact that wood is the only renewable material on the planet! In fact, with more and more responsible forest management, it has become the rule rather than the exception in the United States that major wood producers (Boise-Cascade, Weyerhauser, Georgia-Pacific, etc.) plant 5 to 10 seedlings for every tree they harvest for wood for lumber or pulp for paper. After all, wood is their product, and trees are their source. Hence, it only makes sense that they preserve their own business!

Figure 15-5 Example of the use of Glulam beams in the frame for the roof of a building. This engineered wood or lumber product is remarkably stronger than normal, natural wood timber. (*Source:* Wikipedia Creative Commons, contributed by Mok9 on 9 December 2008.)

or more coarsely crushed granite and an epoxy resin binder. While nearly as strong, and tougher, than natural granite, it is five times cheaper. It is most commonly used instead of cast iron or steel for manufacturing heavy, vibration-damping bases for machine tools and for thermally stable bases for metrology equipment.

More and more elaborate and aesthetically pleasing (more correctly, beautiful) artificial stone is being created every day, with some examples being DuPont Corian and Zodiaq, Silestone Quartz, StoneMark Granite, and Taylor Tere-Stone.

Two examples of what can be achieved are shown in Figure 15–6, one quite old and utilitarian and one very modern and aesthetically pleasing beyond utilitarian.

Synthetic Fibers

While some may not include the synthetic fibers that tend to predominate in modern easy-to-care-for clothing, this author feels they deserve mention, even though it could be questioned

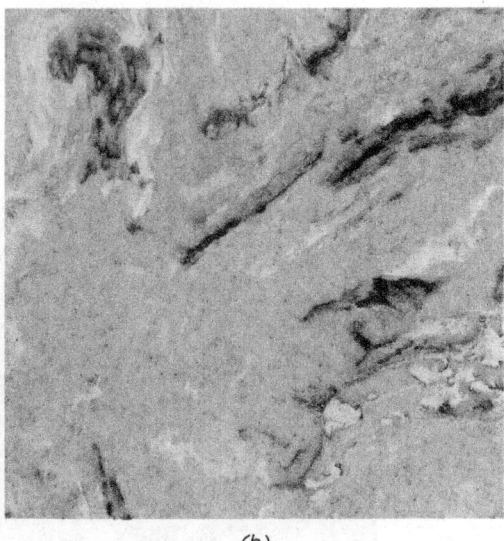

356 Chapter Fifteen

(a) (b)

Figure 15-6 Two quite different examples of artificial or synthetic stone: Vulcanite, a form of concrete used at the turn of the nineteenth century is used in an old walkway (*a*); an example of an extremely attractive and remarkably diverse synthetic stone material used to fabricate countertops by Tere-Stone, consisting of dolomite and a polymeric resin, the one shown here known as "Sierra Latte" (*b*). **Don't miss the color version of this figure, available at www.mhprofessional.com/ReverseEngineering.**
(*Sources:* Wikipedia Creative Commons, originally contributed by PhilaRegion1062 on 11 October 2010 and modified by Rotatebot on 16 December 2011 [*a*]; and a photograph of a sample of Tere-Stone by Taylor Industries, provided by Taylor Industries and used with the permission of Bruce Taylor [*b*].)

whether they were the result of reverse engineering per se. *Synthetic fibers* are man-made materials intended to replace or supplement (in "blends") natural plant-based or animal-based fibers like cotton, wool, and silk. An artificial silk appeared in 1894 as "viscose," rayon (regenerated cellulose) appeared in 1924, and nylon appeared just before the start of World War II (i.e., 1935).

Today, they are many types, including but not limited to acrylic, Ban-Lon (a synthetic yarn), olefin, DuPont Orlon (an acrylic), polyester, and DuPont Spandex or "elastane" (a highly elastic fiber made from a polyurethane-polyurea copolymer).

Synthetic fibers are known for having the following characteristics: heat-sensitive; resistant to most chemicals; resistant to insects and fungi; have low moisture absorbency; do not "breathe" well, which can lead to overheating; dry quickly and without wrinkles; can shrink when heated; can be flame resistant (although they melt easily); are lightweight; are electrostatic; and, most important, are cheap compared to most natural fibers.

Whether created by reverse engineering or not, synthetic fibers most certainly resulted from observations of natural fibers for inspiration to make improvements in one area or another.

15-7 Summary

No area for the application of reverse engineering may be more prone to abuse by the unscrupulous than the creation of materials (i.e., engineering materials) and substances (i.e., consumer substances). Motivations vary widely, as do the ultimate manifestations of the reverse-engineered end product. One can usually justify (as ethical beyond legal) the creation of substitute materials or substances, replacement materials, certain imitations, certain copies, and, perhaps, look-alikes. Impossible to justify under any circumstances are knockoffs, which are solely intended to deceive consumer and casual observer alike.

A major application of reverse engineering, which can easily be shown to be legal and normally justifiable as ethical, is in the creation of generics by reverse engineering originals.

Two exciting applications, and ripe opportunities, for reverse engineering are the creation of synthetic versions of natural materials and substances (in what is known as *biomimicry*) and the creation of imitations of natural materials. Examples of the former include synthetic diamonds, artificial spider silk, and bioadhesives. Examples of the latter include engineered wood, synthetic or artificial stone, and, perhaps arguably, synthetic fibers.

The future for reverse engineering of engineering materials in particular may be the most exciting of all opportunities (see Chapter 18).

15-8 Cited References

Messler, Robert W., Jr., *Engineering Problem-Solving 101: Time-Tested and Timeless Techniques,* McGraw-Hill, 2013).

Messler2, Robert W., Jr., *Joining of Materials* and *Structures: From Pragmatic Process to Enabling Technology,* Elsevier/Butterworth-Heinemann, Burlington, MA, 2004.

15-9 Thought Questions and Problems

15-1 The statement is made near the beginning of Section 15–1: ". . . while inappropriate (and, often, unethical) use of reverse engineering occurs in too many instances . . . nowhere does it seem to be more rampant than in the area of materials and substances."
   a. Using your own words, provide a definition for *materials* and also for *substances,* as used in this chapter, that differentiates these two things in your mind.
   b. Given your definitions from part (a), as well as how each is described in the book, explain why there may be more illegal, illicit, and unethical reverse engineering of these things than of mechanisms, structures, and systems (from the book's title).
   c. Why might it be (and, in fact, is, based on so few successful prosecutions) so common (even though not easy) for *substances* to be "blindly copied" using reverse engineering.

15-2 There are many motivations for reverse engineering *materials* and *substances,* some legal and ethical, some absolutely illegal, and some legal but of questionable ethicality. There are also a variety of uses of reverse engineering of materials, including creation of the following:
   - Substitutes
   - Replacements
   - Imitations

- Copies
- Look-alikes
- Knockoffs

a. Using the definitions and descriptions given for these in Section 15.5, search the Internet for *two* well-known, infamous, and/or interesting examples of each. Discuss each very briefly.

b. Two other terms pertaining to materials and substances to which reverse engineering often applies are *improvements* and *proprietary materials*. Search the Internet for two well-known and/or interesting examples of each. Discuss each very briefly.

15-3 A few very important and legitimate (i.e., legal and ethical) uses of reverse engineering relative to materials and substances are (1) for finding *substitute materials,* (2) for finding *replacement materials,* and (3) for creating *generic materials.*

a. The book gives a few examples of *substitute materials* (in Section 15–3), but there are many others. Use the Internet or your own knowledge and experience to find *three* more examples. Very briefly describe what is sacrificed or compromised in each.

b. The book gives a few examples of *replacement materials* (in Section 15–3), but there are many others. Use the Internet or your own knowledge and experience to find *three* more examples. Very briefly explain the need for each replacement, as well as describe any sacrifice with the replacement.

c. The book described a few areas (or products) for which a *generic* is desired. The usual motivation is to offer a lower-cost version of a brand-name product.

   (1) Use the Internet or your own knowledge and experience to find *two* examples of a generic version in each of *four* different product lines (e.g., over-the-counter medicines, prescription medicines, personal care products, beverages, food products).

   (2) Very briefly describe *two* other reasons of motivations (in addition to lower cost) for the creation of a *generic*.

15-4 The earliest use of materials and substances by human beings all involved materials and substances that came from Nature, i.e., *natural materials* and *natural substances*. These became both our source *and* our inspiration.

a. Use the Internet to help you identify *three* examples of natural materials (i.e., engineering structural or electrical or thermal materials) or natural substances (i.e., nonstructural or electrical or thermal materials) for each of the following:

   (1) fundamentally different natural structural materials for construction
   (2) natural materials for clothing
   (3) natural substances for use as adhesives
   (4) natural substances (not just aromas) for use as fragrances
   (5) natural substances (not just flavors) for use as flavorings
   (6) natural materials used for their electrical or thermal properties

   Natural materials, as well as natural substances, have long been admired by human beings for their special properties (as materials) or characteristics (as substances). Certain of these have inspired *biomimicry* (see Section 15–5).

b. Use the Internet to help you identify examples of a *natural material* or *natural substance* that modern science and engineering are attempting to create or have succeeded in creating for each of the following:

(1) *one* natural adhesive
(2) *two* natural fibers
(3) *two* natural gemstones
c. Recognizing that the goal of biomimicry is not to create an *imitation* of a natural material or substance as much as to *re-create* that material or substance for some unique or special purpose or capability *not* found in existing synthesized materials or substances, prepare a thoughtful essay of about one page that addresses the following:
- Some unique property or quality
- Some especially attractive aesthetic property or quality

15-5 Sometimes the goal of reverse engineering is *not to re-create* a natural material or substance "atom for atom" or "feature for feature" *but*, rather, simply *to imitate* the natural material or substance. This is especially true for imitations of natural materials.

Prepare a brief but thoughtful essay of about one page that addresses the following:
- How some imitations of natural materials are only intended to capture certain selected properties of the natural material, giving an example
- How some imitations are able to surpass certain selected properties of the natural material, giving an example
- What special challenge is involved with trying to imitate the key *structure-property* relationship (i.e., micro- or macrostructure) found in the natural material.

CHAPTER 16

Reverse Engineering Broken, Worn, or Obsolete Parts for Remanufacture

16-1 Necessity Is the Mother of Invention

Aesop (ca. 620–564 BC), the ancient Greek fabulist, told the tale of a crow that, half-dead with thirst, came upon a pitcher only partially filled with water (Figure 16–1).[1] Regrettably, when the crow put its beak into the mouth of the pitcher, it found it was unable to reach far enough down to get to the shallow depth of water there. A thought came to the crow. He took a small pebble in his beak and dropped it into the pitcher, noting that the water rose very slightly. He took another small pebble and dropped it into the pitcher, seeing the water rise a little more. He took pebble after pebble, patiently dropping each one into the pitcher, until, finally, with one more pebble, he was able to reach the water, quench his thirst, and save his life. Aesop's moral for the fable: Little by little does the trick!

What is remarkable is that the crow, even if not Aesop, recognized that necessity is the mother of invention almost 200 years before Plato (428/427–348/347 BC), the great ancient Greek philosopher, scholar, and teacher, is alleged by many to have come up with the famous proverb "Necessity is the mother of invention." Even more remarkable, however, is that the crow, probably through Aesop, as he was Aesop's creation if the story was fictitious and Aesop had not actually seen a crow do such a thing, recognized what it would take the ancient Greek mathematical, scientific, and inventive genius Archimedes (ca. 287–ca. 212 BC) almost 400 years to recognize as he got into a tub of water to take a bath while he contemplated how to tell if the king's crown was made of pure gold or a silver-cheapened alloy! "Eureka!" Indeed, for the crow!

Whatever the origin, the meaning of this age-old proverb is that if someone really needs something or has a problem, he or she will find a way of doing or solving it. The validity of the proverb has been proven over and over again across the history of science and, especially,

---

[1] The fable "The Crow and the Pitcher" is #390 in *Aesop's Fables*.

Figure 16-1 An illustration by Milo Winter in *The Aesop for Children* by Aesop (Children's Press, 1985) with pictures by Milo Winter for "The Crow and the Pitcher," Aesop's fable #390. (*Source:* Wikipedia Creative Commons, contributed by Tagishsimon on 17 February 2007.)

engineering, as it is engineers who are charged with solving problems and coming up with inventions (see Chapter 1).

When it comes to reverse engineering, it was only in the pursuit of substitute or replacement materials or substances (Section 15–2) that necessity was the driving force for the resulting "invention." (Think about it. Did the Soviets *really* have to have a B29 look-alike?) But there are three other situations that *necessitate* the use of reverse engineering to solve a problem that *must* be solved. These three situations are:

1. To allow the remanufacture of a broken part
2. To allow the remanufacture of a deformed or worn part
3. To allow the remanufacture of an obsolete part

Let's look at these three driving forces, motivations, or *necessities* in the remainder of this chapter, as this book comes to its end.

16–2 The Motivation for Reverse Engineering for Remanufacture

Broken (i.e., fractured or cracked), damaged (i.e., deformed or worn), or obsolete (i.e., no longer available) parts are often considered too expensive to replace, if they can even be replaced (ref. Bagci). As a consequence, one of two choices must be made: first, to use a similar, "make-do" part that might require elaborate refitting or compromise functionality or degrade performance of the repaired device, or, second, to discard the broken, damaged, or obsolete model altogether. More often than not, the author can assure the reader from personal experience, a good engineer finds a way to make the needed device work.

Fortunately, reverse engineering, if it is known and properly and fully understood by an engineer, always allows the systematic evaluation of a former model to allow reproduction and, thus, remanufacture, of an identical, similar, better, or cheaper new model. The driving force is necessity!

The author recalls a situation while working in Materials and Processing Engineering at Grumman Aerospace Corporation in the 1970s. The cast-iron frame of a large, 500-ton forming press fractured completely across and through one of the vertical main struts. The press was unique in the company and essential for the fabrication of parts for a major billion-dollar aircraft program. Something had to be done! And quickly!

The vice president of engineering and chief engineer at the company called together, as such senior executives are prone to do, a special "Tiger Team" made up of experts in equipment maintenance, design, manufacturing, and metallurgy and welding. As one of the metallurgy and welding "experts," the 30ish author heard many naysayers—pessimists who preferred to whine and complain before they even tried to work. Some said (erroneously!) that cast iron could not be welded. Others said that no repair would stand up to the forces and stresses required of the press in operation. Despite these detractors, two suggestions emerged: one to weld-repair the strut and another to place a large, structural "Band-Aid" (i.e., splice) across the fracture.

The former approach involved machining a deep V-groove both from the front and from the rear of the strut, deep enough to allow new, tougher, but slightly less strong 55% nickel/45% iron filler metal to rebuild the cross section of the broken strut by shielded metal-arc welding. The latter approach would place 3-inch-thick steel splice plates, one on the front and one on the rear of the strut, spanning the fracture, drilling a pattern of 1-inch-diameter holes at each end of the splice plates and through the cast-iron strut above and below the fracture, and inserting pretensioned high-strength steel bolts with locking washers through all of the drilled holes. Both approaches required a temporary support frame and jack to be erected around the broken strut to act as a "splint" until the repair was completed—not to allow the press to operate, but to keep it stable and prevent it from collapsing.

Ultimately, in what some argued was a "belt-and-suspenders approach," the crack was gouged and welded and the steel "Band-Aid" was applied over the repair-welded strut.[2] The repair worked, and, more important, the press was restored to full operation within two weeks.

Reverse engineering actually played a role, as several of the engineers thought of other repairs they had seen on cast iron, on other failed machine frames, and on other high-load-bearing large structures. All of these experiences helped the team come up with what proved to be a very successful repair.

---

[2] In fact, if one thinks about how tension-loaded bolts and about how fusion welds work, one realizes one or the other of the repairs was unnecessary, as both never would perform together to carry or share loads. If one believed the weld would hold, the stiffness it provided shielded the bolts and plates from carrying much load. On the other hand, if one believed the relatively elastic bolts and plates would hold, the stiffer weld repair was not needed. It is analogous to one wearing a belt and suspenders. However, if the belt is cinched tighter than the suspenders, it carries the entire load. If the suspenders are cinched tighter than the belt, they carry the entire load. They never share the load. What such a system does provide is some level of redundancy. If the belt fails, the suspenders hold up the pants, and vice versa.

So let's look at how reverse engineering can enable remanufacture of broken, damaged, or obsolete parts.

16-3 Reverse Engineering Broken Parts for Remanufacture

The least complicated parts to reverse engineer for remanufacture are those that have simply cracked or fractured into fragments by a brittle mechanism. This is the mode by which parts or other things made from ceramics, glass, inherently brittle metallic materials (e.g., cast iron), or normally ductile metals or alloys that are forced, for some reason, to act in a brittle fashion, fail. For metals and alloys, a brittle cracking or fracture mechanism is promoted by a body-centered cubic (BCC) crystal structure when (1) failure occurs at operating temperatures below the material's ductile-brittle transition temperature (DBTT), (2) loads are applied rapidly or strain occurs at a high rate (i.e., under a high strain rate), (3) in the presence of a severe stress concentration (e.g., from sharp radii at thickness transitions or cutouts, small holes, porosity or other open defects, etc.), or (4) in the presence of certain corrosive environments (e.g., hydrogen for quenched steels inadequately tempered). Hexagonal close-packed (HCP) metals or alloys can exhibit similar, albeit less sensitive, behavior to BCC metals or alloys. Face-centered cubic (FCC) metals and alloys are unaffected by any of the listed factors except certain corrosive environments for certain metals or alloys. All metals and alloys, like all materials, are forced to fail by a brittle mechanism under a state of triaxial stress, which is more prevalent in very thick sections.

The reason brittle fractures are easier to reverse engineer is that the geometry (i.e., shape and dimensions) of the part is retained in the absence of any plastic deformation before fracture or in the immediate vicinity of the fracture during fracture. Thus, by reassembling the fracture fragments, the reassembled part can be measured to capture points in 3D space that define it, for example, using a coordinate measuring machine (CMM) equipped with either a touch probe or laser. The resulting point coordinates are stored in a computer-aided design (CAD) file to create a CAD model as well as enable manufacture. To do the latter, the CAD file is transformed for use in computer-aided manufacturing (CAM) using computer numerical-controlled (CNC) machine tools, for example.

Interested readers are encouraged to seek published or online references on the details of the procedure.

Figure 16–2 provides a nice example of how a broken part can be remanufactured by reverse engineering details.

A really interesting practical example of the technique is to be found in the reconstruction of broken prescriptive eyeglass lenses. Since lenses are made from either inherently brittle glass or more impact resistant, but still somewhat brittle, polycarbonate polymer, when they break, the pieces can be fitted back together to re-form the lens, albeit with unacceptable cracks. But, by providing the fracture fragments to an optician, the key details of the prescription to allow a replacement lens to be made can be extracted using a CMM or comparable measuring system. The question is: What if not all of the fracture fragments were recovered? Could the prescription still be extracted from what is available? And, if so, how little of the lens would actually be needed to allow reverse engineering? Obviously, the answer is less would be needed for lenses that are simpler—not bifocal, trifocal, or progressively variable. More obviously, there is some minimum fraction of some portion of any lens that represents the whole.

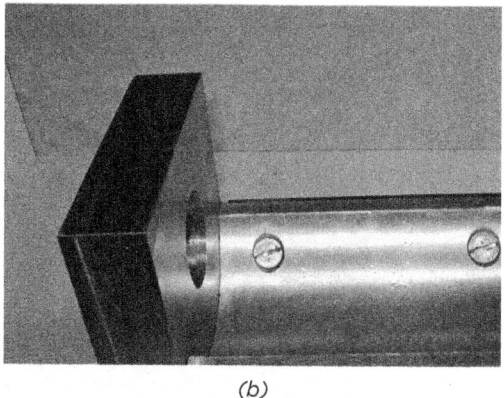

Reverse Engineering Broken, Worn, or Obsolete Parts for Remanufacture **365**

(a)   (b)

Figure 16-2 Photographs of a broken low-alloy steel part (a) that was remanufactured by reverse engineering the original from the broken part (b). The part held a large motor for a cold heading machine. (*Source:* Share Machine, Custom Precision Machining, 2175 Rochester Road, Aurora, IL 60506, used with permission of Zekir Share.)

16-4 Reverse Engineering Deformed or Worn Parts for Remanufacture

When a part has suffered damage from gross plastic deformation (e.g., bending, twisting, bulging, localized section necking or wall thinning, etc.), as is not uncommon in FCC metals or alloys that have experienced overload, the challenge for reverse engineering the original shape to allow remanufacture is greater (Figure 16–3). Likewise for parts which ultimately become useless (i.e., can no longer perform their required function or achieve their required performance) due to the loss of material from wear (i.e., the parts were worn) (Figure 16–4).

For worn parts, missing material—and associated missing part surface and/or volume—provides a gap in the digitized data that can be extracted by a CMM. Fortunately, modern computer technology comes to the rescue! The missing surface and/or volume can be filled in, and needed digital data can be "created," by establishing continuity across curves and/or surface patches using mathematical extrapolation of the curve and/or surface using either polynomial or French-curve splines (ref. Messler).

For deformed parts, the challenge is greater. Once the original geometry (i.e., shape and dimensions) is lost by wear or distorted by deformation, the only recourse (in the absence of engineering drawings) is to attempt to reverse engineer the part from knowledge of the part's role, purpose, and functionality within an overall device, product, or structure (Chapter 7), as well as its form, fit, and function (Section 12–2). By carefully examining all of the other (hopefully, unworn and/or undeformed) parts in contact with the damaged part, data on the original geometry of the damaged part can be extracted.

A search of the Internet under "reverse engineering of broken or worn parts" yields dozens and dozens of companies that specialize in this important area as a service. One example, with an impressive list of clients, offered at random, is 3D ScanCo, a division of Laser Design Inc., located in Atlanta, Georgia.

(a)

(b)

Figure 16-3 The U.S. Navy T2 Tanker USS *Schenectady* (a), which broke in two amidship while lying at the outfitting dock in the constructor's yard in Portland, Oregon. The water temperature in the harbor on January 16, 1943, the day the tanker suddenly fractured was near freezing. The noise could be heard for several miles. Note how the two portions of the hull could be put back together, as the fracture was brittle, without any deformation. Also shown is an American Petroleum Institute (API) steel pipe that ruptured during a hydrostatic proof test well below the proof stress required, due to an intentionally created notch along the pipe's outer surface (b). Note how the pipe bulged, and the pipe wall thinned, before rupturing. The fracture initiated by a brittle mechanism (as evidenced by its initial direction being perpendicular to the primary hoop stress, but turned 45 degrees to terminate in ductile shear. (*Source:* Wikipedia Creative Commons, contributed by Al Rosenfield on 10 January 2006 [a]; and from Donald Wulpi, *How Metal Components Fail*, 2nd edition, ASM International, Materials Park, OH, 1999; used with permission from ASM International [b].)

Reverse Engineering Broken, Worn, or Obsolete Parts for Remanufacture **367**

Figure 16-4 The rear sprocket gear for a bicycle showing the difference between a new gear (*left*) and a severely worn gear (*right*). Wear, which clearly occurred with clockwise rotation, was a combination of metal-to-metal adhesive wear and abrasive wear from sand and other hard particulate materials picked up from the road, and so forth. (*Source:* Wikipedia Creative Commons, contributed by Arc1977 on 10 December 2011.)

16-5 Reverse Engineering Obsolete Parts for Remanufacture

The U.S. Navy has hundreds of ships that have been constructed for it by several large shipyards (General Dynamics Bath Iron Works, Kennebeck, Maine; General Dynamics Electric Boat, Groton, Connecticut; Newport News Shipbuilding, a division of Huntington Ingalls Industries, Newport, Virginia; and several others) over many decades. Complex structural electromechanical systems like naval vessels, not surprisingly, require regular repair, which, not surprisingly increases as the ships age. Eventually, there are ships that require repairs for which either the original manufacturer of the particular part is no longer in business or for which engineering drawings no longer exist. When this occurs, the part may be obsolete, but the ship may not be![3] In such cases, repairs, including remanufacturing of parts, must be done at rework facilities (Figure 16–5).

For this and many other reasons, it is often required to replace obsolete parts. The way it is done often involves reverse engineering.

The cause for the part coming out of service and needing repair or replacement could be, and

---

[3] In the mid-1980s, the U.S. Navy initiated a program it called "RAMP," for which the full title represented by the acronym is not recalled and was not traceable. The objective of RAMP was to allow the navy to rapidly (hence, the "R") repair (i.e., maintain, hence, the "M") its ships—even at sea. The carrier in a fleet is the "mother ship," which both protects and is protected by its supporting ships. Carriers have fully equipped machine shops capable of making whatever is needed for herself or her support ships. The approach in RAMP was to use reverse engineering and CMM $\Rightarrow$ CAD $\Rightarrow$ CAM $\Rightarrow$ CNN.

(a)

(b)

Figure 16-5 A photograph showing modern and immaculately maintained global marine and naval repair shop (*a*) and machining of naval engine cylinder heads at the Zamakona Yards in the Canary Islands, Spain (*b*). (*Source:* Zamakona Yards, Port of Las Palmas, Gran Canaria, Spain. Courtesy of John Roseler, with kind permission.)

often is, wear and/or corrosion (from the harsh seawater and salt air environment), fracture from overload or fatigue, or permanent distortion or deformation (Section 6–3). Thus, the procedure for making the repair or remanufacturing a new part is like that presented in Sections 16–3 and 16–4.

For obsolete parts, the situation can be made even more challenging, however, if the original material(s) used in manufacture are either unknown (which can be dealt with; see Chapter 9) or no longer available. In the latter case, a suitable alternative or substitute (see Section 15–3) is required.

No matter how challenging, a good engineer never quits. The well-informed engineer has reverse engineering in his or her basket of tricks for solving problems.

16-6 Summary

Necessity *is* the mother of invention, and so, too, is having the ability to remanufacture broken, deformed, worn, or obsolete parts. Fortunately, reverse engineering often helps when the full geometry (i.e., shape and dimensions) of a failed part has been altered or lost. More fortunately, modern computer technology helps greatly. Reassembled fracture fragments or unbroken but damaged parts for which material lost due to wear or geometry change due to deformation are first used to construct a 3D CAD model using digitized data obtained from a coordinate measuring machine (CMM). These data allow the creation of a computer-aided design (CAD) model. Necessary adjustments to the CAD model can be made by either filling in data for missing part fragments or material using mathematical extrapolation techniques involving splines or by correcting altered shape and dimensions using the role/purpose/functional and form/fit/function principles underlying reverse engineering. Once the CAD model is deemed representative of the original part, it can be transformed into a data file to allow computer-aided manufacturing (CAM) via computer numerical-controlled (CNC) machines.

The necessity for a damaged, lost, or obsolete part to be remanufactured is often enabled by reverse engineering.

16-7 Cited References

Bagci, Eyup, "Reverse Engineering Applications for Recovery of Broken or Worn Parts and Remanufacture," *Advances in Engineering Software* (Elsevier), Vol. 10, Issue 6 (June 2009), pp. 407–418.

Messler, Robert W., Jr., *Engineering Problem-Solving 101: Time-Tested and Timeless Techniques*, McGraw-Hill, 2013, Chapter 5, pp. 54–56.

16-8 Thought Questions and Problems

16-1 The age-old proverb says: "Necessity is the mother of invention." When it comes to reverse engineering, *necessity*, if one is honest, is rarely the driving force for the resulting "product development" or "invention" (Section 16–1). But there are two general areas for invention, reinvention, or product development that *necessitate* the use of reverse engineering, and within the latter of these, three specific situations:

1. The pursuit of substitute or replacement materials or substances (Section 15–2)

2. Remanufacture
    a. of a broken part
    b. of a deformed or worn part
    c. of an obsolete part for which there are no drawings

In a thoughtful essay of one to two pages, discuss why reverse engineering is an especially effective technique for addressing the aforementioned "necessities." Specifically address why what many consider normal design (e.g., original design or adaptive design) would not suffice. In other words, why do the cited situations necessitate "dissection" of the original to create or re-create the copy?

16-2 When parts that are critical to the operation of a mechanism, device, or system, for which total replacement is impractical if not impossible, break, wear, or otherwise fail, reverse engineering is usually the only viable approach for enabling their remanufacture.
   a. Briefly describe *two or three* situations in which total replacement of a mechanism, device, or system that experienced a failure that prevents proper function is impractical.
   b. Describe at least *one* situation in which total replacement would be impossible.

16-3 Failures involving fracture can occur by either a brittle or a ductile mechanism. In brittle fracture, a part, component, or structural element generally separates completely into two or more pieces or fragments, with very little or no apparent (i.e., gross) plastic deformation prior to fracture. In ductile fracture, on the other hand, a part, component, or structural element always undergoes some (sometimes severe!) plastic deformation before fracture, resulting in cracking without complete separation or separation into two pieces.
   a. Look up "brittle fracture versus ductile fracture" on the Internet or, better yet, in a good introductory materials textbook (Callister, Messler, Shackelford, Askland, Smith, etc.). Prepare a two-column table that lists items for the following:
      (1) factors that contribute to or are associated with brittle fracture (in the left-hand column) and with ductile fracture (in the right-hand column), including crystal structure, rate of loading, etc.
      (2) macroscopic overall and surface features and microscopic surface features (i.e., fractographic features or topography)
   b. Without gross plastic deformation, fragments of a brittle fractured item can be reassembled to re-create the original geometry nearly exactly. This reassembled part can then be "digitized" to create a 3D digital model using a touch-probe or laser coordinate measuring machine, for example.
      Write a very brief (less than one page) description of how this process is accomplished, why it works to allow part remanufacture, and how the measured data are used to create instructions for computer-aided manufacturing of a replacement part (i.e., how CAD data are transformed into CAM data).
   c. Gross plastic deformation associated with ductile fracture complicates remanufacture by the method in part (b) because the original (needed) geometry (i.e., shape and dimensions) is lost.
      Write a very brief (less than one page) explanation of how this very real complication can be managed, if not totally overcome.

16-4 While it may prove to be futile, try to find information in the library or on the Internet pertaining to the U.S. Navy RAMP Program to provide the capability for rapid remanufacture of critical parts for ships via reverse engineering. (This could be very challenging, but worthwhile, as the program was innovative.)

16-5 The reality of the modern world of business and industry is that companies come and go. Even once-giant corporations, leaders in the industrial sectors, stalwarts even, have passed into oblivion. Civil (e.g., Convair, 1943–1996) as well as military (e.g., North American Aircraft, 1928–1996), and Republic Aviation (1931–1965) aerospace manufacturers, automobile manufacturers (e.g., Plymouth, 1926–2001; Pontiac, 1926–2000; and Oldsmobile, 1887–2004), helicopter manufacturers, shipbuilders, consumer electronics manufacturers, computer companies—the list goes on and on, as can be seen from the extensive lists of defunct companies, by industry, on Wikipedia. As a result, there are uncountable examples of products produced by now defunct companies that are still in use day to day. Many of the owners and users of these products want to keep them running, either for nostalgia or because they believe "they don't make them like they used to."

The only way to get replacement parts for obsolete products is via reverse engineering.

a. Use the Internet to find *one* high-profile example of a defunct manufacturer in each of the following industrial sectors, and then a high-profile model produced for which replacements would surely be required for what are inevitably obsolete parts. The industry sectors are:
   (1) military aerospace
   (2) automobiles
   (3) computers
b. Briefly, in one page or less, describe how one might go about re-creating or remanufacturing an obsolete part needing replacement via reverse engineering.

CHAPTER 17

The Law and the Ethics of Reverse Engineering

17-1 Without Morals and Ethics, Law Means Nothing

An anonymous priest serving as chaplain on Holland America's MS *Noordam* during an 11-day Caribbean cruise said the following in his homily the evening before this chapter was begun aboard ship: "Laws are good, but without morals and ethics, they mean nothing." This immediately set the framework in the author's mind for this chapter. There is what is legal pertaining to reverse engineering, and there is what is right.

Professionals, such as doctors, lawyers, and engineers, all serve people or the public, and so have a particular obligation to do what is right beyond what is legal. Most doctors have their own practices, so they are under no pressure from a boss to conduct themselves in accordance with the Hippocratic oath they swore upon entering the profession other than their own personal and professional honor and ethics, the modern version of which states:[1]

> I swear to fulfill, to the best of my ability and judgment, this covenant: I will respect the hard-won scientific gains of those physicians in whose steps I walk, and gladly share such knowledge as is mine with those who are to follow. I will apply, for the benefit of the sick, all measures which are required, avoiding those twin traps of overtreatment and therapeutic nihilism. I will remember that there is art to medicine as well as science, and that warmth, sympathy, and understanding may outweigh the surgeon's knife or the chemist's drug. I will not be ashamed to say "I know not," nor will I fail to call in my colleagues when the skills of another are needed for a patient's recovery. I will respect the privacy of my patients, for their problems are not disclosed to me that the world may know. Most especially must I

---

[1] The modern version of the Hippocratic oath was written in 1964 by Louis Lasagna, academic dean of the School of Medicine at Tufts University, and is used in many medical schools today. While graduates are not required to swear to the oath, 98 percent do!

tread with care in matters of life and death. If it is given me to save a life, all thanks. But it may also be within my power to take a life; this awesome responsibility must be faced with great humbleness and awareness of my own frailty. Above all, I must not play at God. I will remember that I do not treat a fever chart, a cancerous growth, but a sick human being, whose illness may affect the person's family and economic stability. My responsibility includes these related problems, if I am to care adequately for the sick. I will prevent disease whenever I can, for prevention is preferable to cure. I will remember that I remain a member of society, with special obligations to all my fellow human beings, those sound of mind and body as well as the infirm. If I do not violate this oath, may I enjoy life and art, respected while I live and remembered with affection thereafter. May I always act so as to preserve the finest traditions of my calling and may I long experience the joy of healing those who seek my help.

Lawyers, too, often have their own practice or participate with other like-minded lawyers in a firm, so are really only obligated to practice by their code of conduct.[2] But more often than not, in serving the public and their profession via a code of conduct, engineers also have responsibilities to an employer. This has the very real possibility of creating a dilemma for an engineer. Does the engineer mindlessly follow orders from his or her management, even if what is being ordered violates the law or potentially places the consumer or the public at risk? What about if what is being ordered isn't illegal but crosses established lines of proper professional conduct? Ethical conduct says, "No!"

While not every engineer (nor even the majority of engineers) practicing engineering has a professional license, all are bound to conduct their professional careers ethically.[3] The National Society of Professional Engineers (NSPE) *Code of Ethics for Engineers* has the following Preamble:[4]

Engineering is an important and learned profession. As members of this profession, engineers are expected to exhibit the highest standards of honesty and integrity. Engineering has a direct and vital impact on the quality of life for all people. Accordingly, the services provided by engineers require honesty, impartiality, fairness, and equity, and must be dedicated to the protection of the public health, safety, and welfare. Engineers must perform under a standard of professional behavior that requires adherence to the highest principles of ethical conduct.

While it sounds cynical, it is a truism: Doctors treat one patient at a time, so a serious mistake could cost one life at a time. Likewise, lawyers generally defend one client at a time, so a mistake could send one innocent person to prison at a time. But engineers design and build structures that carry hundreds or thousands or more people at a time, so a mistake could hurt or kill hundreds or thousands or more at a time. Perhaps worse, a mistreated patient or misrepresented legal client is known about by only a few close relatives or confidants. But a collapsed bridge, for example, becomes a headline on the nightly national news. In a way, engineers are held to a particularly high standard of conduct because their work directly affects the public.

Besides its Preamble for guiding the conduct of engineers, the NSPE has developed Rules of

---

[2] On August 27, 1908, the American Bar Association adopted the original Canon of Professional Ethics, which has, over the years, undergone changes to become the Modern Rules of Professional Conduct.

[3] The National Society of Professional Engineers (NSPE) in the United States administers examinations for professional licensure and oversees the conduct of licensed engineers. Certain engineering disciplines, most notably civil engineers, are required to have professional licenses if they are engaged in the design of structures for the public, as public safety and health are at stake.

[4] The complete Code of Ethics for Engineers can be found on the NSPE's website at www.nspe.org. Every student and practitioner of engineering should read it—and live by it!

Practice and Professional Obligations, which can be found at their website. It will suffice here to quote only the Fundamental Canons of the society, thus:

Engineers, in the fulfillment of their professional duties, shall:

1. Hold paramount the safety, health, and welfare of the public.
2. Perform services only in areas of their competence.
3. Issue public statements only in an objective and truthful way.
4. Act for each employer or client as faithful agents or trustees.
5. Avoid deceptive acts.
6. Conduct themselves honorably, responsibly, ethically, and lawfully so as to enhance the honor, reputation, and usefulness of the profession.

As for reverse engineering, it must never be conducted to deceive (per Canon #5) and must always be done in an ethical and lawful manner (per Canon #6). As listed in Section 2–3, there are many motivations for engaging in reverse engineering. Most of these motivations are perfectly legal *and* ethical, although there are situations where the line between ethical and unethical can be fuzzy. But, as pointed out in that section, it is never ethical, and can be illegal, to create unlicensed or unapproved duplicates.

17-2 Legal versus Ethical

The dictionary defines the word *legal* as "conduct in conformity with or permitted by law," where the law is established by an official governing body, which might be local, state, national, or international in jurisdiction. The word *ethical,* on the other hand, is defined as "actions or behavior within a set of principles of right conduct," with what is "right" generally arrived at by the consensus of a group, but also guided by higher standards often set by a wider culture.

By these definitions, it is possible for a person to operate within the law (i.e., *legally*), but not in an ethical manner (i.e., *ethically*). It is clear (i.e., unambiguous) what one must not do to operate within the law, and the line between what is legal and illegal is (or is supposed to be) the same for everyone under the particular jurisdiction. There should be no judgment required. On the other hand, it is not always so clear where the line between ethical and unethical conduct is. What is deemed ethical by one person may be deemed unethical by another person. Furthermore, what is deemed ethical in one set of circumstances, may be deemed unethical under another set of circumstances.

Every person in a society is held to the same set of laws, but perhaps not so for ethics or ethical conduct. Professionals, such as medical doctors, lawyers, dentists, veterinarians, architects, and engineers, are all expected to conduct their practices not only within the law but also within a code of ethical conduct established by the profession (e.g., the American Medical Association, the American Bar Association, the American Dental Association, the American Veterinary Medical Association, the American Institute of Architects, and various engineering societies, such as the American Society of Civil Engineers or the NSPE).

The companies and corporations within which most engineers practice their profession and conduct their work are generally quite diligent about having their engineers not break the prevailing law. Unfortunately, while most companies and corporations are careful most of the time to have their engineers operate in an ethical manner, this is not always the case. During a long career, an engineer is far more likely to be faced with a situation where, while not being directed to cross over the line of ethical conduct, there is (or seems to be) an implicit sense or expectation that they are to close their eyes or otherwise blur the line separating ethical from unethical conduct. In the name of "business" in most cases.

Reverse engineering has been the subject of many discussions and papers when it comes to what is considered unethical, even if not illegal. Let's try to clarify this conundrum somewhat, even though the real barometer for ethical conduct may be internal, arising from one's personal system of values, as that little voice in one's head saying, "Don't do that!" The author's mother's suggested barometer was: "If you wouldn't want what you did to appear on the front page of the *New York Times* newspaper for all to see, don't do it!"

17-3 The Legality of Reverse Engineering

Like all law, the law pertaining to reverse engineering depends on the jurisdiction, which is usually national (e.g., for the United States) or, in the case of Europe, established and overseen by a group of countries (i.e., the European Community).

In the United States, even if an entity (known as an *artifact*) or process is protected by trade secrets (e.g., Coca-Cola), reverse engineering the artifact or process is legal as long as the artifact or process was obtained legitimately. For example, it is legal to try to reverse engineer a Mercedes-Benz automobile if one obtains the one to be "dissected" on the open market by legal purchase. Likewise for Coca-Cola, which has been the subject of reverse engineering almost since its introduction in 1886.

Reverse engineering, while legal to perform on artifacts obtained legitimately, does not circumvent patents, however. This notwithstanding, one common motivation for reverse engineering is to determine whether a competitor's product contains patent or copyright infringements for which you or your company holds the rights.

The U.S. law on reverse engineering can be found on the Internet, with an excellent paper on the subject by Samuelson and Scotchmer (ref. Samuelson and Scotchmer).

The situation is different in the European Union, however. Reverse engineering is allowed for the purpose of inoperability but is prohibited for the purpose of creating a competing product, as well as being prohibited for the public release of information obtained through reverse engineering, including software.

It should be obvious that reverse engineering for the purpose of copying a competitor's design, material, or process is absolutely illegal if the resulting copy or clone is misrepresented as the originator's product. This illegal practice is known as *piracy* for software and *knockoffs* for hardware or materials or substances (e.g., the "Rolex" watch one can pick up for a bargain price in Times Square in New York City).

An interesting discussion of the role played by the liberal patent law in the United States from its founding is presented in historian-author Ernest Freeberg's fascinating and (pardon the pun, enlightening) *The Age of Edison: Electric Light and the Invention of Modern America* (ref. Freeberg). In his seventh chapter, "Looking at Inventions, Inventing New Ways of Looking," Freeberg writes of Founding Fathers Benjamin Franklin and Thomas Jefferson, both prolific inventors, and their resentment of any patent. Both felt strongly that, to paraphrase, a free country should allow all citizens an opportunity to share in all benefits of that freedom, including access to new ideas and inventions. Freeberg writes, at length, of various theories as to why Americans, exemplified by Thomas Alva Edison, were such prolific inventors. He reports how Europeans had the edge in scientific education, while American "mechanics" seemed to have the edge as "thinkers."

A favored theory of America's contribution to the exponential growth of inventions is the liberal nature of our patent law. Unlike Europe, where patents require totally new concepts, Freeberg states that " ... under this [U.S.] system, the pursuit of a patent became an abiding preoccupation for many nineteenth-century American men and women. Because the American system granted patents to practical improvements in already established inventions, every framer and mechanic who worked with a machine could see its limitations not just as a source of frustration but also as an opportunity to make improvements, and maybe even a fortune." This surely lent credence and credibility to the technique of reverse engineering. Some, however, thought patents—often based on reverse engineering—contributed to "producing redundant solutions to technical problems that had already been solved," suggesting the evil of mindless copying.

Once again, like so many other things associated with technology, reverse engineering is a two-edged sword.

17-4 The Ethics of Reverse Engineering

In Section 2–3, different motivations were given for why reverse engineering might be used. Different readers may find a few of these motivations questionable in terms of their being ethical (e.g., competitive intelligence gathering). Furthermore, what one reader finds ethical, another reader might find unethical, and vice versa.

To make the question of what is an ethical versus unethical use of reverse engineering even more complicated are considerations of the specific circumstances under which the process is employed, the period in which a decision to reverse engineer is made, and other factors. Few, if any, would even stop to consider whether it is ethical or unethical to use reverse engineering to copy the weapon system of an enemy or potential enemy, as the security of one's country must come first. But is the same true for the security of one's company? Do the ends justify the means?

So what if the People's Republic of China has, in fact (if indeed it did), "stolen" technology for advanced military aircraft, air-to-air or surface-to-air missiles, nuclear warheads, nuclear reactors, or supercomputers? Did not the United States "steal" weapons technology from Germany during or right after World War II (e.g., the V2 rocket)? Did not the United States attempt to "steal" technology—if there was anything worthwhile to steal—during the Cold War with the former Soviet Union (e.g., recall the author's anecdote about examining captured MiG fighter aircraft at Wright-Patterson Air Force Base in the mid- to late 1970s)?

A really good case study on the ethics of reverse engineering is to be found in the cloning of IBM's personal computer in the early 1980s, from which the modern personal computer industry was born.

Prior to the 1970s, the computing field was dominated by multi-million-dollar giant mainframes at one end of the spectrum (e.g., by IBM and Digital Equipment Corporation [DEC]) and by cheap, do-it-yourself home electronic kits on the other end of the spectrum (e.g., Altair 8800). By 1977, three fully assembled personal computer systems appeared: the Apple II, the Commodore PET, and the TRS-80. Together, this "1977 Trinity" played a central role in the emerging computer industry, spawning major subindustries of software and specialized hardware.[5] By the end of the 1970s, two parallel trends emerged: (1) a rapidly growing market for home computers driven by the easy-to-use systems and (2) the fragmentation of the market into mutually incompatible hardware platforms and corresponding software.

Figures 17–1 and 17–2 show the IBM and DEC mainframe computers of the 1970s with the Altair 8800 and the "1977 Trinity," respectively.

---

[5] "The 1977 Trinity" was the title given to the Apple II, Commodore PET, and TRS-80 by *Byte* magazine.

(a)

(b)

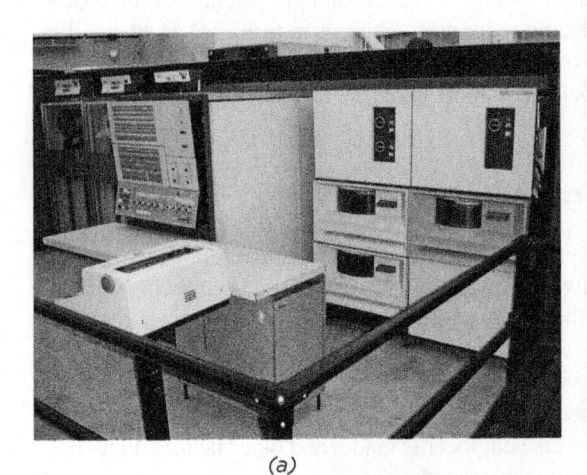

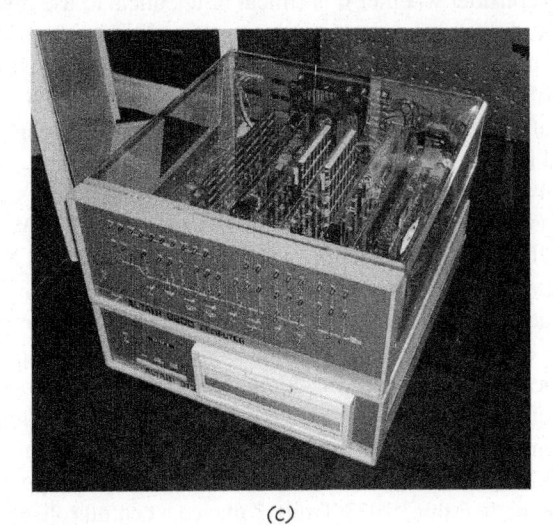

(c)

Figure 17-1 Multi-million-dollar IBM 360 (*a*) and DEC PCP-11 (*b*) mainframe computers dominated the computer marketplace in the 1970s. The MITS Altair 8800 (*c*) was the first personal computer to appear on the market in 1975. It used the Intel 8080 CPU. (*Source:* Wikipedia Creative Commons, contributed by Flickr upload bot on 7 November 2011 [*a*], Stefan King on 18 December 2005 [*b*], and Zzyzx11 on 5 December 2007 [*c*].)

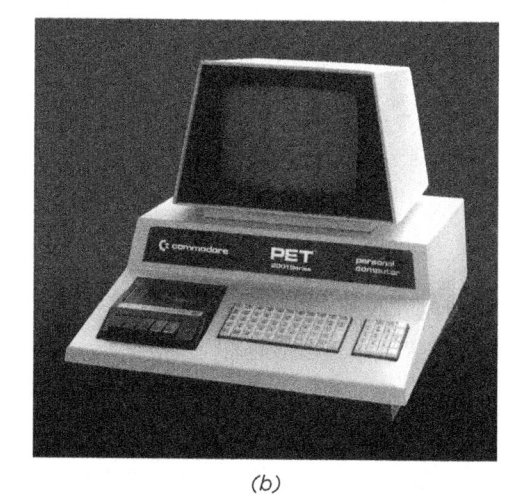

Figure 17-2 The "1977 Trinity" consisted of the Apple II (*a*), Commodore PET (*b*), and TRS-80 (*c*). (*Source:* Wikipedia Creative Commons, contributed by DevonCook on 21 April 2011 [*a*], Rama on 26 October 2011 [*b*], and Ubcule on 25 March 2011 [*c*].)

On August 12, 1981, things changed dramatically with the late arrival of the IBM PC. The industry giant's PC quickly and deeply penetrated and dominated the market. The crucial factor for the new product was the system's open architecture, which allowed software to finally utilize the computer's processor and memory without having to go through the PC's operating system. This spawned a large variety and number of software products that performed extremely well on IBM machines but were incompatible with similar machines from other manufacturers.

Figure 17-3 The IBM 5100 PC introduced in 1977 (*a*) and Compaq's clone (*b*). (*Source:* Wikipedia Creative Commons, contributed by Sandstein on 4 September 2011 [*a*] and Museo8bits on 27 September 2006 [*b*].)

To IBM's dismay, one of the results of their PC's open design was that it was easy to copy by reverse engineering, so the appearance of "clones" was inevitable. And so, the two horns of a dilemma arose: IBM tried to corner the market by placing legal restrictions on their product even as they had built it using openly available parts. Others (e.g., Compaq) felt no compunction about reverse engineering and cloning the IBM 5100 to offer their own entrées to the growing marketplace.

Figure 17–3 shows the IBM 5100 and BIOS clones.

As to what is ethical when it comes to reverse engineering the work of another individual, organization, enterprise, or institution, here are three guidelines:

- It is ethical only if it is legal to perform reverse engineering within the prevailing jurisdiction.
- It is ethical provided there has been no blatant abuse of the rights and intellectual property of the originator or creator in a mindless or artless copy (i.e., with no improvements or other modifications).
- It is ethical if there has been no attempt to mislead, deceive, defraud, or cheat the ultimate user.

A final test may be: Would you be ashamed of what you are doing (or did) if your mother knew about it?

The story—and argument for the right to reverse engineer or not—is brilliantly presented by Basu (ref. Basu). It is left to the interested reader to check it out. It will suffice to capture Basu's closing thoughts and conclusions here:

> A free industrial society cannot thrive if there is insufficient competition. The formation of monopolies results in the concentration of power in the hands of a few which, history has

shown, never results in an overall benefit for society. Only if there are competing products can there be a clear drive to innovate and succeed. Reverse engineering guarantees that no entity can rest on its laurels for more than a short amount of time. The right to reverse engineer is at the heart of modern scientific method and free industry. This right is a must for any technological society dedicated to the ideas of democracy and independence. If scientific advancement is to be considered a natural path for social evolution, then reverse engineering must be viewed as a necessary brick on that road.

In his fascinating *The Age of Edison: Electric Light and the Invention of Modern America,* author Ernest Freeberg notes: "Reflecting on America's remarkable technological creativity, one science educator claimed an invention's 'greatness' was best measured by its power to generate ever more inventions." By this mesaure, there have been few if any inventions greater than Edison's bulb!

Bravo and amen! But play fair, be ethical, and use reverse engineering as a jumping-off point for new innovation, not an easy end goal.

17-5 Summary

There is what is legal and allowed, and there is what is ethical and right. The former has jurisdiction; that is, it may be different in one place than in another. But, at least when it comes to engineering, ethical conduct is pretty much the same everywhere, even though it ties closely to personal values within societal values.

Reverse engineering can be motivated by many purposes, some of which pose little or no conflict with ethical conduct. However, there are some motives for reverse engineering that are not so pure and are not ethical. The most obvious guidepost is: Would I be ashamed if my mother or father knew I did this? Reverse engineering should never be done to deceive and, preferably, should be used not to copy a prior design but to learn from it to stimulate imagination and innovation.

17-6 Cited References

Basu, Shrutarshi, "The Right to Reverse Engineer," http://basus.me/writing/reverse.html.

Freeberg, Ernest, *The Age of Edison: Electric Light and the Invention of Modern America,* Penguin Press, New York, 2013.

Samuelson, Pamela, and Suzanne Scotchmer, *The Law and Economics of Reverse Engineering,* on SamuelsonFINAL.com, April 10, 2002.

17-7 Thought Questions and Problems

17-1 Professionals, such as doctors, lawyers, and engineers, all serve the public and so have a particular obligation to do what is *right,* not just what is legal. All of the people in these professions, like all people in a society, are bound to obey the law and in their practice as professionals are also bound by special laws to do only what is legal for that profession.

But as professionals, doctors, lawyers, and engineers are also honor bound to adhere to the profession's code of ethics.

Reread the *Code of Ethics for Engineers* of the National Society of Professional Engineers (NSPE) given in Section 17–1. Then reread the Society's "Fundamental Canons," of which there are six listed professional duties.

    a. Briefly comment on each of the six listed duties under the NSPE's "Fundamental Canons." Be sure to offer your opinion on the merit of each of these duties.

    b. Without any attempt to dwell on the negative, it is as important to know what *not to do* as what *to do* as one aims to serve (or serves) in an honorable profession like engineering. Therefore, think of (from your personal experiences or readings), find (on the Internet), or make up a reasonable and realistic example of inappropriate conduct by an engineer for each of the six listed professional duties.

17-2 a. Look up several definitions of the words *legal* and *ethical* in several different dictionaries, including business dictionaries. Then, prepare a thoughtful essay of one page or less that clearly elucidates the differences.

    b. Based on your findings in dictionaries and from your essay in part (a), very briefly explain why it is so important for an engineer to do what is *ethical* beyond simply what is *legal*.

17-3 To obey the law, one must know the law, as it is said: "Ignorance of the law is no excuse" [for violating it]. So, as one about to enter—or, perhaps, as one who is new to—the practice of engineering and sure to employ the technique and process of *reverse engineering* many times in your career, for many purposes, you need to know the law pertaining to reverse engineering in the country in which you plan to (or do) practice.

Look up and carefully read the law for conducting reverse engineering legally in your country of planned or actual practice. Then, to embed the key tenets of this law in your mind, prepare a narrative summary, of one page or less, of what is legal.

17-4 In U.S. criminal law, *means, motive, and opportunity* is a popular cultural summation of the three aspects of a crime that must be established before guilt can be determined in a criminal proceeding. In ethics, *motivation* is undoubtedly the differentiating factor for what is ethical and what is not.

The various motivations for conducting reverse engineering are given in Section 2–3.

    a. Briefly comment on how *you* feel about the ethicality of each of the motivations listed. Be sure to indicate for each whether there is no question that it is ethical, no question that it is unethical, or "it depends on the circumstances." For the latter, clarify what you feel are differentiating circumstances.

    b. In a separate brief narrative of about one page, reflect and comment on reverse engineering to create a "clone" versus a "knockoff." Be sure to make it clear whether you believe that one is less ethical than the other.

    c. In a separate brief narrative of about one page, reflect and express your opinion on whether there is—or should be—any limit on what people would or could do for the security of their country. *Before* you form your opinion, be sure to look up the decision in the Nuremburg Trials, in which Nazi officers running "death camps" defended their actions as "only following orders." The judges stated: "Following orders is not an acceptable excuse for what is clearly immoral or unethical conduct."

17-5 In a thoughtful essay of less than one page, comment on the quote (at the end of Section 17–4) from Shrutarshi Batu in his insightful article "The Right to Reverse Engineer" (obtainable via Internet search). *Before* finalizing your thoughts, look into two landmark cases in recent U.S. business history: (1) the forced successful divestiture of the Bell System within AT&T in 1974 and (2) the unsuccessful attempt to break up Microsoft in 2000–2001. (You might even wish to focus your essay on "why the difference?".)

CHAPTER 18

The End of a Book, the Beginning of a New Story: Closing Thoughts

18-1 The First Design

[1] In the beginning God created the heavens and the earth. [2] And the earth was without form, and void, and darkness was upon the face of the deep. And the spirit of God moved upon the face of the waters.

[3] And God said, Let there be light, and there was light.

So says Genesis I, verses 1 through 3, of the creation narrative in the Old Testament in the King James Version of the *Holy Bible*, held as sacred by those of Jewish and Christian faith (Figure 18–1).

The ancient Greeks referred in myths to *Chaos* as the formless void state that preceded the creation of the Universe or *cosmos*, or, more correctly, as the gap between heaven and Earth, with everything thereafter born out of darkness (Figure 18–2). There are as many creation "myths" as there are ancient civilizations, and, somewhat remarkable, almost all have our Earth being the product of a creator, from nothingness. The form of the creator differs, but the common factor is that the Universe, our planet, we and our fellow creatures and plants, are all the product of a grand design. Even the brilliant theoretical physicist and cosmologist Stephen Hawking (1942–) wrote *The Grand Design* with Leonard Mlodinow (ref. Hawking and Mlodinow), although Hawking dodges full acknowledgment of a creator as a supreme architect by talking about "the ultimate miracle."

The myths, legends, stories, and theories may differ in details, but the end product is the same: The Universe. The Earth. Plants and animals in an elegantly balanced ecosystem—and many, including the author, would argue a perfect design!

But what about we human beings? We are not perfect. Some less so than others, but none is perfect. As a logical consequence, engineers are not perfect and, being imperfect, may be incapable of creating anything that is perfect. An interesting subject for philosophical debate,

386 Chapter Eighteen

(a)

(b)

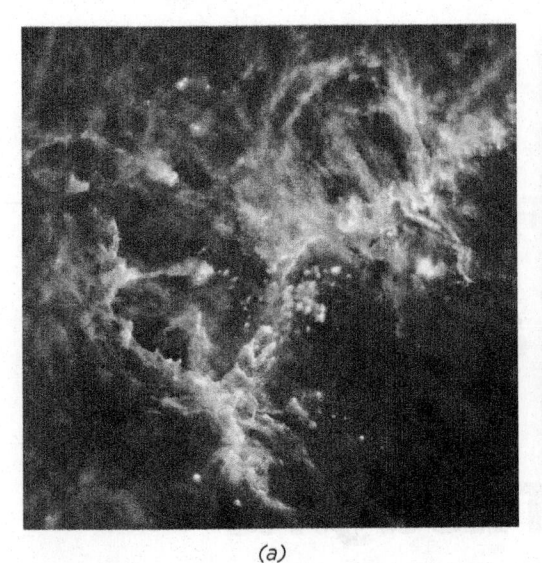

Figure 18-1 A photograph taken from the European Space Agency's Herschel Space Observatory, which is sensitive to near-infrared to submillimeter wavebands, showing the Eagle nebula, with its intensely cold gas and dust (a). "The Pillars of Creation" made famous by a 1995 photograph from NASA's Hubble Space Telescope (b) can be seen at the lower left-center of (a) to lie with the Eagle nebula. Nebulas (or nebulae) are regions of interstellar space where stars are forming, and, so, indicate the formation of a new galaxy. (*Sources:* From www.nasa.gov, with full credit for far-infrared to ESA/Herschel/PACS/SPIRE/Hill, Motte, HOBYS Key Programme Consortium and for x-ray to ESA/XMM-Newton/EPIC/XMM-Newton SOC/Boulanger [a]; and from Wikipedia Creative Commons, originally contributed by watcharakom on 4 May 2005, with modifications by Lokal Profil on 11 June 2006 and Twinsday on 19 March 2012 [b].) **Don't miss the color version of this figure, available at www.mhprofessional.com/ReverseEngineering.**

Figure 18-2 The myth of creation held by the ancient Greeks involving Erebus and Nyx. Erebus was a primordial deity and personification of darkness, while Nyx was a goddess, the personification of night and mother of subsequent personified gods, who, together, created the Universe from the darkness. (*Source:* Wikipedia Creative Commons, contributed by Sailko on 3 February 2011.)

perhaps, but let's accept the premises that (1) humans are not perfect and (2) imperfect beings are incapable of creating anything perfect. What, then, are we faced with?

18-2 Imperfect Humans Need Reverse Engineering

"To err is human (to forgive divine)." So said Alexander Pope (1688–1744), English poet, in "Essay on Criticism," written in 1711. While the more-often-quoted first half of this proverb applies to all of us as humans, even more relevant to those of us who are privileged to be engineers is the tongue-in-cheek-paraphrase title of a wonderful book by Henry Petroski (1942–), *To Engineer Is Human: The Role of Failure in Successful Design* (ref. Petroski). The central cover photograph on Petroski's book makes the point with the most famous—or infamous—engineering blunder ever: the collapse of the months-old Tacoma Narrows Bridge, dubbed "Galloping Gertie" by workers during its construction, on November 7, 1940, due to wind-induced aerodynamic flutter and resonance at the bridge's natural frequency. Amazingly, a homemade movie was taken by a witness that day that shows the bridge swaying and, finally, collapsing. This can be seen on YouTube under "Tacoma Narrows Bridge Collapse—Galloping Gertie." Photographs of the collapsing bridge, while dramatic, are strictly copyrighted and cannot be published without permission.

The moral of the story is this: As hard as they try, engineers don't always get it right, and never get it perfect! While we may let down our Creator by being less than perfect, in a way, we may be unable to achieve perfection because of one of His or Her design rules,[1] that being the *second law of thermodynamics,* which says that things tend to move from order (and perfection) to disorder (and imperfection). Be that as it may, what engineers need to do, as a matter of routine, is *reengineer* less-than-satisfactory designs, attempting to improve what needs improvement and not lessening anything that is working well.

As has, hopefully, been made clear for those who have read this book through, one extremely valuable, but misunderstood, underappreciated, and often disparaged technique for reengineering involves *reverse engineering.* By systematically physically or mentally dissecting an old design, the engineer learns. The intended paramount lesson of this book is that reverse engineering has the potential for much more than artless copying. In using reverse engineering, diligent engineers should always be striving to produce not just a new model but their knowledge and understanding of design for creating other products.

One caveat for those new to engineering is the troubling reality that we all, as engineers, are expected to do, most of the time, the best we can within the allotted time and with the allotted money and other resources.[2] We engineers always feel we could have done better. What we need to learn and believe in our hearts and minds is that we did the best we could within the given constraints. When we've arrived as engineers is when we know that no one else could have done it better. That's not arrogance, it's self-confidence!

---

[1] As an engineer who chose to teach after 16 years in industry, the author believes "the Designer" (as Dr. Wernher von Braun referred to a creator in his quote given in the front matter of this book), like all good designers, set some rules to guide design. The first was to expend as little energy as possible, a variant of the first law of thermodynamics. (This would give the Designer a day off, as suggested in Genesis 1 in the *Holy Bible.*) As long as He or She was at it, the logical second rule would be to make everything perfect. But, as soon as the thought of creating seas from water, which lacks the "perfection" of crystalline ice, and air, which lacks order altogether, arose, the second rule had to be revised to ... accept, even if not prefer, disorder over order (i.e., to accept imperfection). Hence, the second law of thermodynamics. Oh, and good thing about accepting imperfection for most of us!

[2] This is, after all, what our parents told us to do when we were in their charge, that is, the best we can do.

Examples of how engineers could have done better, and how they need to reengineer almost as a matter of routine, abound. They confront us almost every day. Just look and think as you go about your life. What were engineers thinking when they drew plans to clear-cut an ugly scar across a lushly forested verdant mountain to run power transmission lines? What were engineers thinking, or weren't they thinking, when a newly constructed complex interchange is under "road work" a couple of years hence? What were engineers thinking when they built a suspension bridge across a narrow cut between hills in an area near Tacoma, Washington, known for its prevailing winds and didn't take into account lift and flutter and induced resonance, when they learned about all of those things as undergraduate students in engineering school?

For the rest of this chapter, kindly indulge an old engineer who chose to teach halfway through his career as he shares some not-completely-random thoughts that emerged as he contemplated how to close this story. Who knows? What can you lose? Whether you agree or disagree, the author has stimulated your thinking and, hopefully, has encouraged you to take action. That's a teacher's goal!

18-3 Order from Chaos, Light from Darkness, Knowledge from Knowledge

Every autumn in the northeastern United States,[3] Mother Nature "struts her stuff" as the leaves of deciduous trees turn spectacular shades of yellow, gold, orange, and red (Figure 18–3). Then She sends a not-so-subtle reminder that She seems to prefer disorder over order, chaos over perfection. The leaves that filled out such trees fall to litter the ground, day after day, cleanup after cleanup, until either She prevails with the ground covered thick with leaves turned brown or the homeowner prevails with a lawn and garden free of leaves but with aching muscles, a sore back, and a chance to finally rest. Entropy indeed seems to rule!

But there seem to be some serious exceptions. The same Mother Nature from whom the second law of thermodynamics originates seems to suspend it, as opposed to abandon it, on occasion. Sometimes for long occasions!

Regardless of the creation myth, legend, story, or theory, from God to the big bang, order emerged from chaos. Otherwise, we wouldn't be here! Cold gas and dust in nebulae *organize* into stars, planets, and moons (Figure 18–1, again). Order arises from disorder. In the big bang theory, all matter concentrated into an infinitely dense volume at the beginning of time, exploded[4] to give rise to what? Disorder? True, the Universe is still expanding outward from that original explosion, but along the way, we find ourselves living on a planet reordered from dust and gaze out in wonder at uncountable and incomprehensible examples of other reordering from gas and dust! Galaxies. Solar systems. Stars. Planets and planetoids or asteroids. Moons. Each and every one the result of disorder moving to create (or already having created) order. And not for an instant, but for billions of years! More amazingly, if thermodynamics prevails and life on our Planet Earth emerged from

---

[3] The northeastern United States includes the New England states of Maine, New Hampshire, Vermont, Massachusetts, Rhode Island, and Connecticut, and, some would say, upstate New York, as well as the rest of New York, New Jersey, and Pennsylvania. Deciduous trees predominate in these states, and autumn is characterized by increasingly cool and windy weather until winter arrives, officially on December 21, but, in reality, earlier, in November for most.

[4] The "big bang" was not truly an "explosion." Nothing "exploded." Repulsive forces simply overwhelmed attractive forces in the singularity of subatomic pure energy to force new matter outward to fill space.

Figure 18-3 The glorious display of colors in autumn, here near Hogback Mountain, Vermont, in New England, with the triumph of entropy as leaves litter the ground. (*Source:* Wikipedia Creative Commons, contributed by Chensiyuan on 13 October 2009.) **Don't miss the color version of this figure, available at www.mhprofessional.com/ReverseEngineering.**

primordial ooze, and if one accepts evolution (for which there is considerable evidence), most, if not all, forms of life are improving. How did that happen? By becoming *less* ordered or *less* perfect?

So how does and will this specifically affect us, as engineers, even as we concede that it seems to have affected and continues to affect us as human beings? The author suggests we engineers build on what both Genesis and the myths of the ancient Greeks, as but two examples, purport, that is: Light emerged from darkness.[5] With one of the metaphorical meanings of *light* being "knowledge," a consequence would be that knowledge begets knowledge.

Let's look at each suggestion briefly before ending with four opportunities to be enabled by reverse engineering (see Section 18-4).

Following the fall of the Roman Empire, Europe entered what became known as the "Dark Ages" between the fifth and the thirteenth centuries.[6] Intellectual darkness replaced the "light of Rome," and it wasn't until the rise of the Italian Renaissance in the fourteenth century and the general Renaissance in the later sixteenth century that this darkness was replaced with light. It was during this period that the polymath genius of Leonardo da Vinci (1452–1519), the artistic and architectural genius of Michelangelo (1475–1564), and the artistic magnificence of Raffaelo Sanzis, known as Raphael (1483–1520), emerged (Figure 18-4).

---

[5] Another example of something from nothing, even if not order from disorder.
[6] The term "Dark Ages" originated with the Italian scholar Petrarch, actually, Francesco Petrarca (1304–1374) in the 1330s, as it described the loss of the light of curiosity and creativity.

Figure 18-4 Raphael's *The School of Athens*, painted as a fresco between 1509 and 1511, is from the artist's imagination. While it contains undeniably identifiable depictions of some great philosophers of ancient Greece (e.g., Plato and Aristotle), many of the philosophers shown bear the faces of some of Raphael's contemporaries, such as da Vinci, Michelangelo, and others less well known. (*Source:* Wikipedia Creative Commons, originally contributed by Jic on 27 January 2005 and modified by Franks Valli on 11 September 2006, Howcheng on 28 August 2008, and Harpsichord246 on 22 September 2012.) **Don't miss the color version of this figure, available at www.mhprofessional.com/ReverseEngineering.**

A prime enabler for the Renaissance was the invention of a German blacksmith, goldsmith, printer, and publisher, Johannes Gensfleisch zur Laden zum Gutenberg (1395–1488)—the moveable-type printing press in 1450 (Figure 18–5). With the ability to print 3600 pages per day instead of less than half that number by block printing and only a few pages by hand copying, the number of printed books grew exponentially from less than 2 million before Gutenberg's invention to more than 20 million by 1500 and to more than 200 million before the sixteenth century ended, as presses spread beyond Western Europe across the world. Greater and more widespread access to knowledge enabled by the new books directly led to exploration of new worlds beyond Western Europe. Knowledge gained led to a renewed interest in and quest for scientific understanding, proliferation of other inventions, and a general greater sense of purpose among human beings. Knowledge obtained as "light" from "darkness" begat more knowledge.

Another, further emergence of the metaphorical "light" occurred around 1650 to 1700. Known as the Age of Enlightenment, intellectuals of the period attempted to reform society using reason (hence, the alternate name of Age of Reason often heard), challenge ideas grounded in tradition and faith, and advance knowledge though the scientific method based on systematic observation of

Figure 18-5 A 1568 engraving showing an early wooden version of Gutenberg's moveable-type press. Such presses could produce up to 240 impressions per hour. This was a dramatic increase over earlier block presses and, particularly, over hand copying. (*Source:* Wikipedia Creative Commons, contributed by Parhamr on 19 September 2007.)

cause and effect. The period promoted scientific thought, skepticism, and intellectual interchange, and opposed superstition, intolerance, and some abuses of power by the church and the state. Knowledge was attempting to change the world for the better!

The most relevant statement to our purpose to better understand and appreciate reverse engineering to come from the period is: "If there is something you know, communicate it. If there is something you don't know, search for it" (Figure 18–6). "Search for it," just as reverse engineering demands!

Just as the invention of the moveable-type printing press revolutionized the world by

Figure 18-6 An illustration from the frontispiece of a 1772 edition of *Encyclopedie*. *Truth*, in the top center, is surrounded by light and unveiled by the figure to the right, *Philosophy* and *Reason*. (*Source:* Wikipedia Creative Commons, originally contributed by Tomisti on 8 September 2005 and modified by Shizhao on 17 March 2011.)

Internet Users in the World
Distribution by World Regions - 2012 Q2

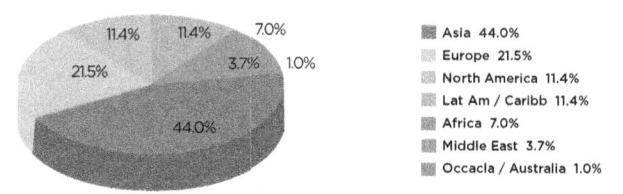

- Asia 44.0%
- Europe 21.5%
- North America 11.4%
- Lat Am / Caribb 11.4%
- Africa 7.0%
- Middle East 3.7%
- Occacia / Australia 1.0%

Source: Internet World Stats - www.internetworldstats.com/stats.htm
Basis: 2,405,518,376 Interrnet users on June 30, 2012
Copyright © 2012 Miniwatts Marketing Group

Figure 18-7 A pie chart depicting the distribution of Internet users around the world by regions as of the second quarter of 2012. The total number of users on June 30, 2012, was a remarkable 2,405,518,376, or approximately 34.3 percent of the world population at that time. (*Source:* Recreated from a pie chart shown on www.internetworldstats.com/stats.html, with copyright held by Miniwatts Marketing Group; used with permission.)

exponentially expanding access to knowledge, the invention of the Internet in the later 1960s promises to make this earlier expansion seem trivial by comparison.[7] As of June 30, 2012, a website at www.internetworldstats.com/stats.html reported 2,405,518,376 users, representing 34.3 percent of the world's population (Figure 18–7). This has occurred in just 40 years since its invention. With a world population estimated at around 500 million in 1650, even if 100 million books had been equally distributed among these people (which they most assuredly were not!), about half that percentage had access to books 200 years after the invention of the printing press. The greater likelihood is that only a small percentage of people had access to books by that time, as very few could read. That's impressive! Exponential, in fact!

For the author, this is a harbinger of a "New Renaissance"—one that will make the old one appear modest, if not trivial, by comparison! At the foundation of this "New Renaissance" will be computer technology. The approach proposed should hinge on reverse engineering as the best and most efficacious way to learn from the old to create anew, as "a necessary brick on the road to scientific advancement," to paraphrase Basu (Section 17–4).

18-4 Learning from the Old to Create Anew: Four Opportunities

If one simply looks at what they or others have done in the past as a guide to what needs to be done in the future, as the authors of predecessors of this book have suggested (ref. Otto and Wood)

---

[7] While there were many key steps along the timeline for the Internet (see "Time-line for the Internet" on Wikipedia.com), the origin is generally attributed to the concept of ARPANET by the military think tank RAND in 1964. ARPANET was to be a means for ensuring that information could be communicated and interchanged among leaders and advisors in the event of a nuclear attack on the United States. While the first attempt at Stanford University crashed when the "g" was typed for "Login," the system was successfully demonstrated in 1971 between National Laboratories across the United States.

and even advocated (ref. Ingle), the company, people, a country, and the world are limited to evolutionary advances in a design at best. In the meantime, recent history in the United States, the European Union, and the rest of the industrialized world abounds with examples where "same-old, same-old" isn't leading to advancement, and old paradigms seem to be failing. New models are needed for running major corporations, wherein better quality at lower cost and with greater agility and responsiveness becomes the mantra and the method. New models are needed for operating local, state, national, and international economies, wherein spending is curbed and cut, while the earnings of those paying the bills are not further reduced by taxes. New models are needed for providing better living standards for both those able-bodied enough to work and for those not so fortunate, but with a moral sense that if one can contribute, one is obliged to contribute by working. And a new model for healthcare and medicine, wherein everyone has access to the best technology and talent available and where the fundamental approaches by which physicians cure is changed to prevent dis-ease and toward treating and curing the underlying cause of any dis-ease that does occur, not simply masking the symptoms or managing the condition.[8]

What is needed in many areas driven by technology is revolution not evolution. Reverse engineering holds the key, by learning from the old to create anew, not simply new! Four opportunities will be briefly described, as follows:

1. Revolution via reverse engineering of energy sources and systems
2. Revolution via reverse engineering of transportation
3. Revolution via reverse engineering of materials
4. Revolution via reverse engineering of medicine

The premise in each case is this: By using reverse engineering principles to see what has been done in the past that has worked well, as well as what has not worked so well or no longer seems to work, totally new designs can emerge, all of which deal with the root-cause(s) to effect desired change(s) with consideration of the entire system, and not simply parts of the whole.

So here we go . . .

Revolution via Reverse Engineering of Energy Sources and Systems

Everyone knows and supports the idea that energy should be sustainable and, within reason, renewable. But, in a reasonable world, sustainable alone may have to do for at least the shorter term, while never losing sight of the greater goal of sources being renewable.

Sustainable energy is the provision of energy that meets the needs of the present without compromising the ability of future generations to meet their needs. *Renewable energy* is limitless and mitigates greenhouse gases and other pollutants to the environment. The focus of sustainable energy is on the ability to continue providing energy, even if some pollution of the environment occurs, as long as it is not sufficient to prohibit heavy use of the source for an indefinite amount of time.

Table 18–1 lists sustainable energy sources, as well first-generation renewable technologies (up to 150-year-old technologies that were used during or emerged from the Industrial Revolution from about 1760 to around 1840 in Europe, and later elsewhere), second-generation renewable

---

[8] The use of prescriptive tranquilizers, sedatives, and painkillers in the United States is growing rampantly among and to disturbing percentages of our population. Doctors are enabling people to cover their dis-ease, with no sense of obligation to seek the underlying cause in too many cases, if statistics are to be believed!

TABLE 18-1 First-, Second-, and Third-Generation Renewable Energy Technologies as Sustainable Energy Sources

*First-generation technologies:**
- Hydropower/water power (from falling or running water)
- Wind power (to provide kinetic energy)
- Biomass combustion (of natural plant matter)
- Geothermal power (from natural steam)

Second-generation technologies:
- Solar heating and cooling (from solar thermal collectors and heat exchangers)
- Wind power (to make electricity)
- Bioenergy (wood, wood waste, straw, sugarcane, manure, and by-products of agricultural processes)
- Solar photovoltaics

Third-generation technologies: †
- Biomass gasification
- Biorefinery technology (to make combustible biofuels)
- Concentrated solar thermal power
- Hot dry rock geothermal energy
- Ocean energy (from waves, currents, tides, heat cycles, and salinity gradients)

*Most of these were used directly to provide kinetic energy or to make steam to provide kinetic energy, and not to make electricity.
†Almost every one of these is intended to make steam to drive electrical generators.

technologies (now active and emerging technologies largely developed after the 1980s), and third-generation renewable technologies (still in the research-and-development stage, albeit with some demonstration systems). Figure 18–8 is a chart showing the relative contributions of various sustainable and/or renewable energy technologies being used to supply the current world.

The keys for the future are these:

- There is no "best" energy source nor is there a "one-size-fits-all" energy source. What is best for one area is different in another area. If one lives near Niagara Falls, one is insane not to use hydroelectric power. People in Phoenix, Arizona, have no such option but can use solar power (combinations of photovoltaic panels and solar concentrators), and would be insane to do otherwise. On the other hand, those who live in Kansas have neither raging rivers nor waterfalls nor blazing sunshine for six to eight hours per day for 350 days per year. Hence, they ought to use wind turbines, as prevailing winds sweep across the prairie. Remember Dorothy in *The Wizard of Oz*?[9]
- Every country, as well as every continent and the world, needs to have a coordinated and integrated long-term energy policy, now, as we didn't have it yesterday when we should have, and we can't wait much longer.
- A key part of an energy policy is to accept that a *system* of different energy sources is the answer, and such systems need to be decided upon and designed with the use of reverse engineering of the past. Technology and technologists, not politics and politicians, often overly influenced by ill-informed and hysterical environmental activists, need to set policy.

[9] *The Wizard of Oz*, a 1939 film released by Metro-Goldwyn-Mayer, was based on the 1900 book *The Wonderful Wizard of Oz*, by L. Frank Baum.

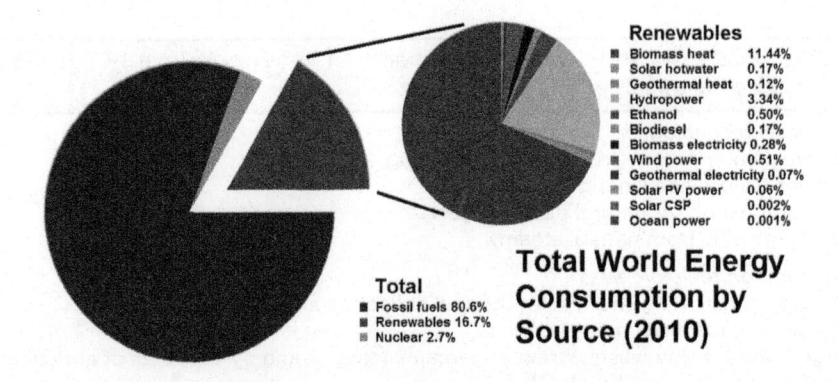

Renewables
Biomass heat	11.44%
Solar hotwater	0.17%
Geothermal heat	0.12%
Hydropower	3.34%
Ethanol	0.50%
Biodiesel	0.17%
Biomass electricity	0.28%
Wind power	0.51%
Geothermal electricity	0.07%
Solar PV power	0.06%
Solar CSP	0.002%
Ocean power	0.001%

Total
- Fossil fuels 80.6%
- Renewables 16.7%
- Nuclear 2.7%

Total World Energy Consumption by Source (2010)

Figure 18-8 A pie chart showing relative contributions to the total consumption of energy for the world in 2010 by source, with a breakout of renewable energy sources. (*Source:* Wikipedia Creative Commons, contributed by Delphi234 on 26 June 2011.) ***Don't miss the color version of this figure, available at www.mhprofessional.com/ReverseEngineering.***

- It must be recognized and made real that new energy sources alone are not the solution to the world's energy needs. Also needed are responsible efforts at conservation, abhorrence of waste, and dramatically improved efficiencies for generation systems.
- New and more effective systems for pollution abatement and distribution are required.

Not incidentally, an objective, scientific, and rational (versus emotional or, worse, hysterical) look must be taken as to (1) whether the world is experiencing climate change or not; (2) if it is, what is the direction and rate; and (3) what is or are the source or sources. Then and only then can we rationally address any problem by attacking its root-cause. And, most important of all, we cannot "let the tail wag the dog" by shutting down our industry and commerce to move back to what would quite literally be a "dark age."

Revolution via Reverse Engineering of Transportation

How we move around our town, state, country, continent, and world is, unless we walk, tied intimately to how we address our energy needs and sources. Realistic assessments have to be made as to the *real* efficiency—and efficacy—of electric vehicles, considering that the electrical energy that allows them to operate overwhelmingly comes from power-generating utilities that burn fossil fuels (see Figure 18–8) and that losses along the way overwhelm any apparent advantage over burning a fossil fuel in the vehicle. And, once we determine whether electric vehicles accomplish what those that advocate and own them believe they accomplish, what do we do for air transportation and ship transportation? Use big extension cords, big batteries, flapping wings, sails, or oars? This is meant as more than facetiousness or cynicism! *What do we use?*

A few keys:

- It's time to look at the recent past, when every city in America had electric trolleys, which were taken out in the 1950s and replaced by personal passenger cars and taxis (Figure 18–9). (But the world cannot operate with pedicabs, either!)

Figure 18-9 A modern "trolleybus," known as *Troiza*, operating in Moscow. These use—and choose between—electrical power and natural gas. Moscow operates the largest trolleybus system in the world. (*Source:* Wikipedia Creative Commons, contributed by Kowkamurka on 19 July 2011.)

- For America, it's time to reflect on how and why we have the finest interstate highway system in the world. It was the brainchild of former General Dwight David Eisenhower (1890–1969), upon returning from the European theater of World War II to become our thirty-fourth president (1953–1961), to provide routes for civil defense. But when gasoline cost around 20 cents per gallon, and diesel fuel cost less, cars and trucks proliferated and Americans and America became addicted. In our collective delirium, we allowed our passenger and freight-hauling railroads to fall into disarray.
- A new look has to be given to mass transportation versus personal transportation.
- New modes of transportation need to be considered (e.g., using the 2- to 3-million-mile network of 1- to 2-meter-diameter pipelines for transporting water, natural gas, oil, and, in some places, powdered coal, grain, and so on, to move cargo-containing pods). Systems have been proposed to move freight in tubes beneath highways (Figure 18–10*a*) and goods beneath urban complexes (Figure 18–10*b*). There are even proposals to move people in evacuated tubes, like those found in drive-up tellers at banks (see YouTube, "Evacuated tube transport . . .").

Then there is the huge question of how we move millions of businesspeople and tourists around the skies every day. How do we do that without burning Jet-A?

Revolution via Reverse Engineering of Materials

The traditional method by which designers select materials in designs has been to match the listed properties of commercially available engineering metals or alloys (e.g., pure copper or a

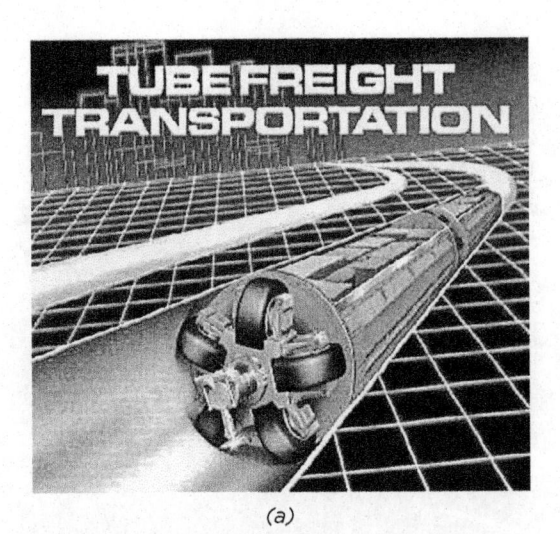

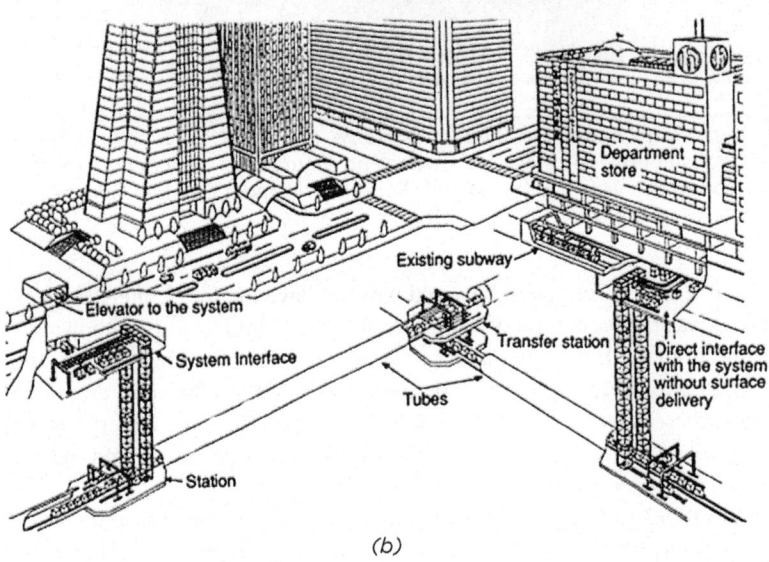

Figure 18-10 Concepts for tube freight transport have received serious study by the U.S. Department of Transportation's Federal Highway Administration (a), and automated underground tube networks have received serious consideration for urban goods transport by the International Association of Traffic and Safety Sciences (IATSS) (ref. Koshi) (b). (*Sources:* U.S. Department of Transportation, Federal Highway Administration, a government agency allowing free use [a]; and the IATSS, with permission [b].)

copper-zinc yellow brass), porous ceramics (e.g., cement and concrete), engineering ceramics (e.g., alumina, silicon carbide), glass (e.g., soda-lime glass, borosilicate glass), or engineering polymers (e.g., polypropylene or polyetherether-ketone/PEEK) to design requirements involving functionality and performance via properties. More recently (see Section 9–6), this approach evolved to a more formal and systematic technique involving a formally derived Material Performance Index (or Indices) as design guidelines on Material Selection Charts. With either approach, it was the rule rather than the exception that engineers settled for the material(s) closest to providing the set of properties they needed. For the most part, they chose the best among comparable options but rarely got the truly optimum set of properties.

Advances in materials science in general, and in nanotechnology in particular, have changed the opportunity. It is rapidly becoming not only a possibility but a reality to have a material that exhibits precisely the set of properties (i.e., the functionally specific properties) needed. These are known as *designer materials* but are probably more appropriately referred to as *materials by design.*[10] Nanotechnology has made feasible, although still at the R&D stage, the possibility of designing and then building a material from the bottom up, rather than choosing and settling for one from the top down.

Materials by design would be conceived, designed, and optimized from first principles of the quantum mechanical theory of electron behavior in solids, using computational modeling and model-driven, computer-controlled manufacturing. Most of these will be like familiar alloys, ceramics, glasses, or polymers, except with exactly the properties desired via a thoughtfully conceived and meticulously developed composition and nano- and/or microstructure, more will be custom-engineered composites, and some will be *meta-materials.* These are meticulously (versus naturally) constructed atom by atom or molecule by molecule—using nanotechnology (e.g., self-assembly)—to create solid materials with qualities and responses not found in existing materials (ref. Messler). Properties could be fairly conventional but precisely matched to need, or nonconventional and seemingly unachievable based on traditional materials.

K. Eric Drexler, the "Father of Nanotechnology," in a new book, *Radical Abundance: How a Revolution in Nanotechnology Will Change Civilization* (ref. Drexler), speaks about how the digital revolution that has driven computing, communication, and information technology for the past four to five decades is "about to give way to a form of production [of materials and devices] that will radically transform the world economy and that could also save the environment." He speaks at length about atomically precise manufacturing (APM), which builds functionally specific properties into materials and devices atom by atom, molecule by molecule, from the bottom up using techniques analogous to and that will evolve from 3D printing or stereolithography.

As stated at the outset of this section, reverse engineering would play a central role in identifying precisely what is needed, compared to what existed before. The details will come from nanotechnology, but the necessary guidance will come from reverse engineering.

Revolution via Reverse Engineering of Medicine

There is prehistoric archaeological evidence of trepanning (i.e., the scraping, chiseling, or drilling of holes in the human skull) as early as 6500 BC in the area of modern France, making it the

---

[10] The subtle change is more about shedding the shallow association of such highly sophisticated materials with designer jeans, designer fragrances, and designer drugs, licit or, worse, illicit, although the parallel with either type of drug is apropos.

oldest surgical technique known to have been performed. The technique reached its peak with the Incas (1430–1533). The purpose was relief from severe headaches or to release "evil spirits" residing in one's head. Physicians in ancient Greece, nearly 2500 years ago, practiced bloodletting, making small incisions in the skin to allow "bad blood" to run off to relieve a variety of diseases. The practice continued until late in the nineteenth century. Similar bloodletting was practiced by the renowned surgeons Charaka and Susruta during the Common Period in ancient India (ca. 300 BC) using leeches applied to wounds, with the practice continuing late into the nineteenth century in America and elsewhere.[11]

While these, and other, once-practiced medical techniques seem crude, if not barbaric, one only needs to reflect on how modern medicine as practiced in the Western World, and elsewhere,[12] can be divided into two schools of thought, in the most simple and unadorned descriptions: (1) the removal and disposal of diseased or severely traumatized organs or other tissue in what is known as "surgery" and (2) the introduction of toxic chemicals[13] into the body of a diseased patient in what is known as "drug therapy" or "medical practice." It is probably safe to say that in the not-too-distant future, your children or grandchildren will ask, "Is it true that when you were little, doctors cut off pieces of someone's body and threw it away or gave them poisons to make them better?" Our current medicine will soon seem as unsophisticated as trepanning, bloodletting, and use of electrical shock therapy or lobotomies for people suffering from severe mental conditions (e.g., schizophrenia).

Three new approaches to medicine with potentially great promise are: (1) immunotherapy, (2) gene therapy and genetic engineering, and (3) tissue engineering. Since it is not the purpose of this book or chapter to address medicine, only to suggest the potential contribution offered by reverse engineering, only a simple definition of each approach will be given and then a brief discussion of tissue engineering will be presented, in particular. Here, then, are the brief definitions, without elaboration:

- *Immunotherapy* is "the treatment of disease by inducing, enhancing, or suppressing an immune response [inherently built into healthy human beings]"[14] (Figure 18–11).
- *Gene therapy* is "the use of DNA as a pharmaceutical agent to treat disease" (Figure 18–12), while *genetic engineering* (or *genetic modification*) is "the direct manipulation of an organism's genome using biotechnology (e.g., the insertion of an isolated and copied segment of genetic material from DNA into the host genome)."

---

[11] Interestingly, modern practitioners of microsurgery for limb reattachment have begun anew to employ specially raised sterile medical leeches to the skin of the reattached part to restore proper blood flow by taking advantage of a natural anticoagulant released by the leech while sucking blood.

[12] In other parts of the world, most notably in the Far East, specifically China, modern medicine routinely also includes herbal treatments, acupuncture, massage, and more holistic approaches. Some of these (e.g., acupuncture) are beginning to receive respect by Western practitioners, and there is far greater and wider appreciation by all modern physicians of the need for holistic treatment, that is, "treatment of the whole person."

[13] In the most severe cases, such as treatment of cancers, highly toxic chemical agents are administered during *chemotherapy*, with unintended and unwanted, but almost inevitable, side effects that extend to damage of healthy tissue and organs while attempting to destroy malignant cells.

[14] *Active immunotherapy* "induces or enhances" while *suppressive immunotherapy* "suppresses" an immune response.

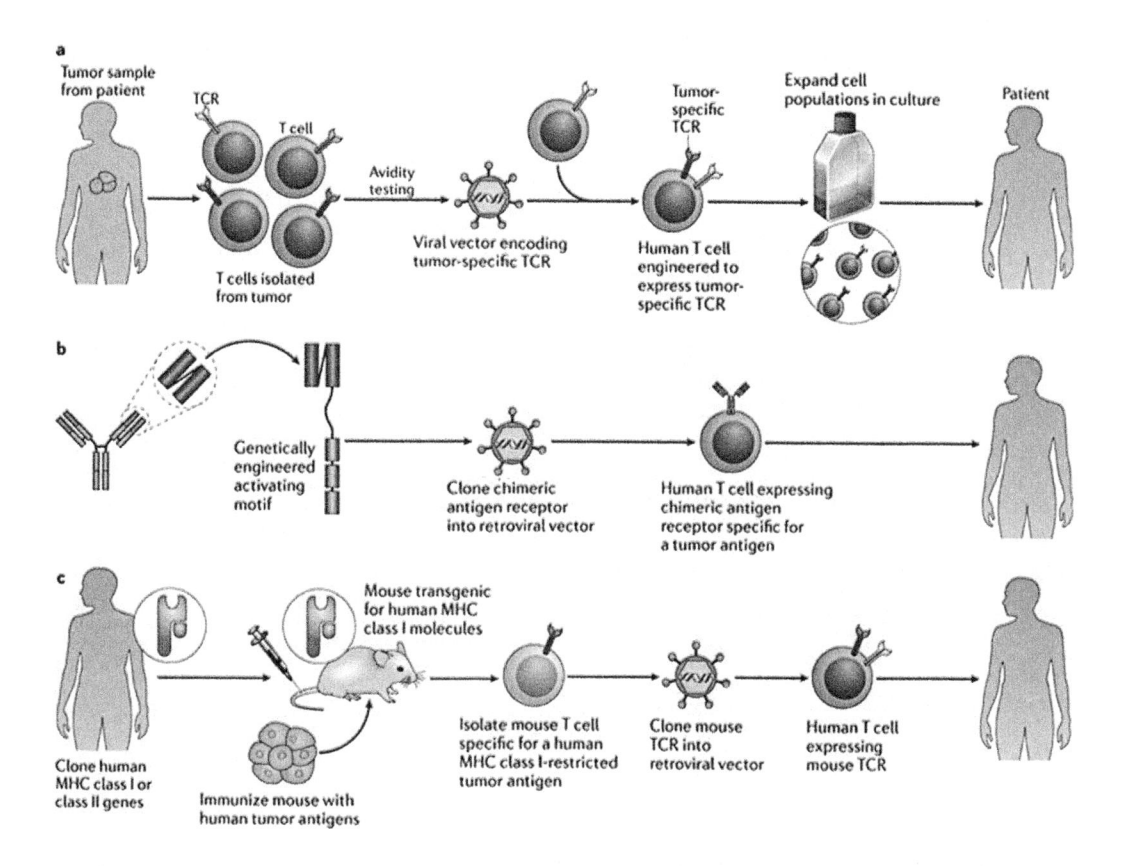

Figure 18-11 Schematic illustration of various approaches to immunotherapy. (*Source:* Nicholas P. Restifo, Mark E. Dudley, and Steven A. Rosenberg, "Adoptive Immunotherapy for Cancer: Harnessing the T Cell Response," *Nature Reviews Immunology,* Vol. 12 (April 2012), pp. 269–281; used with permission.) **Don't miss the color version of this figure, available at www.mhprofessional.com/ReverseEngineering.**

- *Tissue engineering* is "the use of a combination of cells, engineering and materials methods, and suitable biochemical and physio-chemical factors to improve or replace [or, ultimately, restore] biological functions" (Figure 18–13).

From its definition given here, *tissue engineering* is especially interesting for making the point of this entire section. To work at all, knowledge and understanding must be brought together from several sources (here, disciplines), such as engineering, materials, medicine, biology, chemistry or biochemistry, physics or biophysics. It, like genetic engineering, is truly *engineering,* as it involves the application of science and an appreciation of the need for manipulation within the laws of Nature.

One need only understand how *biomedical engineering* or *biotechnology* has helped, and

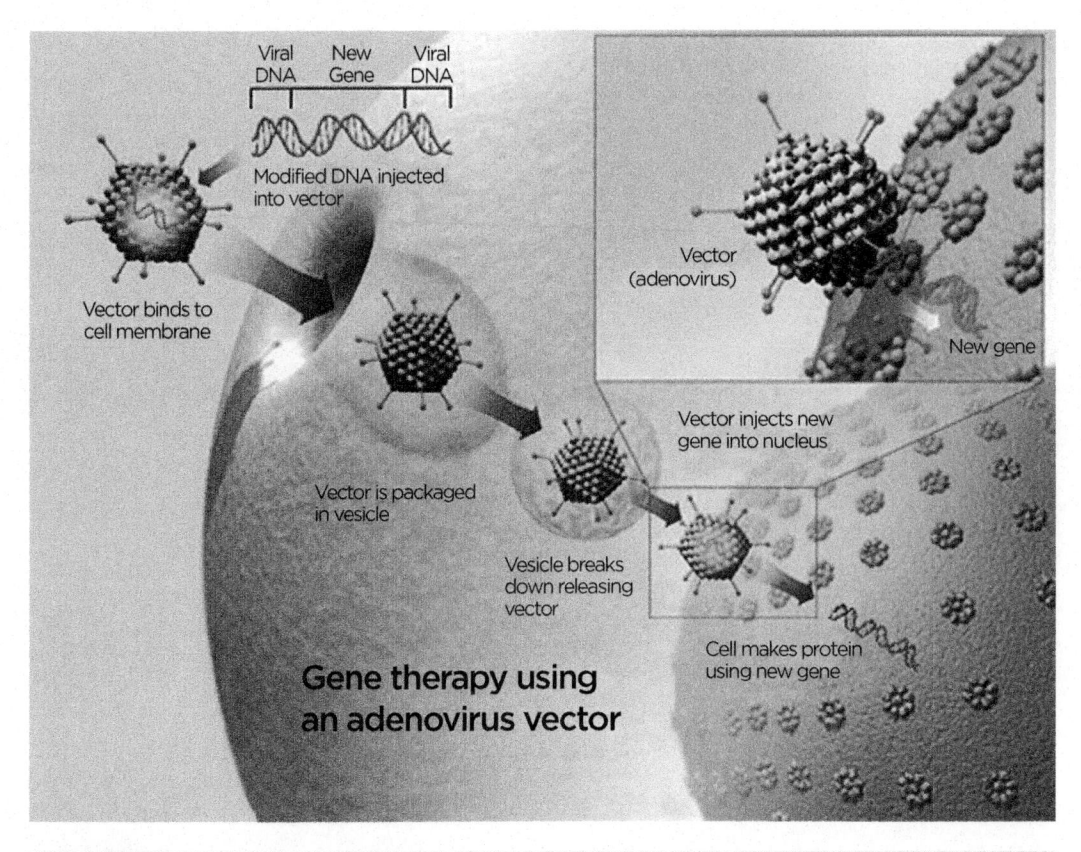

Figure 18-12 Schematic illustration of gene therapy. (*Source:* Wikipedia Creative Commons, originally contributed by Llull on 31 December 2005 and modified by Doectzee on 26 February 2010.) **Don't miss the color version of this figure, available at www.mhprofessional.com/ReverseEngineering.**

will continue to help, medicine, and, in turn, all of our lives, to see the potential offered by reverse engineering in the future. It emerged as a formal discipline (for which one could receive a degree) only about 50 years ago, although one could argue it had been practiced much longer. Mechanical engineers interested in the kinematics of the skeletal, suspensory (i.e., tendon and ligament), and muscle system of human beings; electrical engineers interested in noninvasive imaging techniques for human beings; materials engineers interested in biological and biocompatible materials; and physicians came together, and the new discipline was born. Without question, medicine (and our life) has been made better when doctors learned from engineers, and vice versa. After all, a heart, for example, is just a pump, albeit a complex one. As a consequence, cardiologists learned from engineers expert in fluid dynamics (within mechanical engineering) as well as signal processing (within electrical engineering) how to better detect abnormalities in blood flow through the heart using noninvasive techniques (e.g., MRI and/or ultrasound). Orthopedic surgeons benefitted from engineers expert in structural

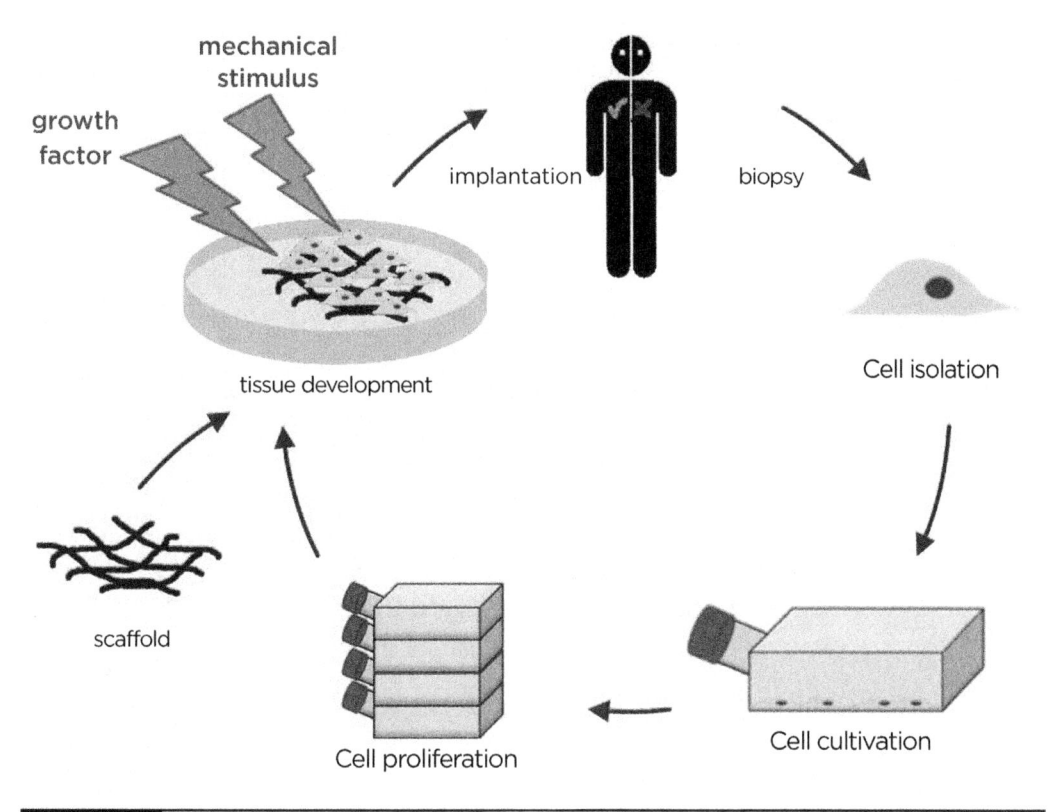

The End of a Book, the Beginning of a New Story: Closing Thoughts   **403**

Figure 18-13   Schematic illustration showing the key steps in tissue engineering. (*Source: Wikipedia Creative Commons, contributed by HIA on 2 December 2010.*) **Don't miss the color version of this figure, available at www.mhprofessional.com/ReverseEngineering.**

mechanics and materials for the design and manufacture of better artificial joint replacements. Synergy succeeds!

18-5   Final Words

As final words, here's the message in a nutshell:

As stated in Ecclesiastes 1:9: "What has been will be again, what has been done will be done again; there is nothing new under the Sun."[15] But this doesn't mean we are doomed to stick with what we have or to accept the status quo. It means we have access to what has always been here for us, if only we look hard enough to see it.

We need to "dissect" the world around us to see how it works, or why some part of it doesn't work. This is the common necessity for reverse engineering. Reverse engineering can be limited

---

[15] From the New International Version of the *Holy Bible*.

Figure 18-14 Our Planet Earth from space; our place "under the Sun." (*Source:* www.earthobservatory.nasa.gov, available for free use courtesy of the National Aeronautics and Space Administration.) **Don't miss the color version of this figure, available at www.mhprofessional.com/ReverseEngineering.**

to copying what was, to make incremental improvements or advancements. It can be used to make truly artless copies. But that is *not* what it is really all about! It is *much* more!

Reverse engineering allows us to observe, to learn, and to create. But we don't have to stop with a creation that is simply a new version of the old version. The greatest opportunity is to create anew something that the world has not yet seen!

18-6 Cited References

Drexler, K. Eric, *Radical Abundance: How a Revolution in Nanotechnology Will Change Civilization,* Public Affairs, New York, 2013.

Hawking, Stephen, and Leonard Mlodinow, *The Grand Design,* Bantam, New York, 2012.

Ingle, Kathryn A., *Reverse Engineering,* McGraw-Hill, New York, 1994.

Koshi, Masaki, "An Automated Underground Tube Network for Urban Goods Transport," *International Association of Traffic and Safety Sciences (IATSS) Research,* Vol. 16, Issue 2, 1999 (Elsevier).

Messler, Robert W., Jr., *The Essence of Materials for Engineers,* Jones & Bartlett Learning, Burlington, MA, 2011, pp. 490–491.

Otto, Kevin, and Kristin Wood, *Product Design: Techniques in Reverse Engineering and New Product Development,* Prentice-Hall, Upper Saddle River, NJ, 2000.

Petroski, Henry, *To Engineer Is Human: The Role of Failure in Successful Design,* Vintage, New York, 1992.

18-7 Recommended Readings

Allwood, Julian M., and Jonathan M. Cullen, *Sustainable Materials with Both Eyes Open,* UIT Cambridge Ltd., Cambridge, UK, 2012. [Materials]

Boyle, Godfrey, *Renewable Energy: Power for a Sustainable Future,* Oxford University Press, Oxford, UK, 2012. [Energy]

Church, George M., and Ed Regis, *Regenesis: How Synthetic Biology Will Reinvent Nature and Ourselves,* Basic Books, New York, 2012. [Biology and medicine]

Cox, Wendell, *21st Century Highways: Innovative Solutions to America's Transportation Needs,* The Heritage Foundation, Washington, DC, 2005. [Transportation]

Ermak, Gennady, *Modern Science and Future Medicine,* 2nd edition, CreateSpace Independent Publishing Platform, 2013. [Medicine]

Kaku, Michio, *Physics of the Future: How Science Will Shape Human Destiny in the Year 2100,* Anchor, New York, 2012.

Peter, Laurie, *Materials for a Sustainable Future,* Royal Society of Chemistry, London, 2012. [Materials]

Scientific American, ed., The Future of Energy: Earth, Wind and Fire, Kindle edition, Scientific American, 2013. [Energy]

Schilperoord, Paul, *Future Tech: Innovations in Transportation,* Black Dog Publishing, London, 2006. [Transportation]

18-8 Thought Questions and Problems

18-1 Section 18.2 is entitled "Imperfect Humans Need Reverse Engineering."

In a thoughtful essay of less than two pages, provide an argument for why imperfect humans need reverse engineering *or,* if you disagree (which you are entitled to do), make your argument for your counterposition.

18-2 Section 18–3 contends that human history, as well as its progress, has been marked by three interlinked ideas: (1) order from chaos, (2) light from darkness (as a metaphor for knowledge from ignorance), and (3) knowledge from (or begets) knowledge. This may seem strictly philosophical (and it surely involves philosophy!), but it is supported by the realities of history.

Pick *one* of the three premises as a *hypothesis*, and write a thoughtful two- to three-page essay, richly supported by biblical and/or historical events. Be especially sure to indicate how certain major leaps forward—if not transformational events—in the progress of a society or all of humankind may have resulted.

18-3 Two interesting observations by the author and others who have looked carefully, and without preconceived notions, are these:

(1) Gutenberg's invention of the moveable-type printing press in 1450 transformed Western Europe, then Eastern Europe, and then Asia, by dramatically increasing access to ideas and knowledge, but this transformational event could be dwarfed by the invention of the "information superhighway" made possible by the Internet.

(2) Evidence supports that the knowledge possessed by our ancient ancestors was far greater than is largely presumed by casual observations, and that willful and systematic destruction or suppression of that knowledge may have set humankind back by centuries, greatly slowing growth of scientific thought and technological progress that could have placed our modern world far ahead of where it is.

Choose one or the other of these and prepare a thoughtful three- to five-page paper on the premise. Support your position with real historical events. Also, indicate the motivation for knowledge destruction as well as suppression in the latter and the further opportunity for even greater proliferation of knowledge in the former.

In Section 18–4, the author suggests four opportunities where we can, as a world people, "learn from the old to create anew," enabled by reverse engineering. The result could well be revolutionary, as opposed to evolutionary, advances. The four opportunities are:

- Revolution via Reverse Engineering of Energy Sources and Systems
- Revolution via Reverse Engineering of Transportation
- Revolution via Reverse Engineering of Materials
- Revolution via Reverse Engineering of Medicine

18-4 and 18-5 Choose any *two* of the four opportunities (one for 18–4 and one for 18–5), and prepare a well-thought-through, well-organized, and articulate essay of four to five pages.

An ever larger and larger number of fertile brains are continually at work in discovery and invention, and these fresh brains start from an ever-widening vantage ground of accumulated research and proven experience. The result must surely be that important inventions and new discoveries will crowd thicker upon the world [in the future than in the past]....

Attributed to an unnamed American engineer near the close of the nineteenth century in Ernest Freeberg, *The Age of Edison: Electric Light and the Invention of Modern America* (Penguin Press, 2013).

APPENDIX A

List of All Material Classes and Major Subtypes, and Major Members of Each

Class
Engineering Metals/Alloys

Members
Al and Al alloys
Co and Co alloys
Cu and Cu alloys (brass, bronze)
Fe and Fe alloys (steels)
Mg and Mg alloys
Mo and Mo alloys
Ni and Ni alloys
Pb and Pb alloys (e.g., Babbitt)
Sn and Sn alloys (e.g., pewter)
Ti and Ti alloys
W and W alloys
Zn and Zn alloys
Zr and Zr alloys

Engineering Polymers
(thermoplastics [TPs] and thermosets [TSs])

Acrylics (TP)
Epoxies (TS)
Melamines (TS)
Nylons (TP)
Polycarbonates (TP)
Polyesters (TP)
Polyethylenes, high density (TP)
Polyethylenes, low density (TP)
Polyethylene teraphthalate (TP)
Polyformaldehyde (TP)
Polymethylmethacrylate (TP)
Polypropylene (TP)
Polystyrene, dense (TP)
Polytetrafluorethylene (TS)
Polyvinylchloride (TP)

Elastomers (natural and synthetic rubbers)	Hard butyl rubber Natural rubber Nitrile rubber Polyurethanes Silicone rubber Soft butyl rubber
Polymer Foam (foamed polymers)	Cork Polyester (TP) Polystyrene (TP) Polyurethane (TP)
Engineering Ceramics Porous ceramics	Brick Cement Common rocks Concrete Porcelain Pottery
Fine ceramics/high-performance ceramics	Alumina (Al_2O_3, sapphire) Borides Carbides Diamond Magnesia (MgO) Mullite (Al_2O_3-SiO_2 eutectic) Sialons Silicon carbide (SiC) Silicon nitride (Si_3N_4) Titania (TiO_2) Zirconia/Partially stabilized zirconia (ZrO_2, PSZ)
Glasses (silicate glasses)	Borosilicate glass (Pyrex) Leaded glasses (Pb-containing) Soda-lime (common) glass Silica/fused silica (SiO_2)
Engineering Composites (composite materials)	Carbon fiber-reinforced polymer (CFRP) Glass fiber-reinforced polymer (GFRP) Kevlar fiber-reinforced polymer (KFRP)
Woods (separate from other composites, with properties parallel and perpendicular to grain direction)	Ash Balsa Fir Oak Pine Wood products (plywood, etc.)
Ice	Ice (frozen water; used by Inuits)

APPENDIX B

Comprehensive List of Specific Manufacturing Methods by Process Class

Casting

Centrifugal casting
- Continuous casting
- Die casting
- Evaporative-pattern casting
  - ✓ Full-mold casting
  - ✓ Lost-foam casting
- Investment casting
  - ✓ Lost-wax casting
  - ✓ Lost-foam casting
- Low pressure
- Permanent mold casting
- Plastic mold casting
- Resin casting
- Rheocasting/thixotropic casting
- Sand casting
- Shell casting
- Slush or slurry casting
- Spray forming (special)

Molding

- Plastics
  - ✓ Blow molding
  - ✓ Compression molding
  - ✓ Dip molding
  - ✓ Expandable bead molding

- ✓ Extrusion
- ✓ Foam molding/foaming
- ✓ Injection molding
- ✓ Laminating
- ✓ Matched mold molding
- ✓ Pressure plug assist molding
- ✓ Pultrusion (composites)
- ✓ Rotational molding
- ✓ Thermoforming
- ✓ Transfer molding
- ✓ Vacuum plug assist molding
- Shrink fitting
- Shrink wrapping

Deformation Processing

- End tube forming
    - ✓ Flaring
    - ✓ Tube beading
- Forging
    - ✓ Cored forging
    - ✓ Drop (drop-hammer) forging
    - ✓ Hammer forging
    - ✓ Heading (fasteners)
    - ✓ High-energy-rate forging
    - ✓ Impact forging (also see Extrusion)
    - ✓ Incremental forging
    - ✓ Isothermal forging
    - ✓ No-draft forging
    - ✓ Powder forging
    - ✓ Press forging
    - ✓ Rotary forging
    - ✓ Upset forging
- Rolling (thick plate and thin sheet)
    - ✓ Cold rolling/cold finishing
    - ✓ Cross-rolling
    - ✓ Cryorolling
    - ✓ Hot rolling
    - ✓ Orbital rolling
    - ✓ Ring rolling
    - ✓ Shape rolling
    - ✓ Texture rolling
    - ✓ Thread rolling
    - ✓ Transverse rolling

Comprehensive List of Specific Manufacturing Methods by Process Class

- Extrusion
  - ✓ Impact extrusion
  - ✓ Pressure extrusion
  - ✓ Tube extrusion
- Pressing/forming
  - ✓ Blanking
  - ✓ Deep drawing (auto body, sinks)
  - ✓ Drawing (bar, wire, tube)
  - ✓ Embossing
  - ✓ Stretch forming
- Bending
  - ✓ Hemming
- Shearing
  - ✓ Piercing
    - Nibbling
    - Notching
    - Perforating
    - Shaving
    - Trimming
  - ✓ Stamping
    - Leather
    - Metal
    - Progressive
  - ✓ Coining
  - ✓ Straight shearing/slitting
- Other
  - ✓ Cold sizing
  - ✓ Decambering
  - ✓ Electroforming (special)
  - ✓ Explosive forming
  - ✓ Flanging
  - ✓ Flattening
  - ✓ Hubbing
  - ✓ Hydroforming
  - ✓ Magnetic pulse-forming
  - ✓ Redrawing
  - ✓ Peening
  - ✓ Seaming
  - ✓ Staking
  - ✓ Straightening
  - ✓ Swaging

Powder Processing
- Cold compacting
- Forging
- Hot pressing
- Hydrostatic isothermal pressing (HIP)

Machining
- Abrasive-jet machining
- Broaching
- Chemical machining (chemical milling)
- Countersinking
- Drilling
- Electric discharge machining
- Electron-beam machining
- Electrochemical machining
- Filing
- Grinding
- Hobbing
- Honing
- Laser machining and cutting
- Milling
- Photochemical machining
- Planing
- Reaming
- Routing
- Sawing
- Tapping (thread tapping)
- Turning
  - ✓ Boring
  - ✓ Cutoff (or parting)
  - ✓ Facing
  - ✓ Knurling
  - ✓ Lathe
  - ✓ Spinning/spin forming/flow turning
- Ultrasonic machining
- Water-jet cutting

Finishing
- Abrasive blasting (sand or grit blasting)
- Buffing
- Burnishing
- Coating (dip)
- Electroplating

Comprehensive List of Specific Manufacturing Methods by Process Class

- Electropolishing
- Etching
- Plating
- Polishing
- Tumbling
- Vibratory finishing
- Wire brushing

Joining

- Adhesive bonding
  - ✓ Adhesive alloys (actually, blends)
  - ✓ Cement and mortar
  - ✓ Elastomeric adhesives
  - ✓ Inorganic adhesives
  - ✓ Natural adhesives
  - ✓ Tapes
  - ✓ Thermoplastic polymer adhesives
  - ✓ Thermosetting polymer adhesives
- Brazing
  - ✓ Dip brazing
  - ✓ Furnace brazing
  - ✓ Induction brazing
  - ✓ Reaction brazing
  - ✓ Resistance brazing
  - ✓ Step-brazing
  - ✓ Torch brazing
- Integral mechanical attachment
  - ✓ Elastic snap-fits (e.g., cantilever hooks)
  - ✓ Formed-in attachments (e.g., crimps)
  - ✓ Press fits
  - ✓ Rigid interlocking features (e.g., dovetail and grooves, mortise-tenons)
- Mechanical fastening
  - ✓ Bolting (with or without nuts/washers)
  - ✓ Keys and keyways
  - ✓ Nailing
  - ✓ Pegging
  - ✓ Pinning (e.g., cotter pins, taper pins)
  - ✓ Retaining clips and rings
  - ✓ Riveting (upset, blind, two-piece)
  - ✓ Screwing (self-tapping screws)
  - ✓ Stapling
  - ✓ Stitching

- Soldering
  - ✓ Hot-plate soldering
  - ✓ Induction soldering
  - ✓ Infrared soldering
  - ✓ Iron soldering
  - ✓ Oven soldering
  - ✓ Step-soldering
  - ✓ Ultrasonic soldering
  - ✓ Wave soldering
- Welding
  - ✓ Fusion arc welding
    - Carbon arc welding
    - Electrogas welding
    - Electroslag welding
    - Flux-cored (open-arc) welding
    - Gas-metal arc (globular, spray, short-circuit, pulsed)
    - Gas-tungsten arc welding
    - Magnetic-impelled arc butt (MIAB) welding
    - Plasma-arc welding
    - Plasma-MIG welding
    - Shield metal-arc welding
    - Stud (arc) welding
    - Submerged arc welding
  - ✓ Fusion beam welding
    - Electron-beam welding
    - Laser-beam welding
  - ✓ Fusion gas welding
    - Oxyacetylene welding
    - Oxyhydrogen welding
    - Methylacetylene propadiene propane (MAPP) welding
  - ✓ Nonfusion (solid-state) welding
    - Cold welding
    - Diffusion welding
    - Explosive welding
    - Forge welding
    - Friction welding
    - Inertia welding
    - Roll welding
    - Ultrasonic welding

- ✓ Other welding
  - Dielectric welding
  - Heated-metal plate welding (thermoplastics)
  - High-frequency welding (thermoplastics)
  - Hot-air welding (for thermoplastics)
  - Magnetic-pulse welding
  - Radio-frequency (RF) welding (thermoplastics)
  - Solvent cementing (an adhesive process for thermoplastics)
- ✓ Resistance welding
  - Flash butt welding
  - Percussion (or capacitor-discharge) welding
  - Projection welding
  - Resistance seam welding
  - Resistance spot welding
  - Upset welding

Index

Page numbers followed by "F" refer to figures, by "FN" refer to footnotes, and by "T" refer to tables.

Academic learning, 18
Actuators, 138
Adaptive design, 22, 23T
Additive processing methods, 203
Adhesives:
    bioadhesives, 350–351
    natural adhesives, 350
    synthetic adhesives, 353
Aesthetics (in design), 166
Alloys, metals and, 173, 175
Analysis, analyzing (approach to engineering), 6, 7
Anomalistic month, 150
Antikythera mechanism, 144, 145
Approaches to engineering, 6
Artificial stone, 350–351
Assembly errors (in failures), 94, 95F
Astrolabe, 145
Atomic bonding:
    covalent, 175, 176T
    ionic, 175, 176T
    metallic, 175, 176T

Backward problem-solving, 17, 17T, 105
Benchmarking, use of reverse engineering for, 24, 25T
Benefits of reverse engineering, 26
Bioadhesives, 350–351

Biomedical engineering, biotechnology, 400–401
Biomimetics, biomimicry, 349
Bottoms-up teardown analysis, 70
Brittle (overload) failures, 98, 99
Built-up details (for manufacturing), 202

Callippic (gear) train, 151
Castability, 170
Casting methods, 205
Castings, identifying, 307–308
Chemical spot test kits for metal ID, 192
Cheops' Pyramid (*see* Khufu's [Great] Pyramid)
Clones, 339–340, 380
Clue(s), definition of, 124
Codes of ethics for engineers, 373–375
Combination failures, 98
Combination properties, 177
Commercial espionage, 18
Competitive technical intelligence gathering, 18, 338
Composite materials, 186, 203FN, 331, 353
Computer-based models, use in engineering, 7, 8
Computer simulation (use in engineering), 7, 8
Conceptual design stage, 7
Conjecture, 156

419

Constructing versus manufacturing, 201
Construction, use of reverse engineering for, 24, 25T
Controls, controllers, 138–139
Converters (of power or motion), 138
Copies, 339
Corrosion failures, 98
Corrosion-fatigue failures, 98
Cost (in design), 166, 317FN
Cost teardown, 65
Cues, definition of, 124
Curiosity (in humans), 1, 2

Deduction:
  of fact versus fiction, 119
  process of, 17
Deformation processing methods, 203, 205
Degradation, 91FN
Design:
  definition of, 19
  failures in, 93
  process of (steps), 20, 21F
  stages of, 7, 8T
  types of, 22, 23T
  use of reverse engineering for, 24, 25T
Design errors (in failures), 94, 95F
Design for assembly, 318, 320–322, 321T
Design for manufacture (DFM), 318–319, 319F
Design for process, 319–320
Designer materials, 399
Detail design stage, 7
Developmental design, 22, 23T
Differential scanning calorimetry (DSC), 193
Dimensionality, 127, 128T
Discovery (of new concepts/technologies via reverse engineering), 25
Dissection, 9, 10, 11
Documentation shortcoming, 18
Ductile (overload) failures, 98, 99

Duplication, unauthorized, 18
Duty cycle, 274–275
Dynamic teardown, 65

Ease of assembly, 165
Ease of fabrication, 165
Ejector pin marks, 288, 295, 301
Electronegative elements, 175
Electropositive elements, 173
Elevated-temperature failures, 98
Embodiment design stage, 7
Energy flow field design/diagram, 76, 85, 292
Energy flow/transport, 129, 129T
Engineered wood/lumber, 354
Engineering composites, 353
Engineering design process (steps), 20, 21F
Enigma, 152
Espionage, 17, 337–338
Ethical conduct, 373–375
Ethicality, 375
Evidence, definition of, 124
Exeligmos (gear) trains, 148, 151
Experience:
  for learning, 4
  value of in reverse engineering, 120
Experiential learning, 4, 5
Experiential Learning Model (ELM), 4, 5F
Experiments/experimentation (use in engineering), 7, 64, 65
Exploded view(s), 58, 59F, 72, 83F, 286, 287, 294, 300, 306

Fabrication errors (in failures), 94, 95F
Failure analysis:
  definition of, 91
  general procedure, 104, 105T
Failures:
  catastrophic, 90, 91
  causes of, 94, 95F

Failures (*Cont.*):
  clues to, 97T
  degradation, 91 FN
  eventual/ultimate, 89
  manifestations of, 91
  mechanisms of, 95, 96T
  premature, 90
  sources of, 92, 93, 94T, 95T
Fastening (in manufacturing), 202
Fatigue failures, 98, 99
  high-cycle/low-stress, 98, 104
  low-cycle/high-stress, 98, 104
Finishing processes, finish processing, 207
Fit, 127, 266
Flow processing methods, 203
Flows, 128, 129T
Fluids:
  flow of, 129, 129T
  movers of, 138
Force flow diagram, 72, 76, 79T, 86F, 128–129, 292
Force flow/transfer, 128, 129T
Form, 127, 266
Form, fit, and function (FFF, F3), 126, 265
Form, fit, and function assessment using observations, 267
  design details, 275, 277
  electrical and/or thermal robustness, 273
  material selection, 270–273
  method-of-manufacture/-construction, 270–273
  precision, 273
  size and robustness, 268, 270
Formability, 170
Formulations of substances, 337, 347, 348
Forensic engineering, 59, 60, 91, 111–112
Forensics, forensic science, 111–112
Forward engineering (*see* Forward problem-solving)

Forward problem-solving, 17, 17T
Fourier transform infrared spectroscopy (FTIR), 193
Fractographic analysis, fractography, 99
Fractures:
  brittle, 364, 366F
  ductile, 365, 366F
Function:
  actual, 64
  intended, 57
  latent, 57
  predicted, 64
Function/functionality of design/entity, 126, 127, 165, 199, 266, 317
Function-Material-Shape-Process interrelationship, 199, 200F
Function units, 130
Functional analysis, 79
Functional diagram, 74
Functional measurement, 63
Functional models, 72, 79, 81F, 136T–137T, 327
Functional structure, 64, 81
Functional tree, 74, 82F
Fundamental approaches to engineering, 6

Gene therapy, 400
Generics, 340, 345–348
Genetic engineering, 400
Geometric measurement, 61
Geometric model, 72
Geometric shape, 127, 128T
Geometric symmetry, 127, 128T
Great Pyramid (of Khufu), 229

Heat treating/treatment, 206, 218–221
High-cycle/low-stress fatigue, 98, 104
Human-caused/-based failures, 93

422   Index

Identification:
   of materials/metals by observation:
      characteristics, 189T–191T
      color, 186
      coolness (from thermal conductivity), 187
      density/heft, 186–187
      flex (for stiffness/modulus), 187
      hardness, 187
      luster, 186
      magnetic attraction/magnetism, 188
      ring, sound, 187
   of polymers by applications, 191T
   as true/false, 119
Imitation:
   definition of, 339
   of natural materials, 352–353
   of Nature (biomimetics), 348–349
Immunotherapy, 400
Improvements of materials, substances, items, 342
Inferring, inference, 119
Injection molding, 288
Inspiration from Nature, 348
Integral mechanical attachments, 322T
Intelligence gathering, 18
Intended function, 57
Internet, growth/impact of, 383

Joining (processes), 207
   adhesive bonding, 207
   brazing and soldering, 207
   integral mechanical attachment, 207
   mechanical fastening, 207
   mechanical joining, 207
   welding, 207
Joints, 139

Khufu's (Great) Pyramid, 30–31, 229
Knockoffs, 341

Knowing (true/false), 119
Kolb ELM (*see* Experiential Learning Model)

Latent function, 57
Learning:
   from experience, 4
   from sensory input, 4
   by taking things apart, 3
Learning styles, 4
Learning Styles Inventory (LSI), 6
Legality, 375
Life-cycle cost, 317FN, 318
Look-alikes, 340–341
Low-cycle/high-stress fatigue, 98, 104

Machinability, 170
Machined parts, identifying, 312
Machining (in manufacturing), 206
Maintenance errors (in failures), 94, 95F
Manifestation of failures, 91
Manufacturability, 165–166, 317
   cost, 166, 318
   of designs, 165
   ease of assembly, 165
   ease of fabrication, 165, 318
Manufacturing:
   versus constructing, 201
   failures in, 93
   methods of, 199
   use of reverse engineering for, 24, 25T
Manufacturing processes taxonomy, 203, 204F
   additive processes, 203
   flow/deformation processes, 203
   processing for finish, 203
   processing for geometry (shape and dimensions), 203

Manufacturing processes taxonomy (*Cont.*):
  processing for properties, 203
  subtractive processes, 203
Market pull (in design), 22
Marketing, use of reverse engineering for, 24, 25T
Material errors (in failures), 94, 95F
Material performance indices, 181
Material properties:
  definition of, 169
  by specific type, 177–181, 182T
Material selection charts, 181–186
Material teardown, 65
Materials, 165, 172, 187F
  amorphous, 336
  composite, 336
  crystalline, 336
  for engineering, 336
  identification of, 186–191
  versus substances (definitions), 335–337
Materials-by-design, 399
Materials-of-construction/-manufacture, 165
Materials revolution, 397–399
Materials science, 163
Matrix teardown, 65
Measurement:
  of function, 63
  use in engineering, 7, 60
Measurement device criteria, 62T
Mechanical dissection, 11, 73
Medicine, revolution in, 399–400
Meta-materials, 399
Metal (identification) test kits, 192
Metalloids (semimetals or semiconductors), 175, 177
Metals (and alloys), 173, 175
Method-of-manufacture/-construction, 199
Method-of-manufacture from observation:
  cast metal parts, 210, 219T
  cold cast ceramic parts, 214

Method-of-manufacture from observation (*Cont.*):
  composite material parts, 214
  deformation processed metal parts, 212, 219T
  forged metal parts, 212–213
  identification of:
    using batch size, 216
    using cost/apparent value, 217
    using geometric complexity, 214
    using joining method, 217–218
    using material class, 210–214
    using production rate, 216
    using roughness/surface finish, 215
    using shape, 214
    using size, 214
    using surface details, 216
    using tolerance/precision, 214–215
    using workmanship, 217
  machined parts, 219T
  molded polymer parts, 214
  powder processed metal (or ceramic) parts, 213–214, 219T
Methodizing, 24, 327–329
Methods engineering, 327
Metonic (gear) train, 151
Military espionage, 18
Misuse (in failures), 94, 95F
Model-centric approach to design, 21
Moldability, 170
Molding parting lines, 303, 303F
Molding/pressure-molding methods, 205
Moon gear, 151
Motion converters, 138
Motivations for reverse engineering, 18, 19T

Net-shape/near-net-shape processes, 204
New paradigms, 384
Nonmetals, 175
Nuclear magnetic resonance (NMR), 193

Observation, use of, 58
Occam's (Ockham's) razor, 242, 242 FN
Olympiad (gear) train, 152
Original design, 22, 23T
Overload failures, 98

Periodic Table of the Elements (Standard), 173, 174F
Physical models (use in engineering), 7
Polymers, 177
  identification by application, 191T
  laboratory techniques for identifying, 192–193
Powder processing methods, 205
Power converters, 138
Power sources, 135
Practice of engineering, 4
Primary properties, 177
Primary shaping processes, 204
Prime movers, 138
Printing press, moveable-type, impact of, 390
Problem statement/formulation, 7
Process, 199
Process attributes, 207, 208T
  cost, 207, 208T
  dimensional accuracy/tolerance/precision, 207, 208T
  geometric complexity, 207, 208T
  material class, 207, 208T
  minimum batch size, 207, 208T
  production rate, 207, 208T
  shape, 207, 208T
  size range, 207, 208T
  speed (of processing), 207, 208T
  surface finish/roughness, 207, 208T
Process selection charts, 208–209
Processing errors (in failures), 94, 95F
Producibility, 24, 318, 330
Product form, 204

Product functional model, 85, 86F
Product security analysis, 18
Product teardown, 56
  benchmarking, 57
  forms of, 65
  procedure for, 71
  process (definition), 69
  purposes, 56, 69T, 70
  subtract-and-operate procedure, 70, 74, 75T
Production, use of reverse engineering for, 24, 25T
Production engineering, 325
Properties of materials:
  acoustical, 170
  biological, 170
  chemical, 169
  combination/complex, 169, 170, 177
  electrical, 169
  magnetic, 169
  mechanical, 169
  optical, 169
  physical, 169
  primary, 177
  radiological, 170
  secondary, 177
  thermal, 169
Property-performance relationship, 171T–172T
Proprietary materials, 342–343
Purpose of design/entity, 126
Pyramids:
  alignment, 241–242
  casing stones, 252
  construction, 244–256
  design layout/measurement, 246–247
  internal details, completion of, 252
  location, 238–239
  materials-of-construction, 247–248
  orientation of Three Pyramids, 242–244
  progress check/plan adherence, 247

Pyramids (*Cont.*):
  purpose, 234–238
  role of reverse engineering, 252–256
  site preparation, 245–246
  site selection, 245
  stone transport/positioning, 248–252
Pyramids of Giza, 237

Quality assurance, using reverse engineering for, 24, 25T

Raman spectroscopy, 193
Remanufacture, using reverse engineering for:
  broken parts, 363, 364
  deformed parts, 365
  obsolete parts, 363, 367, 369
  worn parts, 363, 365
Renewable energy, 394, 395T
Replacements (for materials or substances), 338–339, 344–345
Reverse engineering:
  benefits of, 26
  during Cold War and post–Cold War, 47–49
  definition of, 13, 16, 18, 55
  emergence of, 29
  of Great Pyramid of Khufu, 29–32
  during Industrial Revolution, 35–37, 40, 42
  during Middle Ages, 33, 35
  motivations for, 18, 19T
  procedure for, 73, 117–118, 118T
  for remanufacture, 362
  risks of, 26
  status of (in textbooks), 15, 16
  uses/potential uses, 23, 25T
  during World War II, 42–47
Role of design/entity, 126
Root-cause for failure, 92
Rules of Engineering Practice (NSPE), 375

Saros (gear) train, 151
Scaling/variant design, 22, 23T
Secondary processes/processing methods, 206
Secondary properties, 177
Security analysis of products, 18
Semiconductors, 175, 177
Sensors, 138
Sensory input for learning, 4
Service errors (in failures), 94, 95F
Service failures, 93
Shape, macro-/micro, 199, 200
Sidereal month, 150
Simulation, computer (use in engineering), 7, 8
Special processing methods, 206
Spider silk, artificial, 350
Stages of engineering design, 7, 8T
Statement of problems (in design), 7
Static overload failures, 98
Structure(s), 138
Structure-Property-Processing-Performance interrelationship, 166, 167, 168, 168F
Substances versus materials, definitions of, 335–337
Substitutes (for materials or substances), 338, 343T, 344–345
Subtract-and-operate procedure (SOP), 70, 74, 75T
Subtractive processing methods, 203
Suitability of design for purpose, 265–266
Sun gear, 150
Sustainable energy, 394–396, 395T
Synodic month, 151
Synthesis/synthesizing (approach to engineering), 6, 7
Synthetics:
  bioadhesives, 350–351
  diamonds, 349
  fibers, 355–356
  stone, 350–351

Systems analysis, 81
Systems approach to design, 21, 22

Taking things apart (to learn), 3
Teardown process, 56
  (*See also* Product teardown)
Technical systems, 79, 80F
Technology push (in design), 22
Tissue engineering, 401
Top-down teardown analysis, 70, 73
Topological evidence/features, 125
Trade-off decisions stage in design, 7
Transportation, new concepts for, 396–397, 398
Troubleshooting, use of reverse engineering for, 25, 25T

Unlicensed/unauthorized duplication, 18

Value, definition of, 323
Value analysis, 323
Value analysis teardown, 73
Value engineering (VE), 24, 322–324, 324T
Variant/scaling design, 22, 23T
Visual, auditory, kinesthetic (VAK) learning styles, 4
Vivisection, 9, 10, 74

Wear failures, 98
  corrosion, 98
  fatigue, 98
Weldability, 170
Welding (in manufacturing), 202
Who, What, When, Where, Why, and How ("Five Ws"), 3
Wide-angle x-ray diffraction/scattering, 193
Wood, 354, 354FN
Workmanship, 58, 122, 217

CPSIA information can be obtained
at www.ICGtesting.com
Printed in the USA